PRINCIPLES AND PRACTICE OF
AUTOMATIC PROCESS CONTROL

Carlos A. Smith
University of South Florida

Armando B. Corripio
Louisiana State University

JOHN WILEY & SONS
New York   Chichester   Brisbane   Toronto   Singapore

Library of Congress Cataloging in Publication Data:

Smith, Carlos A.
  Principles and practice of automatic process control.

  Includes bibliographies and index.
  1. Chemical process control—Automation.
I. Corripio, Armando B. II. Title.
TP155.75.S58   1985   660.2'81   84-21982
ISBN 0-471-88346-8

Printed and bound in the United States of America by Braun-Brumfield, Inc.

10

PRINCIPLES AND PRACTICE OF
AUTOMATIC
PROCESS
CONTROL

PREFACE

A most important objective of this book is to present the practice of automatic process control along with the fundamental principles of control theory. To this end we have included in the book a generous number of case studies, problems, and examples taken directly from our experience as industrial practitioners and consultants. It is our belief that, although there are many fine books in the market that cover the principles and theory of automatic process control, most of them do not expose the reader to the practice of these principles.

The notes from which the book was developed have been tested during the last few years in senior-level chemical and mechanical engineering courses at the University of South Florida and Louisiana State University. Also, many parts of the book have been used for the past several years by the authors in teaching short courses to practicing engineers in this country and abroad.

The import of the book is directed toward the process industries. The book can be used by senior engineering students, principally in the fields of chemical, mechanical, metallurgical, petroleum, and environmental engineering, and by technical personnel in the process industries. We believe that in order to control a process, the engineer must first understand the process. This is why throughout this book we base the understanding of process dynamic response on the principles of material and energy balances, fluid flow, heat transfer, separation processes, and reaction kinetics. Most senior engineering students will have the background necessary to understanding these concepts at the level at which we present them. The mathematical level required is covered in any undergraduate engineering curriculum—mainly operational calculus and differential equations.

The definitions of terms and mathematical tools used in the study of process control systems are presented in Chapters 1 and 2. Chapters 3 and 4 present the principles of process dynamic response. In these chapters we use numerous examples to show how to develop simple process models and to learn the physical significance of the parameters that describe the dynamic behavior of the process.

Chapter 5 presents a discussion of some important components of a control system. Namely, sensors, transmitters, control valves, and feedback controllers. The practical operating principles of some common sensors, transmitters, and control valves are presented in Appendix C. Students who are interested in acquiring a working knowledge of process instrumentation are strongly encouraged to study Appendix C.

The design and analysis of feedback control systems is the subject of Chapters 6 and 7, while the rest of the important industrial control techniques are treated in Chapter 8. These are ratio, cascade, feedforward, override, selective, and multivariable control. We use numerous practical examples to illustrate actual industrial applications of these techniques.

The principles of mathematical modeling and computer simulation of processes and their control systems are presented in Chapter 9. In this chapter we present a very useful modular program architecture that can be used to illustrate the principles of dynamic response, stability, and tuning of control systems.

In our experience, a one-semester course should include the first six chapters of the book through Section 6-3, plus the section on feedforward control from Chapter 8. Then, depending upon the availability of time and the instructor's preference, the sections on computing relays, ratio control, cascade control, root locus, and frequency response can be included. These sections are independent of each other. If the course includes a laboratory, the material in Chapter 5 and Appendix C should serve as excellent background for the laboratory experiments. The examples in Chapter 9 can serve as blueprints for computer simulation "experiments" to supplement the actual laboratory experiments.

If two semesters or quarters are available for teaching the course, the entire text can be presented in detail. The course should include a term project using the process control problems of Appendix B. These are actual industrial problems and provide the student with the opportunity to design, from scratch, the control system for a process. We strongly believe that these problems are an important contribution of this book.

In this book we have exclusively used the transfer function approach over the state variable approach for three reasons: First, we strongly believe that transfer functions are more viable for conveying understanding of process control concepts; second, we are not aware of any control schemes used in industry today that require the state variable approach in their design; and finally, the state variable approach requires a stronger mathematical background than transfer functions.

In any work of this type there are numerous people who contribute, encourage, and help the authors in different ways. We are no exception and feel blessed to have these persons around. From industry, both authors would like to thank Charles E. Jones of Dow Chemical USA, Louisiana Division, for supplying the motivation of the industrial practice of process control and for his encouraging us to seek higher education. From academia, our two universities have provided the atmosphere and help necessary for completing this project. We would like to thank the faculty and students of our departments for developing in us a deep appreciation and satisfaction in academic instruction. To serve as agents in the training and development of young minds is certainly a most rewarding profession.

The encouragement of our undergraduate and graduate students (the young minds) will never be forgotten, especially that of Tom M. Brookins, Vanessa Austin, Sterling L. Jordan, Dave Foster, Hank Brittain, Ralph Stagner, Karen Klingman, Jake Martin, Dick Balhoff, Terrell Touchstone, John Usher, Shao-yu Lin, and A. (Jefe) Rovira. From the University of South Florida, Carlos A. Smith would like to thank Dr. L. A. Scott; his friendship and advice during the last ten years have been most helpful. Thanks are also due to Dr. J. C. Busot; his constant question, "When are you going to finish the book?" has certainly helped in providing some of the fuel necessary to continue. From Louisiana State University, Armando B. Corripio would like to acknowledge the role Drs. Paul W. Murrill and Cecil L. Smith played in getting him started in automatic process control. They not only taught him the theory, they instilled in him their love for the subject and for teaching it.

Finally, the authors would like to thank the secretarial staffs of both universities for their care, efficiency, and patience in typing the manuscript. From USF we thank Phyllis Johnson and Lynn Federspeil. From LSU we thank Janet Easley, Janice Howell, and Jimmie Keebler.

Carlos A. Smith
Tampa, Florida

Armando B. Corripio
Baton Rouge, Louisiana
1984

CONTENTS

Chapter 1   Introduction   **1**

  1-1 A Process Control System   1
  1-2 Important Terms and Objective of Automatic Process Control   3
  1-3 Regulatory and Servo Control   4
  1-4 Transmission Signals   4
  1-5 Control Strategies   5

      Feedback Control   5
      Feedforward Control   6

  1 6 Underlying Reasons for Process Control   8
  1-7 Background Needed for Process Control   8
  1-8 Summary   9

Chapter 2   Mathematical Tools for Control Systems Analysis   **10**

  2-1 The Laplace Transform   10

      Definition   10
      Properties of the Laplace Transform   13

  2 2 Solution of Differential Equations Using the Laplace Transform   22

      Laplace Transform Solution Procedure   22
      Inversion of Laplace Transforms by Partial Fractions Expansion   24
      Eigenvalues and Stability   37
      Finding the Roots of Polynomials   38
      Outline of Laplace Transform Method for Solving Differential
      Equations   42

  2-3 Linearization and Deviation Variables   43

      Deviation Variables   43
      Linearization of Functions of One Variable   45
      Linearization of Functions of Two or More Variables   48

  2-4 Review of Complex Number Algebra   52

      Complex Numbers   52
      Operations with Complex Numbers   54

  2-5 Summary   57
      References   57
      Problems   58

Chapter 3   First-Order Dynamic Systems   **64**

  3-1 Thermal Process   64
  3-2 Gas Process   71
  3-3 Transfer Functions and Block Diagrams   74

      Transfer Functions   74
      Block Diagrams   76

3-4   Dead Time   83
3-5   Level Process   85
3-6   Chemical Reactor   89
3-7   Response of First-Order Processes to Different Types of Forcing
      Functions   93

         Step Function   93
         Ramp Function   94
         Sinusoidal Function   94

3-8   Summary   96
      Problems   97

Chapter 4   Higher-Order Dynamic Systems   104

4-1   Tanks in Series—Noninteracting System   104
4-2   Tanks in Series—Interacting System   111
4-3   Thermal Process   115
4-4   Response of Higher-Order Systems to Different Types of Forcing
      Functions   122

         Step Function   123
         Sinusoidal Function   128

4-5   Summary   130
      References   131
      Problems   131

Chapter 5   Basic Components of Control Systems   136

5-1   Sensors and Transmitters   136
5-2   Control Valves   138

         Control Valve Action   138
         Control Valve Sizing   139
         Selection of Design Pressure Drop   144
         Control Valve Flow Characteristics   147
         Control Valve Gain   152
         Control Valve Summary   154

5-3   Feedback Controllers   154

         Actions of Controllers   157
         Types of Feedback Controllers   158
         Reset Windup   169
         Feedback Controller Summary   171

5-4   Summary   172
      References   172
      Problems   172

Chapter 6   Design of Single-Loop Feedback Control Systems   176

6-1   The Feedback Control Loop   176
         Closed-Loop Transfer Function   179

Characteristic Equation of the Loop | 181
Steady-State Closed-Loop Response | 187

6-2 Stability of the Control Loop | 198

Criterion of Stability | 198
Routh's Test | 200
Effect of Loop Parameters on the Ultimate Gain | 203
Direct Substitution Method | 205
Effect of Dead Time | 208

6-3 Tuning of Feedback Controllers | 210

Quarter Decay Ratio Response by Ultimate Gain | 211
Process Characterization | 214
Process Step Testing | 216
Quarter Decay Ratio Response | 225
Tuning for Minimum Error Integral Criteria | 226
Tuning Sampled-Data Controllers | 234
Summary | 237

6-4 Synthesis of Feedback Controllers | 237

Development of the Controller Synthesis Formula | 237
Specification of the Closed-Loop Response | 238
Controller Modes and Tuning Parameters | 239
Derivative Mode for Dead-Time Processes | 243
Summary | 249

6-5 Prevention of Reset Windup | 249
6-6 Summary | 253
References | 253
Problems | 254

Chapter 7 Classical Single-Loop Feedback Control Design | **273**

7-1 Root Locus Technique | 277

Examples | 277
Rules for Plotting Root Locus Diagrams | 282
Root Locus Summary | 292

7-2 Frequency Response Techniques | 292

Bode Plots | 300
Polar Plots | 320
Nichols Plot | 328
Frequency Response Summary | 329

7-3 Pulse Testing | 329

Performing the Pulse Test | 330
Derivation of the Working Equation | 331
Numerical Evaluation of the Fourier Transform Integral | 333

7-4 Summary | 336
References | 337
Problems | 337

Chapter 8   Additional Control Techniques   **344**

8-1   Computing Relays   344
8-2   Ratio Control   354
8-3   Cascade Control   362
8-4   Feedforward Control   369

    A Process Example   370
    Lead/Lag Unit   378
    Block Diagram Design of Linear Feedforward Control   379
    Two Other Examples   384
    Inverse Response   390
    Feedforward Control Summary   391

8-5   Override and Selective Control   392
8-6   Multivariable Process Control   396

    Signal Flow Graphs (SFG)   399
    Pairing Controlled and Manipulated Variables   407
    Interaction and Stability   419
    Decoupling   420

8-7   Summary   429
    References   429
    Problems   430

Chapter 9   Modeling and Simulation of Process Control Systems   **448**

9-1   Development of Complex Process Models   449
9-2   Dynamic Model of a Distillation Column   450

    Tray Equations   452
    Feed and Top Trays   454
    Reboiler   455
    Condenser Model   458
    Condenser Accumulator Drum   460
    Initial Conditions   463
    Input Variables   464
    Summary   464

9-3   Dynamic Model of a Furnace   464
9-4   Solution of Partial Differential Equations   469
9-5   Computer Simulation of Dynamic Process Models   471

    Continuous Stirred Tank Reactor Simulation—An Example   471
    Numerical Integration by Euler's Method   475
    Duration of Simulation Runs   475
    Selection of Integration Interval   478
    Display of Simulation Results   479
    Example Results for Euler's Method   480
    Modified Euler Method   482
    Runge-Kutta-Simpson Method   489
    Summary   491

9-6 Special Simulation Languages and Subroutines 491
9-7 Control Simulation Examples 493
9-8 Stiffness 505

 Sources of Stiffness in a Model 505
 Numerical Integration of Stiff Systems 512

9-9 Summary 516
 References 516
 Problems 516

Appendix A Instrumentation Symbols and Labels 527

Appendix B Case Studies 533

 Case I Ammonium Nitrate Prilling Plant Control System 533
 Case II Natrual Gas Dehydration Control System 535
 Case III Sodium Hypochlorite Bleach Preparation Control
   System 536
 Case IV Control Systems in the Sugar-Refining Process 537
 Case V CO_2 Removal from Synthesis Gas 538
 Case VI Sulfuric Acid Process 543

Appendix C Sensors, Transmitters, and Control Valves 545

 Pressure Sensors 545
 Flow Sensors 546
 Level Sensors 557
 Temperature Sensors 560
 Composition Sensors 567
 Transmitters 567

  Pneumatic Transmitter 569
  Electronic Transmitter 570

 Types of Control Valves 570

  Reciprocating Stem 571
  Rotating Stem 574

 Control Valve Actuator 575

  Pneumatically Operated Diaphragm Actuator 575
  Piston Actuator 576
  Electrohydraulic and Electromechanical Actuators 577
  Manual-Handwheel Actuator 577

 Control Valve Accessories 578

  Positioners 578
  Boosters 579
  Limit Switches 580

Control Valves—Additional Considerations   581
  Viscosity Corrections   581
  Flashing and Cavitation   586
Summary   592
References   595

Appendix D   Computer Program to Solve for Roots of Polynomials   **597**

Index   **605**

PRINCIPLES AND PRACTICE OF

AUTOMATIC
PROCESS
CONTROL

CHAPTER
1
Introduction

The principal purpose of this chapter is to present you, the reader, with the need for automatic process control and to motivate you to study it. Automatic process control is concerned with maintaining process variables, temperatures, pressures, flows, compositions, and the like at some desired operating value. As we shall see in the ensuing pages, processes are dynamic in nature. Changes are always occurring, and if actions are not taken, the important process variables—those related to safety, product quality, and production rates—will not achieve design conditions.

This chapter also introduces two control systems, takes a look at some of their components, and defines some terms used in the field of process control. Finally, the background needed for the study of process control is discussed.

1-1. A PROCESS CONTROL SYSTEM

In order to fix ideas, let us consider a heat exchanger in which a process stream is heated by condensing steam. The process is sketched in Fig. 1-1.

The purpose of this unit is to heat the process fluid from some inlet tempertature $T_i(t)$, up to a certain desired outlet temperature, $T(t)$. As mentioned, the heating medium is condensing steam. The energy gained by the process fluid is equal to the heat released by the steam, provided there are no heat losses to the surroundings, that is, the heat

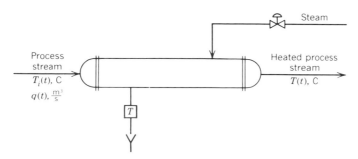

Figure 1-1. Heat exchanger.

exchanger and piping are well insulated. In this case the heat released is the latent heat of condensation of the steam.

In this process there are many variables that can change, causing the outlet temperature to deviate from its desired value. If this happens some action must be taken to correct for this deviation. That is, the objective is to control the outlet process temperature to maintain its desired value.

One way to accomplish this objective is by first measuring the temperature $T(t)$, then comparing it to its desired value, and, based on this comparison, deciding what to do to correct for any deviation. The flow of steam can be used to correct for the deviation. That is, if the temperature is above its desired value, then the steam valve can be throttled back to cut the steam flow (energy) to the heat exchanger. If the temperature is below its desired value, then the steam valve could be opened some more to increase the steam flow (energy) to the exchanger. All of this can be done manually by the operator, and since the procedure is fairly straightforward, it should present no problem. However, since in most process plants there are hundreds of variables that must be maintained at some desired value, this correction procedure would require a tremendous number of operators. Consequently, we would like to accomplish this control automatically. That is, we want to have instruments that control the variables without requiring intervention from the operator. This is what we mean by *automatic process control*.

To accomplish this objective a *control system* must be designed and implemented. A possible control system and its basic components are shown in Fig. 1-2. (Appendix A presents the symbols and identifications for different instruments.) The first thing to do is to measure the outlet temperature of the process stream. This is done by a *sensor* (thermocouple, resistance temperature device, filled system thermometers, thermistors, etc.). This sensor is connected physically to a *transmitter*, which takes the output from the sensor and converts it to a signal strong enough to be transmitted to a *controller*. The controller then receives the signal, which is related to the temperature, and compares it with the desired value. Depending on this comparison, the controller decides what to do to maintain the temperature at its desired value. Based on this decision, the controller then sends another signal to the *final control element*, which in turn manipulates the steam flow.

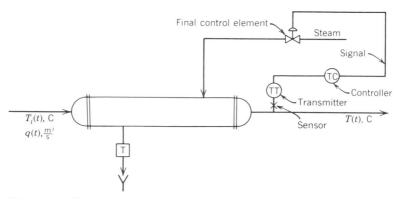

Figure 1-2.   Heat exchanger control system.

The preceding paragraph presents the four basic components of all control systems. They are

1. *Sensor*, also often called the primary element.

2. *Transmitter*, also called the secondary element.

3. *Controller*, the "brain" of the control system.

4. *Final control element*, often a control valve but not always. Other common final control elements are variable speed pumps, conveyors, and electric motors.

The importance of these components is that they perform the three basic operations that *must* be present in *every* control system. These operations are

1. *Measurement (M):* Measuring the variable to be controlled is usually done by the combination of sensor and transmitter.

2. *Decision (D):* Based on the measurement, the controller must then decide what to do to maintain the variable at its desired value.

3. *Action (A):* As a result of the controller's decision, the system must then take an action. This is usually accomplished by the final control element.

As mentioned, these three operations, M, D, and A, *must* be present in *every* control system. The decision-making operation in some systems is rather simple, while in others it is more complex; we will look at many of them in this book. The engineer designing a control system must be sure that the action taken affects the variable to be controlled, that is, that the action taken affects the measured value. Otherwise, the system is not controlling and will probably do more harm than good.

1-2. IMPORTANT TERMS AND OBJECTIVE OF AUTOMATIC PROCESS CONTROL

At this time it is necessary to define some terms used in the field of automatic process control. The first term is *controlled variable*. This is the variable that must be maintained or controlled at some desired value. In the preceding example, the process outlet temperature, $T(t)$, is the controlled variable. The second term is *set point*, the desired value of the controlled variable. The *manipulated variable* is the variable used to maintain the controlled variable at its set point. In the example, the flow of steam is the manipulated variable. Finally, any variable that can cause the controlled variable to deviate away from set point is defined as a *disturbance or upset*. In most processes there are a number of different disturbances. As an example, in the heat exchanger shown in Fig. 1-2, possible disturbances are the inlet process temperature, $T_i(t)$, the process flow, $q(t)$, the quality of the energy of the steam, ambient conditions, process fluid composition, fouling, and so on. What is important here is to understand that in the process industries, most often it is because of these disturbances that automatic process control is needed. If there were no disturbances, design operating conditions would prevail and there would be no necessity of continuously "policing" the process.

The following additional terms are also important. *Open-loop* refers to the condition in which the controller is disconnected from the process. That is, the controller is not making the decision of how to maintain the controlled variable at set point. Another instance in which open-loop control exists is when the action (A) taken by the controller does not affect the measurement (M). This is indeed a major flaw in the control system design. *Closed-loop control* refers to the condition in which the controller is connected to the process, comparing the set point to the controlled variable and determining corrective action.

With these terms defined, the objective of an automatic process control system can be stated as follows:

The objective of an automatic process control system is to use the manipulated variable to maintain the controlled variable at its set point in spite of disturbances.

1-3. REGULATORY AND SERVO CONTROL

In some processes the controlled variable deviates from a constant set point because of disturbances. *Regulatory control* refers to systems designed to compensate for these disturbances. In some other instances the most important disturbance is the set point itself. That is, the set point may be changed as a function of time (typical of this are batch processes), and therefore the controlled variable must follow the set point. *Servo control* refers to control systems designed for this purpose.

Regulatory control is by far more common than servo control in the process industries. However, the basic approach to designing either of them is essentially the same. Thus, the principles learned in this book apply to both cases.

1-4. TRANSMISSION SIGNALS

Let us now say a few words about the signals used to provide communication between instruments of a control system. There are three principal types of signals in use in the process industry today. The *pneumatic signal*, or air pressure, ranges normally between 3 and 15 psig. Less often, signals of 6 to 30 psig or 3 to 27 psig are used. The usual representation in piping and instrument diagrams (P&ID) for pneumatic signals is —#——#—. The *electrical*, or *electronic, signal* ranges normally between 4 and 20 mA. Less often 10 to 50 mA, 1 to 5 V or 0 to 10 V are used. The usual representation in P&ID's for this signal is --------. The third type of signal, which is becoming common, is the *digital*, or *discrete, signal* (zeros and ones). The use of process-control systems based on large-scale computers, minicomputers, or microprocessors is forcing increased use of this type of signal.

It is often necessary to change one type of signal into another type. This is done by a *transducer*. For example, there may be a need to change from an electrical signal, mA, to a pneumatic signal, psig. This is done by the use of a current (I) to pneumatic (P) transducer (I/P). This is shown graphically in Fig. 1-3. The input signal may be 4 to 20 mA and the output 3 to 15 psig. There are many other types of transducers: pneumatic-to-current (P/I), voltage-to-pneumatic (E/P), pneumatic-to-voltage (P/E), and so on.

Figure 1-3.   I/P transducer.

1-5.   CONTROL STRATEGIES

Feedback Control

The control scheme shown in Fig. 1-2 is referred to as feedback control, also called a feedback control loop. This technique was first applied to control of an industrial process by James Watt about 200 years ago. The application consisted of maintaining constant speed of a steam engine under variable load; this was a regulatory control application. In this scheme the controlled variable is obtained and fed back to the controller so that it can make a decision. One must understand the working principles of feedback control to recognize its advantages and disadvantages; the heat exchanger control loop shown in Fig. 1-2 is presented to foster this understanding.

If the inlet process temperature increases, thus creating a disturbance, its effect must propagate through the entire heat exchanger before the outlet temperature changes. Once the outlet temperature changes, the signal from the transmitter to the controller also changes. It is then that the controller becomes aware that it must compensate for the disturbance by changing the steam flow. The controller then signals the valve to close its opening and thus decrease the steam flow. Fig. 1-4 shows graphically the effect of the disturbance and the action of the controller.

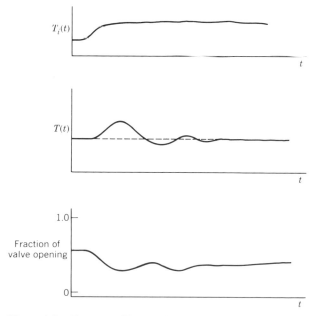

Figure 1-4.   Response of heat exchanger control system.

It is interesting to note that at first the outlet temperature increases, because of the increase in inlet temperature, but it then decreases even below set point and continues to oscillate around set point until the temperature finally stabilizes. This oscillatory response shows that the operation of a feedback control system is essentially a trial-and-error operation. That is, when the controller notices that the outlet temperature has increased above the set point, it signals the valve to close, but the closure is more than required. Therefore, the outlet temperature is brought down below the set point. Noticing this, the controller signals the valve to open again somewhat to bring the temperature back up. This trial and error continued until the temperature reached and stayed at set point.

The *advantage* of feedback control is that it is a very simple technique, as shown in Fig. 1-2, that compensates for *all* disturbances. Any disturbance will affect the controlled variable, and once this variable deviates from set point, the controller will change its output to return it to set point. The feedback control loop does not know, nor does it care, which disturbance enters the process. It tries only to maintain the controlled variable at set point and in so doing compensates for all disturbances. The *disadvantage* of feedback control is that it can compensate for a disturbance only after the controlled variable has deviated from set point. That is, the disturbance must propagate through the entire process before the feedback control scheme can compensate for it.

The job of the engineer is to design a control scheme that will maintain the controlled variable at its set point. Once this is done, he must then tune the controller so that it minimizes the trial-and-error operation required to control. To do a creditable job, the engineer must know the characteristics, or ''personality,'' of the process to be controlled. Once this ''process personality'' is known, the engineer can design the control system and obtain the best ''controller personality'' to match that of the process. What is meant by ''personality'' is explained in the next few chapters. To help you now, however, imagine that you are trying to persuade someone to behave in a certain way i.e., to control someone's behavior. You are the controller and that someone is the process. The wisest thing for you to do is to learn that someone's personality and then adapt yourself to that personality to do a good job of persuading or controlling. This is what is meant by ''tuning the controller.'' That is, the controller is adapted, or tuned, to the process. Most controllers have up to three parameters used to tune them, as presented in Chapters 5 and 6.

Feedforward Control

Feedback control is the most common control strategy in the process industries. Its simplicity accounts for its popularity. In some processes, however, feedback control may not provide the required control performance. For these processes other types of control may have to be designed. Chapter 8 presents additional control strategies that have proven to be profitable. One such strategy is feedforward control. The objective of feedforward cotnrol is to measure the disturbances and compensate for them before the controlled variable deviates from set point. If applied correctly the controlled variable will not deviate from set point.

A concrete example of feedforward control is the heat exchanger shown in Fig. 1-1. Suppose that the "major" disturbances are the inlet temperature, $T_i(t)$, and the process flow, $q(t)$. To implement feedforward control these two disturbances must first be measured and then a decision must be made about how to manipulate the steam flow to compensate for these disturbances. Figure 1-5 shows this control strategy. The feedforward controller makes the decision about how to manipulate the steam flow to maintain the controlled variable at set point depending on the inlet temperature and process flow.

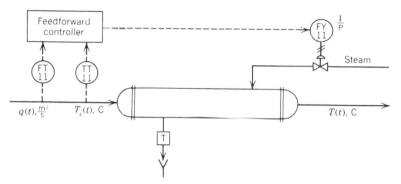

Figure 1-5. Heat exchanger feedforward control system.

Earlier in Section 1-2 we learned that there are a number of different disturbances. The feedforward control system shown in Fig. 1-5 compensates for only two of them. If any of the other ones enter the process, this strategy will not compensate for it, and the result will be a permanent deviation from set point of the controlled variable. To avoid this deviation some feedback compensation must be added to feedforward control; this is shown in Fig. 1-6. Feedforward control now compensates for the "major" disturbances, $T_i(t)$ and $q(t)$, while feedback control compensates for all other disturbances.

Chapter 8 presents the development of the feedforward controller and the instrumentation required to implement it. Actual industrial cases are used to discuss in detail this important strategy.

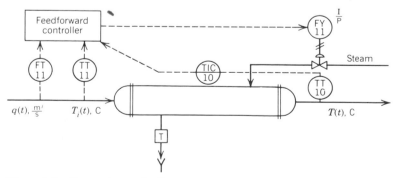

Figure 1-6. Heat exchanger feedforward control with feedback compensation.

It is important to notice that the three basic operations, M, D, A, are still present in this more "advanced" control strategy. Measurement is performed by the sensors and transmitters. Decision is made by the feedforward controller and the feedback controller, TIC-10. Action is taken by the steam valve.

In general, the control strategies presented in Chapter 8 are more costly, in hardware and manpower necessary to design, implement, and maintain them, than feedback control. Therefore, they must be justified (monetarily) before they can be implemented. The best procedure is to first design and implement a simple control strategy, keeping in mind that if it does not prove satisfactory then a more "advanced" strategy may be justifiable. It is important, however, to recognize that these advanced strategies still require some feedback compensation.

1-6. UNDERLYING REASONS FOR PROCESS CONTROL

Earlier in this chapter the objective of automatic process control was defined as "maintaining the controlled variable at set point in spite of disturbances." It is wise at this time to enumerate some of the "reasons" why this is important. These reasons are based on our industrial experience and we would like to pass them on to the reader. They may not be the only ones, but again we think that they are the most important.

1. Prevent injury to plant personnel or damage to equipment. *Safety* must always be in everyone's mind; it is the single most important consideration.
2. Maintain product quality (composition, purity, color, etc.) on a continuous basis and with minimum cost.
3. Maintain plant production rate at minimum cost.

So it can be said that the reasons for automation of process plants are to provide a safe environment and at the same time to maintain desired product quality and high plant throughput and reduce the demand on human labor.

1-7. BACKGROUND NEEDED FOR PROCESS CONTROL

To be successful in the practice of automatic process control, the engineer must first understand the principles of process engineering. Therefore, this book assumes that the reader is familiar with the basic principles of thermodynamics, fluid flow, heat transfer, separation processes, reaction processes, and the like.

For the study of process control it is also important to understand how processes behave dynamically. Consequently, it is necessary to develop the set of equations that describe different processes. This is called *modeling*. To do this, the knowledge of the basic principles mentioned in the previous paragraph and of mathematics through differential equations is needed. In process control the Laplace transforms are used heavily. This greatly simplifies the solution of differential equations and the dynamic analysis of processes and their control systems. Chapter 2 of this book is devoted to the development and usage of the Laplace transforms along with a review of complex number algebra.

Another important "tool" for the study and practice of process control is computer simulation. Many of the equations developed to describe processes are nonlinear in nature and, consequently, the most exact way to solve them is by numerical methods; this means computer solution. The computer solution of process models is called *simulation*. Chapters 3 and 4 present an introduction to the modeling of some simple processes. Chapter 9 develops models for more complex processes and also presents an introduction to simulation.

1-8. SUMMARY

In this chapter the need for automatic process control has been discussed. Industrial processes are not static but rather very dynamic; they are continuously changing because of many types of disturbances. It is principally because of this dynamic nature that control systems are needed to continuously and automatically watch over the variables that must be controlled.

The working principles of a control system can be summarized with the three letters M, D, A. M refers to the measurement of process variables. D refers to the decision to be made based on the measurements of the process variables. Finally, A refers to the action to be taken based on the decision.

The basic components of a process control system were also presented: sensor, transmitter, controller, and final control element. The most common types of signals—pneumatic, electronic or electrical, and digital—were introduced along with the purpose of transducers.

Two control strategies were presented: feedback and feedforward control. The advantages and disadvantages of both strategies were briefly discussed. Chapters 6 and 7 present the subject of design and analysis of feedback control loops. Feedforward control is presented in more detail, along with other control strategies, in Chapter 8.

In writing this book we have been constantly aware that to be successful, the engineer must be able to apply the principles learned. Consequently, the book covers the necessary principles of automatic process control for its successful practice. The book is full of actual cases drawn from our several years of industrial experience as full-time practitioners or part-time consultants. We sincerely hope that you get excited about studying automatic process control. It is a very dynamic, challenging, and rewarding area of process engineering.

CHAPTER
2

Mathematical Tools for Control Systems Analysis

The techniques of Laplace transforms and linearization have been found to be particularly useful in analyzing process dynamics and designing control systems. This is because they allow us to gain a general insight into the behavior of a wide variety of processes and instruments. In contrast, the technique of computer simulation provides us with a more accurate and detailed analysis of the dynamic behavior of specific systems, but seldom allows us to generalize our findings to other processes.

In this chapter we will review the Laplace transform method of solving linear differential equations. By this method we can convert a linear differential equation into an algebraic equation. This in turn allows us to develop the useful concept of transfer functions, which will be introduced in this chapter and extensively used in the ones that follow. Since the differential equations that represent most processes are nonlinear, we will introduce the method of linearization to approximate the nonlinear differential equations with linear ones so that the technique of Laplace transforms can be applied. A familiarity with complex numbers is required to work with Laplace transforms. Because of this we have included a brief review of complex number algebra as a separate section. We firmly believe that a knowledge of Laplace transforms is essential for understanding the fundamentals of process dynamics and control systems design.

2-1. THE LAPLACE TRANSFORM

Definition

The Laplace transform of a function of time, $f(t)$, is defined by the following formula:

$$F(s) = \mathscr{L}[f(t)] = \int_0^\infty f(t)e^{-st} \, dt \qquad (2-1)$$

where

$f(t)$ is a function of time

$F(s)$ is the corresponding Laplace transform

s is the Laplace transform variable

t is time

In the application of Laplace transforms to the design of control systems, the functions of time are the variables in the system, including the manipulated and controlled variables, the transmitter signals, the disturbances, the control valve positions, the flow through the control valves, and any other intermediate variables or signals. It is therefore very important to realize that the Laplace transforms apply to the variables and signals and not to the processes or the instruments.

In order to gain a familiarity with the definition of Laplace transforms, let us find the transforms of a number of common input signals.

Example 2-1: In the analysis of control systems, signals are applied as inputs to the system (e.g., disturbances, set point changes, etc.) in order to study its response. Although some of the types of signals are usually difficult or even impossible to achieve in practice, they provide useful tools for the comparison of responses. In this example we will derive the Laplace transforms of

    (a)   A unit step function
    (b)   A pulse
    (c)   A unit impluse function
    (d)   A sine wave

Solution.

(a)   Unit Step Function

This is a sudden change of unity magnitude at time equals zero. It is sketched graphically in Fig. 2-1a and can be represented algebraically by

$$u(t) = \begin{cases} 0 & t < 0 \\ 1 & t \geq 0 \end{cases}$$

Its Laplace transform is given by

$$\mathcal{L}[u(t)] = \int_0^\infty u(t)e^{-st}\,dt = -\frac{1}{s}e^{-st}\Big|_0^\infty = -\frac{1}{s}(0-1)$$

$$\mathcal{L}[u(t)] = \frac{1}{s}$$

(b)   A Pulse of Magnitude H and Duration T

The pulse is sketched in Fig. 2-1b and can be represented algebraically by

$$f(t) = \begin{cases} 0 & t < 0,\, t \geq T \\ H & 0 \leq t < T \end{cases}$$

Its Laplace transform is given by

$$\mathcal{L}[f(t)] = \int_0^\infty f(t)\,e^{-st}\,dt = \int_0^T He^{-st}\,dt$$

$$= -\frac{H}{s} e^{-st} \Big|_0^T = -\frac{H}{s}(e^{-sT} - 1)$$

$$\mathscr{L}[f(t)] = \frac{H}{s}(1 - e^{-sT})$$

(c)   A Unit Impulse Function

This is an ideal pulse of infinite amplitude and zero duration with an area of unity—in other words, a pulse of unit area with all of its area concentrated at time equals zero. A sketch of this function is given in Fig. 2-1c. The symbol $\delta(t)$ is commonly used to represent it and it is known as the "Dirac delta" function. An algebraic expression could be obtained by taking limits on the pulse function of part (b):

$$\delta(t) = \lim_{T \to 0} f(t)$$

with

$$HT = 1 \text{ (the area) or } H = 1/T$$

The Laplace transform can be obtained by taking the limit of the result of part (b):

$$\mathscr{L}[\delta(t)] = \lim_{T \to 0} \frac{1}{Ts}(1 - e^{-sT}) = \frac{1}{0}(1 - 1) = \frac{0}{0}$$

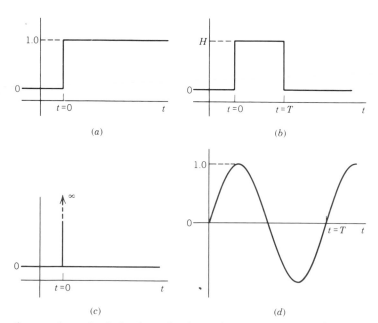

Figure 2-1.   Common input signals for the study of control system response. (*a*) Unit step function, $u(t)$. (*b*) Pulse. (*c*) Impulse function, $\delta(t)$. (*d*) Sine wave, $\sin \omega t$ ($\omega = 2\pi/T$).

This requires the application of L'Hopital's rule for undefined limits:

$$\mathcal{L}[\delta(t)] = \lim_{T \to 0} \frac{\dfrac{d}{dT}(1 - e^{-sT})}{\dfrac{d}{dT}(Ts)}$$

$$= \lim_{T \to 0} \frac{s\,e^{-sT}}{s}$$

$$\mathcal{L}[\delta(t)] = 1$$

This is a very significant result, as it tells us that the unit impulse has a Laplace transform of unity.

(d)   A Sine Wave of Unity Amplitude and Frequency ω

The sine wave is sketched in Fig. 3-1d and can be represented in exponential form by

$$\sin \omega t = \frac{e^{i\omega t} - e^{-i\omega t}}{2i}$$

where $i = \sqrt{-1}$ is the unit of imaginary numbers.
   Its Laplace transform is given by

$$\mathcal{L}[\sin \omega t] = \int_0^\infty \sin \omega t \; e^{-st}\, dt$$

$$= \int_0^\infty \frac{e^{i\omega t} - e^{-i\omega t}}{2i}\, e^{-st}\, dt$$

$$= \frac{1}{2i}\left[\int_0^\infty e^{-(s-i\omega)t}\, dt - \int_0^\infty e^{-(s+i\omega)t}\, dt \right]$$

$$= \frac{1}{2i}\left[-\frac{e^{-(s-i\omega)t}}{s - i\omega} + \frac{e^{-(s+i\omega)t}}{s + i\omega} \right]\Bigg|_0^\infty$$

$$= \frac{1}{2i}\left[-\frac{0 - 1}{s - i\omega} + \frac{0 - 1}{s + i\omega} \right]$$

$$= \frac{1}{2i}\frac{2i\omega}{s^2 + \omega^2}$$

$$\mathcal{L}[\sin \omega t] = \frac{\omega}{s^2 + \omega^2}$$

The preceding example illustrates some of the algebraic manipulations involved in deriving the Laplace transform of a signal. Tables of Laplace transforms are available in most mathematics and engineering handbooks. A short list of the transforms of some common functions is given in Table 2-1.

Properties of the Laplace Transform

In this section we will look at some important properties of the Laplace transform. These properties are useful in that they allow us to derive the transforms of some functions from those of simpler functions such as those listed in Table 2-1. They also allow us to

Table 2-1　Laplace Transforms of Common Functions

$f(t)$	$F(s) = \mathscr{L}[f(t)]$
$\delta(t)$	1
$u(t)$	$\dfrac{1}{s}$
t	$\dfrac{1}{s^2}$
t^n	$\dfrac{n!}{s^{n+1}}$
e^{-at}	$\dfrac{1}{s+a}$
te^{-at}	$\dfrac{1}{(s+a)^2}$
$t^n e^{-at}$	$\dfrac{n!}{(s+a)^{n+1}}$
$\sin \omega t$	$\dfrac{\omega}{s^2 + \omega^2}$
$\cos \omega t$	$\dfrac{s}{s^2 + \omega^2}$
$e^{-at} \sin \omega t$	$\dfrac{\omega}{(s+a)^2 + \omega^2}$
$e^{-at} \cos \omega t$	$\dfrac{s+a}{(s+a)^2 + \omega^2}$

develop the relationships between the transforms of a function and its derivatives and integrals, and to determine the initial and final values of a function from its transform.

Linearity.　This is a most important property; it states that the Laplace transform is linear. This means that if k is a constant

$$\mathscr{L}[kf(t)] = k\mathscr{L}[f(t)] = kF(s) \tag{2-2}$$

Because it is linear, the distributive property also holds for the Laplace transform:

$$\mathscr{L}[f(t) + g(t)] = \mathscr{L}[f(t)] + \mathscr{L}[g(t)] \tag{2-3}$$
$$= F(s) + G(s)$$

Both of these properties can be easily proven by application of Eq. (2-1), the definition of the Laplace transform.

Real Differentiation Theorem.　This theorem establishes the relationship between the Laplace transform of a function and that of its derivatives. It states that

$$\mathscr{L}\left[\frac{df(t)}{dt}\right] = sF(s) - f(0) \tag{2-4}$$

Proof.　From the definition of the Laplace transform, Eq. (2-1)

$$\mathscr{L}\left[\frac{df(t)}{dt}\right] = \int_0^\infty \frac{df(t)}{dt} e^{-st}\, dt$$

Integrate by parts:

$$u = e^{-st} \qquad\qquad dv = \frac{df(t)}{dt} dt$$

$$du = -se^{-st} dt \qquad v = f(t)$$

$$\mathcal{L}\left[\frac{df(t)}{dt}\right] = f(t)e^{-st} \Big|_0^\infty - \int_0^\infty f(t)(-se^{-st} dt)$$

$$= [0 - f(0)] + s \int_0^\infty f(t)e^{-st} dt$$

$$= -f(0) + s\,\mathcal{L}[f(t)]$$

$$= sF(s) - f(0) \qquad\qquad\qquad \text{q.e.d.}$$

The extension to higher derivatives is straightforward:

$$\mathcal{L}\left[\frac{d^2 f(t)}{dt^2}\right] = \mathcal{L}\left[\frac{d}{dt}\left(\frac{df(t)}{dt}\right)\right]$$

$$= s\mathcal{L}\left[\frac{df(t)}{dt}\right] \quad \frac{df}{dt}(0)$$

$$- s\,[sF(s) - f(0)] - \frac{df}{dt}(0)$$

$$- s^2 F(s) - sf(0) - \frac{df}{dt}(0)$$

In general

$$\mathcal{L}\left[\frac{d^n f(t)}{dt^n}\right] = s^n F(s) - s^{n-1} f(0) - s^{n-2}\frac{df}{dt}(0) - \cdots$$

$$\cdots \quad s\,\frac{d^{n-2}f}{dt^{n-2}}(0) - \frac{d^{n-1}f}{dt^{n-1}}(0) \qquad (2\text{-}5)$$

For the very important case of the function and its derivatives having zero initial conditions, this expression simplifies to

$$\mathcal{L}\left[\frac{d^n f(t)}{dt^n}\right] = s^n F(s) \qquad (2\text{-}6)$$

You can see that, for the case of zero initial conditions, taking the Laplace transform of a derivative of a function can be done simply by replacing variable s for the "d/dt" operator, and $F(s)$ for $f(t)$.

Real Integration Theorem. This theorem establishes the relationship between the transform of a function and that of its integral. It states that

$$\mathcal{L}\left[\int_0^t f(t)\,dt\right] = \frac{1}{s} F(s) \qquad (2\text{-}7)$$

Proof. From the definition of the Laplace transform, Eq. (2-1), we

$$\mathcal{L}\left[\int_0^t f(t)\,dt\right] = \int_0^\infty \left[\int_0^t f(t)\,dt\right] e^{-st}\,dt$$

Integrate by parts:

$$u = \int_0^t f(t)\,dt \qquad dv = e^{-st}\,dt$$

$$du = f(t)\,dt \qquad v = -\frac{1}{s}e^{-st}$$

$$\mathcal{L}\left[\int_0^t f(t)\,dt\right] = -\frac{1}{s}e^{-st}\int_0^t f(t)\,dt \,\Big|_0^\infty - \int_0^\infty f(t)\,dt\left(-\frac{1}{s}e^{-st}\right)$$

$$= -\frac{1}{s}\left[0 - \int_0^0 f(t)\,dt\right] + \frac{1}{s}\int_0^\infty f(t)\,e^{-st}\,dt$$

$$= \frac{1}{s}\,\mathcal{L}\left[f(t)\right] = \frac{1}{s}F(s) \qquad\qquad \text{q.e.d.}$$

Notice that this derivation assumes that the initial value of the integral is zero. Under these conditions the transform of the nth integral of a function is the transform of the function divided by s^n.

Complex Differentiation Theorem. This theorem facilitates the evaluation of the transforms involving the time variable t. It is stated by

$$\mathcal{L}[t\,f(t)] = -\frac{d}{ds}F(s) \qquad\qquad (2\text{-}8)$$

Proof. From the definition of the Laplace transform, Eq. (2-1), we see that

$$F(s) = \int_0^\infty f(t)e^{-st}\,dt$$

Take the derivative of this equation with respect to s:

$$\frac{dF(s)}{ds} = \int_0^\infty f(t)\,(-te^{-st})\,dt$$

$$= -\int_0^\infty t\,f(t)e^{-st}\,dt$$

$$= -\mathcal{L}[t\,f(t)] \qquad\qquad \text{q.e.d.}$$

Rearranging this result we obtain the statement of the theorem.

Real Translation Theorem. This theorem deals with the translation of a function in the time axis, as illustrated in Fig. 2-2. The translated function is the original function delayed in time. As we shall see in Chapter 3, process time delays are caused by transportation lag. This phenomenon is commonly known as *dead time*.

Because the Laplace transform does not contain information about the original func-

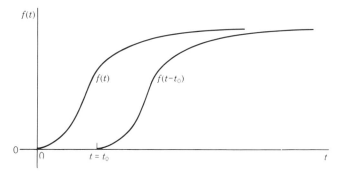

Figure 2-2. Function translated in time is zero for all times less than the time delay t_0.

tion for negative time, the delayed function must be assumed to be zero for all times less than the time delay (see Fig. 2-2). The theorem is stated by the following formula:

$$\mathscr{L}\left[f(t - t_0)\right] - e^{-st_0}F(s) \tag{2-9}$$

Proof. From the definition of the Laplace transform, Eq. (2-1), we get

$$\mathscr{L}[f(t - t_0)] = \int_0^{\infty} f(t - t_0)e^{-st}\,dt$$

Let $\tau = t - t_0$ (or $t = t_0 + \tau$) and substitute:

$$\mathscr{L}\left[f(t - t_0)\right] = \int_{\tau=-t_0}^{\tau=\infty} f(\tau)e^{-s(t_0+\tau)}\,d(t_0 + \tau)$$

$$= \int_{\tau=0}^{\tau=\infty} f(\tau)\,e^{-st_0}\,e^{-s\tau}\,d\tau$$

$$= e^{-st_0}\int_0^{\infty} f(\tau)\,e^{-s\tau}\,d\tau$$

$$= e^{-st_0}F_{(s)} \qquad\qquad \text{q.e.d.}$$

Notice that in this proof we made use of the fact that $f(\tau) = 0$ for $\tau < 0$ ($t < t_0$).

Complex Translation Theorem. This theorem facilitates the evaluation of the transforms of functions involving exponentials of time.

$$\mathscr{L}\left[e^{at}f(t)\right] = F(s - a) \tag{2-10}$$

Proof. From the definition of the Laplace transform, Eq. (2-1), we obtain

$$\mathscr{L}\left[e^{at}f(t)\right] = \int_0^{\infty} e^{at}f(t)\,e^{-st}\,dt$$

$$= \int_0^{\infty} f(t)\,e^{-(s-a)t}\,dt$$

Let $\sigma = s - a$ and substitute:

$$\mathcal{L}\left[e^{at}f(t)\right] = \int_0^\infty f(t)e^{-\sigma t}\, dt$$

$$= F(\sigma) = F(s - a) \qquad\qquad\qquad \text{q.e.d.}$$

Final Value Theorem.   This theorem allows us to calculate the final or steady-state value of a function from its transform. It is also useful in checking the validity of derived transforms. If the limit of $f(t)$ as $t \to \infty$ exists, it can be determined from its Laplace transform as follows:

$$\lim_{t\to\infty} f(t) = \lim_{s\to 0} sF(s) \qquad\qquad (2\text{-}11)$$

Initial Value Theorem.   This theorem is useful in calculating the initial value of a function from its transform and in providing another check on the validity of derived transforms.

$$\lim_{t\to 0} f(t) = \lim_{s\to\infty} sF(s) \qquad\qquad (2\text{-}12)$$

The proofs of these last two theorems add little to understanding them. In the examples that follow we shall check the validity of our derived transforms by applying the initial and final value theorems.

Example 2-2.   Derive the transforms of the following expression by application of the properties of Laplace transforms:

$$\frac{d^2x(t)}{dt^2} + 2\xi\omega_n \frac{dx(t)}{dt} + \omega_n^2 x(t) = Kr(t)$$

where K, ω_n, and ξ are constants, and $x(0) = dx/dt(0) = 0$

Solution.   By application of the linearity property of Laplace transforms, Eqs. (2-2) and (2-3), we can take the transform of each term. Then we apply the real differentiation theorem, Eq. (2-5):

$$\mathcal{L}\left[\frac{d^2x(t)}{dt^2}\right] = s^2X(s) - sx(0) - \frac{dx}{dt}(0) = s^2X(s)$$

$$\mathcal{L}\left[2\xi\omega_n \frac{dx(t)}{dt}\right] = 2\xi\omega_n \mathcal{L}\left[\frac{dx(t)}{dt}\right]$$

$$= 2\xi\omega_n \left[sX(s) - x(0)\right]$$

$$= 2\xi\omega_n\, sX(s)$$

$$\mathcal{L}\left[\omega_n^2 x(t)\right] = \omega_n^2\, \mathcal{L}\left[x(t)\right] = \omega_n^2 X(s)$$

$$\mathcal{L}\left[Kr(t)\right] = K\, \mathcal{L}\left[r(t)\right] = KR(s)$$

Substituting into the original equation yields

$$s^2X(s) + 2\xi\omega_n sX(s) + \omega_n^2 X(s) = KR(s)$$

solving for $X(s)$ gives

$$X(s) = \frac{K}{s^2 + 2\xi\omega_n s + \omega_n^2} R(s)$$

The preceding example illustrates the fact that the Laplace transform converts the original differential equation into an algebraic equation. Herein lies the great usefulness of Laplace transforms, since algebraic equations are a lot easier to manipulate than differential equations. The price we pay for this advantage is the need to transform and then invert the transform to obtain the solution in the "time domain," that is, with time as the independent variable.

Example 2-3. Derive the Laplace transform of the following function:

$$y(t) = te^{-at}$$

where a is a constant.

Solution. We can evaluate the transform $Y(s)$ by using one of two properties: the complex differentiation theorem or the complex translation theorem.
Apply the complex differentiation theorem, Eq. (2-8).
Let

$$f(t) = e^{-at} \qquad F(s) = \frac{1}{s + a} \qquad \text{(from Table 2-1)}$$

$$Y(s) = \mathcal{L}\,[te^{-at}] = \mathcal{L}\,[tf(t)] = \frac{d}{ds}F(s) =$$

$$= -\frac{d}{ds}\left(\frac{1}{s + a}\right) = -\left[-\frac{1}{(s + a)^2}\right]$$

$$Y(s) = \frac{1}{(s + a)^2}$$

We can obtain the same answer by applying the complex translation theorem, Eq. (2-10):
Let

$$f(t) = t \qquad F(s) = \frac{1}{s^2} \qquad \text{(from Table 2-1)}$$

then

$$\mathcal{L}\,[e^{-at}\,t] = \mathcal{L}\,[e^{-at}f(t)] = F\,[s - (-a)]$$

$$= F(s + a) = \frac{1}{(s + a)^2}$$

This checks our previous answer.

We now check the validity of the transform by application of the initial and final value theorems.

Initial Value

$$\lim_{t \to 0} y(t) = 0e^{-a(0)} = 0$$

$$\lim_{s \to \infty} sY(s) = \lim_{s \to \infty} \frac{s}{(s + a)^2} = \frac{\infty}{\infty}$$

From L'Hopital's rule

$$\lim_{s \to \infty} Y(s) = \lim_{s \to \infty} \frac{1}{2(s + a)} = 0 \qquad \text{(check)}$$

Final value

$$\lim_{t \to \infty} y(t) = \lim_{t \to \infty} \frac{t}{e^{at}} = \frac{\infty}{\infty}$$

From L'Hopital's rule

$$\lim_{t \to \infty} y(t) = \lim_{t \to \infty} \frac{1}{ae^{at}} = 0$$

$$\lim_{s \to 0} sY(s) = \lim_{s \to 0} \frac{s}{(s + a)^2} = \frac{0}{a^2} = 0 \qquad \text{(check)}$$

Unfortunately zero is not a very good check because it does not allow us to detect sign errors.

Example 2-4.   Derive the Laplace transform of the following time-delayed function:

$$m(t) = u(t - 3) e^{-(t-3)} \cos \omega(t - 3)$$

where ω is a constant.

Note: The term $u(t - 3)$ is included in this expression to make it explicitly clear that the function is zero for $t < 3$. We recall that $u(t - 3)$ is the unit step function at $t = 3$, which can be written as

$$u(t - 3) = \begin{cases} 0 & t < 3 \\ 1 & t \geq 3 \end{cases}$$

Thus, the presence of the $u(t - 3)$ does not alter the rest of the function for $t \geq 3$.

Solution.   Let

$$f(t - 3) = e^{-(t-3)} \cos \omega(t - 3)$$

then

$$f(t) = e^{-t} \cos \omega t$$

$$F(s) = \frac{(s + 1)}{(s + 1)^2 + \omega^2} \qquad \text{(from Table 2-1)}$$

By applying the real translation theorem, Eq. (2-9)

$$M(s) = \mathcal{L}\left[u(t - 3) f(t - 3)\right]$$
$$= e^{-3s} F(s)$$
$$M(s) = \frac{(s + 1)e^{-3s}}{(s + 1)^2 + \omega^2}$$

We now check the validity of this answer by applying the initial and final value theorems.

Initial Value

$$\lim_{t \to 0} m(t) = u(-3) (e^{+3}) \cos (-3\omega) = 0$$

because $u(-3) = 0$

$$\lim_{s \to \infty} sM(s) = \lim_{s \to \infty} \frac{s(s + 1) e^{-3s}}{(s + 1)^2 + \omega^2} = \frac{\infty(0)}{\infty}$$

We can separate the undefined part and apply L'Hopital's rule:

$$\lim_{s \to \infty} sM(s) = \left[\lim_{s \to \infty} \frac{s(s + 1)}{(s + 1)^2 + \omega^2}\right]\left[\lim_{s \to \infty} e^{-3s}\right] = \begin{bmatrix} \infty \\ \infty \end{bmatrix} [0]$$

$$= \left[\lim_{s \to \infty} \frac{2s + 1}{2(s + 1)}\right]\left[\lim_{s \to \infty} e^{-3s}\right] = \begin{bmatrix} \infty \\ \infty \end{bmatrix} [0]$$

$$= \left[\lim_{s \to \infty} \frac{2}{2}\right] [0] = 0 \qquad \text{(check)}$$

Final Value

In principle the final value theorem cannot be applied to periodic functions. This is because these functions oscillate forever without ever reaching a steady state. Nevertheless, application of the final value theorem to periodic functions results in the final value around which the function oscillates.

$$\lim_{t \to \infty} m(t) = (1) (0) \cos (\infty) = 0$$

$$\lim_{s \to 0} sM(s) = \frac{(0) (1) e^0}{1^2 + \omega^2} = 0 \qquad \text{(check)}$$

Example 2-5. Derive the Laplace transform of the following function:

$$c(t) = [u(t) - e^{-t/\tau}]$$

where τ is a constant.

Solution. Applying the linearity property, Eq. (2-3):

$$C(s) = \mathcal{L}\left[u(t) - e^{-t/\tau}\right]$$

$$= \mathcal{L}\left[u(t)\right] - \mathcal{L}\left[e^{-t/\tau}\right]$$

$$= \frac{1}{s} - \frac{1}{s + 1/\tau} \qquad \text{(from Table 2-1)}$$

$$C(s) = \frac{1}{s(\tau s + 1)}$$

Initial Value

$$\lim_{t \to 0} c(t) = 1 - e^0 = 0$$

$$\lim_{s \to \infty} sC(s) = \lim_{s \to \infty} \frac{s}{s(\tau s + 1)} = \frac{1}{\infty} = 0 \qquad \text{(check)}$$

Final Value

$$\lim_{t \to \infty} c(t) = 1 - e^{-\infty} = 1$$

$$\lim_{s \to 0} sC(s) = \lim_{s \to 0} \frac{s}{s(\tau s + 1)} = 1 \qquad \text{(check)}$$

2-2. SOLUTION OF DIFFERENTIAL EQUATIONS USING THE LAPLACE TRANSFORM

In order to illustrate the use of the Laplace transform to solve linear ordinary differential equations, let us consider the following second-order differential equation:

$$a_2 \frac{d^2 y(t)}{dt^2} + a_1 \frac{dy(t)}{dt} + a_0 y(t) = bx(t) \qquad (2\text{-}13)$$

The problem of solving this equation can be stated as follows: Given the coefficients a_0, a_1, a_2, and b, the appropriate initial conditions, and the function $x(t)$, find the function $y(t)$ that satisfies Eq. (2-13).

The function $x(t)$ is usually referred to as the "forcing function" or input variable and $y(t)$ is the "output function" or dependent variable. Variable t is the independent variable, time. In the design of control systems, a differential equation like Eq. (2-13) usually represents how a particular process or instrument relates its output signal, $y(t)$, to its input signal, $x(t)$.

Laplace Transform Solution Procedure

There are basically three steps involved in the solution of a differential equation using the Laplace transform.

Step 1. Transformation of the differntial equation into an algebraic equation in the Laplace transform variable s. This is done by taking the Laplace transform of each side of the equation:

$$\mathcal{L}\left[a_2 \frac{d^2y(t)}{dt^2} + a_1 \frac{dy(t)}{dt} + a_0\, y(t)\right] = \mathcal{L}\left[bx(t)\right] \tag{2-14}$$

Then, making use of the distributive property of the transform, Eq. (2-2), and of the real differentiation theorem, Eq. (2-5), we see that

$$\mathcal{L}\left[a_2 \frac{d^2y(t)}{dt^2}\right] = a_2\left[s^2Y(s) - sy(0) - \frac{dy}{dt}(0)\right]$$

$$\mathcal{L}\left[a_1 \frac{dy(t)}{dt}\right] = a_1\left[sY(s) - y(0)\right]$$

$$\mathcal{L}\left[a_0\, y(t)\right] = a_0\, Y(s)$$

$$\mathcal{L}\left[b\, x(t)\right] = b\, X(s)$$

Next we substitute these terms into Eq. (2-14) and rearrange:

$$(a_2s^2 + a_1s + a_0)\, Y(s) - (a_2s + a_1)\, y(0) - a_2\frac{dy}{dt}(0) - b\,X(s)$$

Notice that this is an algebraic equation and that the Laplace transform variable s can be manipulated like any other algebraic quantity.

Step 2. Manipulation of the algebraic equation to solve for the output variable $Y(s)$ in terms of the input variable and the initial conditions:

$$Y(s) = \frac{bX(s) + (a_2s + a_1)\, y(0) + a_2 \dfrac{dy}{dt}(0)}{a_2s^2 + a_1s + a_0} \tag{2-15}$$

Step 3. Inversion of the resulting equation to obtain the output variable as a function of time $y(t)$:

$$y(t) = \mathcal{L}^{-1}\left[Y(s)\right] \tag{2-16}$$

$$= \mathcal{L}^{-1}\left[\frac{b\,X(s) + (a_2s + a_1)\, y(0) + a_2 \dfrac{dy}{dt}(0)}{a_2s^2 + a_1s + a_0}\right]$$

In this procedure the first two steps are relatively easy and straightforward, with all of the difficulty concentrated in the third step. The usefulness of Laplace transforms in the design of control systems stems from the fact that the inversion step is seldom necessary. This is so because all of the important characteristics of the time response $y(t)$ can be recognized in the terms of $Y(s)$. In other words, the entire analysis can be made in the Laplace or "s-domain" without ever having to invert the transform back into the "time domain." The term "domain" is used in the preceding sentence to denote the independent variable of the field in which the analysis and design are carried out.

The inversion step establishes the relationship between the Laplace transform $Y(s)$ and its inverse $y(t)$. We shall demonstrate the inversion step by the method of partial fractions expansion. However, let us first generalize Eq. (2-15) for the case of an nth-order equation.

For the nth-order linear ordinary differential equation with constant coefficients

$$a_n \frac{d^n y(t)}{dt^n} + a_{n-1} \frac{d^{n-1} y(t)}{dt^{n-1}} + \ldots + a_0 y(t) = $$

$$b_m \frac{d^m x(t)}{dt^m} + b_{m-1} \frac{d^{m-1} x(t)}{dt^{m-1}} + \ldots + b_0 x(t) \qquad (2\text{-}17)$$

with zero initial conditions.

$$y(0) = 0; \quad \frac{dy}{dt}(0) = 0 ; \ldots ; \frac{d^{n-1}y}{dt^{n-1}}(0) = 0$$

$$x(0) = 0; \quad \frac{dx}{dt}(0) = 0 ; \ldots ; \frac{d^{m-1}x}{dt^{m-1}}(0) = 0$$

it is easy to show that the Laplace transformed equation is given by

$$Y(s) = \left[\frac{b_m s^m + b_{m-1} s^{m-1} + \ldots + b_0}{a_n s^n + a_{n-1} s^{n-1} + \ldots + a_0} \right] X(s) \qquad (2\text{-}18)$$

The case of zero initial conditions is the most common in control systems design because the signals are usually defined as deviations from some initial steady state (see Section 2-3). When this is done, the initial value of the perturbation is, by definition, zero; the initial values of the time derivatives are also zero because of the assumption that the system is initially at steady state, that is, not changing with time.

Transfer Function.   If the variables $X(s)$ and $Y(s)$ in Eq. (2-18) are the transforms of the input and output signals, respectively, of a process, instrument, or control system, the term in brackets is by definition the *transfer function* of that process, instrument, or control system. The transfer function of a system is the expression that when multiplied by the transform of its input signal results in the transform of its output signal. Transfer functions provide a useful mechanism for analyzing dynamic behavior and designing control systems. They will be discussed in detail in Chapter 3.

Inversion of Laplace Transforms by Partial Fractions Expansion

The last step in the process of solving a differential equation by Laplace transforms is the inversion of the algebraic equation for the output variable $Y(s)$. The inversion can be represented by

$$y(t) = \mathcal{L}^{-1} [Y(s)] \qquad (2\text{-}19)$$

As this is the most difficult step of the solution procedure, our objective in this section is to establish a general relationship between the transform of the output variable $Y(s)$ and its inverse $y(t)$. This will allow us to carry out our analysis of the response of the system by analyzing its transform function $Y(s)$ without having to actually invert it. We shall establish the relationship between $Y(s)$ and $y(t)$ by the method of partial fractions expansion. This method was first introduced by the British physicist Oliver Heaviside (1850–1925) as part of his revolutionary "operational calculus."

As we saw in the preceding section, the Laplace transform of the output or dependent variable of an nth-order linear differential equation with constant coefficients can be expressed by

$$Y(s) = \left[\frac{b_m s^m + b_{m-1} s^{m-1} + \ldots + b_0}{a_n s^n + a_{n-1} s^{n-1} + \ldots + a_0} \right] X(s) \tag{2-18}$$

where

$Y(s)$ is the Laplace transform of the output variable
$X(s)$ is the Laplace transform of the input variable
$a_0, a_1, \ldots, a_n$ are the constant coefficients of the output variable and its derivatives
$b_0, b_1, \ldots, b_m$ are the constant coefficients of the input variable and its derivatives.

A glance at Table 2-1 shows that the Laplace transforms of the most common functions are ratios of polynomials in the Laplace transform variables. Assuming that this is the case for $X(s)$, it can be easily shown that $Y(s)$ is also the ratio of two polynomials:

$$Y(s) = \frac{(b_m s^m + b_{m-1} s^{m-1} + \ldots + b_0)}{(a_n s^n + a_{n-1} s^{n-1} + \ldots + a_0)} \frac{[\text{numerator of } X(s)]}{[\text{denominator of } X(s)]} \tag{2-20}$$

$$- \frac{N(s)}{D(s)}$$

where

$$N(s) = \beta_j s^j + \beta_{j-1} s^{j-1} + \ldots + \beta_1 s + \beta_0$$

$$D(s) = s^k + \alpha_{k-1} s^{k-1} + \ldots + \alpha_1 s + \alpha_0$$

$\beta_0, \beta_1, \ldots, \beta_j$ are the constant coefficients for the numerator polynomial $N(s)$ of jth degree $(j \geq m)$
$\alpha_0, \alpha_1, \ldots, \alpha_{k-1}$ are the constant coefficients of the denominator polynomial $D(s)$ of kth degree $(k \geq n)$

Notice that we have assumed that the coefficient of s^k in $D(s)$ is unity. We can do this without loss of generality because we can always divide numerator and denominator by the coefficient of s^k and thus match Eq. (2-20).

It can be shown that Eq. (2-20) can also represent the case in which the output variable responds to more than one input forcing function. However, it does not represent the case in which either the system or the input signal contains time delays (transportation lags or dead times). For the sake of simplicity we shall ignore this very important case for the time being and consider it as a special case at the end of this section.

The first step in the partial fractions expansion of $Y(s)$ is to factor the denominator polynomial $D(s)$:

$$D(s) = s^k + \alpha_{k-1} s^{k-1} + \ldots + \alpha_1 s + \alpha_0 \tag{2-21}$$

$$= (s - r_1)(s - r_2) \ldots (s - r_k)$$

where $r_1, r_2, \ldots, r_k$ are the *roots* of the polynomial, that is, the values of s that satisfy the equation

$$D(s) = s^k + \alpha_{k-1} s^{k-1} + \ldots + \alpha_1 s + \alpha_0 = 0 \tag{2-22}$$

We recall that a kth-degree polynomial can have as many as k distinct roots. You can see that a polynomial can always be factored as shown in Eq. (2-21) by noting that setting s equal to any of the roots results in one of the factors $(s - r)$ being zero and thus $D(s) = 0$.

Substitution of Eq. (2-21) into Eq. (2-20) results in

$$Y(s) = \frac{N(s)}{(s - r_1)(s - r_2) \ldots (s - r_k)} \tag{2-23}$$

From this equation it can be shown that the transform $Y(s)$ can be expressed as a sum of k fractions:

$$Y(s) = \frac{A_1}{s - r_1} + \frac{A_2}{s - r_2} + \ldots + \frac{A_k}{s - r_k} \tag{2-24}$$

where $A_1, A_2, \ldots, A_k$ are a set of constant coefficients to be evaluated by a set procedure. This step is called "expansion in partial fractions."

Once the transform of the output is expanded as in Eq. (2-24), we can use the distributive property of the inverse transform to obtain the inverse function:

$$
\begin{aligned}
y(t) &= \mathscr{L}^{-1}\left[Y(s)\right] \\
&= \mathscr{L}^{-1}\left[\frac{A_1}{s - r_1} + \frac{A_2}{s - r_2} + \ldots + \frac{A_k}{s - r_k}\right] \\
&= A_1 \mathscr{L}^{-1}\left[\frac{1}{s - r_1}\right] + A_2 \mathscr{L}^{-1}\left[\frac{1}{s - r_2}\right] + \ldots A_k \mathscr{L}^{-1}\left[\frac{1}{s - r_k}\right]
\end{aligned} \tag{2-25}
$$

The individual inverses can usually be determined by using a table of Laplace transforms such as Table 2-1.

In order to evaluate the coefficients of the partial fractions and complete the inversion process we must consider four cases:

1. Unrepeated real roots.
2. Unrepeated pairs of complex conjugate roots.
3. Repeated roots.
4. Presence of dead time.

We shall consider each of these cases in turn.

Case 1. Unrepeated Real Roots

To evaluate the coefficient A_i of a fraction involving an unrepeated real root r_i, we multiply both sides of Eq. (2-24) by the factor $(s - r_i)$. This results in the following equation after rearranging:

$$(s - r_i)\, Y(s) = \frac{A_1(s - r_i)}{s - r_1} + \ldots + A_i + \ldots + \frac{A_k(s - r_i)}{s - r_k} \tag{2-26}$$

Notice that, since the root r_i is not repeated, there is no cancellation of the numerator and denominator factors except for the ith fraction. By letting $s = r_i$ in Eq. (2-26), we obtain the formula for coefficient A_i:

$$A_i = \lim_{s \to r_i} (s - r_i) Y(s) = \lim_{s \to r_i} (s - r_i) \frac{N(s)}{D(s)} \tag{2-27}$$

This formula is used to evaluate the coefficients of all fractions involving unrepeated real roots. The inverse of the corresponding terms of $Y(s)$ in Eq. (2-25) are, from Table 2-1

$$\mathcal{L}^{-1} \left[\frac{A_i}{s - r_i} \right] = A_i e^{r_i t} \tag{2-28}$$

If all of the roots of $D(s)$ are unrepeated real roots, the inverse function is

$$y(t) = A_1 e^{r_1 t} + A_2 e^{r_2 t} + \ldots + A_k e^{r_k t} \tag{2-29}$$

Let us illustrate this procedure by means of an example.

Example 2-6.   Given the second-order differential equation

$$\frac{d^2 c(t)}{dt^2} + 3 \frac{dc(t)}{dt} + 2c(t) = 5u(t)$$

where $u(t)$ is the unit step function (see Example 2-1a), find the function $c(t)$ that satisfies the equation for the case in which the initial conditions are zero:

$$c(0) = 0; \qquad \frac{dc}{dt}(0) = 0$$

Solution.

Step 1.   Take the Laplace transform of the equation.

$$s^2 C(s) + 3sC(s) + 2C(s) = 5U(s)$$

Step 2.   Solve for $C(s)$.

$$C(s) = \frac{5}{s^2 + 3s + 2} U(s)$$

$$= \frac{5}{s^2 + 3s + 2} \frac{1}{s}$$

where $U(s) = 1/s$ is obtained from Table 2-1.

Step 3.   Invert $C(s)$.
The roots of the denominator polynomial are

$$s(s^2 + 3s + 2) = 0$$

$$r_1 = 0$$

$$r_{2,3} = \frac{-3 \pm \sqrt{9 - 8}}{2} = -2, -1$$

or

$$s(s^2 + 3s + 2) = s(s + 1)(s + 2)$$

Expanding $C(s)$ in partial fractions gives

$$C(s) = \frac{5}{s(s + 1)(s + 2)} = \frac{A_1}{s} + \frac{A_2}{s + 1} + \frac{A_3}{s + 2}$$

From Eq. (2-27) we obtain

$$A_1 = \lim_{s \to 0} s \frac{5}{s(s + 1)(s + 2)} = \frac{5}{(1)(2)} = \frac{5}{2}$$

$$A_2 = \lim_{s \to -1} (s + 1) \frac{5}{s(s + 1)(s + 2)} = \frac{5}{(-1)(1)} = -5$$

$$A_3 = \lim_{s \to -2} (s + 2) \frac{5}{s(s + 1)(s + 2)} = \frac{5}{(-2)(-1)} = \frac{5}{2}$$

or

$$C(s) = \frac{5/2}{s} - \frac{5}{s + 1} + \frac{5/2}{s + 2}$$

Inverting with the help of Table 2-1 yields

$$c(t) = \frac{5}{2} u(t) - 5e^{-t} + \frac{5}{2} e^{-2t}$$

Case 2.   Unrepeated Pairs of Complex Conjugate Roots

We recall that if the coefficients of a polynomial are real numbers, its roots are either real numbers or pairs of complex conjugate numbers. In other words, if r_i is a complex root of $D(s)$, there must be another complex root that is the conjugate of r_i, that is, has the same real part and the same imaginary part but with opposite sign. For simplicity, let us say that these two roots are r_1 and r_2:

$$r_1 = r + iw \qquad r_2 = r - iw$$

where

$i = \sqrt{-1}$ is the unit of imaginary numbers

r is the real part of r_1 and r_2

w is the imaginary part of r_1

The partial fractions expansion of $Y(s)$ is then

$$Y(s) = \frac{N(s)}{(s - r - iw)(s - r + iw) \ldots (s - r_k)} \tag{2-30}$$

$$= \frac{A_1}{s - r - iw} + \frac{A_2}{s - r + iw} + \ldots + \frac{A_k}{s - r_k}$$

By the use of complex number algebra (see Section 2-4) we can apply Eq. (2-27) to evaluate A_1 and A_2:

$$A_1 = \lim_{s \to r + iw} (s - r - iw) \, Y(s) \tag{2-31}$$

$$A_2 = \lim_{s \to r - iw} (s - r + iw) \, Y(s)$$

It can be shown that A_1 and A_2 constitute a pair of complex conjugate numbers:

$$A_1 = B + iC \qquad A_2 = B - iC \tag{2-32}$$

where B and C are, respectively, the real and imaginary parts of A_1.

Having determined the values of coefficients A_1 and A_2 and ignoring for the moment that they are complex, let us look into the inverses of these terms. These can be obtained by application of Eq. (2-28):

$$\mathcal{L}^{-1} \left[\frac{A_1}{s - r - iw} \right] = A_1 e^{(r + iw)t} = A_1 e^{rt} e^{iwt}$$

$$= A_1 e^{rt}(\cos wt + i \sin wt) \tag{2-33}$$

$$\mathcal{L}^{-1} \left[\frac{A_2}{s - r + iw} \right] = A_2 e^{(r - iw)t} = A_2 e^{rt} e^{-iwt}$$

$$= A_2 e^{rt} (\cos wt - i \sin wt) \tag{2-34}$$

We have made the use here of the identity for the exponential of a pure imaginary number:

$$e^{ix} = \cos x + i \sin x \tag{2-35}$$

Combining Eqs. (2-33) and (2-34) yields

$$\mathcal{L}^{-1} \left[\frac{A_1}{s - r - iw} + \frac{A_2}{s - r + iw} \right]$$

$$= e^{rt} \left[(A_1 + A_2) \cos wt + i(A_1 - A_2) \sin wt \right] \tag{2-36}$$

$$= e^{rt} (2B \cos wt - 2C \sin wt)$$

where we have made use of Eq. (2-32). Notice that this shows that the solution $y(t)$ contains only real coefficients, since the numbers B and C are real numbers.

A simpler form of Eq. (2-36) is given by

$$\mathcal{L}^{-1} \left[\frac{A_1}{s - r - iw} + \frac{A_2}{s - r + iw} \right] = 2\sqrt{B^2 + C^2} \, e^{rt} \cos (wt + \theta) \tag{2-37}$$

where

$$\theta = \tan^{-1} \left(\frac{C}{B} \right)$$

Equations (2-36) and (2-37) can be shown to be equivalent by the substitution of the following trigonometric identities:

$$\sin (wt - \theta) = \sin wt \cos \theta - \cos wt \sin \theta$$

$$B = \sqrt{B^2 + C^2} \cos \theta$$

$$C = \sqrt{B^2 + C^2} \sin \theta$$

It is important to note that the arguments of the sine and cosine functions in Eqs. (2-36) and (2-37) must be in *radians*, not degrees. This is because the units of w are radians per unit time.

The real part of the complex roots, r, appears in the exponential of time e^{rt} in the final solution, while the imaginary part, w, appears in the argument of the sine and cosine functions. The two complex conjugate factors can be combined into a single "quadratic" (second-order) factor as follows:

$$\frac{B + iC}{s - r - iw} + \frac{B - iC}{s - r + iw} = \frac{2B(s - r) - 2Cw}{s^2 - 2rs + r^2 + w^2}$$

$$= \frac{2B(s - r) - 2Cw}{(s - r)^2 + w^2}$$

(2-38)

Notice that the denominator of this quadratic factor can be matched to the one in the last two entries of Table 2-1 by simply letting $a = -r$. How do the numerators match?

Example 2-7. Given the differential equation

$$\frac{d^2 c(t)}{dt^2} + 2\frac{dc(t)}{dt} + 5c(t) = 3u(t)$$

with zero initial conditions

$$c(0) = 0; \qquad dc/dt\,(0) = 0$$

find, by Laplace transforms, the function $c(t)$ that satisfies the equation.

Solution.

Step 1. Laplace transform the equation.

$$s^2C(s) + 2sC(s) + 5C(s) = 3U(s)$$

Step 2. Solve for $C(s)$ and substitute $U(s)$ from Table 2-1.

$$C(s) = \frac{3}{s^2 + 2s + 5}\frac{1}{s}$$

Step 3. Invert $C(s)$.
The roots of $(s^2 + 2s + 5)s = 0$ are

$$r_{1,2} = \frac{-2 \pm \sqrt{4 - 20}}{2} = -1 \pm i2$$

$$r_3 = 0$$

Expanding in partial fractions gives

$$C(s) = \frac{3}{(s + 1 - i2)(s + 1 + i2)s} = \frac{A_1}{s + 1 - i2} + \frac{A_2}{s + 1 + i2} + \frac{A_3}{s}$$

$$A_1 = \lim_{s \to -1+i2} (s + 1 - i2) \frac{3}{(s + 1 - i2)(s + 1 + i2)s} = \frac{3}{i4(-1 + i2)}$$

$$= \frac{3(-2 + i)}{4(-2 - i)(-2 + i)} = \frac{-6 + 3i}{20}$$

$$A_2 = \lim_{s \to -1-i2} (s + 1 + i2) \frac{3}{(s + 1 - i2)(s + 1 + i2)s} = \frac{3}{-i4(-1 - i2)}$$

$$= \frac{-3(2 + i)}{4(2 - i)(2 + i)} = \frac{-6 - 3i}{20}$$

$$A_3 = \lim_{s \to 0} s \frac{3}{(s + 1 - i2)(s + 1 + i2)s} = \frac{3}{5}$$

or

$$C(s) = \frac{(6 + 3i)/20}{s + 1 - i2} + \frac{(-6 - 3i)/20}{s + 1 + i2} + \frac{3/5}{s}$$

Inverting with the help of Eqs. (2-36) and (2-37) and Table 2-1 gives

$$c(t) = e^{-t} \left(\frac{-3}{5} \cos 2t - \frac{3}{10} \sin 2t \right) + \frac{3}{5} u(t) = \frac{3\sqrt{5}}{10} e^{-t} \cos (2t + 2.678) + \frac{3}{5} u(t)$$

with $r = 1$, $w = 2$, $B = -3/10$, $C = 3/20$, $\theta = 2.678$ radians.

Case 3. Repeated Roots

The formula presented for the first two cases cannot be used to evaluate the coefficients of fractions involving repeated roots. The procedure presented here applies equally to the case when the repeated roots are real or complex.

The partial fractions expansion of a transform for which a root r_1 is repeated m times is given by

$$Y(s) = \frac{N(s)}{(s - r_1)^m \dots (s - r_k)} \tag{2-39}$$

$$= \frac{A_1}{(s - r_1)^m} + \frac{A_2}{(s - r_1)^{m-1}} + \dots + \frac{A_m}{s - r_1} + \dots + \frac{A_k}{s - r_k}$$

To evaluate the coefficients $A_1, A_2, \dots, A_m$ we apply the following formulas in the order given:

$$A_1 = \lim_{s \to r_1} [(s - r_1)^m Y(s)]$$

$$A_2 = \lim_{s \to r_1} \frac{d}{ds} [(s - r_1)^m Y(s)]$$

$$A_3 = \lim_{s \to r_1} \frac{1}{2!} \frac{d^2}{ds^2} [(s - r_1)^m Y(s)] \tag{2-40}$$

$$\vdots \qquad \vdots \qquad \vdots$$

$$A_m = \lim_{s \to r_1} \frac{1}{(m - 1)!} \frac{d^{m-1}}{ds^{m-1}} [(s - r_1)^m Y(s)]$$

Once the coefficients are evaluated, inversion of Eq. (2-39), using Table 2-1, results in the following:

$$y(t) = \left[\frac{A_1 t^{m-1}}{(m-1)!} + \frac{A_2 t^{m-2}}{(m-2)!} + \ldots + A_m \right] e^{r_1 t} + \ldots + A_k e^{r_k t} \tag{2-41}$$

For the rare case of repeated pairs of complex conjugate roots, some effort can be saved by using the fact that the coefficients are complex conjugate pairs. From Eq. (2-32) we obtain

$$A_1 = B_1 + iC_1 \qquad A_1^c = B_1 - iC_1$$

where A_1^c is the conjugate of A_1.

Then, by combination of Eqs. (2-36) and (2-41) we can write

$$y(t) = e^{rt} \left\{ \left[\frac{2B_1 t^{m-1}}{(m-1)!} + \frac{2B_2 t^{m-2}}{(m-2)!} + \ldots + 2B_m \right] \cos wt \right.$$
$$\left. - \left[\frac{2C_1 t^{m-1}}{(m-1)!} + \frac{2C_2 t^{m-2}}{(m-2)!} + \ldots + 2C_m \right] \sin wt \right\} + \ldots + A_k w^{r_k t} \tag{2-42}$$

Example 2-8.   Given the differential equation

$$\frac{d^3 c(t)}{dt^3} + 3 \frac{d^2 c(t)}{dt} + 3 \frac{dc(t)}{dt} + c(t) = 2u(t)$$

with zero initial conditions

$$c(0) = 0; \qquad \frac{dc}{dt}(0) = 0; \qquad \frac{d^2 c}{dt^2}(0) = 0$$

find, using the method of Laplace transforms, the function $c(t)$ that satisfies the equation.

Solution.

Step 1.   Transform the equation.

$$s^3 C(s) + 3s^2 C(s) + 3s C(s) + C(s) = 2U(s)$$

Step 2.   Solve for $C(s)$ and substitute $U(s)$ from Table 2-1.

$$C(s) = \frac{2}{(s^3 + 3s^2 + 3s + 1)} \frac{1}{s}$$

Step 3.   Invert back to get $c(t)$.
The roots are:

$$(s^3 + 3s^2 + 3s + 1) s = 0$$
$$r_{1,2,3} = -1, -1, -1$$
$$r_4 = 0$$

Expand in partial fractions:

$$C(s) = \frac{2}{(s+1)^3 s} = \frac{A_1}{(s+1)^3} + \frac{A_2}{(s+1)^2} + \frac{A_3}{(s+1)} + \frac{A_4}{s}$$

Evaluate the coefficients by Eq. (2-40):

$$A_1 = \lim_{s \to -1} \left[(s + 1)^3 \frac{2}{(s + 1)^3 s} \right] = -2$$

$$A_2 = \lim_{s \to -1} \frac{d}{ds} \left[(s + 1)^3 \frac{2}{(s + 1)^3 s} \right] = \lim_{s \to -1} \frac{d}{ds} \left[\frac{2}{s} \right]$$

$$= \lim_{s \to -1} \left[-\frac{2}{s^2} \right] = -2$$

$$A_3 = \lim_{s \to -1} \frac{1}{2} \frac{d^2}{ds^2} \left[(s + 1)^3 \frac{2}{(s + 1)^3 s} \right] = \lim_{s \to -1} \frac{1}{2} \frac{d}{ds} \left[-\frac{2}{s^2} \right]$$

$$= \lim_{s \to -1} \frac{1}{2} \left[\frac{4}{s^3} \right] = -2$$

$$A_4 = \lim_{s \to 0} s \frac{2}{(s + 1)^3 s} = 2$$

$$C(s) = -\frac{2}{(s + 1)^3} - \frac{2}{(s + 1)^2} - \frac{2}{(s + 1)} + \frac{2}{s}$$

Inverting with the help of Eq. (2-41) and Table 2-1 results in

$$c(t) = -[t^2 + 2t + 2] e^{-t} + 2u(t)$$

Notice that the same result can be obtained by inverting each term directly using Table 2-1.

Case 4. Presence of Dead Time

The technique of partial fractions expansion is restricted for use with Laplace transforms that can be expressed as ratios of two polynomials. As we learned from the real translation theorem, Eq. (2-9), when the Laplace transform contains dead time (transportation lag or time delay), the exponential term e^{-st_0}, where t_0 is the dead time, appears in the transform function. Since the exponential is a transcendental function, the inversion procedure must be appropriately modified.

If the exponential function appears in the denominator of the Laplace transform, it is not possible to invert it by partial fractions expansion. This is because we no longer have a finite number of roots and thus there would be an infinite number of fractions in the expansion. On the other hand, exponential terms in the numerator can be handled, as we shall now see.

Let us first consider the case in which the Laplace transform consists of an exponential term multiplied by the ratio of two polynomials:

$$Y(s) = \left[\frac{N(s)}{D(s)} \right] e^{-st_0} = [Y_1(s)] e^{-st_0} \tag{2-43}$$

The procedure is to expand in partial fractions only the ratio of the polynomials.

$$Y_1(s) = \frac{N(s)}{D(s)} = \frac{A_1}{s - r_1} + \frac{A_2}{s - r_2} + \ldots + \frac{A_k}{s - r_k} \tag{2-44}$$

This expansion may require the application of any of the first three cases. Next we invert Eq. (2-44) to obtain

$$Y_1(t) = A_1 e^{r_1 t} + A_2 e^{r_2 t} + \ldots + A_k e^{r_k t}$$

To invert Eq. (2-43) we make use of the real translation theorem, Eq. (2-9).

$$Y(s) = e^{-s t_0} Y_1(s) = \mathcal{L}[y_1(t - t_0)] \tag{2-45}$$

Inverting this equation results in

$$
\begin{align}
y(t) = \mathcal{L}^{-1}[Y(s)] &= y_1(t - t_0) \\
&= A_1 e^{r_1(t - t_0)} + A_2 e^{r_2(t - t_0)} \ldots + A_k e^{r_k(t - t_0)}
\end{align}
\tag{2-46}
$$

It is important to notice the effect of removing the exponential term from the partial fractions expansion procedure. Had we expanded the original function, Eq. (2-43), we would have obtained

$$Y(s) = \frac{A_1 e^{-r_1 t_0}}{s - r_1} + \frac{A_2 e^{-r_2 t_0}}{s - r_2} + \ldots + \frac{A_k e^{-r_k t_0}}{s - r_k}$$

Although this may appear to work in certain cases, it is fundamentally *incorrect*.

Next let us consider the case of multiple delays. This introduces more than one exponential function in the numerator of the Laplace transform. The procedure is then to algebraically manipulate the function into a sum of terms, each involving the product of an exponential and the ratio of two polynomials:

$$
\begin{align}
Y(s) &= \left[\frac{N_1(s)}{D_1(s)}\right] e^{-s t_{01}} + \left[\frac{N_2(s)}{D_2(s)}\right] e^{-s t_{02}} + \ldots \\
&= [Y_1(s)] e^{-s t_{01}} + [Y_2(s)] e^{-s t_{02}} + \ldots
\end{align}
\tag{2-47}
$$

Then each polynomial ratio is expanded in partial fractions and inverted to produce a result of the form

$$y(t) = y_1(t - t_{01}) + y_2(t - t_{02}) + \ldots \tag{2-48}$$

Multiple delays may occur when the system is subjected to different forcing functions, each delayed by a different period of time.

Example 2-9. Given the differential equation

$$\frac{dc(t)}{dt} + 2c(t) = f(t)$$

with $c(0) = 0$, find the response of the output for
    (a)   A unit step change at $t = 1$: $f(t) = u(t - 1)$
    (b)   A staircase function of unit steps at every unit of time

$$f(t) = u(t - 1) + u(t - 2) + u(t - 3) + \ldots$$

These functions are sketched in Fig. 2-3.

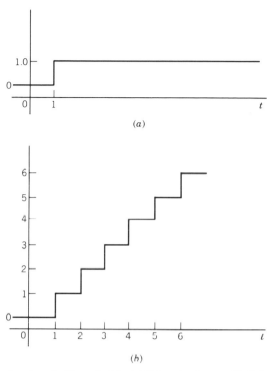

Figure 2-3. Input functions for Example 2-9. (*a*) Delayed unit step. (*b*) Staircase of unit steps.

Solution.

Step 1. Transform the differential equation and input function.

$$sC(s) + 2C(s) = F(s)$$

Applying the real translation theorem and Table 2-1 yields

(a) $F(s) = \mathcal{L}[u(t - 1)] = e^{-s}\dfrac{1}{s}$

(b) $F(s) = \mathcal{L}[u(t - 1) + u(t - 2) + u(t - 3) + \ldots]$

$$= \frac{1}{s}(e^{-s} + e^{-2s} + e^{-3s} + \ldots)$$

Step 2. Solve for $C(s)$.

$$C(s) = \frac{1}{s + 2}F(s)$$

Step 3. Invert to obtain $c(t)$.

(a) $C(s) = \dfrac{1}{s + 2}\dfrac{1}{s}e^{-s} = C_1(s)e^{-s}$

$$C_1(s) = \frac{1}{(s+2)s} = \frac{A_1}{s+2} + \frac{A_2}{s}$$

$$A_1 = \lim_{s \to -2} (s+2)\frac{1}{(s+2)s} = \frac{1}{-2} = -\frac{1}{2}$$

$$A_2 = \lim_{s \to 0} s\frac{1}{(s+2)s} = \frac{1}{2}$$

$$C_1(s) = -\frac{1/2}{s+2} + \frac{1/2}{s}$$

Invert to obtain

$$c_1(t) = -\frac{1}{2}e^{-2t} + \frac{1}{2}u(t)$$

Apply Eq. (2-46):

$$c(t) = c_1(t-1) = \frac{1}{2}u(t-1)[1 - e^{-2(t-1)}]$$

Note that unit step $u(t-1)$ must also be multiplied by the exponential term to indicate that $c(t) = 0$ for $t < 1$.

    (b)   For the staircase function we see that

$$C(s) = \left[\frac{1}{s+2} \cdot \frac{1}{s}\right](e^{-s} + e^{-2s} + e^{-3s} + \ldots)$$

$$= C_1(s)e^{-s} + C_1(s)e^{-2s} + C_1(s)e^{-3s} + \ldots$$

We note that $C_1(s)$ is the same as for part (a), and therefore the result of the expansion and inversion steps is the same. Applying Eq. (2-46) to each term results in

$$c(t) = c_1(t-1) + c_1(t-2) + c_1(t-3) + \ldots$$

$$= \frac{1}{2}u(t-1)[1 - e^{-2(t-1)}] + \frac{1}{2}u(t-2)[1 - e^{-2(t-2)}]$$

$$+ \frac{1}{2}u(t-3)[1 - e^{-2(t-3)}] + \ldots$$

Suppose that we now wanted to evaluate this function at $t = 2.5$. The answer is

$$c(2.5) = \frac{1}{2}(1)[1 - e^{-2(2.5-1)}] + \frac{1}{2}(1)[1 - e^{-2(2.5-2)}]$$

$$+ \frac{1}{2}(0)[1 - e^{-2(2.5-3)}] + \ldots$$

$$= \frac{1}{2}(0.950) + \frac{1}{2}(0.632) = 0.791$$

Notice that all of the terms after the first two are zero.

    The preceding four cases are summarized in Table 2-2. As these cases cover essentially all of the possibilities in the solution of linear differential equations with constant coefficients, the one-to-one correspondence of the entries in Table 2-2 makes it unnecessary to actually invert the Laplace transform of the dependent variable. This is because we can usually recognize the terms of the time function $y(t)$ in the Laplace transform $Y(s)$.

Table 2-2 Relationship Between the Laplace Transform $Y(s)$ and Its Inverse $y(t)$

Denominator of $Y(s)$	Partial fraction term	Term of $y(t)$
1. Unrepeated real root	$\dfrac{A}{s - r}$	Ae^{rt}
2. Pair of complex conjugate roots	$\dfrac{B(s - r) + Cw}{(s - r)^2 + w^2}$	$e^{rt}(B \cos wt + C \sin wt)$
3. Real root repeated m times	$\displaystyle\sum_{j=1}^{m} \frac{A_j}{(s - r)^j}$	$\displaystyle e^{rt}\sum_{j=1}^{m} A_j \frac{t^{j-1}}{(j-1)!}$

Dead-time term in the numerator of $Y(s)$	Term of $y(t)$
4. $Y_1(s)e^{-t_0 s}$	$y_1(t - t_0)$

Eigenvalues and Stability

It is evident from inspection of Table 2-2 that the roots of the denominator of the Laplace transform $Y(s)$ determine the response $y(t)$. We can also see from Eq. (2-20) that some of the roots of the denominator of $Y(s)$ are the roots of the following equation:

$$a_n s^n + a_{n-1} s^{n-1} + \ldots + a_1 s + a_0 = 0 \qquad (2\text{-}49)$$

where $a_0, a_1, \ldots, a_n$ are the coefficients of the dependent variable and its derivatives in differential equation Eq. (2-17). The balance of the roots of the denominator polynomial come from the input or forcing function $X(s)$.

Equation (2-49) is said to be the *characteristic equation* of the differential equation and of the system whose dynamic response it represents. The roots of the characteristic equation are called the *eigenvalues* (German for "characteristic" or "proper" values) of the differential equation. The significance of the eigenvalues is that they are by definition characteristic of the differential equation and independent of the input forcing function. From Table 2-2 we can see that the eigenvalues determine whether the time response is going to be monotonic (cases 1 and 3) or oscillatory (case 2), independent of whether or not the forcing function has these characteristics. Note that the sine and cosine functions of case 2 cause an oscillatory response (see Fig. 2-1). The eigenvalues also determine whether or not the response is *stable*.

A differential equation is said to be *stable* if its time response remains bound (finite) for a bound forcing function. From Table 2-2 we see that in order for this condition of stability to hold, *all of the eigenvalues must have negative real parts r*. This is because the term e^{rt} appears in each of the possible response terms and, in order for this exponential term to remain finite as time increases, r must be negative (r is either the real root or the real part of the root in each case). We will discuss stability in more detail in Chapter 6 when we look at the response of feedback control systems.

Finding the Roots of Polynomials

The most time-consuming operation in the inversion step is the finding of the roots of the denominator polynomial $D(s)$ when its degree is three or higher. This is because root finding is an iterative or trial-and-error procedure. As time is valuable, it is important to use efficient methods of polynomial root finding. Three of the most efficient methods are:

1. Newton's method for real roots.
2. The Newton-Bairstow method for complex conjugate and real roots.
3. Müller's method for complex and real roots.

Of these, Newton's method is the most convenient for manual calculation, specially when combined with the method of nested multiplications to evaluate the polynomial and its derivative. This method works as long as there is no more than one pair of complex conjugate roots.

The Newton-Bairstow method is commonly used to find quadratic factors (second-degree polynomial factors) of the polynomial on programmable calculators. From the coefficients of each quadratic factor we can find two roots; these can be real or a complex conjugate pair.

Müller's[1] method is the most efficient for finding the roots, real or complex, of any function. With this method, however, the necessary computations are involved, and complex number arithmetic must be used to find the complex roots. For this reason this method is indicated when a computer is to be used to calculate the roots. A FORTRAN program to find the roots of a polynomial by Müller's method is listed in Appendix D. The algorithm for the Newton-Bairstow method is described in any good text on numerical methods[2,3]. We will restrict our presentation here to Newton's method.

The problem of finding the roots of a polynomial can be formulated as follows: Given the polynomial of the nth degree

$$f_n(s) = a_n s^n + a_{n-1} s^{n-1} + \ldots + a_1 s + a_0 \tag{2-50}$$

find all of its roots, that is, all of the values of s that satisfy the equation

$$f_n(s) = 0$$

The three basic steps required to solve this problem by iteration or trial and error are as follows:

1. Assume an initial approximation s_0 to the root.
2. Calculate an improved approximation s_k.
3. Check for convergence within a specified error tolerance. If not within this tolerance, repeat steps 2 and 3 until convergence is attained.

After the procedure has converged to a value r_1, this value is taken as the first root and the polynomial of degree $n - 1$ is determined by

$$f_{n-1}(s) = \frac{f_n(s)}{s - r_1} \tag{2-51}$$

Then the iterative procedure is repeated for $f_{n-1}(s)$ to find the second root r_2, then for $f_{n-2}(s)$ to find r_3, and so on until all n roots have been found. The polynomial $f_{n-2}(s)$ is known as the "reduced polynomial."

The basic difference between different iteration methods is the formula in step 2 of the iteration procedure that is used to improve the approximation to the root sought.

Newton's Method.   The iteration formula for Newton's method is given by

$$s_{k+1} = s_k - \frac{f(s_k)}{f'(s_k)} \tag{2-52}$$

where

s_k is the previous approximation to the root
s_{k+1} is the new approximation
$f(s_k)$ is the value of the polynomial at s_k
$f'(s_k) = \dfrac{df}{ds}(s_k)$

This formula is very efficient in terms of the number of iterations required to approximate the root within a given error tolerance. However, it requires the evaluation of the polynomial and its derivative at each iteration, while other methods (e.g., secant, Müller) require only the evaluation of the function at each iteration. Nevertheless, Newton's method can be efficient for polynomials when the method of nested multiplications is used to evaluate the polynomial and its derivative.

Nested Multiplications.[3]   To evaluate the nth-degree polynomial of Eq. (2-49) and its derivative at $s = s_k$, we introduce the method of nested multiplications or synthetic division, which consists of the following procedure in which a_i are the coefficients of the polynomial, and b_i and c_i are two sets of variables to be used in the calculations:

1.   Let $b_n = a_n$ and $c_n = b_n$.
2.   For $i = n - 1, n - 2, \ldots, 1, 0$, let $b_i = a_i + b_{i+1}s_k$.
3.   For $i = n - 1, n - 2, \ldots, 1$, let $c_i = b_i + c_{i+1}s_k$.

Then

$$f(s_k) = b_0$$

$$f'(s_k) = c_1$$

We have thus calculated the values of the polynomial and its derivative doing just $2n - 1$ multiplications and additions. In comparison, the direct evaluation of a polynomial and its derivative requires n^2 multiplications and $2n - 1$ additions. For a fifth-degree polynomial, the nested multiplications method requires 9 multiplications per iteration instead of 25!

In addition to the dramatic savings in computations, the method of nested multiplications offers the additional advantage that, once the root is found, the reduced polynomial of degree $(n - 1)$ is given by

$$f_{n-1}(s) = b_n s^{n-1} + b_{n-1} s^{n-2} + \ldots + b_2 s + b_1 \tag{2-53}$$

This eliminates the need to perform the division indicated by Eq. (2-51) after finding each root.

Example 2-10. Given the differential equation

$$\frac{d^3 c(t)}{dt^3} + 2 \frac{d^2 c(t)}{dt^2} + 3 \frac{dc(t)}{dt} + 4c(t) = 4\delta(t)$$

where $\delta(t)$ is the unit impulse function at $t = 0$, and given that all initial conditions are zero:

$$c(0) = 0; \qquad \frac{dc}{dt}(0) = 0; \qquad \frac{d^2 c}{dt^2}(0) = 0$$

find the response $c(t)$ by Laplace transforms.

Solution.

Step 1. Transform the differential equation and the input function.

$$s^3 C(s) + 2s^2 C(s) + 3sC(s) + 4C(s) = 4\mathcal{L}[\delta(t)]$$
$$\mathcal{L}[\delta(t)] = 1 \qquad \text{(from Table 2-1)}$$

Step 2. Solve for $C(s)$.

$$C(s) = \frac{4}{s^3 + 2s^2 + 3s + 4}$$

To factor the polynomial we need to find the roots of the polynomial

$$f(s) = s^3 + 2s^2 + 3s + 4$$

Newton Iteration Procedure

Initial approximation $s_0 = -1$

Iteration 1

	(3)	(2)	(1)	(0)
	$a_3 = 1$	$a_2 = 2$	$a_1 = 3$	$a_0 = 4$
$s_0 = -1$	$b_3 s = -1$	$b_2 s = -1$	$b_1 s = -2$	
	$b_3 = 1$	$b_2 = 1$	$b_1 = 2$	$b_0 = 2 = f(-1)$
$s_0 = -1$	$c_3 s = -1$	$c_2 s = 0$		
	$c_3 = 1$	$c_2 = 0$	$c_1 = 2 = f'(-1)$	
$s_2 = -1 - (2/2) = -2$				

Iteration 2

$$
\begin{array}{cccc}
(3) & (2) & (1) & (0) \\
a_3 = 1 & a_2 = 2 & a_1 = 3 & a_0 = 4
\end{array}
$$

$s_1 = -2$ $\qquad\qquad$ $\underline{-2}$ $\qquad$ 0 $\qquad\qquad$ $\underline{-6}$

$\qquad\qquad\quad b_3 = 1 \qquad b_2 = 0 \qquad b_1 = 3 \qquad b_0 = -2 = f(-2)$

$s_1 = -2$ $\qquad\qquad$ $\underline{-2}$ $\qquad$ 4

$\qquad\qquad\quad c_3 = 1 \qquad c_2 = -2 \qquad c_1 = 7 = f'(-2)$

$s_2 = -2 - (-2/7) = -1.714$

Iteration 3

$$
\begin{array}{cccc}
(3) & (2) & (1) & (0) \\
a_3 = 1 & a_2 = 2 & a_1 = 3 & a_0 = 4
\end{array}
$$

$s_2 = -1.714$ $\qquad\qquad$ $\underline{-1.714}$ $\qquad$ $\underline{-0.490}$ $\qquad$ $\underline{-4.303}$

$\qquad\qquad\quad b_3 = 1 \quad b_2 = 0.286 \quad b_1 = 2.510 \quad b_0 = -0.303 = f(-1.714)$

$s_2 = -1.714$ $\qquad\qquad$ $\underline{-1.714}$ $\qquad$ 2.448

$\qquad\qquad\quad c_3 = 1 \quad c_2 = -1.428 \quad c_1 = 4.958 = f'(-1.714)$

$s_3 = -1.714 - (-0.303/4.958) = -1.653$

Iteration 4

$$
\begin{array}{cccc}
(3) & (2) & (1) & (0) \\
a_3 = 1 & a_2 = ? & a_1 = 3 & a_0 = 4
\end{array}
$$

$s_3 = -1.653$ $\qquad\qquad$ $\underline{-1.653}$ $\qquad$ $\underline{-0.573}$ $\qquad$ $\underline{4.012}$

$\qquad\qquad\quad b_3 = 1 \quad b_2 = 0.347 \quad b_1 = 2.427 \quad b_0 = -0.012 = f(-1.653)$

$s_3 = -1.653$ $\qquad\qquad$ $\underline{-1.653}$ $\qquad$ 2.159

$\qquad\qquad\quad c_3 = 1 \quad c_2 = -1.306 \quad c_1 = 4.586 = f'(-1.653)$

$s_4 = -1.653 - (-0.012/4.586) = -1.651$

Iteration 5

$$
\begin{array}{cccc}
(3) & (2) & (1) & (0) \\
a_3 = 1 & a_2 = ? & a_1 = 3 & a_0 = 4
\end{array}
$$

$s_4 = -1.651$ $\qquad\qquad$ $\underline{-1.651}$ $\qquad$ $\underline{-0.577}$ $\qquad$ $\underline{-4}$

$\qquad\qquad\quad b_3 = 1 \quad b_2 = 0.349 \quad b_1 = 2.423 \quad b_0 = 0 = f(-1.651)$

Since the polynomial function is zero, there is no need to calculate the derivative after this last iteration.

Thus, the first root is $r_1 = -1.651$

From Eq. (2-53), the reduced polynomial is given by

$$
f_2(s) = s^2 + 0.349s + 2.423
$$

Notice that the coefficients are the b's calculated in iteration 5. The roots of this polynomial can be found using the quadratic formula

$$
r_{2,3} = \frac{-0.349 \pm \sqrt{0.1218 - 4(2.423)}}{2} = -0.174 \pm i1.547
$$

Then

$$C(s) = \frac{4}{(s + 1.651)(s + 0.174 - i1.547)(s + 0.174 + i1.547)}$$

$$= \frac{A_1}{s + 1.651} + \frac{A_2}{s + 0.174 - i1.547} + \frac{A_3}{s + 0.174 + i1.547}$$

$$A_1 = 0.875$$

$$A_2 = -0.437 - i0.417$$

$$A_3 = -0.437 + i0.417$$

$$C(s) = \frac{0.875}{s + 1.651} + \frac{-0.437 - i0.417}{s + 0.174 - i1.547} + \frac{-0.437 + i0.417}{s + 0.174 + i1.547}$$

Inverting yields

$$c(t) = 0.875 e^{-1.651t} + e^{-0.174t}$$
$$\cdot \, [-0.874 \cos (1.547t) + 0.834 \sin (1.547t)]$$

Outline of Laplace Transform Method for Solving Differential Equations

The procedure for solving differential equations using the Laplace transform and partial fractions expansion is outlined as follows.

Given an nth-order differential equation in the output variable $y(t)$ with input variable $x(t)$ in the form of Eq. (2-17).

Step 1. Laplace transform the equation term by term into an algebraic equation in $Y(s)$ and $X(s)$.

Step 2. Algebraically solve for the transform of the output variable $Y(s)$ and substitute the transform of the input variable $X(s)$ to obtain a ratio of two polynomials:

$$Y(s) = \frac{N(s)}{D(s)} \tag{2-20}$$

Step 3. Invert by expansion in partial fractions as follows:
(a) Find the roots of the denominator of $Y(s)$ by a method such as Newton's (previous subsection) or a computer program (Appendix D).
(b) Factor the denominator.

$$D(s) = (s - r_1)(s - r_2) \ldots (s - r_k) \tag{2-21}$$

(c) Expand the transform in partial fractions.

$$Y(s) = \frac{A_1}{s - r_1} + \frac{A_2}{s - r_2} + \ldots + \frac{A_k}{s - r_k} \tag{2-24}$$

      where

$$A_i = \lim_{s \to r_i} (s - r_i) \frac{N(s)}{D(s)} \tag{2-27}$$

if all of the roots are unrepeated, real, or complex. If there are repeated roots, the corresponding coefficients must be evaluated using the set of formulas that constitute Eq. (2-40).

(d) Invert Eq. (2-24) with the help of a table of Laplace transforms (Tables 2-1 or 2-2). For unrepeated roots the solution is of the form

$$y(t) = A_1 e^{r_1 t} + A_2 e^{r_2 t} + \ldots + A_k e^{r_k t} \tag{2-29}$$

For the case of complex conjugate pairs of roots, the solution is of the form of Eqs. (2-36) or (2-37). For the case of repeated roots, the solution is of the form of Eqs. (2-41) or (2-42).

If dead-time terms are present in the numerator, the procedure must be altered as indicated in Eqs. (2-47) and (2-48).

Fortunately, most of this inversion procedure can be avoided when analyzing and designing control systems. This is because, as indicated in Table 2-2, we can recognize the terms of the time response in the terms of the denominator of the Laplace transform. However, in order to use Laplace transforms at all, the equations that represent the process and instruments must be linear. As this is not usually the case, we will look at this problem next.

2-3. LINEARIZATION AND DEVIATION VARIABLES

One of the major difficulties of analyzing the dynamic response of industrial processes is the fact that they are nonlinear, that is, they cannot be represented by linear equations. (In order for an equation to be linear each of its terms must contain no more than one variable or derivative, and it must appear to the first power.) Unfortunately, only linear systems can be analyzed by the powerful tool of Laplace transforms that we learned in the preceding section. Another difficulty is that there is no convenient technique to analyze the dynamics of a nonlinear system in such a way that it can be generalized to a wide variety of physical systems.

In this section we will learn the technique of *linearization*. With linearization we can approximate the nonlinear equations that represent a process with linear equations that can then be analyzed by Laplace transforms. Our basic assumption is that the response of the linear approximation represents the response of the process in the region near the operating point around which the linearization has been performed.

The manipulation of linearized equations is greatly facilitated by the use of deviation or perturbation variables. We will define these next.

Deviation Variables

A deviation variable is defined as the difference between the value of a variable or signal and its value at the operating point:

$$X(t) = x(t) - \bar{x} \tag{2-54}$$

where

> $X(t)$ is the deviation variable
> $x(t)$ is the corresponding absolute variable
> $\bar{x}$ is the value of x at the operating point (base value)

In other words, the deviation variable is the deviation of a variable from its operating or base value. As illustrated in Fig. 2-4, the transformation from absolute to deviation value of a variable is equivalent to moving the zero on the axis for that variable to the base value.

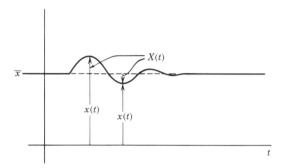

Figure 2-4.   Definition of deviation variable.

As the base value of the variable is a constant, the derivatives of the deviation variables are always equal to the corresponding derivatives of the variables:

$$\frac{d^n X(t)}{dt^n} = \frac{d^n x(t)}{dt^n} \qquad \text{for } n = 1, 2, \text{ etc.} \qquad (2\text{-}55)$$

The major advantage of using deviation variables derives from the fact that the base value $\bar{x}$ is usually the *initial value* of the variable. In addition, the operating point is usually at *steady state*. This means that the initial conditions of the deviation variables and their derivatives are all zero:

$$x(0) = \bar{x} \qquad X(0) = 0$$

Also

$$\frac{d^n X}{dt^n}(0) = 0 \qquad \text{for } n = 1, 2, \text{ etc.}$$

Then, when taking the Laplace transform of any of the derivatives of deviation variables, Eq. (2-6) applies:

$$\mathcal{L}\left[\frac{d^n X(t)}{dt^n}\right] = s^n X(s)$$

where $X(s)$ is the Laplace transform of the deviation variable.

Another important feature of the case when all the deviation variables are deviations from the initial steady-state conditions is that all the constant terms drop out of the linearized differential equations. We shall demonstrate this shortly.

Linearization of Functions of One Variable

Consider the first-order differential equation

$$\frac{dx(t)}{dt} = f[x(t)] + k \tag{2-56}$$

where $f[x(t)]$ is a nonlinear function of x and k is a constant. The Taylor series expansion of $f[x(t)]$ around a value $\bar{x}$ is given by

$$f[x(t)] = f(\bar{x}) + \frac{df}{dx}(\bar{x})[x(t) - \bar{x}] + \frac{1}{2!}\frac{d^2f}{dx^2}(\bar{x})[x(t) - \bar{x}]^2$$

$$+ \frac{1}{3!}\frac{d^3f}{dx^3}(\bar{x})[x(t) - \bar{x}]^3 + \ldots \tag{2-57}$$

The linear approximation consists of dropping all of the terms of the series except the first two:

$$f[x(t)] = f(\bar{x}) + \frac{df}{dx}(\bar{x})[x(t) - \bar{x}] \tag{2-58}$$

or, substituting the definition of deviation variable $X(t)$ from Eq. (2-54),

$$f[x(t)] \doteq f(\bar{x}) + \frac{df}{dx}(\bar{x})X(t) \tag{2-59}$$

The graphical interpretation of this approximation is given in Fig. 2-5. The linear approximation is a straight line passing through the point $[\bar{x}, f(\bar{x})]$ with slope $df/dx\ (\bar{x})$. This line is by definition the tangent to the curve $f(x)$ at $\bar{x}$. Notice that the difference between the linear approximation and the actual function is small near the operating point $\bar{x}$ and large away from it. The region where the linear approximation is accurate enough to represent the nonlinear function is difficult to assess. The more nonlinear a function is, the smaller the region over which the linear approximation is accurate.

Substituting the linear approximation Eq. (2-59) into Eq. (2-56) results in

$$\frac{dx(t)}{dt} = f(\bar{x}) + \frac{df}{dx}(\bar{x})X(t) + k \tag{2-60}$$

If the initial conditions are

$$x(0) = \bar{x} \qquad \frac{dx}{dt}(0) = 0 \qquad X(0) = 0$$

then

$$0 = f(\bar{x}) + \frac{df}{dx}(\bar{x})(0) + k$$

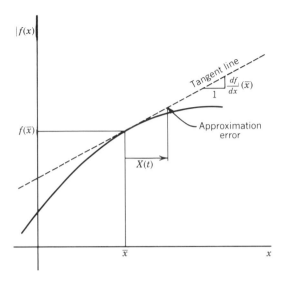

Figure 2-5.   Linear approximation is the tangent to the function at the operating point.

or

$$f(\bar{x}) + k = 0$$

Substituting into Eq. (2-58) yields

$$\frac{dX(t)}{dt} = \frac{df}{dx}(\bar{x})X(t) \tag{2-61}$$

This shows how the constant terms drop out of the linearized equation when the base value is the initial steady-state condition. Notice that it is possible to omit all of the intermediate steps and go directly from Eq. (2-56) to Eq. (2-61).

The following are examples of some nonlinear functions that are common in process models:

1. Arrhenius temperature dependence of reaction rate.
    $$k(T) = k_0 e^{-(E/RT)}$$
    where k_0, E, and R are constants.
2. Vapor pressure of a pure substance (Antoine equation).
    $$p^0(T) = e^{[A - B/(T + C)]}$$
    where A, B, and C are constants.
3. Vapor–liquid equilibrium by relative volatility.
    $$y(x) = \frac{\alpha x}{1 + (\alpha - 1)x}$$
    where α is a constant.
4. Pressure drop through fittings and pipes.
    $$\Delta P(F) = kF^2$$
    where k is a constant.

5. Heat transfer rate by radiation.
    $$q(T) = \varepsilon \sigma A T^4$$
    where ε, σ, and A are constants.
6. Enthalpy as a function of temperature.
    $$H(T) = H_0 + AT + BT^2 + CT^3 + DT^4$$
    where H_0, A, B, C, and D are constant.

Example 2-11. Linearize the Arrhenius equation for the dependence of chemical reaction rates on temperature:

$$k(T) = k_0 e^{-(E/RT)}$$

where k_0, E, and R are constants.

Solution. From Eq. (2-56) we find

$$k(T) \doteq k(\overline{T}) + \frac{dk}{dT}(\overline{T})(T - \overline{T})$$

$$\frac{dk}{dT}(\overline{T}) = k_0 e^{(E/R\overline{T})}\left(\frac{E}{R\overline{T}^2}\right) = k(\overline{T})\frac{E}{R\overline{T}^2}$$

Substitute to obtain

$$k(T) \doteq k(\overline{T}) + k(\overline{T})\frac{E}{R\overline{T}^2}(T - \overline{T})$$

In terms of deviation variables we see that

$$K(T) \doteq \left[k(\overline{T})\frac{E}{R\overline{T}^2}\right]\mathbf{T}$$

where $K(T) = k(T) - k(\overline{T})$ and $\mathbf{T} = T - \overline{T}$.
    To illustrate that the only variables in the linearized equation are K and $\mathbf{T}$, let us consider the following numerical problem:

$$k_0 = 8 \times 10^9 \ s^{-1}$$
$$E = 22000 \ cal/g \ mole$$
$$\overline{T} = 373K \ (100C)$$

and

$$R = 1.987 \ cal/g \ mole \ K$$
$$k(\overline{T}) = 8 \times 10^9 \ e^{-[22000/(1.987)(373)]} = 1.0273 \times 10^{-3} \ s^{-1}$$
$$\frac{dk}{dt}(\overline{T}) = (1.0273 \times 10^{-3})\frac{22000}{(1.987)(373)^2}$$
$$= 8.175 \times 10^{-5} \ s^{-1}K^{-1}$$

This results in the following linearized equations:

$$k(T) = 1.0273 \times 10^{-3} + 8.175 \times 10^{-5} (T - 373)$$
$$K(T) = 8.175 \times 10^{-5} \text{ T}$$

Linearization of Functions of Two or More Variables

Consider the nonlinear function of two variables $f[x(t), y(t)]$. The Taylor series expansion around a point $(\bar{x}, \bar{y})$ is given by

$$
\begin{aligned}
f[x(t), y(t)] = f(\bar{x}, \bar{y}) &+ \frac{\partial f}{\partial x}(\bar{x}, \bar{y})[x(t) - \bar{x}] \\
&+ \frac{\partial f}{\partial y}(\bar{x}, \bar{y})[y(t) - \bar{y}] + \frac{1}{2!} \frac{\partial^2 f}{\partial x^2}(\bar{x}, \bar{y})[x(t) - \bar{x}]^2 \\
&+ \frac{1}{2!} \frac{\partial^2 f}{\partial y^2}(\bar{x}, \bar{y})[y(t) - \bar{y}]^2 \\
&+ \frac{\partial^2 f}{\partial x \partial y}(\bar{x}, \bar{y})[x(t) - \bar{x}][y(t) - \bar{y}] + \dots
\end{aligned}
\tag{2-62}
$$

The linear approximation consists of dropping the second- and higher-order terms to obtain

$$
f[x(t), y(t)] \doteq f(\bar{x}, \bar{y}) + \frac{\partial f}{\partial x}(\bar{x}, \bar{y})[x(t) - \bar{x}] + \frac{\partial f}{\partial y}(\bar{x}, \bar{y})[y(t) - \bar{y}]
\tag{2-63}
$$

The error of this linear approximation is small for x and y in the neighborhood of $\bar{x}$ and $\bar{y}$. As an example that graphically illustrates this point, let us consider the area of a rectangle as a function of its sides h and w:

$$a(h,w) = hw$$

The partial derivatives are

$$\frac{\partial a}{\partial h} = w \qquad \frac{\partial a}{\partial w} = h$$

The linear approximation is given by

$$a(h,w) \doteq a(\bar{h}, \bar{w}) + \bar{w}(h - \bar{h}) + \bar{h}(w - \bar{w})$$

As shown graphically in Fig. 2-6, the error of this approximation is a small rectangle of area $(h - \bar{h})(w - \bar{w})$. This error is small when h and w are near $\bar{h}$ and $\bar{w}$. In terms of deviation variables

$$A(h,w) \doteq \bar{w}H + \bar{h}W$$

where $A(h,w) = a(h,w) - a(\bar{h}, \bar{w})$.

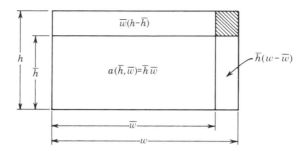

Figure 2-6.   The cross-hatched area is the error of the linear approximation to the function $a(h,w) = hw$.

In general, a function of n variables $x_1, x_2, \ldots, x_n$, is linearized by the formula

$$f(x_1, x_2, \ldots, x_n) \doteq f(\bar{x}_1, \bar{x}_2, \ldots, \bar{x}_n) + \overline{\frac{\partial f}{\partial x_1}}(x_1 - \bar{x}_1)$$

$$+ \overline{\frac{\partial f}{\partial x_2}}(x_2 - \bar{x}_2) + \ldots + \overline{\frac{\partial f}{\partial x_n}}(x_n - \bar{x}_n) \tag{2-64}$$

$$= f(\bar{x}_1, \bar{x}_2, \ldots, \bar{x}_n) + \sum_{k=1}^{n} \overline{\frac{\partial f}{\partial x_k}}(x_k - \bar{x}_k)$$

where

$$\overline{\frac{\partial f}{\partial x_k}}$$

denotes the partial derivatives evaluated at $(\bar{x}_1, \bar{x}_2, \ldots, \bar{x}_n)$.

Example 2-12.   Find the linear approximation to the nonlinear function

$$f(x,y,z) = 2x^2 + xy^2 - 3\frac{y}{z}$$

at the point $\bar{x} = 1, \bar{y} = 2, \bar{z} = 3$.

Solution.   From Eq. (2-64) the linear approximation is given by

$$f(x,y,z) = f(\bar{x},\bar{y},\bar{z}) + \overline{\frac{\partial f}{\partial x}}(x - \bar{x}) + \overline{\frac{\partial f}{\partial y}}(y - \bar{y}) + \overline{\frac{\partial f}{\partial z}}(z - \bar{z})$$

Taking the indicated partial derivatives of the function gives

$$\frac{\partial f}{\partial x} = 4x + y^2$$

$$\frac{\partial f}{\partial y} = 2xy - \frac{3}{z}$$

$$\frac{\partial f}{\partial z} = \frac{3y}{z^2}$$

Evaluating the function and its partial derivatives at the base point results in

$$\bar{f} = 2(1)^2 + 1(2)^2 - 3\frac{2}{3} = 4$$

$$\frac{\overline{\partial f}}{\partial x} = 4(1) + (2)^2 = 8$$

$$\frac{\overline{\partial f}}{\partial y} = 2(1)(2) - \frac{3}{3} = 3$$

$$\frac{\overline{\partial f}}{\partial z} = \frac{3(2)}{(3)^2} = \frac{2}{3}$$

Substituting these values, the linearized function is given by

$$f(x,y,z) \doteq 4 + 8(x - 1) + 3(y - 2) + \frac{2}{3}(z - 3)$$

or in terms of the deviation variables F, X, Y, and Z

$$F \doteq 8X + 3Y + \frac{2}{3}Z$$

Example 2-13.   The density of ideal gas is given by the following formula:

$$\rho = \frac{Mp}{RT}$$

where M is the molecular weight and R is the ideal gas constant.

Find the linear approximation to the density as a function of T and p and evaluate the coefficients for air ($M = 29$) at 300K and atmospheric pressure (101,300 N/m^2). In SI units the ideal gas constant is $R = 8,314$ N-m/kgmole-K.

Solution.   From Eq. (2-64) the linear approximation is given by

$$\rho = \bar{\rho} + \frac{\overline{\partial \rho}}{\partial T}(T - \bar{T}) + \frac{\overline{\partial \rho}}{\partial p}(p - \bar{p})$$

The partial derivatives of the density function are

$$\frac{\partial \rho}{\partial T} = -\frac{Mp}{RT^2} = -\frac{\rho}{T}$$

$$\frac{\partial \rho}{\partial p} = \frac{M}{RT} = -\frac{\rho}{p}$$

Evaluating at the base condition, we obtain

$$\bar{\rho} = \frac{M\bar{p}}{R\bar{T}} = \frac{(29)(101300)}{(8314)(300)} = 1.178 \text{ kg/m}^3$$

$$\frac{\overline{\partial \rho}}{\partial T} = -\frac{\bar{\rho}}{\bar{T}} = -\frac{1.178}{300} = -0.00393 \text{ kg/m}^3\text{K}$$

$$\frac{\overline{\partial \rho}}{\partial p} = \frac{\bar{\rho}}{\bar{p}} = \frac{1.178}{101300} = 1.163 \times 10^{-5} \text{ kg/m-N}$$

Substituting we obtain the linearized function:

$$\rho \doteq 1.178 - 0.00393\,(T - \overline{T}) + 1.163 \times 10^{-5}\,(p - \overline{p})$$

or, in terms of the deviation variables **R**, **T**, **P**

$$\mathbf{R} \doteq -0.00393\mathbf{T} + 1.163 \times 10^{-5}\mathbf{P}$$

Example 2-14. The response of the composition of reactant A in a continuous stirred tank reactor (see Fig. 2-7) can be calculated by the following equations:

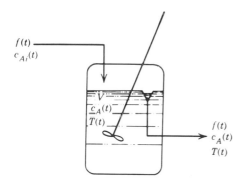

Figure 2-7. Continuous stirred tank reactor.

$$V\frac{dc_A(t)}{dt} = f(t)[c_{Ai}(t) - c_A(t)] - Vk[T(t)]c_A(t)$$

$$k[T(t)] = k_0 e^{-[E/RT(t)]}$$

where k_0, V, E, and R are constant. In deriving these equations we have assumed that the reactor is adiabatic and of constant volume and that the heat of reaction is negligible.

Write the linearized equation in terms of the deviation variables around the initial steady-state conditions, and in terms of the Laplace transform variables.

Solution. Linearizing the two nonlinear terms of the equation in turn results in

$$f(t)[c_{Ai}(t) - c_A(t)] \doteq \overline{f}(\overline{c}_{Ai} - \overline{c}_A) + \overline{f}[C_{Ai}(t) - C_A(t)]$$
$$+ [\overline{c}_{Ai} - \overline{c}_A]F(t)$$

$$Vk[T(t)]c_A(t) \doteq Vk(\overline{T})\overline{c}_A + V\frac{dk}{dT}(\overline{T})\overline{c}_A T(t) + Vk(\overline{T})C_A(t)$$

Substituting the result from Example 2-11, we get

$$Vk[T(t)]c_A(t) \doteq Vk(\overline{T})\overline{c}_A + Vk(\overline{T})\frac{E}{R\overline{T}^2}\overline{c}_A T(t) + Vk(\overline{T})C_A(t)$$

Since the base values represent a steady state, we can see that

$$0 = \overline{f}(\overline{c}_{Ai} - \overline{c}_A) - Vk(\overline{T})\overline{c}_A$$

If we now substitute these last two equations into the original differential equation, we obtain

$$V\frac{dC_A(t)}{dt} \doteq \bar{f}C_{Ai}(t) + (\bar{c}_{Ai} - \bar{c}_A)F(t) - [\bar{f} + Vk(\bar{T})]C_A(t)$$

$$- Vk(\bar{T})\frac{E}{R\bar{T}^2}\bar{c}_A\mathbf{T}(t)$$

Taking advantage of the fact that the base values are the initial values (i.e., $C_A(0)$ = 0), we can now take the Laplace transform of the linearized equation:

$$VsC_A(s) = \bar{f}C_{Ai}(s) + (\bar{c}_{Ai} - \bar{c}_A)F(s) - [\bar{f} + Vk(\bar{T})]C_A(s)$$

$$- Vk(\bar{T})\frac{E}{R\bar{T}^2}\bar{c}_A T(s)$$

Rearranging yields

$$C_A(s) = \frac{K_A}{\tau s + 1}C_{Ai}(s) + \frac{K_F}{\tau s + 1}F(s) - \frac{K_T}{\tau s + 1}T(s)$$

where the constant parameters are

$$\tau = \frac{V}{\bar{f} + Vk(\bar{T})} \qquad K_A = \frac{\bar{f}}{\bar{f} + Vk(\bar{T})}$$

$$K_F = \frac{\bar{c}_{Ai} - \bar{c}_A}{\bar{f} + Vk(\bar{T})} \qquad K_T = \frac{Vk(\bar{T})E\bar{c}_A}{R\bar{T}^2[\bar{f} + Vk(\bar{T})]}$$

The result of this last example shows that the parameters of the linearized equation depend on the base values of the system variables. It follows that for a nonlinear system the dynamic response will be different at different operating conditions. This point will be illustrated further in Chapters 3 and 4.

2-4. REVIEW OF COMPLEX NUMBER ALGEBRA

In the preceding sections we have learned that linearization and Laplace transforms are powerful tools for establishing general relationships among the variables and signals that constitute process control systems. Unfortunately, the manipulation of Laplace transforms requires familiarity with complex number algebra. In this section we will review some of the fundamental operations of complex numbers. Our objective is to provide a ready reference for those readers who might not feel comfortable with complex numbers.

Complex Numbers

A number is said to be complex when it cannot be represented as a pure real number or a pure imaginary number; an *imaginary number* is one that contains the square root of negative unity ($i = \sqrt{-1}$). One way to write a complex number is as follows:

$$c = a + ib \tag{2-65}$$

where

    a is the *real part* of complex number c
    b is the *imaginary part*, itself a real number.

A complex number can be represented graphically in a plane by plotting the imaginary part of the vertical or *imaginary axis*, and the real part on the horizontal or *real axis*. Such a plane is known as the *complex plane* and is represented in Fig. 2-8. Each point on this plane represents a number that can be real if on the real axis, imaginary if on the imaginary axis, or complex if anywhere else.

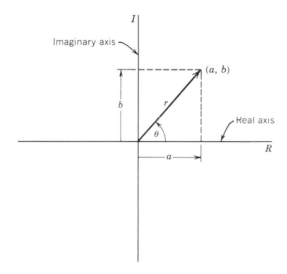

Figure 2-8. Complex plane.

An alternative way to represent a complex number is the *polar notation*. As shown in Fig. 2-8, the distance from the origin to the point (a,b) is given by

$$r = \sqrt{a^2 + b^2} = |c| \qquad (2\text{-}66)$$

This is known as the *magnitude* of complex number c given in Eq. (2-65). The other part required to represent number c in polar notation is angle θ, which is given by (see Fig. 2-8)

$$\theta = \tan^{-1}\frac{b}{a} = \angle c \qquad (2\text{-}67)$$

This is the angle that the line from the origin to point (a,b) forms with the positive real axis. It is known as the *argument* of the complex number. Equations (2-66) and (2-67) can be used to convert a complex number from its cartesian form (a,b) to its polar form (r,θ). The reverse operation can be done by using the following equations, which you can easily verify from inspection of Fig. 2-8:

$$a = r\cos\theta \qquad (2\text{-}68)$$
$$b = r\sin\theta \qquad (2\text{-}69)$$

If we substitute Eqs. (2-68) and (2-69) into Eq. (2-65) and factor the magnitude r, we obtain

$$c = r(\cos \theta + i \sin \theta) = re^{i\theta} \tag{2-70}$$

where we have made use of the trigonometric identity

$$e^{i\theta} = \cos \theta + i \sin \theta \tag{2-71}$$

Equation (2-70) is the polar form of complex number c.

Operations with Complex Numbers

Given two complex numbers

$$c = a + ib$$
$$p = v + iw$$

The sum is given by

$$c + p = (a + v) + i(b + w) \tag{2-72}$$

The difference is

$$c - p = (a - v) + i(b - w) \tag{2-73}$$

The multiplication is given by

$$\begin{aligned} cp &= (a + ib)(v + iw) \\ &= av + i^2bw + ibv + iaw \\ &= (av - bw) + i(bv + aw) \end{aligned} \tag{2-74}$$

where we have substituted $i^2 = -1$.

Thus, addition, subtraction, and multiplication of complex numbers follow the same rules of general algebra. So does division, except that in order to clear the denominator we must make use of the *conjugate*.

The conjugate of a complex number is defined as a number having the same real part and an imaginary part that is equal in magnitude and opposite in sign. In other words

$$\text{conj. } (a + ib) = a - ib \tag{2-75}$$

The product of a complex number and its conjugate is a real number given by

$$(a + ib)(a - ib) = a^2 + b^2 \tag{2-76}$$

With this, the division of complex numbers is carried out by multiplying numerator and denominator by the conjugate of the denominator:

$$\begin{aligned} \frac{c}{p} &= \frac{a + ib}{v + iw} \cdot \frac{v - iw}{v - iw} \\ &= \frac{(av + bw) + i(bv - aw)}{v^2 + w^2} \\ &= \left(\frac{av + bw}{v^2 + v^2} \right) + i\left(\frac{bv - aw}{v^2 + w^2} \right) \end{aligned} \tag{2-77}$$

Multiplication and division can be performed more easily in polar notation. Let c and p be given by

$$c = re^{i\theta}$$
$$p = qe^{i\beta}$$

Then the product is given by

$$cp = rqe^{i(\theta + \beta)} \tag{2-78}$$

and the ratio by

$$\frac{c}{p} = \frac{r}{q}e^{i(\theta - \beta)} \tag{2-79}$$

Raising to a power is also simpler in polar notation:

$$c^n = r^n e^{in\theta} \tag{2-80}$$

and so is extracting the nth root. A number has n nth roots when all its real and complex roots are considered:

$$\sqrt[n]{c} = \sqrt[n]{r}\, e^{i\frac{\theta + 2k\pi}{n}} \qquad k = 0, \pm 1, \pm 2, \ldots \tag{2-81}$$

where the value of k is changed until n distinct roots have been calculated.

The above constitute the fundamental operations of complex number algebra. It is important to realize that FORTRAN and other modern computer programming languages can perform operations and calculate functions involving complex numbers.

Example 2-15. Given the numbers

$$a = 3 + i4 \qquad b = 8 - i6 \qquad c = -1 + i$$

(a) Convert them to polar form:

$$|a| = \sqrt{9 + 16} = 5 \qquad |b| = \sqrt{64 + 36} = 10 \qquad |c| = \sqrt{1 + 1} = \sqrt{2}$$

$$\sphericalangle a = \tan^{-1}\frac{4}{3} = 0.9273 \qquad \sphericalangle b = \tan^{-1}\frac{-6}{8} = -0.6435 \qquad \sphericalangle c = \tan^{-1}\frac{1}{-1} = \frac{3\pi}{4}$$

$$a = 5e^{i0.9273} \qquad b = 10e^{-i0.6435} \qquad c = \sqrt{2}e^{(i3\pi/4)}$$

Notice that b is in the fourth quadrant and c is in the second quadrant of the complex plane (see Fig. 2-9). Notice also that the angles are in radians.

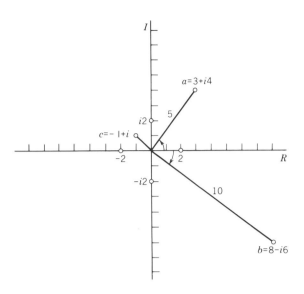

Figure 2-9.   Complex numbers for Example 2-15.

(b)   The following are examples of operations with complex numbers:

$$a + b = (3 + 8) + i(4 - 6) = 11 - i2$$
$$a - b = (3 - 8) + i(4 + 6) = -5 + i10$$
$$ac = (-3 - 4) + i(3 - 4) = -7 - i$$
$$bc = (-8 + 6) + i(8 + 6) = -2 + i14$$

The following illustrates the distributive property of multiplication:

$$(a + b)c = (11 - i2)(-1 + i) = (-11 + 2) + i(11 + 2) = -9 + i13$$
$$ac + bc = (-7 - i) + (-2 + i14) = -9 + i13$$

The following exemplifies division:

$$\frac{a}{b} = \frac{3 + i4}{8 - i6} \cdot \frac{8 + i6}{8 + i6} = \frac{(24 - 24) + i(18 + 32)}{64 + 36} = i0.50$$

In polar notation

$$ac = 5e^{i0.9273}\sqrt{2}e^{i(3\pi/4)} = 5\sqrt{2}e^{i3.2834}$$
$$= 5\sqrt{2}\cos 3.2834 + i5\sqrt{2}\sin 3.2834 = -7 - i$$
$$\frac{a}{b} = \frac{5e^{i0.9273}}{10e^{-i0.6435}} = 0.50e^{i1.5708}$$
$$= 0.50\cos 1.5708 + i0.50\sin 1.5708 = i0.50$$

These results match those obtained earlier.

(c) Find the fourth roots of 16. In polar coordinates

$$16 = 16e^{i0}$$
$$x = \sqrt[4]{16}\,e^{i0} = \sqrt[4]{16}\,e^{i[(0 + 2k\pi)/4]} = 2e^{i(2k\pi/4)}$$

The roots are

For $k = 0$ $x = 2$
$k = 1$ $x = 2e^{i(\pi/2)} = i2$
$k = -1$ $x = 2e^{-i(\pi/2)} = -i2$
$k = 2$ $x = 2e^{i\pi} = -2$

These roots are plotted in Fig. 2-9.

2-5. SUMMARY

In this chapter we have learned the techniques of Laplace transforms and linearization. By combining these tools we shall be able to represent the dynamic response of processes and instruments by sets of algebraic equations in the Laplace transform variable, s. This will lead us, as we shall see in the next chapter, to the important concept of transfer functions. In the rest of this book we shall use transfer functions for designing and analyzing process control systems.

REFERENCES

1. Müller, D. E., "A Method for Solving Algebraic Equations Using an Automatic Computer," *Mathematical Tables and Other Aids to Computation*, Vol. 10, 1956, pp. 208–215.

2. Ketter, R. L., and S. P. Prawel, Jr., *Modern Methods of Engineering Computation*, McGraw-Hill, New York, 1969.

3. Conte, Samuel D., and Carl de Boor, *Elementary Numerical Analysis*, 3rd ed., McGraw-Hill, New York, 1980.

4. D'Azzo, John J., and C. H. Houpis, *Feedback Control System Analysis and Synthesis*, 2nd ed., McGraw-Hill, New York, 1966, Chapter 4.

5. Murrill, Paul W., *Automatic Control of Processes*, Intext, Scranton, Pa., 1967, Chapters 4, 5, and 9.

6. Luyben, William L., *Process Modeling, Simulation, and Control for Chemical Engineers*, McGraw-Hill, New York, 1973, Chapters 6 and 7.

PROBLEMS

2-1.   Using the definition of the Laplace transform, derive the transforms $F(s)$ of the following functions:

   a. $f(t) = t$
   b. $f(t) = e^{-at}$ where a is constant.
   c. $f(t) = \cos \omega t$ where ω is constant.
   d. $f(t) = e^{-at} \cos \omega t$ where a and ω are constant.

   Note: In parts (c) and (d) you will need the trigonometric identity

$$\cos x = \frac{e^{ix} + e^{-ix}}{2}$$

Check your answer against the entries in Table 2-1.

2-2.   Using a table of Laplace transforms and the properties of the transform, find the transforms $F(s)$ of the following functions:

   a. $f(t) = u(t) + 2t + 3t^2$
   b. $f(t) = e^{-2t}[u(t) + 2t + 3t^2]$
   c. $f(t) = u(t) + e^{-2t} - 2e^{-t}$
   d. $f(t) = u(t) - e^{-t} + te^{-t}$
   e. $f(t) = u(t - 2)[1 - e^{-2(t-2)} \sin(t - 2)]$

2-3.   Check the validity of your results to Problem 2-2 by application of the initial and final value theorems. Do these theorems apply in all of the cases?

2-4.   A useful function to use as a forcing function in the analysis of process control systems is the partial ramp sketched in Fig. 2-10. Derive the Laplace transform of the partial ramp function by each of the following methods:

   a. Use direct application of the definition of the Laplace transform, considering the following sections of the function:

$$f(t) = \begin{cases} \dfrac{H}{t_1}t & t < t_1 \\ H & t \geq t_1 \end{cases}$$

   where H is the height and t_1 the duration of the ramp.
   b. Use a table of Laplace transforms, the linearity property, and the real translation theorem, considering the function as the sum or superposition of the following functions:

$$f(t) = \frac{H}{t_1}t - u(t - t_1)\frac{H}{t_1}t + Hu(t - t_1)$$

   Check your answer by application of the initial and final value theorems.

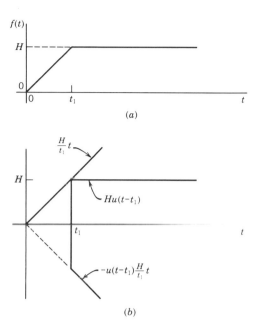

Figure 2-10. Forcing function for Problem 2-4. (*a*) Partial ramp function. (*b*) Functions that, when superimposed, produce the partial ramp function.

2-5. In the statement of the real translation theorem we pointed out that for the theorem to apply, the delayed function has to be zero for all times less than the delay time. Demonstrate this by calculating the Laplace transform of the function

$$f(t) = e^{-(t-t_0)/\tau}$$

where t_0 and τ are constants,

a. Assuming that it holds for all times greater than zero, that is, that it can be rearranged as

$$f(t) = e^{t_0/\tau}e^{-t/\tau}$$

b. Assuming that it is zero for $t \le t_0$, that is, that it should be properly written as

$$f(t) = u(t - t_0)e^{-(t-t_0)/\tau}$$

Are the two answers the same? Which one agrees with the result of the real translation theorem?

2-6. Find the solution $y(t)$ of the following differential equations using the method of Laplace transforms and partial fractions expansion. The initial conditions of $y(t)$ and its derivatives are zero, and the forcing function is the unit step function.

$$x(t) = u(t)$$

a. $2\dfrac{dy(t)}{dt} + y(t) = 5x(t)$

b. $\dfrac{d^2y(t)}{dt^2} + 9\dfrac{dy(t)}{dt} + 9y(t) = x(t)$

c. $\dfrac{d^2y(t)}{dt^2} + 3\dfrac{dy(t)}{dt} + 9y(t) = x(t)$

d. $\dfrac{d^2y(t)}{dt^2} + 6\dfrac{dy(t)}{dt} + 9y(t) = x(t)$

e. $2\dfrac{d^3y(t)}{dt^3} + 7\dfrac{d^2y(t)}{dt^2} + 21\dfrac{dy(t)}{dt} + 9y(t) = x(t)$

2-7. Repeat Problem 2-6(d) using as the forcing function

a. $x(t) = e^{-3t}$
b. $x(t) = u(t - 1)e^{-3(t-1)}$

2-8. The standard form of the differential equation for a first-order lag with dead time is

$$\tau\frac{dy(t)}{dt} + y(t) = Kx(t - t_0)$$

where

τ is the time constant
K is the gain
t_0 is the dead time

Assuming that the initial condition is $y(0) = y_0$ find, by Laplace transforms, the solution $y(t)$ for each of the following forcing functions:
a. Unit impulse $x(t) = \delta(t)$
b. A unit step $x(t) = u(t)$
c. The partial ramp of Problem 2-4
d. A sine wave of frequency ω: $x(t) = \sin \omega t$

2-9. A standard form of the differential equation for a second-order lag is given by

$$\tau^2\frac{d^2y(t)}{dt^2} + 2\zeta\tau\frac{dy(t)}{dt} + y(t) = Kx(t)$$

where

τ is the characteristic time constant
ζ is the damping ratio
K is the gain

Assuming that all initial conditions are zero, use the method of Laplace transforms to find the unit step response of $y(t)$ for the following cases:

a. $\zeta > 1$, known as *overdamped*.
b. $\zeta = 1$, known as *critically damped*.
c. $0 < \zeta < 1$, known as *underdamped*.
d. $\zeta = 0$, known as *undamped*.
e. $\zeta < 0$, known as *unstable*.

Based on your results, which terms in the function $y(t)$ are characteristic of an underdamped and undamped lag? Can you verify the definition of stability by inspection of your result to part (e)?

2-10. Using a computer program to calculate the roots of the polynomial in the denominator of the following Laplace transforms, find the inverse $y(t)$ by the method of partial fractions expansion:

a. $Y(s) = \dfrac{s^2 + 1}{(s^5 + 1)} \cdot \dfrac{1}{s}$

b. $Y(s) = \dfrac{2}{24s^4 + 50s^3 + 35s^2 + 10s + 1} \cdot \dfrac{1}{s}$

c. $Y(s) = \dfrac{1}{4s^4 + 4s^3 + 17s^2 + 16s + 4}$

A root-finding program is listed in Appendix D.

2-11. Linearize the following functions with respect to the indicated variable. Give your results in terms of the deviation variables.

a. Antoine equation for vapor pressure:

$$p^\circ(T) = e^{[A - B/(T + C)]}$$

where A, B, and C are constant.

b. Vapor composition in equilibrium with liquid:

$$y(x) = \frac{\alpha x}{1 + (\alpha - 1)x}$$

where α, the relative volatility, is constant.

c. Flow through a valve:

$$f(\Delta p_v) = C_v \sqrt{\frac{\Delta p_v}{G}}$$

where C_v and G are constant.

d. Heat transfer by radiation:

$$q(T) = \varepsilon \sigma A T^4$$

where ε, σ, and A are constant.

2-12. As pointed out in the text, the range of applicability of the linearized equation depends on the degree of nonlinearity of the original function at the base point.

Demonstrate this point by calculating the range of the liquid mole fraction x in Problem 2-11(b) over which the linearized function matches the actual function within $\pm 5\%$ of the vapor mole fraction y. Also calculate the range of values of x over which the parameter of the linearized function remains within $\pm 5\%$ of its base value. Use the following numerical values:

a. $\alpha = 1.10$ $\bar{x} = 0.10$
b. $\alpha = 1.10$ $\bar{x} = 0.90$
c. $\alpha = 5.0$ $\bar{x} = 0.10$
d. $\alpha = 5.0$ $\bar{x} = 0.90$

2-13. Repeat Problem 2-12 for the Arrhenius function of Example 2-11. Calculate the range of values of the temperature over which the reaction rate coefficient $k(T)$ is matched within $\pm 5\%$ by the linearized function. Calculate also the range of values of T over which the parameter of the linearized equation remains within $\pm 5\%$ of its base value. What would be the results if the value of E doubled?

2-14. Evaluate the parameters of the transform obtained in Example 2-14 using the kinetic parameters from Example 2-11 and the following reactor parameters.

$$V = 2.6 \text{ m}^3 \qquad \bar{f} = 2 \times 10^{-3} \text{ m}^3/\text{s} \qquad \bar{c}_{Ai} = 12 \text{ kgmoles/m}^3$$

Notice that the value of $\bar{c}_A$ can be calculated from the steady-state relationship given in Example 2-14.

2-15. Raoult's law gives the vapor mole fraction y of a component at equilibrium by the following relationship:

$$y(x,T,p) = \frac{p^\circ(T)}{p} x$$

where

x is the liquid mole fraction
T is the temperature
p is the total pressure

and $p^\circ(T)$, the vapor pressure, is given by the Antoine equation, as in Problem 2-11(a).

a. Linearize the formula for the vapor mole fraction and express the result in terms of deviation variables.
b. Evaluate the parameters of the linear equation for the following conditions:

$\bar{x} = 0.50$ (benzene mole fraction in the liquid)
$\bar{T} = 95C$ $\bar{p} = 760$ mm Hg

Antoine constants for benzene: $A = 15.9008$
$B = 2788.51C$
$C = 220.80C$

2-16. Write, in terms of deviation variables, the linear approximations to the following functions:

    a. $f(x,y) = y^2 x + 2x + \ln y$

    b. $f(x,y) = \dfrac{3\sqrt{x}}{y} + 2 \sin xy$

    c. $f(x,y) = y^x$

2-17. The tank shown in Fig. 2-11 is put in a line to smooth out the flow under variations in inlet pressure $p_i(t)$ and in discharge pressure $p_o(t)$. At the steady-state base conditions the flow through the system is 25.0 kgmoles/s and the pressures are

$$\bar{p}_i = 2000 \text{ kN/m}^2$$
$$\bar{p} = 1800 \text{ kN/m}^2$$
$$\bar{p}_o = 1600 \text{ kN/m}^2$$

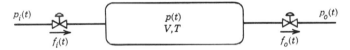

Figure 2-11. Pressure surge tank for Problem 2-17.

The volume of the tank is $V = 10 \text{ m}^3$. A mole balance on the tank, assuming ideal gas behavior and constant temperature of 400K, is given by

$$\frac{V}{RT} \frac{dp(t)}{dt} = f_i(t) - f_o(t)$$

where $R = 8314 \text{ N-m/kgmole K}$, is the ideal gas law constant. The inlet and outlet flow rates are given by

$$f_i(t) = k_i \sqrt{p_i(t)[p_i(t) - p(t)]}$$
$$f_o(t) = k_o \sqrt{p(t)[p(t) - p_o(t)]}$$

where k_i and k_o are the (constant) conductance coefficients of the inlet and outlet valves, respectively. These valves are adjusted to obtain the base pressures at the base flow given above. Linearize the equations given above and solve them to obtain the response of the pressure in the tank $p(t)$ to the following forcing functions:

a. A step change in inlet pressure, constant discharge pressure

$$P_i(t) = u(t) \qquad p_o(t) = \text{constant}$$

b. A step change in outlet pressure, constant inlet pressure

$$P_o(t) = u(t) \qquad p_i(t) = \text{constant}$$

Use the Laplace transform method to solve the differential equation.

CHAPTER
3
First-Order Dynamic Systems

This chapter has two principal objectives. The first is to present an introduction to the development of simple process models. This modeling is necessary whenever analysis of control systems is required. The second objective, a byproduct of the first, is to learn the physical meaning of some process parameters that describe the "personality" of processes. As explained in Chapter 1, once the personality of the process is known, then the required control system can be designed. We will also learn some new terms and mathematical manipulations important in our study of automatic process control. The tools learned in Chapter 2 will be used extensively in this chapter.

All of the above will be done by process examples. We start with some simple examples and build from these to more complex and realistic ones.

The modeling of industrial processes usually starts with a balance on a conserved quantity: mass or energy. This balance can be written as:

$$
\begin{array}{ccccc}
\text{Rate of mass/energy} & & \text{Rate of mass/energy} & & \text{Rate of accumulation} \\
\text{into process} & - & \text{out of process} & = & \text{of mass/energy in process.}
\end{array}
$$

As we can imagine, in writing these balances and all other auxiliary equations, we must make use of almost every area of process engineering, such as thermodynamics, heat transfer, fluid flow, mass transfer, reaction engineering, etc. This makes the modeling of industrial processes most interesting and challenging.

3-1. THERMAL PROCESS

Consider the well-stirred tank shown in Fig. 3-1. We are interested in knowing how the outlet temperature, $T(t)$, responds to changes in inlet temperature, $T_i(t)$.

In this example constant inlet and outlet volumetric flows, liquid densities, and liquid heat capacities are assumed. All of these properties are known. The liquid in the tank is well mixed and the tank is well insulated; that is, this is an adiabatic process.

An unsteady-state energy balance on the contents of the tank gives us the desired relation between the inlet and outlet temperatures:

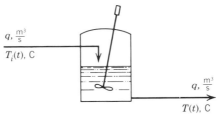

Figure 3-1. Thermal process.

$$qp_i h_i(t) - qph(t) = \frac{d(V\rho u(t))}{dt} \tag{3-1}$$

Or in terms of temperature

$$qp_i C_{pi} T_i(t) - qpC_p T(t) = \frac{d(V\rho C_v T(t))}{dt}$$

where

ρ_i, ρ = inlet and outlet liquid densities, respectively, kg/m^3

C_{pi}, C_p = inlet and outlet liquid heat capacities at constant pressures, respectively, J/kg-C

C_v = liquid heat capacity at constant volume, J/kg-C

V = volume of liquid in tank, m^3

h_i, h = inlet and outlet liquid enthalpies, respectively, J/kg

u = internal energy of liquid in tank, J/kg.

Since the density and heat capacity are assumed to be constant over the operating temperature range, the last equation can be written as

$$qpC_p T_i(t) - qpC_p T(t) = V\rho C_v \frac{dT(t)}{dt} \tag{3-2}$$

This equation is a first-order linear ordinary differential equation that provides the relationship between the inlet and outlet temperatures. It is important to notice that in this equation there is only one unknown, $T(t)$. The inlet temperature, $T_i(t)$, is an input variable and thus is not considered an unknown because it is up to us to specify how it will change, for example, a step change or a ramp change. To show that there is one equation with one unknown, we write

$$qpC_p T_i(t) - qpC_p T(t) = V\rho C_v \frac{dT(t)}{dt} \tag{3-2}$$

$$1 \text{ eq., } 1 \text{ unk. } (T(t))$$

The solution of this differential equation for a certain inlet temperature yields the response of the outlet temperature as a function of time. The inlet temperature is called the *input variable* or *forcing function*, since it is the variable that forces the outlet temperature to change. The outlet temperature is called the *output variable*, or *responding variable*, since it is the variable that responds to the forcing function.

Before solving the previous equation, a variable change is made that somewhat simplifies the solution. Write a steady-state energy balance on the contents of the tank:

$$q\rho C_p \overline{T}_i - q\rho C_p \overline{T} = 0 \tag{3-3}$$

Subtracting Eq. (3-3) from Eq. (3-2) yields

$$q\rho C_p(T_i(t) - \overline{T}_i) - q\rho C_p(T(t) - \overline{T}) = V\rho C_v \frac{d(T(t) - \overline{T})}{dt} \tag{3-4}$$

As discussed in Chapter 2, we now define the following *deviation variables* as

$$\mathbf{T}(t) = T(t) - \overline{T} \tag{3-5}$$

$$\mathbf{T}_i(t) = T_i(t) - \overline{T}_i \tag{3-6}$$

where

$\overline{T}, \overline{T}_i$ = steady state values of outlet and inlet temperatures, respectively, C

$\mathbf{T}(t), \mathbf{T}_i(t)$ = deviation variables of outlet and inlet temperatures, respectively, C

Substituting Eqs. (3-5) and (3-6) into (3-4) yields

$$q\rho C_p \mathbf{T}_i(t) - q\rho C_p \mathbf{T}(t) = V\rho C_v \frac{d\mathbf{T}(t)}{dt} \tag{3-7}$$

Eq. (3-7) is the same as Eq. (3-2) except that it is in terms of deviation temperatures. The solution of this equation yields $\mathbf{T}(t)$, the deviation temperature, versus time for a certain forcing function, $\mathbf{T}_i(t)$. If the actual outlet temperature, $T(t)$, is desired, the steady-state value $\overline{T}$ must be added to $\mathbf{T}(t)$ as per Eq. (3-5).

The definition and use of *deviation variables* in the analysis and design of process control systems is *most* important. These variables are almost exclusively used throughout control theory. Thus, the meaning and importance of deviation variables must be well understood. As explained in Chapter 2, their use offers the advantage that their value indicates the degree of deviation from some operating steady-state value. In practice, this steady-state value may be the desired value of the variable. Another advantage in the use of these variables is that their initial value is zero, assuming we start from steady state, simplifying the solution of differential equations such as Eq. (3-7). As mentioned, these variables are extensively used throughout this book.

Equation (3-7) can be rearranged as follows:

$$\frac{V\rho C_v}{q\rho C_p} \frac{d\mathbf{T}(t)}{dt} + \mathbf{T}(t) = \mathbf{T}_i(t)$$

and let

$$\tau = \frac{V\rho C_v}{q\rho C_p} \tag{3-8}$$

so

$$\tau \frac{d\mathbf{T}(t)}{dt} + \mathbf{T}(t) = \mathbf{T}_i(t) \tag{3-9}$$

Since this is a linear differential equation, the use of Laplace transform yields

$$\tau s \mathbf{T}(s) - \tau \mathbf{T}(0) + \mathbf{T}(s) = \mathbf{T}_i(s)$$

But $\mathbf{T}(0) = 0$; therefore, using algebra

$$\mathbf{T}(s) = \frac{1}{\tau s + 1} \mathbf{T}_i(s) \qquad (3\text{-}10)$$

or

$$\frac{\mathbf{T}(s)}{\mathbf{T}_i(s)} = \frac{1}{\tau s + 1} \qquad (3\text{-}11)$$

Equation (3-11) is called a *transfer function*. It is a first-order transfer function because it is developed from a first-order differential equation. Processes described by this transfer function are called *first-order processes, first-order systems*, or *first-order lags*. Sometimes, they are also referred to as single capacitance systems because this is the same type of transfer function that describes a one resistor–one capacitor (R-C) electrical system.

The name "transfer function" arises from the fact that the solution of the equation *transfers* the input, or forcing function, $\mathbf{T}_i(t)$, to the output, or responding variable, $\mathbf{T}(t)$. Transfer functions are discussed in more detail in Section 3-3.

Let us assume that the inlet temperature, $T_i(t)$, to the tank increases by A degrees C. That is, it experiences a step change of A degrees in magnitude. Mathematically, this is written as follows:

$$T_i(t) = \overline{T}_i \qquad t < 0$$
$$T_i(t) = \overline{T}_i + A \qquad t \geq 0$$

or as shown in Chapter 2

$$\mathbf{T}_i(t) = A u(t)$$

Taking the Laplace transforms, we obtain

$$\mathbf{T}_i(s) = \frac{A}{s}$$

Substituting into Eq. (3-10) yields

$$\mathbf{T}(s) = \frac{A}{s(\tau s + 1)}$$

and using partial fractions to obtain the inverse transform yields

$$\mathbf{T}(t) = A(1 - e^{-t/\tau}) \qquad (3\text{-}12)$$

or

$$T(t) = \overline{T} + A(1 - e^{-t/\tau}) \qquad (3\text{-}13)$$

The solutions of Eqs. (3-12) and (3-13) are shown graphically in Fig. 3-2.

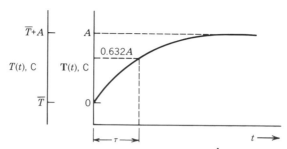

Figure 3-2. Response of a first-order process to a step change in forcing function—time constant.

Equation (3-12) provides the physical significance of τ, which is called the *process time constant*. Letting $t = \tau$, yields

$$\mathbf{T}(\tau) = A(1 - e^{-\tau/\tau}) = A(1 - e^{-1})$$

$$\mathbf{T}(\tau) = 0.632A$$

That is, in one time constant, 63.2% of the total change is reached. This shown graphically in Fig. 3-2. Consequently, *the time constant is related to the speed of response of the process*. The slower a process responds to a forcing function, or input, the larger the value of τ. The faster the process responds to a forcing function, the smaller the value of τ. The units of τ must be those of time. From Eq. (3-8) we see that

$$\tau = \frac{[m^3] \, [kg/m^3] \, [J/kg\text{-}C]}{[m^3/s] \, [kg/m^3] \, [J/kg\text{-}C]} = \text{seconds}$$

It is also very important to realize that the time constant is composed of the different physical properties and operating parameters of the process, as shown by Eq. (3-8). That is, the time constant depends on the volume of liquid in the tank (V), the heat capacities (C_p and C_v), and the process flow (q). If any of these characteristics change, the time constant will also change. Another way of saying this is that if any of the above process conditions change, the "personality" of the process will also change and will be reflected in the speed of response of the process, or time constant.

Another important factor is that in this example, the value of the time constant is constant over the operating range of $T(t)$. This is a property of linear systems and is not the case for nonlinear systems, as we will see sortly.

Up to now, the tank has been assumed well insulated, yielding an adiabatic process; that is, there are no heat losses to the atmosphere. Consequently, there is no heat-loss term in the energy balance. Removing this assumption of adiabatic operation and taking heat loss into account in the energy balance yields the following equation:

$$q\rho C_p T_i(t) - Q(t) - q\rho C_p T(t) = V\rho C_v \frac{dT(t)}{dt}$$

or

$$q\rho C_p T_i(t) - UA[T(t) - T_s(t)] - q\rho C_p T(t) = V\rho C_v \frac{dT(t)}{dt} \qquad (3\text{-}14)$$

1 eq., 1 unk. ($T(t)$)

where

$$U = \text{overall heat transfer coefficient, J/m}^2\text{-K-s}$$
$$A = \text{heat transfer area, m}^2$$
$$T_s(t) = \text{temperature of surroundings, } C, \text{ an input variable}$$

The overall heat transfer coefficient, U, is a function of several things, one of them being temperature. However, in this particular example, it is assumed to be constant. Since the mass of liquid in the tank and its density are also assumed to be constant, then the height of liquid is constant, and consequently, the heat transfer area, A, is also constant.

To obtain the deviation variables, start by writing a steady-state energy balance for this process;

$$q\rho C_p \overline{T}_i - UA(\overline{T} - \overline{T}_s) - q\rho C_p \overline{T} = 0 \tag{3-15}$$

Subtracting Eq. (3-15) from Eq. (3-14) yields

$$q\rho C_p(T_i(t) - \overline{T}_i) - UA[(T(t) - \overline{T}) - (T_s(t) - \overline{T}_s)]$$

$$- q\rho C_p(T(t) - \overline{T}) = V\rho C_v \frac{d(T(t) - \overline{T})}{dt} \tag{3-16}$$

Define a new deviation variable as

$$\mathbf{T}_s(t) = T_s(t) - \overline{T}_s \tag{3-17}$$

Substituting Eqs. (3-5), (3-6), and (3-17) into Eq. (3-16) yields

$$q\rho C_p \mathbf{T}_i(t) - UA(\mathbf{T}(t) - \mathbf{T}_s(t)) - q\rho C_p \mathbf{T}(t) = V\rho C_v \frac{d\mathbf{T}(t)}{dt} \tag{3-18}$$

Equation (3-18) is the same as Eq. (3-14) except that it is written in terms of deviation variables.

Equation (3-18) is also a first-order linear ordinary differential equation. In this case, there is still one equation with one unknown, $\mathbf{T}(t)$. The new variable $\mathbf{T}_s(t)$ is another forcing function. As the surrounding temperature changes, $T_s(t)$, this will affect the heat losses and consequently the process liquid temperature. Equation (3-18) can be rearranged as follows:

$$\frac{V\rho C_v}{q\rho C_p + UA} \frac{d\mathbf{T}(t)}{dt} + \mathbf{T}(t) = \frac{q\rho C_p}{q\rho C_p + UA} \mathbf{T}_i(t) + \frac{UA}{q\rho C_p + UA} \mathbf{T}_s(t)$$

or

$$\tau \frac{d\mathbf{T}(t)}{dt} + \mathbf{T}(t) = K_1 \mathbf{T}_i(t) + K_2 \mathbf{T}_s(t) \tag{3-19}$$

where

$$\tau = \frac{V\rho C_v}{q\rho C_p + UA}, \qquad \text{seconds} \tag{3-20}$$

$$K_1 = \frac{q\rho C_p}{q\rho C_p + UA}, \qquad \text{dimensionless } (C/C) \tag{3-21}$$

$$K_2 = \frac{UA}{q\rho C_p + UA}, \quad \text{dimensionless } (C/C). \tag{3-22}$$

The right-hand side of Eq. (3-19) shows the two forcing functions, $T_i(t)$ and $T_s(t)$, acting upon the response or left-hand side of the equation, $T(t)$.

Taking the Laplace transform of Eq. (3-19) gives

$$\tau s\mathbf{T}(s) - \tau\mathbf{T}(0) + \mathbf{T}(s) = K_1\mathbf{T}_i(s) + K_2\mathbf{T}_s(s)$$

But $T(0) = 0$. Rearranging this equation yields

$$\mathbf{T}(s) = \frac{K_1}{\tau s + 1}\,\mathbf{T}_i(s) + \frac{K_2}{\tau s + 1}\,\mathbf{T}_s(s) \tag{3-23}$$

If the surrounding temperature remains constant, $T_s(t) = \overline{T}_s$ and $\mathbf{T}_s(t) = 0$, the transfer function relating the process temperature to the inlet water temperature is

$$\frac{\mathbf{T}(s)}{\mathbf{T}_i(s)} = \frac{K_1}{\tau s + 1} \tag{3-24}$$

If the inlet liquid temperature remains constant, $T_i(t) = \overline{T}_i$ and $\mathbf{T}_i(t) = 0$, the transfer function relating the process temperature to the surrounding temperature is

$$\frac{\mathbf{T}(s)}{\mathbf{T}_s(s)} = \frac{K_2}{\tau s + 1} \tag{3-25}$$

If both the inlet liquid temperature and the surrounding temperature change, then Eq. (3-23) provides the correct relationship.

In the last three equations we have encountered a new and very important parameter, K. This parameter is called *process gain* or *steady-state gain*. To obtain the physical significance of this gain, let us assume that the inlet temperature to the tank increases by A degrees C. The response of the temperature to this forcing function is given by

$$\mathbf{T}(s) = \frac{K_1 A}{s(\tau s + 1)}$$

from which

$$T(t) = K_1 A(1 - e^{-t/\tau}) \tag{3-26}$$

or

$$T(t) = \overline{T} + K_1 A(1 - e^{-t/\tau}) \tag{3-27}$$

The response is shown graphically in Fig. 3-3. The total amount of change is given by $K_1 A$, the gain times the change in forcing function. We can say that *the gain tells us how much the output variable will change per unit change in forcing function or input variable.* That is, the gain defines the sensitivity of the process.

We define the gain, mathematically, as follows:

$$K = \frac{\Delta O}{\Delta I} = \frac{\Delta \text{ output variable}}{\Delta \text{ input variable}} \tag{3-28}$$

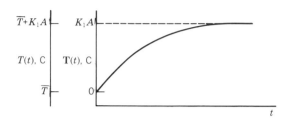

Figure 3-3. Response of a first-order process to a step change in forcing function—process gain.

The gain is another parameter that is related to the ''personality'' of the process to be controlled. Consequently, it depends on the physical properties and operating parameters of the process, as shown by Eqs. (3-21) and (3-22). The gains of this process depend upon the flow, density, and heat capacity of the process liquid (q, ρ, C_p), the overall heat transfer coefficient (U), and the heat transfer area (A). If any of these change, the personality of the process will change and this will be reflected in the gain. As with the time constant, the gains in this particular example are constant over the operating range.

There are two gains in this example. The first one, K_1, relates the outlet temperature to the inlet temperature. The other gain, K_2, relates the outlet temperature to the surrounding temperature. The units of gain must be the units of the output variable divided by the units of the forcing function, or input variable. This can be seen from Eq. (3-28).

Equation (3-23) shows that there is only one time constant in this process. That is, the time it takes the outlet temperature to reach a certain percentage of its total change due to a change in inlet temperature is equal to the time it takes to reach the same percentage when the surrounding temperature changes. This is not always the case. In this example there is more than one gain, one for each forcing function. Some processes may also have more than one time constant, possibly one for each forcing function. We will see examples of this as we go along.

It is always important during the analysis of any process to stop at some point to check the development for possible errors. A convenient point is usually after the development of Eq. (3-23). A quick check can be made by examining the signs of the equations to see whether they make sense in the real world. In this equation, both gains are positive. The equation indicates that if the inlet temperature increases, the outlet temperature also increases. This makes sense for this process. Equation (3-23) also shows that if the surrounding temperature increases, the outlet temperature increases. This makes sense because if the surrounding temperature increases, the rate of heat losses from the tank decreases, thereby increasing the temperature of the contents of the tank. This quick check builds our confidence and permits us to proceed with the analysis with a renewed hope of possible success.

3-2. GAS PROCESS

Consider the gas vessel shown in Fig. 3-4. The vessel acts as a surge tank, or damper, in a process. Assume that the process occurs isothermally, at some temperature, T, and that the flow through the outlet valve is expressed by

$$q_o(t) = \frac{\Delta p(t)}{R_v} = \frac{p(t) - p_2(t)}{R_v} \tag{3-29}$$

where R_v = resistance to flow across the valve, psi/scfm. We are interested in knowing how the pressure in the tank responds to changes in inlet flow, $q_i(t)$, and in downstream pressure, $p_2(t)$.

For this process, an unsteady-state mass balance will give the required relation:

$$\rho q_i(t) - \rho q_o(t) = \frac{dm(t)}{dt} \tag{3-30}$$

$$1 \, \text{eq.}, 2 \, \text{unk.} \, (qo(t), m(t))$$

where

$m(t)$ = mass of gas in vessel, lbm
ρ = density of gas at standard conditions of 14.7 psia and 60°F, lbm/ft^3.

If the pressure in the tank is low, the ideal gas equation of state relates the mass of gas in the vessel to the pressure:

$$p(t) = \frac{RT}{VM} m(t) \tag{3-31}$$

$$2 \, \text{eq.}, 3 \, \text{unk.} \, (p(t))$$

where

T = absolute temperature in tank, °R
V = volume of tank, ft^3
M = molecular weight of gas
R = ideal gas law constant = $10.73 \dfrac{\text{ft}^3\text{-psia}}{\text{lb moles-°R}}$

The expression for the flow through the outlet valve, Eq. (3-29), provides another equation:

$$q_o(t) = \frac{p(t) - p_2(t)}{R_v} \tag{3-29}$$

$$3 \, \text{eq.}, 3 \, \text{unk.}$$

Substituting Eqs. (3-29) and (3-31) into Eq. (3-30) gives

$$\rho q_i(t) - \rho \frac{[p(t) - p_2(t)]}{R_v} = \frac{VM}{RT} \frac{dp(t)}{dt} \tag{3-32}$$

To obtain the deviation variables and transfer function, we follow the same development as in the previous example. Writing the steady-state mass balance gives

Figure 3-4. Gas vessel.

$$\rho\bar{q}_i - \rho\bar{q}_o = 0$$

or

$$\rho\bar{q}_i - \rho\frac{(\bar{p} - \bar{p}_2)}{R_v} = 0 \tag{3-33}$$

Subtracting Eq. (3-33) from Eq. (3-32), we obtain

$$\rho(q_i(t) - \bar{q}_i) - \frac{\rho}{R_v}[(p(t) - \bar{p}) - p_2(t) - \bar{p}_2)] = \frac{VM}{RT}\frac{d(p(t) - \bar{p})}{dt} \tag{3-34}$$

We define the deviation variables as

$$Q(t) = q_i(t) - \bar{q}_i$$
$$P(t) = p(t) - \bar{p}$$

and

$$P_2(t) = p_2(t) - \bar{p}_2$$

Substituting these deviation variables into Eq. (3-34) and rearranging the equation algebraically gives

$$\tau\frac{dP(t)}{dt} + P(t) = K_1 Q_i(t) + P_2(t) \tag{3-35}$$

where

$$\tau = \frac{VM\,R_v}{RT\rho} \quad \text{minutes}$$

$$K_1 = R_v, \quad \text{psia/scfm}$$

Taking the Laplace transform yields the relationship between the response or output variable $P(t)$ and the forcing functions $Q_i(t)$ and $P_2(t)$:

$$P(s) = \frac{K_1}{\tau s + 1}Q_i(s) + \frac{1}{\tau s + 1}P_2(s) \tag{3-36}$$

From this equation we obtain the transfer function between $P(s)$ and $Q_i(s)$

$$\frac{P(s)}{Q_i(s)} = \frac{K_1}{\tau s + 1} \tag{3-37}$$

and between $P(s)$ and $P_2(s)$:

$$\frac{P(s)}{P_2(s)} = \frac{1}{\tau s + 1} \tag{3-38}$$

Both transfer functions are first order.

By now we can start building a feeling for the complete response of any first-order system. We know, for example, by analyzing Eq. (3-37), that if the inlet flow to the vessel changes by $+10$ scfm, the pressure in the vessel will ultimately change by $10K_1$ psi. This is true if no other upset enters the process. We also know that a change of $0.632\,(10K_1)$

in pressure will occur in one time constant, τ. This is shown graphically in Fig. 3-5. Remember that K_1 is the gain that $Q_1(t)$ has on $P(t)$ and that τ gives the speed of response of $P(t)$ once it responds to a change in $Q_i(t)$.

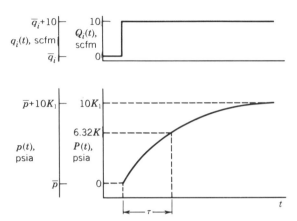

Figure 3-5.   Response of pressure to step change in inlet gas flow.

Similarly, by looking at Eq. (3-38), we know that when the downstream pressure from the valve changes by -2 psi the pressure in the vessel will ultimately change by -2 psi. The gain of $P_2(t)$ on $P(t)$ is $+1$ psi/psi. If the downstream pressure decreases, there will be a higher pressure drop aross the valve resulting in a larger outflow from the tank and, consequently, reduced pressure in the tank. We also know that 63.2% of the total change in pressure, or $0.632(-2)$ psi, will occur in τ minutes.

3-3.   TRANSFER FUNCTIONS AND BLOCK DIAGRAMS
Transfer Functions

The transfer function concept is a most important one in the study of process dynamics and of automatic process control. It is wise at this time to consider some of its properties and characteristics.

We have already defined the transfer function as the ratio of the Laplace transform of the output variable divided by the Laplace transform of the input variable.

The transfer function is usually represented by

$$G(s) = \frac{Y(s)}{X(s)} = \frac{K(a_m s^m + a_{m-1}s^{m-1} + \ldots + a_1 s + 1)}{(b_n s^n + b_{n-1}s^{n-1} + \ldots + b_1 s + 1)} \tag{3-39}$$

where

$G(s)$ = the general representation of a transfer function
$Y(s)$ = the Laplace transform of the output variable
$X(s)$ = the Laplace transform of the forcing function or input variable
K, a's, and b's = constants.

Equation (3-39) shows the best way to write a transfer function. When written in this way, K represents the gain of the system and will have as units the units of $Y(s)$ over the units of $X(s)$. The other constants, a's and b's, will have as units (time)^i, where i is the power of the Laplace variable, s, associated with the particular constant; this will render a dimensionless term inside the parentheses since the unit of s is 1/time.

Note: In general the unit of s is the reciprocal of the unit of the independent variable used in the definition of Laplace transform, Eq. (2-1). In process dynamics and control the independent variable is time and, therefore, the unit of s is 1/time.

Notice that the coefficient of s^0 is 1.

The transfer function *completely* defines the steady-state and dynamic characteristics, the total response, of a system described by a *linear* differential equation. It is characteristic of the system and its terms determine whether the system is stable or unstable and whether its response to a nonoscillatory input is oscillatory or not. The system, or process, is said to be stable when its output remains bound (finite) for a bound input. Chapters 6 and 7 treat in detail the subject of stability of process systems.

The following are some important properties of transfer functions:

1. In the transfer functions of real physical systems the highest power of s in the numerator is never higher than that of the denominator. In other words, $n \geq m$

2. The transfer function relates the transforms of the deviation of the input and output variables from some initial steady state. Otherwise the nonzero initial conditions would contribute additional terms to the transform of the output variable.

3. For stable systems the steady-state relationship between the change in output variable and the change in input variable can be obtained by

$$\lim_{s \to 0} G(s)$$

This stems from the final value theorem, presented in Chapter 2:

$$\lim_{t \to \infty} Y(T) = \lim_{s \to 0} sY(s)$$

$$= \lim_{s \to 0} sG(s)X(s)$$

$$= \left[\lim_{s \to 0} G(s) \right]\left[\lim_{s \to 0} sX(s) \right]$$

$$= \left[\lim_{s \to 0} G(s) \right] \lim_{t \to \infty} X(T)$$

This means that the change in the output variable after a very long time, if bound, can be obtained by multiplying the transfer function with $s = 0$ times the final value of the change in input.

Block Diagrams

A very useful tool in process control is the graphical representation of transfer functions by means of block diagrams. These block diagrams were first introduced by James Watt when he applied the concept of feedback control to the steam engine, as mentioned in Chapter 1. The steam engine consisted of several linkages and other mechanical devices complex enough that Watt decided to show graphically the interaction of all these devices in his control scheme. In this section an introduction to block diagrams and block diagram algebra is presented.

In general, block diagrams consist of four basic elements: arrows, summing points, branch points, and blocks; Fig. 3-6 shows these elements. All block diagrams are formed by a combination of these elements. The *arrows* in general indicate flow of information; they represent process variables or control signals. Each arrowhead indicates the direction of the flow of information. The *summing points* represent the algebraic summation of the input arrows $(E(s) = R(s) - C(s))$. A *branch point* is the position on an arrow at which the information branches out and goes concurrently to other summing points or blocks. The *blocks* represent the mathematical operation, in transfer function form such as $G_C(s)$, which is performed on the input signal (arrow) to produce the output signal. The arrows and block shown in Fig. 3-6 represent the following mathematical expression:

$$M(s) = G_C(s)E(s) = G_C(s)(R(s) - C(s))$$

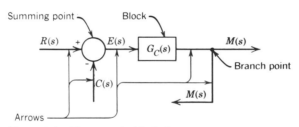

Figure 3-6.   Elements of a block diagram.

Any block diagram can be handled, or manipulated, algebraically. Table 3-1 shows some rules of block-diagram algebra. These rules are important any time a complicated block diagram is simplified.

Let us look at some examples of block diagram algebra.

Example 3-1.   Draw the block diagram depicting Eqs. (3-11) and (3-23).

Equation (3-11) is shown in Fig. 3-7. Equation (3-23) may be drawn in two different ways, as shown in Fig. 3-8.

The block diagrams of Eq. (3-23) show graphically that the total response of the system is obtained by algebraically adding the response due to a change in inlet temperature to the response due to a change in surrounding temperature. *This property of algebraic additions of responses due to several inputs to obtain the final response is a property of linear systems and is called the principle of superposition.* This principle also serves as the basis for defining linear systems. That is, we say that a system is linear if it obeys the principle of superposition.

Table 3-1   Rules for Block Diagram Algebra

1. $Y = A - B - C$

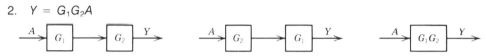

2. $Y = G_1G_2A$

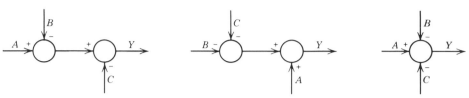

3   $Y = G_1(A - B)$

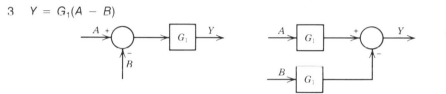

4. $Y = (G_1 + G_2)A$

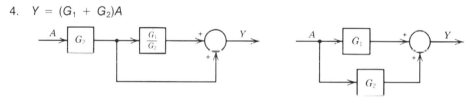

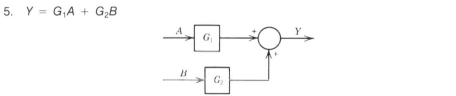

5. $Y = G_1A + G_2B$

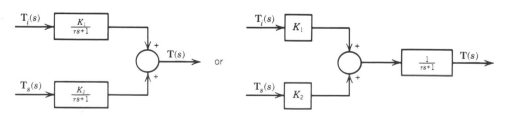

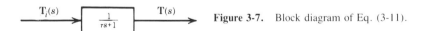

Figure 3-7.   Block diagram of Eq. (3-11).

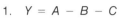

Figure 3-8.   Block diagram of Eq. (3-23).

77

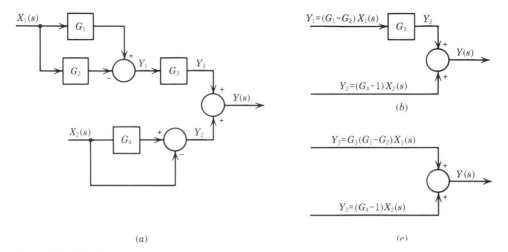

Figure 3-9. Block diagram algebra.

Example 3-2. Determine the transfer functions relating $Y(s)$ to $X_1(s)$ and $X_2(s)$ from the block diagram shown in Fig. 3-9a. That is, obtain

$$\frac{Y(s)}{X_1(s)} \quad \text{and} \quad \frac{Y(s)}{X_2(s)}$$

Using rule 4 the block diagram shown in Fig. 3-9a can be reduced to that of Fig. 3-9b. Then using rule 2 this figure can be further reduced to that of Fig. 3-9c. Then

$$Y(s) = G_3(G_1 - G_2)X_1(s) + (G_4 - 1)X_2(s)$$

from which the two desired transfer functions can be determined:

$$\frac{Y(s)}{X_1(s)} = G_3(G_1 - G_2)$$

and

$$\frac{Y(s)}{X_2(s)} = G_4 - 1$$

This example has shown a procedure to reduce a block diagram to a transfer function. This reduction of block diagrams is necessary in the study of process control, as shown in chapters 6, 7, and 8. In these chapters numerous examples of block diagrams of feedback, cascade, and feedforward control systems will be developed. Let us look at the reduction to transfer functions of some of these block diagrams.

Example 3-3. Figure 3-10 shows the block diagram of a typical feedback control system. From this diagram determine

$$\frac{C(s)}{L(s)} \quad \text{and} \quad \frac{C(s)}{C^{\text{set}}(s)}$$

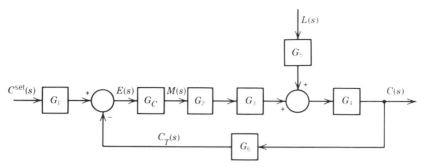

Figure 3-10. Block diagram of a feedback control system.

Using rules 2, 3, and 5, we find

$$C(s) = G_C G_2 G_3 G_4 E(s) + G_5 G_4 L(s) \tag{3-40}$$

Also, using rule 5, we obtain

$$E(s) = G_1 C^{set}(s) - G_6 C(s) \tag{3-41}$$

Substituting Eq. (3-41) into (3-40) yields

$$C(s) = G_1 G_C G_2 G_3 G_4 C^{set}(s) - G_C G_2 G_3 G_4 G_6 C(s) + G_5 G_4 L(s)$$

and after some algebraic manipulation we get

$$C(s) = \frac{G_1 G_C G_2 G_3 G_4}{1 + G_C G_2 G_3 G_4 G_6} C^{set}(s) + \frac{G_5 G_4}{1 + G_C G_2 G_3 G_4 G_6} L(s) \tag{3-42}$$

The individual transfer functions are now obtained from Eq. (3-42):

$$\frac{C(s)}{C^{set}(s)} = \frac{G_1 G_C G_2 G_3 G_4}{1 + G_C G_2 G_3 G_4 G_6} \tag{3-43}$$

and

$$\frac{C(s)}{L(s)} = \frac{G_5 G_4}{1 + G_C G_2 G_3 G_4 G_6} \tag{3-44}$$

Example 3-3 shows how to reduce a simple feedback control loop block diagram to transfer functions. These types of block diagrams and transfer functions will become useful in Chapters 6 and 7, when feedback control is discussed.

The transfer functions given by Eqs. (3-43) and (3-44) are referred to as "closed-loop transfer functions." The reason for this term will become evident in Chapter 6. Looking at Eq. (3-43) notice that the numerator is the multiplication of all of the transfer functions in the forward path between the two variables related by the transfer function, $C^{set}(s)$ and $C(s)$. The denominator of this equation is one (1) plus the multiplication of all the transfer functions in the control loop. Figure 3-11 shows the same block diagram as the one shown in Fig. 3-10 indicating what is meant by the control loop. An inspection of Eq. (3-44) shows that the numerator is again the multiplication of the transfer functions

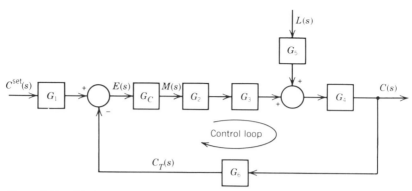

Figure 3-11. Block diagram showing control loop.

in the forward path between $L(s)$ and $C(s)$. The denominator is the same as the one of Eq. (3-43).

With Eqs. (3-43) and (3-44) as background, the form of a closed-loop transfer function reduced from block diagrams similar to that shown in Fig. 3-10 can be generalized.

$$G(s) = \frac{Y(s)}{X(s)} = \frac{\sum_{l=1}^{L} \left[\prod_{j=1}^{J} G_j \right]_l}{1 + \sum_{k=1}^{K} \left[\prod_{i=1}^{I} G_i \right]_k} \qquad (3\text{-}45)$$

where

L = number of forward paths between $X(s)$ and $Y(s)$
J = number of transfer functions in each forward path between $X(s)$ and $Y(s)$
G_j = transfer function in each forward path
K = number of nested loops in block diagram
I = number of transfer functions in each loop
G_i = transfer function in each loop

The term in the numerator of Eq. (3-45) indicates that the J transfer functions, G_j, in each forward path are multiplied together and then all L forward paths are added. The denominator is one (1) plus the summation of the products of the I transfer functions, G_i, in each loop for all K loops.

Example 3-4. Consider another typical block diagram as shown in Fig. 3-12. Chapter 8 shows that this block diagram depicts a cascade control system. Determine the following transfer functions:

$$\frac{C(s)}{R(s)} \quad \text{and} \quad \frac{C(s)}{L(s)}$$

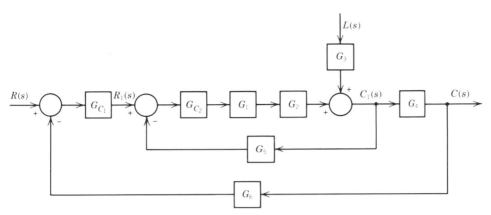

Figure 3-12. Block diagram of cascade control system.

The block diagram of Fig. 3-12 can be thought of as being composed of two closed-loop systems, one inside the other (in practice this is exactly what it is). The first step is then to reduce the inside loop; Fig. 3-13 shows this loop by itself. Using Eq. (3-45) the following two transfer functions for this inside loop can be determined:

$$\frac{C_1(s)}{R_1(s)} = \frac{G_{C_2}G_1G_2}{1 + G_{C_2}G_1G_2G_5} \tag{3-46}$$

and

$$\frac{C_1(s)}{L(s)} = \frac{G_3}{1 + G_{C_2}G_1G_2G_5} \tag{3-47}$$

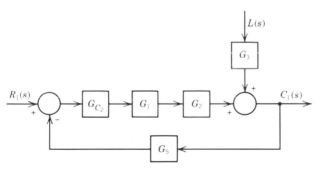

Figure 3-13. Block diagram of inside loop.

Substituting these two transfer functions, Eqs. (3-46) and (3-47), into the original block diagram results in a new reduced block diagram, as shown in Fig. 3-14. By using this block diagram and Eq. (3-45) the desired transfer functions can be determined.

$$\frac{C(s)}{R(s)} = \frac{\dfrac{G_{C_1}G_{C_2}G_1G_2G_4}{1 + G_{C_2}G_1G_2G_5}}{1 + \dfrac{G_{C_1}G_{C_2}G_1G_2G_4G_6}{1 + G_{C_2}G_1G_2G_5}} = \frac{G_{C_1}G_{C_2}G_1G_2G_4}{1 + G_{C_2}G_1G_2G_5 + G_{C_1}G_{C_2}G_1G_2G_4G_6} \tag{3-48}$$

and

$$\frac{C(s)}{L(s)} = \frac{\dfrac{G_3 G_4}{1 + G_{C_2} G_1 G_2 G_5}}{1 + \dfrac{G_{C_1} G_{C_2} G_1 G_2 G_4 G_6}{1 + G_{C_2} G_1 G_2 G_5}} = \frac{G_3 G_4}{1 + G_{C_2} G_1 G_2 G_5 + G_{C_1} G_{C_2} G_1 G_2 G_4 G_6}$$

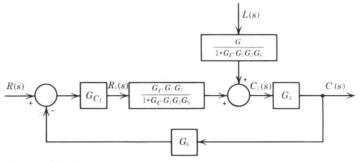

Figure 3-14.   Reduced block diagram.

Example 3-5.   Determine the transfer functions $C(s)/R(s)$ and $C(s)/L(s)$ from the block diagram shown in Fig. 3-15.

The first required transfer function, $C(s)/R(s)$, is easy to obtain since there is only one forward path between the two variables involved.

$$\frac{C(s)}{R(s)} = \frac{G_C G_1 G_2 G_3}{1 + G_C G_1 G_2 G_3 G_6}$$

To obtain the second transfer function we must realize that there are two forward paths between $L(s)$ and $C(s)$. Using Eq. (3-45) we obtain

$$\frac{C(s)}{L(s)} = \frac{(G_4 - G_5 G_1 G_2)G_3}{1 + G_C G_1 G_2 G_3 G_6}$$

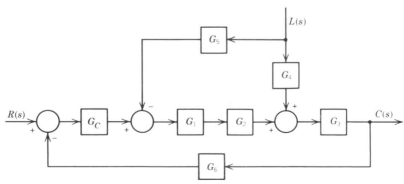

Figure 3-15.   Block diagram of feedforward control system.

A useful recommendation is to write next to each arrow the units of the process variable or control signal that the arrow represents. If this is done, then it is fairly simple to recognize the units of the gain of a block, which are the units of the output arrow over the units of the input arrow. This procedure also helps in avoiding the algebraic summation of arrows with different units.

As mentioned at the beginning of this section, block diagrams are a very helpful tool in process control. We will learn and practice more about the logic of drawing them as ge wo along in our study of process dynamics and control. Chapters 6, 7, and 8 make great use of block diagrams to help analyze and design control systems.

3-4. DEAD TIME

Consider the process shown in Fig. 3-16. This is essentially the same process as the one shown in Fig. 3-1. The difference is that in this case we are interested in knowing how $T_1(t)$ responds to changes in inlet and surrounding temperatures.

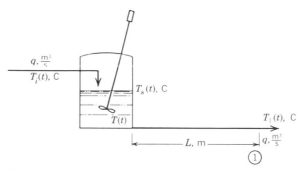

Figure 3-16. Thermal process.

Let us make the following two assumptions about the exit pipe between the tank and point 1. First, the pipe is well insulated. Second, the flow of liquid through the pipe is highly turbulent (plug flow) so that there is essentially no backmixing in the liquid.

Under these assumptions, the response of $T_1(t)$ to the disturbances $T_i(t)$ and $T_s(t)$ will be the same as $T(t)$ except that it will be delayed by some amount of time. That is, there will be a finite amount of time between the response of $T(t)$ and the response of $T_1(t)$. This is shown graphically in Fig. 3-17 for a step change in inlet temperature $T_i(t)$. The amount of time between the time the disturbance enters the process and the time temperature $T_1(t)$ starts to respond is called *dead time*, time delay, or transportation lag. It is respresented by the term t_0.

In this particular example, the dead time can easily be calculated as follows:

$$t_0 = \frac{\text{distance}}{\text{velocity}} = \frac{L}{q/A_p} = \frac{A_p L}{q} \qquad (3\text{-}49)$$

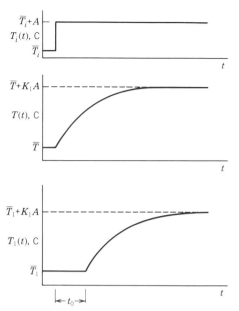

Figure 3-17.   Response of thermal process to a step change in inlet temperature.

where

t_0 = dead time, seconds
A_p = cross-sectional area of pipe, m^2
L = length of pipe, m

The dead time is an integral part of the process, and consequently, it must be accounted for in the transfer function relating $T_1(t)$ to $T_i(t)$ and $T_s(t)$. Equation (2-9) indicates that the Laplace transform of a delayed function is equal to the Laplace transform of the non-delayed function times the term $e^{-t_0 s}$. The term $e^{-t_0 s}$ is the Laplace transform of a pure dead time; thus, if we are interested in the response of $T_1(t)$ to changes in $T_i(t)$ and $T_s(t)$, we must multiply the transfer functions, Eqs. (3-24) and (3-25), by $e^{-t_0 s}$ or

$$\frac{\mathbf{T}_1(s)}{\mathbf{T}_i(s)} = \frac{K_1 e^{-t_0 s}}{\tau s + 1} \qquad (3\text{-}50)$$

and

$$\frac{\mathbf{T}_1(s)}{\mathbf{T}_s(s)} = \frac{K_2 e^{-t_0 s}}{\tau s + 1} \qquad (3\text{-}51)$$

In this example, the dead time has developed because of the time it takes the liquid to move from the exit of the tank to point 1. In most processes, however, the dead time is not as easily defined. It is usually inherent and distributed throughout the process, that is, the tank, reactor, column, and so on. In these cases, its numerical evaluation is not as easily obtained as in the case at hand, but requires a very detailed model or empirical evaluation. Chapter 6 shows how to perform this empirical evaluation.

At this point, it must be recognized that the dead time is another parameter that helps define the personality of the process. Equation (3-49) shows that t_0 depends on some physical properties and operating characteristics of the process, as are K and τ. If any condition of the process changes, this change may be reflected in a change in t_0.

Before concluding this section, we must say that a significant amount of dead time in the process is the *worst* thing that can happen to a control system. The performance of a control system is severely affected by dead time. This will be shown in Chapters 6 and 7.

3-5. LEVEL PROCESS

Consider the process shown in Fig. 3-18. In this process we are interested in knowing how the level, $h(t)$, of liquid in the tank will respond to changes in inlet flow, $q_i(t)$, and to changes in the exit valve opening, $vp(t)$.

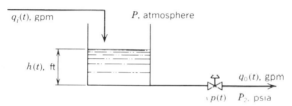

Figure 3-18. Level process.

The flow of liquid through a valve is given, as will be shown in Chapter 5, by

$$q(t) = C_v(vp(t)) \sqrt{\frac{\Delta P(t)}{G}}$$

where

$q(t)$ = flow, gpm
C_v = valve coefficient, gpm/(psi)$^{1/2}$
$vp(t)$ = valve position. This term represents the fraction of valve opening. A value of 0 indicates that the valve is closed. A value of 1 indicates that the valve is wide open
$\Delta P(t)$ = pressure drop across the valve, psi
G = specific gravity of liquid flowing through valve, dimensionless.

For this process, the pressure drop across the valve is given by

$$\Delta P(t) = P + \frac{\rho g h(t)}{144 g_c} - P_2$$

where

P = pressure over the liquid, psia
ρ = density of liquid, lbm/ft^3

g = acceleration due to gravity, 32.2 ft/sec^2
g_c = conversion factor, 32.2 lbm-ft/lbf-sec^2
$h(t)$ = level in tank, ft
P_2 = downstream pressure from valve, psia

In this equation we have assumed that the friction losses through the pipe from the tank to the valve are negligible.

An unsteady-state mass balance around the tank will yield the desired relationship:

$$\frac{\rho q_i(t)}{7.48} - \frac{\rho q_o(t)}{7.48} = \frac{dm(t)}{dt}$$

or

$$\frac{\rho q_i(t)}{7.48} - \frac{\rho q_o(t)}{7.48} = A\rho \frac{dh(t)}{dt}$$

where

A = cross-sectional area of tank, ft^2
7.48 = conversion factor between gal and ft^3

Assuming equal inlet and outlet densities, we have

$$q_i(t) - q_o(t) = 7.48A \frac{dh(t)}{dt} \tag{3-52}$$

1 eq., 2 unk. ($q_o(t)$, $h(t)$)

We now have one equation with two unknowns and therefore must find another independent equation to describe the process. The valve equation provides the other required equation:

$$q_o(t) = C_v(vp(t)) \sqrt{\frac{\left(P + \dfrac{\rho g h(t)}{144 g_c} - P_2\right)}{G}} \tag{3-53}$$

2 eq., 2 unk.

This set of equations, (3-52) and (3-53), describes the process. To simplify this description, we can substitute Eq. (3-53) into (3-52):

$$q_i(t) - C_v(vp(t)) \sqrt{\frac{\left(P + \dfrac{\rho g h(t)}{144 g_c} - P_2\right)}{G}} = 7.48A \frac{dh(t)}{dt} \tag{3-54}$$

The analytical solution of this equation is not possible because of the nonlinear nature of the second term on the left-hand side of the equation. The only way to solve this equation analytically is to linearize the nonlinear term. The only other way to solve Eq. (3-54) is by numerical methods (computer solution).

In Chapter 2 we learned how to linearize nonlinear terms using Taylor's Series expansion. Let us now use this technique to linearize the nonlinear term in Eq. (3-54).

Since this term must be linearized with respect to h and vp, the linearization will be done about the values $\bar{h}$ and $\overline{vp}$, the nominal steady-state values:

$$q_o(t) \simeq \bar{q}_o + \left.\frac{\partial q_o}{\partial vp}\right|_{ss} (vp(t) - \overline{vp}) + \left.\frac{\partial q_o}{\partial h}\right|_{ss} (h(t) - \bar{h})$$

or

$$q_o(t) \simeq \bar{q}_o + C_v \sqrt{\frac{P + \dfrac{\rho g\bar{h}}{144 g_c} - P_2}{G}} (vp(t) - \overline{vp})$$

$$+ \frac{C_v \rho g\overline{vp}}{288 g_c G} \left[\frac{P + \dfrac{\rho g h}{144 g_c} - P_2}{G}\right]^{-1/2} (h(t) - \bar{h})$$

To simplify the notation, let

$$C_1 = C_v \sqrt{\frac{P + \dfrac{\rho g\bar{h}}{144 g_c} - P_2}{G}} \tag{3-55}$$

and

$$C_2 = \frac{C_v \rho g\overline{vp}}{288 G g_c} \left[\frac{P + \dfrac{\rho g\bar{h}}{144 g_c} - P_2}{G}\right]^{-1/2} \tag{3-56}$$

so

$$q_o(t) \simeq \bar{q}_o + C_1(vp(t) - \overline{vp}) + C_2(h(t) - \bar{h}) \tag{3-57}$$

Substituting this last equation into Eq. (3-54) gives a linear differential equation:

$$q_i(t) - \bar{q}_o - C_1(vp(t) - \overline{vp}) - C_2(h(t) - \bar{h}) = 7.48A \frac{dh(t)}{dt} \tag{3-58}$$

We must never lose sight of the fact, as explained in Chapter 2, that this is a linearized version of Eq. (3-54). Equation (3-58), as explained in Section 2-2, will yield an accurate solution around the point of linearization, $\bar{h}$ and $\overline{vp}$. Outside a certain range around this point, the linearization will break down, giving erroneous results.

Now that we have a linear differential equation, the desired transfer functions can be obtained. To obtain them we continue with the same procedure as before. Writing a steady-state mass balance around the tank, we see that

$$\rho \bar{q}_i - \rho \bar{q}_o = 0$$

or

$$\bar{q}_i - \bar{q}_o = 0$$

Subtracting this equation from Eq. (3-58) yields

$$(q_i(t) - \bar{q}_i) - C_1(vp(t) - \overline{vp}) - C_2(h(t) - \bar{h}) = 7.48A \frac{d(h(t) - \bar{h})}{dt}$$

We define the following deviation variables:

$$Q_i(t) = q_i(t) - \bar{q}_i$$
$$VP(t) = vp(t) - \overline{vp}$$
$$H(t) = h(t) - \bar{h}$$

Substituting these deviation variables into the linearized differential equation

$$Q_i(t) - C_1 VP(t) - C_2 H(t) = 7.48A \frac{dH(t)}{dt} \qquad (3\text{-}59)$$

and rearranging this equation algebraically gives

$$\tau \frac{dH(t)}{dt} + H(t) = K_1 Q_i(t) - K_2 VP(t)$$

where

$\tau = 7.48A/C_2$, minutes
$K_1 = 1/C_2$, ft/gpm
$K_2 = C_1/C_2$, ft/valve position

Finally, we take the Laplace transform

$$H(s) = \frac{K_1}{\tau s + 1} Q_i(s) - \frac{K_2}{\tau s + 1} VP(s)$$

from which the two transfer functions are

$$\frac{H(s)}{Q_i(s)} = \frac{K_1}{\tau s + 1} \qquad (3\text{-}60)$$

and

$$\frac{H(s)}{VP(s)} = \frac{-K_2}{\tau s + 1} \qquad (3\text{-}61)$$

The reader should convince himself that the units of the time constant and gains are correct. He should also remember the meaning of these three parameters. K_1 is the gain, or sensitivity, that $Q_i(t)$ has in relation to $H(t)$. It gives the amount of change of level in the tank per unit change of flow into the tank. This change takes place while a constant opening is kept in the outlet valve. K_2 gives the amount of change of level in the tank per unit change in valve position. Notice the negative sign in front of this gain, which means that as the valve position changes positively and the valve opens, the level changes negatively, or drops; this makes sense physically.

The block diagram for this process is shown in Fig. 3-19. In this example the linearization of the nonlinear term was choosen to be done around the $\bar{h}$ and $\overline{vp}$ values, which are the nominal steady-state values. These values, $\bar{h}$ and $\overline{vp}$, are part of the gains and time constant expressions. If a different steady-state value had been choosen for linearization, say $\bar{h}_1$ and $\overline{vp}_1$, the numerical values of the gains and time constant would have been different. *This is an indication of the nonlinearity of the process.* The parameters

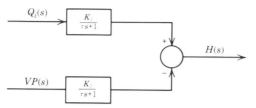

Figure 3-19. Block diagram for level process.

that describe the "personality" of the process are functions of the level of operation, or operating conditions. This is in contrast to linear systems, in which these parameters are constant over the whole range of operation. *The fact that most processes are nonlinear by nature is very important in process control.* The more nonlinear a process is, the more difficult it is to control. For now it is important to understand the meaning of nonlinearities, where they come from, and how they affect the personality of processes.

3-6. CHEMICAL REACTOR

A chemical reactor is a typical example of a highly nonlinear process. Consider the reactor shown in Fig. 3-20. The reactor is a vessel where the "well known" highly exothermic reaction $A \rightarrow B$ occurs. To remove the heat of reaction, the reactor is surrounded by a jacket in which saturated steam is produced from saturated liquid. The temperature of the jacket, T_s, can be assumed to be constant. It can also be assumed that the jacket is well insulated, the reactants and products are liquid, and their densities and heat capacities do not vary much with temperature or composition.

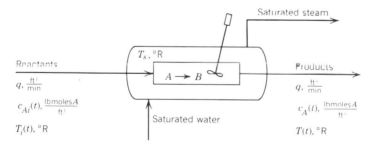

Figure 3-20. Chemical reactor.

The rate of reaction is given by the following expression:

$$r_A(t) = k_0 e^{-E/RT(t)} c_A(t), \qquad \frac{\text{lb moles } A \text{ produced}}{\text{ft}^3\text{-min}}$$

The frequency factor, k_0, and the energy of activation, E, are constants specific for each reaction. The heat of reaction is considered constant and given by ΔHr in Btu/lb mole of A reacted.

We are interested in knowing the effect on the outlet concentration of A, $c_A(t)$, to changes in inlet concentration of A, $c_{Ai}(t)$, and inlet temperature, $T_i(t)$.

Writing an unsteady-state mole balance around the reactor, we have

$$qc_{Ai}(t) - Vr_A(t) - qc_A(t) = V\frac{dc_A(t)}{dt} \tag{3-62}$$

$$1 \text{ eq., } 2 \text{ unk. } (c_A(t), r_A(t))$$

where V = volume of reactor, ft³.

The rate of reaction gives us another relationship:

$$r_A(t) = k_0 e^{-E/RT(t)} c_A(t) \tag{3-63}$$

$$2 \text{ eq., } 3 \text{ unk. } (T(t))$$

We are still short one independent equation to fully describe this process. This relationship should involve the temperature of the reaction. The required relationship is an energy balance around the reactor:

$$q\rho C_p T_i(t) - Vr_A(t)(\Delta H_r) - UA(T(t) - T_s) - q\rho C_p T(t) = V\rho C_v \frac{dT(t)}{dt} \tag{3-64}$$

$$3 \text{ eq., } 3 \text{ unk.}$$

where

    ρ = density of products and reactants, assumed constant, lbm/ft³

    C_p = heat capacity at constant pressure of products and reactants, assumed constant, Btu/lbm-°R

    C_v = heat capacity at constant volume, assumed constant, Btu/lbm-°R

    U = overall heat transfer coefficient, assumed constant, Btu/°R-ft²-min

    A = heat transfer area, ft²

    ΔH_r = heat of reaction, assumed constant, $-$Btu/lb mole of A reacted

    T_s = temperature of saturated steam, °R

Equations (3-62), (3-63), and (3-64) describe the chemical reactor. This set of equations is quite nonlinear, mainly because of Eq. (3-63). Again, the solution to these equations can be obtained either by numerical methods or by linearizing the nonlinear term so that an approximate analytical solution can be obtained. Let us follow the last method to determine the desired transfer functions.

Linearizing Eq. (3-63) around the operating conditions $\bar{T}$ and $\bar{c}_A$ yields

$$r_A(t) = \bar{r}_A + C_1 T(t) + C_2 C_A(t) \tag{3-65}$$

where

$$\bar{r}_A = k_0 e^{-E/R\bar{T}} \bar{c}_A$$

$$C_1 = \left.\frac{\partial r_A(t)}{\partial T(t)}\right|_{ss} = \frac{k_0 E \bar{c}_A}{R\bar{T}^2} e^{-E/R\bar{T}}$$

$$C_2 = \left.\frac{\partial r_A(t)}{\partial C_A(t)}\right|_{ss} = k_0 e^{-E/R\bar{T}}$$

and

$$\mathbf{T}(t) = T(t) - \bar{T}$$
$$C_A(t) = c_A(t) - \bar{c}_A$$

Substituting Eq. (3-65) into Eqs. (3-62) and (3-64) yields two linear differential equations with two unknowns:

$$qc_{Ai}(t) - V\bar{r}_A - VC_1\mathbf{T}(t) - VC_2C_A(t) - qc_A(t) = V\frac{dc_A(t)}{dt} \tag{3-66}$$

and

$$q\rho C_p T_i(t) - V(\Delta H_r)\bar{r}_A - V(\Delta H_r)C_1\mathbf{T}(t) - V(\Delta H_r)C_2C_A(t)$$
$$- UA(T(t) - T_s) - q\rho C_p T(t) = V\rho C_v\frac{dT(t)}{dt} \tag{3-67}$$

Using the method learned in the previous examples, we can obtain the two differential equations in terms of deviation variables:

$$qC_{Ai}(t) - VC_1\mathbf{T}(t) - VC_2C_A(t) - qC_A(t) = V\frac{dC_A(t)}{dt} \tag{3-68}$$

and

$$q\rho C_p\mathbf{T}_i(t) - V(\Delta H_r)C_1\mathbf{T}(t) - V(\Delta H_r)C_2C_A(t) - UA\mathbf{T}(t) - q\rho C_p\mathbf{T}(t) = V\rho C_v\frac{dT(t)}{dt} \tag{3-69}$$

where

$$C_{Ai}(t) = c_{Ai}(t) - \bar{c}_{Ai}$$
$$\mathbf{T}_i(t) = T_i(t) - \bar{T}_i$$

Rearranging Eqs. (3-68) and (3-69) algebraically yields

$$\tau_1\frac{dC_A(t)}{dt} + C_A(t) = K_1C_{Ai}(t) - K_2\mathbf{T}(t) \tag{3-70}$$

and

$$\tau_2\frac{d\mathbf{T}(t)}{dt} + \mathbf{T}(t) = K_3\mathbf{T}_i(t) - K_4C_A(t) \tag{3-71}$$

where

$$\tau_1 = \frac{V}{VC_2 + q}, \qquad \text{minutes}$$

$$K_1 = \frac{q}{VC_2 + q}, \qquad \text{dimensionless}$$

$$K_2 = \frac{VC_1}{VC_2 + q}, \qquad \frac{\text{lb moles } A}{\text{ft}^3\text{-}°\text{R}}$$

$$\tau_2 = \frac{V\rho C_v}{V(\Delta H_r)C_1 + UA + q\rho C_p}, \qquad \text{minutes}$$

$$K_3 = \frac{q\rho C_p}{V(\Delta H_r)C_1 + UA + q\rho C_p}, \qquad \text{dimensionless}$$

$$K_4 = \frac{V(\Delta H_r)C_2}{V(\Delta H_r)C_1 + UA + q\rho C_p}, \qquad \frac{°\text{R-ft}^3}{\text{lb moles } A}$$

Taking the Laplace transforms of Eqs. (3-70) and (3-71) yields

$$C_A(s) = \frac{K_1}{\tau_1 s + 1} C_{Ai}(s) - \frac{K_2}{\tau_1 s + 1} T(s) \qquad (3\text{-}72)$$

and

$$T(s) = \frac{K_3}{\tau_2 s + 1} T_i(s) - \frac{K_4}{\tau_2 s + 1} C_A(s) \qquad (3\text{-}73)$$

The block diagram representation of this process is shown in Fig. 3-21.

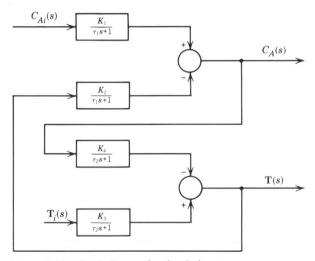

Figure 3-21.   Block diagram for chemical reactor.

This reacting process brings in a new characteristic that we have not seen before. This is the first example in which there is an interaction between two variables. These two variables, the outlet concentration, $C_A(t)$, and the outlet temperature, $T(t)$, are said to be "coupled." Any disturbance that affects one of them will also affect the other. As an example, if the inlet temperature of the reactants changes, this will affect the outlet temperature by the transfer function

$$\frac{T(s)}{T_i(s)} = \frac{K_3}{\tau_2 s + 1}$$

But once the outlet temperature changes, this will also affect the outlet concentration by the transfer function

$$\frac{C_A(s)}{T(s)} = \frac{K_2}{\tau_1 s + 1}$$

Processes with which this occurs are called interacting processes. The control of both variables presents a very interesting challenge to the process, or control, engineer.

As with the previous example, we should note that the gains and time constant depend on the linearization point. Consequently, they are not constant over the whole range. This is due to the nonlinear characteristic of the process.

Before concluding this example, the reader should verify the units and signs of the gains and time constants. As mentioned in an earlier example, this will help in checking our model development before going through the solution.

3-7. RESPONSE OF FIRST-ORDER PROCESSES TO DIFFERENT TYPES OF FORCING FUNCTIONS

We have seen that the transfer function of a first-order process without dead time is of the form

$$G(s) = \frac{Y(s)}{X(s)} = \frac{K}{\tau s + 1}$$

where

$Y(s)$ = transform of output variable
$X(s)$ = transform of forcing function or input variable

In this section we will study the response of this type of process to different types of forcing functions. These functions are the most common ones in the study of automatic process control.

Step Function

A step change of A units in magnitude in the forcing function is shown in the time domain as

$$X(t) = Au(t) \tag{3-74}$$

and in the Laplace domain as

$$X(s) = \frac{A}{s}$$

Then

$$Y(s) = \frac{KA}{s(\tau s + 1)}$$

Using the methods learned in Chapter 2, we can invert this function back to the time domain:

$$Y(t) = KA(1 - e^{-t/\tau}) \tag{3-75}$$

Fig. 3-22 shows the forcing function and responding variable graphically. We notice that the steepest slope of a response curve occurs at the beginning of the response. This is the *typical* response of all first-order processes to a step change in forcing function.

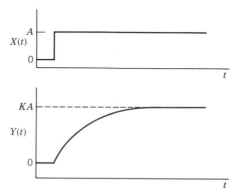

Figure 3-22. Response of first-order process to a step change in forcing function.

Ramp Function

The ramp function in the time domain is represented as

$$X(t) = Atu(t) \tag{3-76}$$

and in the Laplace domain as

$$X(s) = \frac{A}{s^2}$$

Then

$$Y(s) = \frac{KA}{s^2(\tau s + 1)}$$

Again, using the methods learned in Chapter 2 we can invert this function back to the time domain:

$$Y(t) = KA(t + \tau e^{-t/\tau} - \tau) \tag{3-77}$$

Fig. 3-23 shows the forcing function and responding variable graphically. As time increases, the transients die out, $e^{-t/\tau}$ becomes negligible, and the response also becomes a ramp. That is

$$Y(t)|_{t \to \infty} = KA(t - \tau) \tag{3-78}$$

Sinusoidal Function

Suppose the forcing function to the process is a sine function such as

$$X(t) = A \sin \omega t\, u(t) \tag{3-79}$$

In the Laplace domain it is

$$X(s) = \frac{A\omega}{s^2 + \omega^2}$$

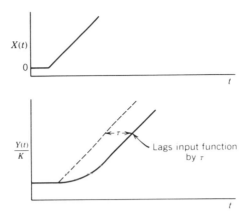

Figure 3-23. Response of first-order process to a ramp forcing function.

Then

$$Y(s) = \frac{KA\omega}{(\tau s + 1)(s^2 + \omega^2)}$$

Inverting this function back to the time domain gives

$$Y(t) = \frac{KA\omega\tau}{\tau^2\omega^2 + 1} e^{-t/\tau} + \frac{KA}{\sqrt{\tau^2\omega^2 + 1}} \sin(\omega t + \theta) \qquad (3\text{-}80)$$

where

$$\theta = \tan^{-1}(-\omega\tau). \qquad (3\text{-}81)$$

The angle θ is called the phase angle. As time increases, the term $e^{-t/\tau}$ becomes negligible and the response reaches a steady oscillation given by

$$Y(t)\Big|_{t\to\infty} = \frac{KA}{\sqrt{\tau^2\omega^2 + 1}} \sin(\omega t + \theta) \qquad (3\text{-}82)$$

This steady oscillation is called the frequency response of the system. It is interesting to note that the amplitude of this function is equal to the amplitude of the forcing function times the gain of the process but attenuated by the factor $1/\sqrt{\tau^2\omega^2 + 1}$. The ratio of the amplitude of $Y(t)\big|_{t\to\infty}$ to the amplitude of $X(t)$, $K/\sqrt{\tau^2\omega^2 + 1}$, is defined as the amplitude ratio. This amplitude ratio and the phase angle are both functions of the frequency of the input function, ω. The study of how they vary as ω varies is an important part of automatic process control and is covered in Chapter 7.

Figure 3-24 shows the forcing function and responding variables after the transients have vanished.

There are two other important forcing functions—pulse and impulse functions. The study of the response of first-order processes to these functions will be assigned as problems at the end of the chapter.

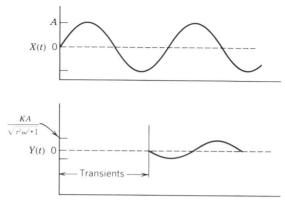

Figure 3-24. Response of first-order process to a sine wave forcing function.

3-8. SUMMARY

In this chapter we have seen the development of simple mathematical models of processes that are described by first-order differential equations. The starting point for these models is usually a balance equation of a conserved quantity—mass or energy. The procedure is to first develop all of the equations that describe the process, as shown in Section 3-6 for the chemical reactor. In this particular example, it took three equations to describe the process. We must note two important points in this development. First, to be able to describe any process, we must end up with the same number of independent equations as unknowns. This is why we have stressed the unknowns next to each equation. This should help us in keeping track of the equations needed to describe the process and develop the model. We shall call this method of writing equations and unknowns the "description method."* Second, the solution method depends upon the linear or nonlinear nature of the equations, on what is required, and on the degree of accuracy that is required. If the set of equations is linear, the solution can be obtained analytically by the Laplace transform or any other classical technique. If any equation is nonlinear, then the best technique for obtaining a fairly accurate solution is the application of numerical methods, which usually means computer work. The other technique available is the linearization of the nonlinear equation to obtain an analytical solution. The solution obtained in this manner is usually accurate over only a small region around the linearization point. This linearization technique permits us to obtain a transfer function of the process.

Several other concepts were presented and explained in this chapter. Transfer functions were defined as the ratio of the Laplace transform of the output variable to the Laplace transform of the input variable. The meaning of the transfer function was explained. The transfer function fully describes the steady-state and dynamic characteristics of the two variables. The variables are used in deviation form.

---

*This term was developed by Dr. J. C. Busot at the University of South Florida in his thermodynamics courses.

The transfer functions developed and studied in this chapter are of the general form

$$\frac{Y(s)}{X(s)} = \frac{Ke^{-t_0 s}}{\tau s + 1} \tag{3-83}$$

They are called first-order-plus-dead-time transfer functions, or first-order lags or single capacitances. This transfer function contains three parameters: the gain, K, time constant, τ, and dead time, t_0. The understanding of these parameters is fundamental to our study of process control. The gain, K, or sensitivity, specifies the amount of change of the output variable per unit change in the input variable. As shown previously K can be defined mathematically as follows:

$$K = \frac{\Delta Y}{\Delta X} = \frac{\Delta \text{ output variable}}{\Delta \text{ input variable}} \tag{3-84}$$

The time constant, τ, is related to the speed of response of the process *once the process starts to respond* to a forcing function. The time constant was shown to be, from Eq. (3-12), the time required to reach 63.2% of the total change, ΔY. The slower the process is to respond to an input, the larger the value of τ. The dead time, t_0, is the time interval between the appearance of the forcing function and the time the process, or responding variable, starts to respond. As mentioned, t_0 is the worst thing that can happen to any control system.

It is important to remember that these three parameters, K, τ, and t_0, define the "personality" of the process; they are functions of the physical parameters of the process. It was also shown that for a linear system, these parameters are constant over the complete operating range of the process. For a nonlinear process, the parameters were shown to be functions of the operating condition, and consequently, not constant over the operating range of the process. This chapter showed how to obtain these parameters starting from balance equations. Chapter 6 shows how to evaluate them from process data.

Finally, the chapter concluded with a study of the response of first-order processes to forcing functions of different types. This knowledge will prove helpful later in our study of process control.

PROBLEMS

3-1. Consider the mixing process shown in Figure 3-25. You may assume that the density of the input and output streams are very similar and that the flow rates F_1 and F_2 are constant. Obtain the transfer functions relating the outlet concentration to each of the inlet concentrations. Show the units of all gains and time constants.

3-2. Consider the isothermal reactor shown in Figure 3-26. The rate of reaction is given by

$$r_A(t) = kC_A(t), \qquad \text{moles of } A/\text{ft}^3\text{-min}$$

where k is constant.

You may assume that the density and all other physical properties of products

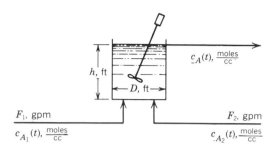

Figure 3-25.   Sketch for Problem 3-1.

and reactants are similar. You may also assume that the flow regime between points 2 and 3 is very turbulent (plug flow), minimizing backmixing.

Obtain the transfer functions relating:

a. The concentration of A at 2 to the concentration of A at 1.
b. The concentration of A at 3 to the concentration of A at 2.
c. The concentration of A at 3 to the concentration of A at 1.

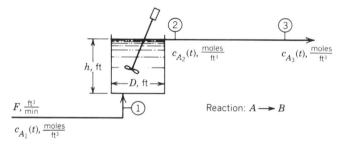

Figure 3-26.   Sketch for Problem 3-2.

3-3. Consider the process shown in Figure 3-27. The tank is spherical with a radius of 4 ft. The nominal flow into and out of the tank is 30,000 lbm/hr, the density of the liquid is 70 lbm/ft^3, and the steady-state level is 5 ft. The volume of a sphere is given by $4\pi r^3/3$. The relation between volume and height is given by

$$V(t) = V_T \left[\frac{h^2(t)(3r - h(t))}{4r^3} \right]$$

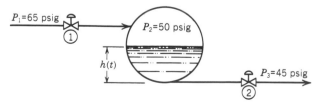

Figure 3-27.   Sketch for Problem 3-3.

and the flows through the valves by

$$\omega(t) = 500 \, C_v \, vp(t) \sqrt{G_f \Delta p}$$

where

r = radius of sphere, ft
$V(t)$ = volume of liquid in tank, ft^3
V_T = total volume of tank, ft^3
$h(t)$ = height of liquid in tank, ft
$\omega(t)$ = flow rate, lbm/hr
C_v = valve coefficient, gpm/psi$^{1/2}$
$\quad C_{v_1} = 20.2 \quad$ and $\quad C_{v_2} = 28.0$
Δp = pressure drop across valve, psi
G_f = specific gravity of fluid
$vp(t)$ = valve position, a fraction of valve opening

The pressure above the liquid level is maintained constant at a value of 50 psig. Obtain the transfer functions that relate the level of liquid in the tank to changes in valve positions of valves 1 and 2. Also, plot the gains and time constants versus different operating levels while keeping the valve positions constant.

3-4. Consider the heating tank shown in Figure 3-28. A process fluid is being heated in the tank by a heating medium flowing through the tubes. The rate of heat transfer, $q(t)$, to the process fluid is related to the pneumatic signal, $m(t)$ by the following:

$$q(t) = a + b(m(t) - 9)$$

You may assume that the process is adiabatic, the fluid is well mixed in the tank, and the heat capacity and density of the fluid are constant. Obtain the transfer functions that relate the outlet fluid temperature to the inlet temperature, $T_i(t)$, the process flow rate, $F(t)$, and the pneumatic signal, $m(t)$. Draw the complete block diagram for this process.

3-5. Consider the mixing process shown in Figure 3-29. The purpose of this process is to blend a stream, weak in component A, with another stream, pure A. The

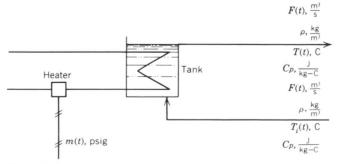

Figure 3-28. Sketch for Problem 3-4.

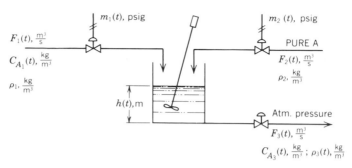

Figure 3-29. Sketch for Problem 3-5.

density of stream 1, ρ_1, can be considered constant since the amount of A in this stream is small. The density of the outlet stream is, of course, a function of the concentration and is given by

$$\rho_3(t) = a_3 + b_3 C_{A_3}(t)$$

The flow through valve 1 is given by

$$F_1(t) = C_{v_1} vp_1(t) \sqrt{\frac{\Delta p_1}{G_1}}$$

The flow through valve 2 is given by

$$F_2(t) = C_{v_2} vp_2(t) \sqrt{\frac{\Delta p_2}{G_2}}$$

Finally, the flow through valve 3 is given by

$$F_3(t) = C_{v_3} \sqrt{\frac{\Delta p_3(t)}{G_3(t)}}$$

The relationship between the valve position and the pneumatic signal is given by

$$vp_1(t) = a_1 + b_1(m_1(t) - d_1)$$

and

$$vp_2(t) = a_2 + b_2(m_2(t) - d_2)$$

where a_1, b_1, d_1, a_2, b_2, d_2, a_3, b_3 = known constants.

C_{v_1}, C_{v_2}, C_{v_3} = valve coefficients of valve 1, 2, and 3, respectively, m³/ (s-psi$^{1/2}$)

$vp_1(t)$, $vp_2(t)$ = valve position of valves 1 and 2, respectively, a dimensionless fraction

Δp_1, Δp_2 = pressure drop across valves 1 and 2, respectively, constants, psi

$\Delta p_3(t)$ = pressure drop across valve 3, psi

G_1, G_2 = specific gravity of streams 1 and 2, respectively, constants, dimensionless

$G_3(t)$ = specific gravity of stream 3, dimensionless

Develop the block diagram for this process showing all transfer functions and
how the forcing functions $m_1(t)$, $m_2(t)$, and $C_{A_1}(t)$ affect the responding variables
$h(t)$ and $C_{A_3}(t)$.

3-6.   Determine the transfer function $C(s)/R(s)$ for the system shown in Figure 3-30.

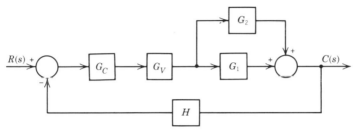

Figure 3-30.   Sketch for Problem 3-6.

3-7.   Determine the transfer function $C(s)/L(s)$ for the system shown in Figure 3-31.

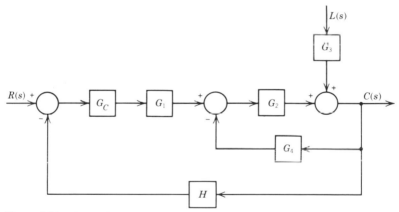

Figure 3-31.   Sketch for Problem 3-7.

3-8.   Determine the transfer function $C(s)/R(s)$ for the system shown in Figure 3-32.

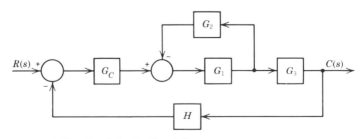

Figure 3-32.   Sketch for Problem 3-8.

3-9.   Obtain the response of a process described by a first-order plus dead-time transfer function to the forcing function shown in Figure 3-33.

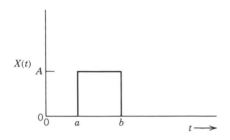

Figure 3-33.   Sketch for Problem 3-9.

3-10.   Assume that the following equation is the description of a certain process.

$$\frac{Y(s)}{X(s)} = \frac{3e^{-0.5s}}{5s + 0.2}$$

a.   Obtain the steady-state gain, time constant, and dead time of this process.
b.   The initial condition of the variable y is $y(0) = 2$. For a forcing function as shown in Figure 3-34, what is the final value of $y(t)$?

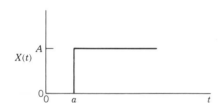

Figure 3-34.   Sketch for Problem 3-10.

3-11.   Consider the chemical reactor shown in Figure 3-35. In this reactor, an endo-

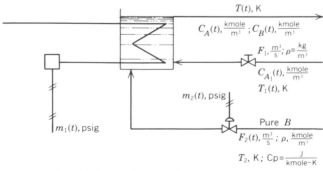

Figure 3-35.   Sketch for Problem 3-11.

thermic reaction of the type $A + 2B \rightarrow C$ takes place. The rate of appearance of A, in kmole/m^3-s, is given by

$$r_A(t) = -k_0 e^{-E/RT(t)} C_A(t) C_B(t)$$

where

$r_A(t)$ = rate of reaction, kmole/m^3-s
k_0 = frequency factor, constant, m^3/kmole-s
E = energy of activation, constant, cal/gmole
R = gas law constant, 1.987 cal/gmole-K
$T(t)$ = temperature in reactor, K
$C_A(t)$ = concentration of A in reactor, kmole/m^3
$C_B(t)$ = concentration of B in reactor, kmole/m^3
ΔH_r = heat of reaction, J/kmole

The heat input to the reactor is related to the pneumatic signal to the heater by

$$q(t) = r + s(m_1(t) - 9)$$

where

$q(t)$ = heat input to reactor, J/s
s, r = constants

The flow of pure B through the valve is given by

$$F_2(t) = C_{v_2} vp_2(t) \sqrt{\frac{\Delta p_2}{G_2}}$$

where

C_{v_2} = valve coefficient, constant, m^3/s-psi$^{1/2}$
Δp_2 = pressure drop across valve, constant, psi
G_2 = specific gravity of B, constant, dimensionless
$vp_2(t)$ = valve position, a fraction

You may assume that the operation is adiabatic and that the physical properties of the reactants and products are similar. The flow rate F_1 can be assumed to be constant.

Develop the block diagram, giving all transfer functions, to show graphically the interactions between the forcing functions $m_1(t)$, $m_2(t)$, $C_{A_1}(t)$, and $T_1(t)$ and the responding variables $C_A(t)$, $C_B(t)$ and $T(t)$. Give the units of all gains and time constants.

3-12. Obtain the response of a process described by a first-order transfer function to an impulse forcing function.

CHAPTER
4
Higher-Order Dynamic Systems

The previous chapter presented several examples of processes described by first-order ordinary differential equations. This chapter is concerned with processes that are described by higher-order ordinary differential equations. In general, the objectives of the chapter follow closely those of the previous chapter. The mathematical models of more complex systems are presented and the meanings of the parameters that describe the characteristics of these processes are explained.

4-1. TANKS IN SERIES—NONINTERACTING SYSTEM

Higher-order systems are either interacting or noninteracting. Examples of both types are presented in this chapter using actual processes. The meanings of the terms interacting and noninteracting are also explained.

A typical example of a noninteracting system is the set of tanks shown in Fig. 4-1. Determine the transfer functions relating the level in the second tank to the inlet flow to the first tank, $q_i(t)$, and to the pump flow, $q_0(t)$.

In this example all tanks are open to the atmosphere, and the process is isothermal.

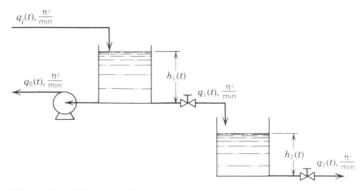

Figure 4-1. Tanks in series—noninteracting system.

The openings of the valves will remain constant and the flow of liquid through the valves is given by

$$q(t) = \frac{C_V}{7.48} \sqrt{\frac{\Delta P(t)}{G}} = \frac{C_V}{7.48} \sqrt{\frac{\rho g h(t)}{144 \, g_c G}} = C_V' \sqrt{h(t)}$$

where

C_V = valve coefficient, $\dfrac{\text{gal/min}}{\sqrt{\text{psi}}}$

7.48 = conversion factor from gal to ft^3

Writing an unsteady-state mass balance around the first tank gives

$$\rho q_i(t) - \rho q_1(t) - \rho q_0(t) = \rho A_1 \frac{dh_1(t)}{dt} \tag{4-1}$$

1 eq., 2 unk. ($q_1(t)$, $h_1(t)$)

where

ρ = density of liquid, lbm/ft^3

A_1 = cross-sectional area of tank 1, ft^2

The valve expression provides another equation:

$$q_1(t) = C_{V_1}' \sqrt{h_1(t)} \tag{4-2}$$

2 eq., 2 unk.

Equations (4-1) and (4-2) describe the first tank. Now we proceed to the second tank. An unsteady state mass balance around the second tank gives

$$\rho q_1(t) - \rho q_2(t) - \rho A_2 \frac{dh_2(t)}{dt} \tag{4-3}$$

3 eq., 4 unk. ($q_2(t)$, $h_2(t)$)

Again the valve expression provides another equation:

$$q_2(t) = C_{V_2}' \sqrt{h_2(t)} \tag{4-4}$$

4 eq., 4 unk.

Equations (4-1) through (4-4) describe the process. Since Eqs. (4-2) and (4-4) are nonlinear, a computer simulation will provide the most exact solution. However, since it is desired to determine transfer functions, these equations must be linearized before we can proceed in a fashion similar to that used in Chapter 3.

Substituting Eq. (4-2) into Eq. (4-1) and Eqs. (4-2) and (4-4) into Eq. (4-3) and dividing each of the resulting equations by the density yields

$$q_i(t) - C_{v1}' \sqrt{h_1(t)} - q_0(t) = A_1 \frac{dh_1(t)}{dt} \tag{4-5}$$

and

$$C_{v1}' \sqrt{h_1(t)} - C_{v2}' \sqrt{h_2(t)} = A_2 \frac{dh_2(t)}{dt} \tag{4-6}$$

From Eq. (4-5) we get, after linearizing and defining deviation variables

$$Q_i(t) - C_1 H_1(t) - Q_o(t) = A_1 \frac{dH_1(t)}{dt} \qquad (4\text{-}7)$$

where

$$C_1 = \left. \frac{\partial q_1(t)}{\partial h_1(t)} \right|_{ss} = \frac{1}{2} C_{v1}' (\bar{h}_1)^{-1/2}$$

and the deviation variables

$$Q_i(t) = q_i(t) - \bar{q}_i$$
$$Q_o(t) = q_o(t) - \bar{q}_o$$
$$H_1(t) = h_1(t) - \bar{h}_1$$

From Eq. (4-6) we get

$$C_1 H_1(t) - C_2 H_2(t) = A_2 \frac{dH_2(t)}{dt} \qquad (4\text{-}8)$$

where

$$C_2 = \left. \frac{\partial q_2(t)}{\partial h_2(t)} \right|_{ss} = \frac{1}{2} C_{v2}' (\bar{h}_2)^{-1/2}$$

and

$$H_2(t) = h_2(t) - \bar{h}_2$$

Rearranging Eqs. (4-7) and (4-8) yields

$$\tau_1 \frac{dH_1(t)}{dt} + H_1(t) = K_1 Q_i(t) - K_1 Q_o(t) \qquad (4\text{-}9)$$

and

$$\tau_2 \frac{dH_2(t)}{dt} + H_2(t) = K_2 H_1(t) \qquad (4\text{-}10)$$

where

$$\tau_1 = \frac{A_1}{C_1}, \qquad \text{minutes}$$

$$\tau_2 = \frac{A_2}{C_2}, \qquad \text{minutes}$$

$$K_1 = \frac{1}{C_1}, \qquad \text{ft-min/ft}^3$$

$$K_2 = \frac{C_1}{C_2}, \qquad \text{dimensionless}$$

Taking the Laplace transform of Eqs. (4-9) and (4-10) and rearranging, we get

$$H_1(s) = \frac{K_1}{\tau_1 s + 1} Q_i(s) - \frac{K_1}{\tau_1 s + 1} Q_o(s) \qquad (4\text{-}11)$$

$$H_2(s) = \frac{K_2}{\tau_2 s + 1} H_1(s) \qquad (4\text{-}12)$$

Equation (4-11) relates the level in the first tank to the inlet and outlet flows. Equation (4-12) relates the level in the second tank to the level in the first tank.

To determine the desired transfer functions, substitute Eq. (4-11) into Eq. (4-12)

$$H_2(s) = \frac{K_1 K_2}{(\tau_1 s + 1)(\tau_2 s + 1)} (Q_i(s) - Q_o(s)) \qquad (4\text{-}13)$$

or the individual transfer functions

$$\frac{H_2(s)}{Q_i(s)} = \frac{K_1 K_2}{(\tau_1 s + 1)(\tau_2 s + 1)} \qquad (4\text{-}14)$$

and

$$\frac{H_2(s)}{Q_o(s)} = \frac{-K_1 K_2}{(\tau_1 s + 1)(\tau_2 s + 1)} \qquad (4\text{-}15)$$

At this point the reader should convince himself that the units of $K_1 K_2$ are ft-min/ft^3 and that the units of τ_1 and τ_2 are minutes. But what about the signs of Eqs. (4-14) and (4-15)? Do they make sense?

The transfer functions given by Eqs. (4-14) and (4-15) are called *second-order transfer functions* or *second-order lags*. It is fairly simple to see from their development that they are "formed" by two first-order transfer functions in series.

The block diagram for this system can be represented in different forms as shown in Fig. 4-2. The block diagram of Fig. 4-2a is developed by "chaining" Eqs. (4-11) and (4-12). The diagram shows that the inlet and outlet flows initially affect the level in the first tank, $H_1(s)$. A change in this level then affects the level in the second tank, $H_2(s)$. Figures 4-2b and 4-2c show other more compact diagrams. Even though the block diagram in Fig. 4-2a gives a better description of how things really happen, all three diagrams are used without any preference.

Now extend the process shown in Fig. 4-1 by another tank, as shown in Fig. 4-3. For this new process, determine the transfer functions relating the level in the third tank to the inlet flow to the first tank and to the pump flow.

Since the first two tanks have already been modeled, in Eqs. (4-1) through (4-4), the third tank is now modeled. Writing an unsteady-state mass balance around the third tank results in

$$\rho q_2(t) - \rho q_3(t) = \rho A_3 \frac{dh_3(t)}{dt} \qquad (4\text{-}16)$$

5eq., 6 unk. ($q_3(t)$, $h_3(t)$)

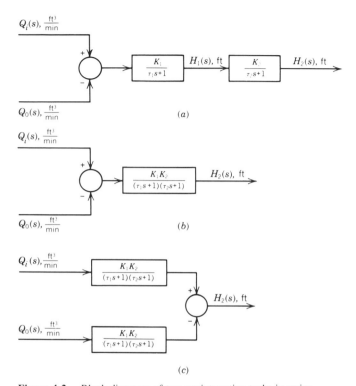

Figure 4-2. Block diagrams of two noninteracting tanks in series.

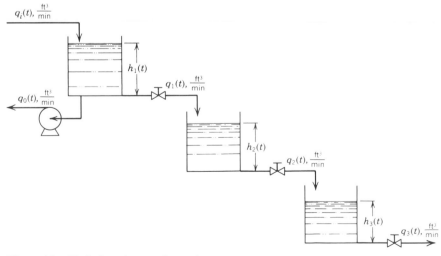

Figure 4-3. Tanks in series—noninteracting system.

The valve expression provides the next required equation:

$$q_3(t) = C'_{V_3} \sqrt{h_3(t)} \tag{4-17}$$

6 eq., 6 unk.

The new process, Fig. 4-3, is now modeled by Eqs. (4-1), (4-2), (4-3), (4-4), (4-16) and (4-17).

Substituting Eqs. (4-4) and (4-17) into Eq. (4-16) and dividing the resulting equation by the density yields

$$C'_{v2} \sqrt{h_2(t)} - C'_{v3} \sqrt{h_3(t)} = A_3 \frac{dh_3(t)}{dt} \tag{4-18}$$

from which we get

$$C_2 H_2(t) - C_3 H_3(t) = A_3 \frac{dH_3(t)}{dt} \tag{4-19}$$

where

$$C_3 = \left. \frac{\partial q_3(t)}{\partial h_3(t)} \right|_{ss} = \frac{1}{2} C'_{v3} \, (\bar{h}_3)^{-1/2}$$

and the deviation variable $H_3(t) - h_3(t) - \bar{h}_3$.

Rearranging Eq. (4-19) and taking the Laplace transform yields

$$H_3(s) - \frac{K_3}{\tau_3 s + 1} H_2(s) \tag{4-20}$$

where

$$\tau_3 = \frac{A_3}{C_3}, \quad \text{minutes}$$

$$K_3 = \frac{C_2}{C_3}, \quad \text{dimensionless}$$

Finally, substituting Eq. (4-13) into Eq. (4-20) gives

$$H_3(s) = \frac{K_1 K_2 K_3}{(\tau_1 s + 1)(\tau_2 s + 1)(\tau_3 s + 1)} (Q_i(s) - Q_o(s)) \tag{4-21}$$

from which the following transfer functions are determined:

$$\frac{H_3(s)}{Q_i(s)} = \frac{K_1 K_2 K_3}{(\tau_1 s + 1)(\tau_2 s + 1)(\tau_3 s + 1)} \tag{4-22}$$

and

$$\frac{H_3(s)}{Q_o(s)} = \frac{-K_1 \dot{K}_2 K_3}{(\tau_1 s + 1)(\tau_2 s + 1)(\tau_3 s + 1)} \tag{4-23}$$

These two transfer functions are called *third-order transfer functions* or *third-order lags*. Figure 4-4 shows three possible ways to represent Eq. (4-21) in block diagram form.

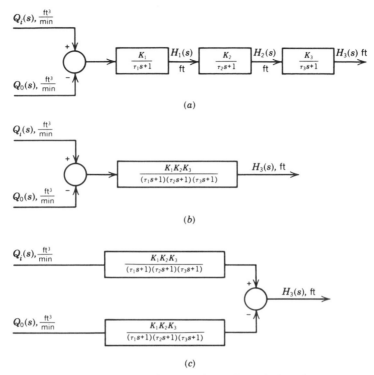

Figure 4-4. Block diagrams of three noninteracting tanks in series.

Figure 4-4a is especially interesting because it shows how the forcing functions, $Q_i(s)$ and $Q_o(s)$, affect the different levels.

Note that these transfer functions were obtained by multiplying the first-order transfer functions, that is

$$\frac{H_3(s)}{Q_i(s)} = \frac{H_1(s)}{Q_i(s)} \cdot \frac{H_2(s)}{H_1(s)} \cdot \frac{H_3(s)}{H_2(s)}$$

This is the case for noninteracting systems in series. This statement can be generalized by writing

$$G(s) = \prod_{i=1}^{n} G_i(s) \tag{4-24}$$

where

$n = $ number of noninteracting systems in series
$G(s) = $ transfer function relating the output from the last system, the nth system, to the input to the first system.
$G_i(s) = $ individual transfer function for each system.

The processes shown in Figs. 4-1 and 4-3 are referred to as noninteracting systems because there is no full interaction between the variables. The level of the first tank affects the level of the second tank, but this level does not in turn affect the level of the first tank. The same is true for the levels in the second and third tanks. Examples of interacting systems are presented in the next few sections.

Finally, it is important to recognize one more point. The set of equations that describe the processes shown so far in this chapter is more complex, because more equations are involved than those presented in Chapter 3. The way the previous examples were worked out was *first* by writing the model, the set of equations that describes the process, and *then* by deciding how to solve this model. Developing the model is always the first step; the solution of the model follows the modeling.

4-2. TANKS IN SERIES—INTERACTING SYSTEM

Rearranging the tanks of Fig. 4-1 results in an interacting system as shown in Fig. 4-5.

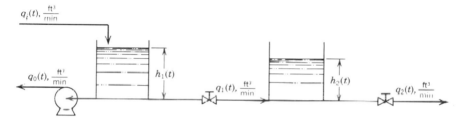

Figure 4-5. Tanks in series—interacting system.

The interaction between the tanks is clearly shown from the valve equation for the flow, $q_1(t)$, that is

$$q_1(t) = \frac{C_{V_1}}{7.48} \sqrt{\frac{\Delta P(t)}{G}} = \frac{C_{V_1}}{7.48} \sqrt{\frac{\rho g(h_1(t) - h_2(t))}{144 \, g_c G}}$$

$$= C'_{V_1} \sqrt{h_1(t) - h_2(t)}$$

This equation shows that the flow between the two tanks depends on the levels in *both* tanks, each affecting the other.

Determine the same two transfer functions as for the noninteracting system

$$\frac{H_2(s)}{Q_i(s)} \quad \text{and} \quad \frac{H_2(s)}{Q_o(s)}$$

Start by writing an unsteady-state mass balance around the first tank; this is given by Eq. (4-1):

$$\rho q_i(t) - \rho q_1(t) - \rho q_o(t) = \rho A_1 \frac{dh_1(t)}{dt} \tag{4-1}$$

1 eq., 2 unk. ($q_1(t)$, $h_1(t)$)

The valve expression provides the next equation:

$$q_1(t) = C'_{V_1} \sqrt{h_1(t) - h_2(t)} \tag{4-25}$$

2 eq., 3 unk. ($h_2(t)$)

Another independent equation is still needed. An unsteady-state mass balance around the second tank helps in obtaining this equation, which is Eq. (4-3).

$$\rho q_1(t) - \rho q_2(t) = \rho A_2 \frac{dh_2(t)}{dt} \tag{4-3}$$

3 eq., 4 unk. ($q_2(t)$)

The expression for the flow through the last valve is given by Eq. (4-4)

$$q_2(t) = C'_{V_2} \sqrt{h_2(t)} \tag{4-4}$$

4 eq., 4 unk.

There is now the same number of independent equations as unknowns; the process is described (modeled). The solution now follows.

Substituting Eq. (4-25) into Eq. (4-1) and dividing the resulting equation by the density yields

$$q_i(t) - C'_{v1} \sqrt{h_1(t) - h_2(t)} - q_o(t) = A_1 \frac{dh_1(t)}{dt}$$

From which we get

$$Q_i(t) - C_4 H_1(t) + C_4 H_2(t) - Q_o(t) = A_1 \frac{dH_1(t)}{dt} \tag{4-26}$$

where

$$C_4 = \left. \frac{\partial q_1(t)}{\partial h_1(t)} \right|_{ss} = - \left. \frac{\partial q_1(t)}{\partial h_2(t)} \right|_{ss} = \frac{1}{2} C'_{v1} (\bar{h}_1 - \bar{h}_2)^{-1/2}$$

Rearranging Eq. (4-26) and taking the Laplace transform gives

$$H_1(s) = \frac{K_4}{\tau_4 s + 1} Q_i(s) + \frac{1}{\tau_4 s + 1} H_2(s) - \frac{K_4}{\tau_4 s + 1} Q_o(s) \tag{4-27}$$

where

$$K_4 = \frac{1}{C_4}, \qquad \text{ft-min/ft}^3$$

$$\tau_4 = \frac{A_1}{C_4}, \qquad \text{minutes}$$

Following the same procedure for the second tank gives

$$H_2(s) = \frac{K_5}{\tau_5 s + 1} H_1(s) \tag{4-28}$$

where

$$K_5 = \frac{C_4}{C_4 + C_2}, \quad \text{dimensionless}$$

$$\tau_5 = \frac{A_2}{C_4 + C_2}, \quad \text{minutes}$$

Finally, substituting Eq. (4-27) into Eq. (4-28) yields

$$H_2(s) = \frac{K_4 K_5}{(\tau_4 s + 1)(\tau_5 s + 1)}(Q_i(s) - Q_o(s))$$

$$+ \frac{K_5}{(\tau_4 s + 1)(\tau_5 s + 1)} H_2(s)$$

$$H_2(s) = \frac{K_4 K_5}{\tau_4 \tau_5 s^2 + (\tau_4 + \tau_5)s + (1 - K_5)}(Q_i(s) - Q_o(s))$$

or

$$H_2(s) = \frac{\dfrac{K_4 K_5}{1 - K_5}}{\left(\dfrac{\tau_4 \tau_5}{1 - K_5}\right)s^2 + \left(\dfrac{\tau_4 \tau_5}{1 - K_5}\right)s + 1}(Q_i(s) - Q_o(s)) \tag{4-29}$$

from which the desired transfer functions are obtained:

$$\frac{H_2(s)}{Q_i(s)} = \frac{\dfrac{K_4 K_5}{1 - K_5}}{\left(\dfrac{\tau_4 \tau_5}{1 - K_5}\right)s^2 + \left(\dfrac{\tau_4 + \tau_5}{1 - K_5}\right)s + 1} \tag{4-30}$$

and

$$\frac{H_2(s)}{Q_o(s)} = \frac{-\dfrac{K_4 K_5}{1 - K_5}}{\left(\dfrac{\tau_4 \tau_5}{1 - K_5}\right)s^2 + \left(\dfrac{\tau_4 + \tau_5}{1 - K_5}\right)s + 1} \tag{4-31}$$

These transfer functions are of second order. Block diagrams depicting this interacting process are shown in Fig. 4-6.

There are several things that can be learned by comparing the transfer functions for the interacting and noninteracting systems. Comparing Eqs. (4-14) and (4-30) reveals that the time constants are different. Furthermore, for the interacting case the larger time constant is greater than for the noninteracting case, resulting in a slower responding system.

To prove this last statement we consider the case in which both individual time constants are equal, that is

$$\tau_4 = \tau_5 = \tau$$

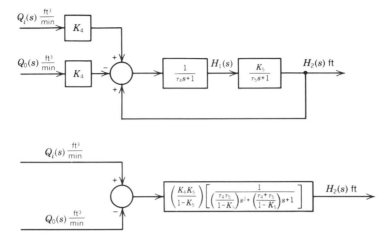

Figure 4-6. Block diagrams of two interacting tanks system.

and for this to be true

$$A_2 = 2A_1$$
$$C_2 = C_4$$

Then $K_5 = 0.5$.

With this information Eq. (4-31) becomes

$$\frac{H_2(s)}{Q_o(s)} = \frac{\dfrac{-K_4K_5}{(\tau s + 1)^2}}{1 - \dfrac{K_5}{(\tau s + 1)^2}} = \frac{-K_4K_5}{\tau^2 s^2 + 2\tau s + (1 - K_5)}$$

The roots of the denominator are

$$\text{Roots} = \frac{-(1 + \sqrt{K_5})}{\tau}, \quad \frac{-(1 - \sqrt{K_5})}{\tau}$$

from which we obtain the two "effective" time constants for this interacting system as

$$\tau_{4\text{eff}} = \frac{\tau}{1 + \sqrt{K_5}} = \frac{\tau}{1.707} = 0.58\tau$$

$$\tau_{5\text{eff}} = \frac{\tau}{1 - \sqrt{K_5}} = \frac{\tau}{0.293} = 3.41\tau$$

and the ratio of these two "effective" time constants as

$$\frac{\tau_{5\text{eff}}}{\tau_{4\text{eff}}} = 5.8$$

even though $\tau_4 = \tau_5$! This clearly shows that the larger time constant for an interacting system is greater than the larger for a noninteracting system.

Another fact about interacting systems is that the "effective" time constants are real. To prove this statement let us equate the denominator of Eq. (4-31) to zero:

$$\left(\frac{\tau_4\tau_5}{1 - K_5}\right)s^2 + \left(\frac{\tau_4 + \tau_5}{1 - K_5}\right)s + 1 = 0$$

Based on the definition of τ_4, τ_5, and K_5 we get

$$\left(\frac{A_1A_2}{C_2C_4}\right)s^2 + \left(\frac{A_1(C_2 + C_4) + A_2C_4}{C_2C_4}\right)s + 1 = 0$$

The roots of this equation are obtained by the use of the quadratic expression. For these roots to be real the following must be true:

$$b^2 \quad 4ac = \frac{[A_1(C_2 + C_4) + A_2C_4]^2}{C_2^2C_4^2} - \frac{4A_1A_2}{C_2C_4} > 0$$

or

$$(A_1C_2 - A_2C_4)^2 + A_1C_4(2A_1C_2 + A_1C_4 + 2A_2C_4) > 0$$

and since all constants are positive, this inequality is always true. Therefore, we can say that the time constants of interacting systems are always real. This becomes important when we study the response of these systems to different forcing functions.

By far the great majority of processes are described by higher order transfer functions. Both interacting and noninteracting processes are found in industry. Of the two, the interacting process is the most common. More examples of interacting processes are presented in the following sections.

4-3. THERMAL PROCESS

Consider the unit shown in Fig. 4-7. The objective of this unit is to cool a hot process fluid. The cooling medium, water, goes through a jacket. For this process assume that the water in the cooling jacket is well mixed as is the fluid in the tank and that the density and heat capacity of the process fluid and of the cooling water do not change significantly

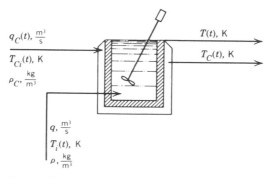

Figure 4-7. Cooling of a hot process fluid.

with temperature. Since the process fluid flows out by overflow, the level and the heat transfer area in the tank are constant. Finally, it can also be assumed that the tank is well insulated.

Determine the transfer functions that relate the outlet process fluid temperature, $T(t)$, to the inlet cooling water temperature, $T_{C_i}(t)$, to the cooling water flow rate, $q_C(t)$, and to the inlet process temperature, $T_i(t)$.

An unsteady-state energy balance on the process fluid is

$$q\rho C_p T_i(t) - UA[T(t) - T_C(t)] - q\rho C_p T(t) = V\rho C_V \frac{dT(t)}{dt} \tag{4-32}$$

$$\text{1 eq., 2 unk. } (T(t), T_C(t))$$

where

$$U = \text{overall heat transfer coefficient, assumed constant, J/m}^2\text{-K-s}$$
$$A = \text{heat transfer area, m}^2$$
$$V = \text{volume of tank, m}^3$$
$$C_p, C_V = \text{heat capacities of process fluid, J/kg-K}$$

Another independent equation is still needed. The equation is obtained from an unsteady-state energy balance around the cooling jacket:

$$q_C(t)\rho_c C_{pc} T_{C_i}(t) + UA[T(t) - T_C(t)] - q_C(t)\,\rho_c C_{pc} T_C(t) \tag{4-33}$$

$$= V_C \rho_c C_{V_C} \frac{dT_C(t)}{dt} \qquad\qquad 2\,\text{eq.}, 2\,\text{unk.}$$

where

$$C_{p_C}, C_{V_C} = \text{heat capacities of cooling water, J/kg-K}$$
$$V_C = \text{volume of cooling jacket, m}^3$$

From Eq. (4-32) and (4-33) the desired transfer functions can be obtained. However, before this is done Eq. (4-33) must be linearized. Specifically, the nonlinear terms are the first and last terms of the left-hand side. Equation (4-32) is already linear.

Following the procedure already learned we obtain from Eqs. (4-32) and (4-33)

$$q\rho C_p \mathbf{T}_i(t) - UA[\mathbf{T}(t) - \mathbf{T}_C(t)] - q\rho C_p \mathbf{T}(t) = V\rho C_v \frac{d\mathbf{T}(t)}{dt} \tag{4-34}$$

and

$$C_1 Q_C(t) + C_2 \mathbf{T}_{ci}(t) + UA\,[\mathbf{T}(t) - \mathbf{T}_C(t)] - C_3 Q_C(t) - C_2 \mathbf{T}_C(t) = V_C \rho_c C_{vc} \frac{d\mathbf{T}_C(t)}{dt} \tag{4-35}$$

where

$$C_1 = \rho_c C_{p_C} \overline{T}_{C_i}, \qquad \text{J/m}^3$$
$$C_2 = \overline{q}_c \rho_c C_{p_c}, \qquad \text{J/s} - \text{K}$$
$$C_3 = \rho_c C_{p_C} \overline{T}_C, \qquad \text{J/m}^3$$

and the deviation variables are

$$\mathbf{T}_i(t) = T_i(t) - \bar{T}_i$$
$$\mathbf{T}(t) = T(t) - \bar{T}$$
$$\mathbf{T}_C(t) = T_C(t) - \bar{T}_C$$
$$\mathbf{Q}_C(t) = q_C(t) - \bar{q}_C$$
$$\mathbf{T}_{C_i}(t) = T_{C_i}(t) - \bar{T}_{C_i}$$

Rearranging Eq. (4-34) and taking the Laplace transform yields

$$\mathbf{T}(s) = \frac{1}{\tau_1 s + 1} [K_1 \mathbf{T}_i(s) + K_2 \mathbf{T}_c(s)] \tag{4-36}$$

where

$$\tau_1 = \frac{V \rho C_V}{UA + q\rho C_p}, \qquad \text{seconds}$$

$$K_1 = \frac{q\rho C_p}{UA + q\rho C_p}, \qquad \text{dimensionless}$$

$$K_2 = \frac{UA}{UA + q\rho C_p}, \qquad \text{dimensionless}$$

Similarly from Eq. (4-35) we obtain

$$\mathbf{T}_c(s) = \frac{1}{\tau_2 s + 1} [K_3 \mathbf{T}_{ci}(s) - K_4 \mathbf{Q}_c(s) + K_5 \mathbf{T}(s)] \tag{4-37}$$

where

$$\tau_2 = \frac{V_c \rho_c C_{V_c}}{C_2 + UA}, \qquad \text{seconds}$$

$$K_3 = \frac{C_2}{C_2 + UA}, \qquad \text{dimensionless}$$

$$K_4 = \frac{C_3 - C_1}{C_2 + UA}, \qquad \text{K/m}^3\text{-s}$$

$$K_5 = \frac{UA}{C_2 + UA}, \qquad \text{dimensionless}$$

By substituting Eq. (4-37) into Eq. (4-36) and performing some algebraic manipulations the following transfer functions are determined:

$$\frac{\mathbf{T}(s)}{\mathbf{T}_i(s)} = \left(\frac{K_1}{1 - K_2 K_5}\right) \left[\frac{\tau_2 s + 1}{\left(\dfrac{\tau_1 \tau_2}{1 - K_2 K_5}\right) s^2 + \left(\dfrac{\tau_1 + \tau_2}{1 - K_2 K_5}\right) s + 1}\right] \tag{4-38}$$

$$\frac{T(s)}{T_{C_i}(s)} = \left(\frac{K_2 K_3}{1 - K_2 K_5}\right) \left[\frac{1}{\left(\dfrac{\tau_1 \tau_2}{1 - K_2 K_5}\right) s^2 + \left(\dfrac{\tau_1 + \tau_2}{1 - K_2 K_5}\right) s + 1} \right] \tag{4-39}$$

$$\frac{T(s)}{Q_C(s)} = \left(\frac{-K_4 K_2}{1 - K_2 K_5}\right) \left[\frac{1}{\left(\dfrac{\tau_1 \tau_2}{1 - K_2 K_5}\right) s^2 + \left(\dfrac{\tau_1 + \tau_2}{1 - K_2 K_5}\right) s + 1} \right] \tag{4-40}$$

The block diagram for this system is shown in Fig. 4-8 and is obtained by chaining Eqs. (4-36) and (4-37).

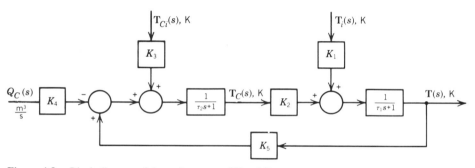

Figure 4-8.   Block diagram of thermal process of Fig. 4-7.

It can be seen that these three transfer functions, Eqs. (4-38), (4-39), and (4-40), are of second order. Equation (4-38) is a bit different from the other two. Specifically, it has the term $(\tau_2 s + 1)$ in the numerator. This type of transfer function is discussed later in this chapter. At this moment it is important to understand the significance of these three second-order transfer functions. Consider the inlet cooling water temperature, for example. If $T_C(t)$ changes, it will affect first the jacket temperature and then the process fluid temperature; there are two first-order systems in series. The same dynamic behavior is true for a change in cooling water flow, as seen from the transfer functions, Eqs. (4-39) and (4-40). The dynamic terms in these two equations are exactly the same. The first transfer function, Eq. (4-38), tells that the dynamics of a change in inlet process fluid temperature, $T_i(t)$, on $T(t)$ are different from the other two disturbances. As mentioned previously, these differences are explained later in this chapter.

In this example the heat transfer rate expression $UA(T(t) - T_C(t))$ has been used. In so doing, however, the dynamics of the wall of the tank have been neglected. We have implied that once the temperature of the cooling water, $T_C(t)$, changes the process fluid will feel a change in heat transfer immediately. In reality, however, this is not so. When the temperature of the cooling water changes, the heat transfer to the wall of the tank will change and, consequently, the temperature of the wall will start to change. It is then that the heat transfer rate from the wall to the process fluid will change. So the wall of the tank presents another capacitance in the system. The magnitude of this capacitance depends, among other things, on thickness, density, heat capacity, and other physical properties of the material of construction of the wall.

Taking this wall into consideration gives a better understanding of this capacitance. It will be assumed that both tank wall surfaces, the one next to the process fluid and the one next to the cooling water, are at the same temperature. This is a good assumption when the wall is not too thick and it has a large thermal conductivity.

The energy balance on the process fluid is thus changed to

$$q\rho C_p T_i(t) - h_i A_i(T(t) - T_m(t)) - q\rho C_p T(t) = V\rho C_V \frac{dT(t)}{dt} \tag{4-41}$$

$$\text{1 eq., 2 unk. } (T(t), T_m(t))$$

where

$$h_i = \text{inside film heat transfer coefficient, assumed constant, J/m}^2\text{-s-K}$$
$$A_i = \text{inside heat transfer area, m}^2$$
$$T_m(t) = \text{temperature of the metal wall, K}$$

Proceeding with an unsteady-state energy balance on the wall of the tank, we can write

$$h_i A_i(T(t) - T_m(t)) - h_o A_o(T_m(t) - T_C(t)) = V_m \rho_m C_{V_m} \frac{dT_m(t)}{dt} \tag{4-42}$$

$$\text{2 eq., 3 unk. } (T_C(t))$$

where

$$h_o = \text{outside film heat transfer coefficient, assumed constant, J/m}^2\text{-s-K}$$
$$A_o = \text{outside heat transfer area, m}^2$$
$$V_m = \text{volume of the metal wall, m}^3$$
$$\rho_m = \text{density of the metal wall, kg/m}^3$$
$$C_{V_m} = \text{heat capacity of the metal wall, J/kg-K}$$

Finally, an unsteady-state energy balance on the cooling water gives the other needed equation:

$$q_C(t)\rho_C C_{p_C} T_{C_i}(t) + h_o A_o(T_m(t) - T_C(t)) - q_C(t)\rho_C C_{p_C} T_C(t) = V_C \rho_C C_{V_C} \frac{dT_C(t)}{dt} \tag{4-43}$$

$$\text{3 eq., 3 unk.}$$

Three differential equations are now required to fully describe the system. Using the procedure already learned, we can develop the following transfer function from Eq. (4-41):

$$T(s) = \frac{K_6}{\tau_3 s + 1} T_i(s) + \frac{K_7}{\tau_3 s + 1} T_m(s) \tag{4-44}$$

The block diagram depicting this equation is shown in Fig. 4-9.

From Eq. (4-42) we obtain

$$T_m(s) = \frac{K_8}{\tau_4 s + 1} T(s) + \frac{K_9}{\tau_4 s + 1} T_C(s) \tag{4-45}$$

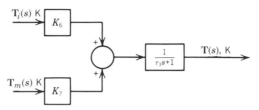

Figure 4-9. Block diagram representation of Eq. (4-44).

and finally from Eq. (4-43) we get

$$\mathbf{T}_C(s) = \frac{K_{10}}{\tau_5 s + 1} \mathbf{T}_{C_i}(s) + \frac{K_{11}}{\tau_5 s + 1} \mathbf{T}_m(s) - \frac{K_{12}}{\tau_5 s + 1} Q_C(s) \qquad (4\text{-}46)$$

where

$$\mathbf{T}_m(t) = T_m(t) - \bar{T}_m$$

$$\tau_3 = \frac{V \rho C_p}{q \rho C_p + h_i A_i}, \quad \text{seconds} \qquad\qquad K_6 = \frac{q \rho C_p}{q \rho C_p + h_i A_i}, \quad \text{dimensionless}$$

$$K_7 = \frac{h_i A_i}{q \rho C_p + h_i A_i}, \quad \text{dimensionless} \qquad\qquad \tau_4 = \frac{V_m \rho_m C_{p_m}}{h_i A_i + h_o A_o}, \quad \text{seconds}$$

$$K_8 = \frac{h_i A_i}{h_i A_i + h_o A_o}, \quad \text{dimensionless} \qquad\qquad K_9 = \frac{h_o A_o}{h_i A_i + h_o A_o}, \quad \text{dimensionless}$$

$$\tau_5 = \frac{V_C \rho_C C_{p_C}}{h_o A_o + \bar{q}_C \rho_C C_{p_C}}, \quad \text{seconds} \qquad K_{10} = \frac{\bar{q}_C \rho_C C_{p_C}}{h_o A_o + \bar{q}_C \rho_C C_{p_C}}, \quad \text{dimensionless}$$

$$K_{11} = \frac{h_o A_o}{h_o A_o + \bar{q}_C \rho_C C_{p_C}}, \quad \text{dimensionless} \qquad K_{12} = \frac{\rho_C C_{p_C}(\bar{T}_C - \bar{T}_{C_i})}{h_o A_o + \bar{q}_C \rho_C C_{p_C}}, \quad \frac{K}{m^3/s}$$

Chaining Eqs. (4-45) and (4-46) to Eq. (4-44) gives the block diagram shown in Fig. 4-10.

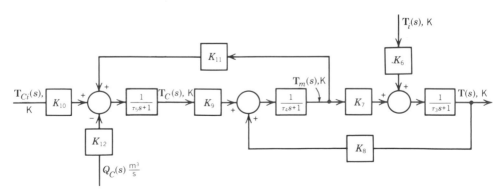

Figure 4-10. Block diagram of thermal process.

In order to determine the desired transfer functions, we substitute Eq. (4-46) into (4-45):

$$\mathbf{T}_m(s) = \frac{K_8}{\tau_4 s + 1} \mathbf{T}(s) + \frac{K_9}{(\tau_4 s + 1)(\tau_5 s + 1)} (K_{10} \mathbf{T}_{C_i}(s)$$

$$+ K_{11} \mathbf{T}_m(s) - K_{12} Q_C(s))$$

or

$$\mathbf{T}_m(s) = \frac{K_8(\tau_5 s + 1)}{(\tau_4 s + 1)(\tau_5 s + 1) - K_9 K_{11}} \mathbf{T}(s)$$

$$+ \frac{1}{(\tau_4 s + 1)(\tau_5 s + 1) - K_9 K_{11}} (K_9 K_{10} \mathbf{T}_{C_i}(s)$$

$$- K_9 K_{12} Q_C(s))$$

Substituting this last equation into Eq. (4-44) yields

$$\mathbf{T}(s) = \frac{K_6}{\tau_3 s + 1} \mathbf{T}_i(s)$$

$$+ \frac{K_7 K_8(\tau_5 s + 1)}{(\tau_3 s + 1)[(\tau_4 s + 1)(\tau_5 s + 1) - K_9 K_{11}]} \mathbf{T}(s)$$

$$+ \frac{K_7 K_9}{(\tau_3 s + 1)[(\tau_4 s + 1)(\tau_5 s + 1) - K_9 K_{11}]} (K_{10} \mathbf{T}_{C_i}(s) - K_{12} Q_C(s))$$

and after some algebra we get

$$\mathbf{T}(s) = \left(\frac{K_6(1 - K_9 K_{11})}{1 - K_9 K_{11} - K_7 K_8} \right) \left[\frac{\tau_6^2 s^2 + \tau_7 s + 1}{\tau_8^3 s^3 + \tau_9^2 s^2 + \tau_{10} s + 1} \right] \mathbf{T}_i(s)$$

$$+ \left(\frac{K_7 K_9}{1 - K_9 K_{11} - K_7 K_8} \right) \left[\frac{1}{\tau_8^3 s^3 + \tau_9^2 s^2 + \tau_{10} s + 1} \right] (K_{10} \mathbf{T}_{C_i}(s) \qquad (4\text{-}47)$$

$$- K_{12} Q_C(s))$$

where

$$\tau_6 = \left(\frac{\tau_4 \tau_5}{1 - K_9 K_{11}} \right)^{1/2}, \qquad \text{seconds}$$

$$\tau_7 = \frac{\tau_4 - \tau_5}{1 - K_9 K_{11}}, \qquad \text{seconds}$$

$$\tau_8 = \left(\frac{\tau_3 \tau_4 \tau_5}{1 - K_9 K_{11} - K_7 K_8} \right)^{1/3}, \qquad \text{seconds}$$

$$\tau_9 = \left(\frac{\tau_4 \tau_5 + \tau_3 \tau_4 + \tau_3 \tau_5}{1 - K_9 K_{11} - K_7 K_8} \right)^{1/2}, \qquad \text{seconds}$$

$$\tau_{10} = \frac{\tau_3(1 - K_9 K_{11}) + \tau_4 + \tau_5 - K_7 K_8 K_5}{1 - K_9 K_{11} - K_7 K_8}, \qquad \text{seconds}$$

From Eq. (4-47) the desired transfer functions are obtained:

$$\frac{\mathbf{T}(s)}{\mathbf{T}_i(s)} = \frac{K_6(1 - K_9K_{11})}{1 - K_9K_{11} - K_7K_8} \left[\frac{\tau_6^2 s^2 + \tau_7 s + 1}{\tau_8^3 s^3 + \tau_9^2 s^2 + \tau_{10} s + 1} \right] \qquad (4\text{-}48)$$

$$\frac{\mathbf{T}(s)}{\mathbf{T}_{C_i}(s)} = \left(\frac{K_7 K_9 K_{10}}{1 - K_9K_{11} - K_7K_8} \right) \left[\frac{1}{\tau_8^3 s^3 + \tau_9^2 s^2 + \tau_{10} s + 1} \right] \qquad (4\text{-}49)$$

and

$$\frac{\mathbf{T}(s)}{Q_C(s)} = \left(\frac{-K_7 K_9 K_{12}}{1 - K_9K_{11} - K_7K_8} \right) \left[\frac{1}{\tau_8^3 s^3 + \tau_9^2 s^2 + \tau_{10} s + 1} \right] \qquad (4\text{-}50)$$

The transfer functions given by Eqs. (4-49) and (4-50) are of the third order; that is, the denominator is a polynomial in s of the third order. The block diagram of Fig. 4-10 shows graphically that if the inlet cooling water temperature changes, it first affects the cooling water jacket temperature, $T_C(t)$, then this temperature affects the metal wall temperature, $T_m(t)$, and finally this temperature affects the process fluid temperature, $T(t)$. There are three first-order systems in series. These transfer functions are analogous to the transfer functions given by Eqs. (4-39) and (4-40). The transfer function given by Eq. (4-48), also of third order, is a bit different from Eqs. (4-49) and (4-50) and similar to Eq. (4-38).

This thermal process is another example of an interacting process. The interaction develops through the heat transfer rate expression: $UA(T(t) - T_C(t))$ in Eqs. (4-32) and (4-33), and $h_i A_i(T(t) - T_m(t))$ and $h_o A_o(T_m(t) - T_C(t))$ in Eqs. (4-41), (4-42), and (4-43). The block diagrams of Figs. 4-8 and 4-10 show this interaction graphically.

4-4. RESPONSE OF HIGHER-ORDER SYSTEMS TO DIFFERENT TYPES OF FORCING FUNCTIONS

Two types of higher-order transfer functions have been developed:

$$G(s) = \frac{Y(s)}{X(s)} = \prod_{i=1}^{n} G_i(s) = \frac{K}{\displaystyle\prod_{i=1}^{n} (\tau_i s + 1)} \qquad (4\text{-}51)$$

and

$$G(s) = \frac{Y(s)}{X(s)} = \frac{K \displaystyle\prod_{j=1}^{m} (\tau_{ld_j} s + 1)}{\displaystyle\prod_{i=1}^{n} (\tau_{lg_i} s + 1)} \qquad (4\text{-}52)$$

where $n > m$.

This section presents the response of these higher-order systems to different types of forcing functions, specifically, step functions and sinusoidal functions. From these studies some generalizations about the responses can be made.

Step Function

The responses of systems described by Eq. (4-51) are shown first. A second-order transfer function is usually written in either of the following two forms:

$$G(s) = \frac{Y(s)}{X(s)} = \frac{K}{(\tau_1 s + 1)(\tau_2 s + 1)} = \frac{K}{\tau_1 \tau_2 s^2 + (\tau_1 + \tau_2)s + 1} \tag{4-53}$$

or

$$G(s) = \frac{Y(s)}{X(s)} = \frac{K}{\tau^2 s^2 + 2\tau\xi s + 1} \tag{4-54}$$

where

τ — characteristic time constant, time
ξ = damping ratio, dimensionless

and the relations between the parameters of the two forms are

$$\tau = \sqrt{\tau_1 \tau_2} \tag{4-55}$$

and

$$\xi = \frac{\tau_1 + \tau_2}{2\sqrt{\tau_1 \tau_2}} \tag{4-56}$$

The response of a second-order transfer function to a step change of unit magnitude in forcing function, $X(s) = 1/s$, is obtained as follows. From Eq. (4-54)

$$Y(s) = \frac{K}{s(\tau^2 s^2 + 2\tau\xi s + 1)} = \frac{K}{s(s - r_1)(s - r_2)} \tag{4-57}$$

where

$$r_1 = -\frac{\xi}{\tau} + \frac{\sqrt{\xi^2 - 1}}{\tau} \tag{4-58}$$

$$r_2 = -\frac{\xi}{\tau} - \frac{\sqrt{\xi^2 - 1}}{\tau} \tag{4-59}$$

From these last two equations it can be noted that the response of this system depends on the value of the damping ratio, ξ.

For a value of $\xi < 1$ the roots, r_1 and r_2 are complex and the response of the system obtained using the procedure learned in Chapter 2 is given by the following equation:

$$Y(t) = K\left[1 - \frac{1}{\sqrt{1 - \xi^2}} e^{-\xi t/\tau} \sin\left(\sqrt{1 - \xi^2}\, \frac{t}{\tau} + \tan^{-1}\frac{\sqrt{1 - \xi^2}}{\xi}\right)\right] \tag{4-60}$$

The response of this type of system is shown graphically in Fig. 4-11. As can be seen, the response is oscillatory; systems of this type are said to be *underdamped*.

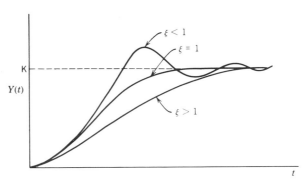

Figure 4-11.   Response of second-order system to a unit step change in forcing function.

For a value of $\xi = 1$ the roots are real and equal and the response of the system is given by

$$Y(t) = K\left[1 - \left(1 + \frac{t}{\tau}\right)e^{-t/\tau}\right]$$ (4-61)

The response of this system is also shown in Fig. 4-11. This is the fastest approach to the final value without overshooting it, and consequently there is no oscillation. Systems where $\xi = 1$ are called *critically damped*.

For a value of $\xi > 1$ the roots are real and different and the response of the system is given by

$$Y(t) = K\left[1 - 0.5e^{-\frac{\xi t}{\tau}}\left[e^{\frac{\sqrt{\xi^2 - 1}\,t}{\tau}}\left(1 + \frac{\xi}{\sqrt{\xi^2 - 1}}\right) + e^{\frac{-\sqrt{\xi^2 - 1}\,t}{\tau}}\left(1 - \frac{\xi}{\sqrt{\xi^2 - 1}}\right)\right]\right]$$ (4-62)

The response of this type of system is also shown in Fig. 4-11. The response never overshoots the final value and approaches it more slowly than the critically damped systems. This type of system is said to be *overdamped*.

All three types of second-order systems are very important in the study of automatic process control. The responses of control systems look similar to one of the above responses. The open-loop response of most industrial processes is similar to a critically damped or overdamped response. That is, they usually do not oscillate. Oscillation may occur once the loop is closed, however. The response of closed-loop systems is shown in Chapter 6.

It is important to recognize the differences between the response of second-order systems and that of first-order systems when subject to step changes in forcing function. The most important difference is that for second-order systems the steepest slope does not occur at the beginning of the response but, rather, at a later time. For first-order systems, as shown in the previous chapter, the steepest slope occurs at the beginning of the response. Another difference is that first-order systems will not oscillate while second-order systems may.

As seen in Fig. 4-11, the amount of damping of a second-order system is given by the damping ratio, ξ. This ratio, as well as the characteristic time constant, τ, depends on the physical parameters of the process. If any physical parameter changes, this will be reflected in a change in either ξ or τ or both.

The analysis of the response of the underdamped system is of particular interest in our study of automatic process control. This stems from the fact that, as mentioned above, the response of most closed loops is similar to the underdamped response. Because of this similarity, there are some important terms related to the underdamped response that must be defined. Referring to Fig. 4-12, these terms are defined below.

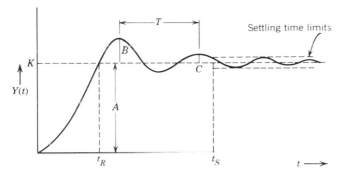

Figure 4-12. Response of an underdamped second-order process.

Overshoot. The "overshoot" is the amount by which the response overshoots the final steady-state value. It is generally expressed as the ratio B/A and given by

$$\frac{B}{A} = e^{-\pi\xi/\sqrt{1-\xi^2}} \tag{4-63}$$

Decay Ratio. The decay ratio is defined as

$$\frac{C}{B} = e^{-2\pi\xi/\sqrt{1-\xi^2}} \tag{4-64}$$

This is an important term because it serves as a criterion for establishing satisfactory response of control systems.

Rise Time, t_R. This is the time it takes the response to first reach the final value.

Settling Time, t_S. This is the time it takes the response to come within some prescribed limits of the final value and remain within these limits. These limits are arbitrary. Typical values are $\pm 5\%$ or $\pm 3\%$.

Period of Oscillation, T. The period of oscillation is given by

$$T = \frac{2\pi\tau}{\sqrt{1-\xi^2}}, \qquad \text{time/cycle} \tag{4-65}$$

Another term related to the period of oscillation is the cyclical frequency, f, and is defined as follows:

$$f = \frac{1}{T} = \frac{\sqrt{1 - \xi^2}}{2\pi\tau}, \qquad \text{cycle/time} \tag{4-66}$$

Still another two terms are the natural period of oscillation and the natural cyclical frequency. These are defined when $\xi = 0$ as

$$T_n = 2\pi\tau \tag{4-67}$$

and

$$f_n = \frac{1}{2\pi\tau} \tag{4-68}$$

Often the following expression for a second-order transfer function is also used:

$$G(s) = \frac{Y(s)}{X(s)} = \frac{K}{\dfrac{s^2}{\omega_n^2} + \dfrac{2\xi}{\omega_n}s + 1} \tag{4-69}$$

The term ω_n is referred to as the natural frequency. Comparing Eq. (4-69) and (4-54) it is easily seen that

$$\omega_n = \frac{1}{\tau} \tag{4-70}$$

The radian frequency, ω, and the cyclical frequency, f, are related by

$$\omega = 2\pi f \tag{4-71}$$

and by substituting Eq. (4-66) into Eq. (4-71) the radian frequency and the natural frequency can be related:

$$\omega = 2\pi f = \frac{2\pi\sqrt{1 - \xi^2}}{2\pi\tau} = \frac{\sqrt{1 - \xi^2}}{\tau} = \omega_n \sqrt{1 - \xi^2} \tag{4-72}$$

All of the above presentation has been for second-order systems. For third-order or higher-order systems with real and distinct time constants, the response to a step change of unit magnitude is given by Eq. (4-73) and shown in Fig. 4-13.

$$Y(t) = K\left[1 - \sum_{i=1}^{n} \frac{\tau_i^{n-1} e^{-t/\tau_i}}{\prod_{\substack{j=1 \\ j \neq i}}^{n} (\tau_i - \tau_j)} \right] \tag{4-73}$$

The general method for solution of other types of time constants is presented in Chapter 2.

Probably the most important characteristic of the responses shown in Fig. 4-13 is that they look very *similar* to the response of a second-order overdamped system with some amount of dead time. As the order of the system increases, the apparent dead time also increases. This is important in the study of automatic process control because most

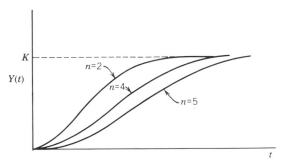

Figure 4-13. Response of overdamped higher-order systems to a unit step change in forcing function.

industrial processes are composed of a number of first-order systems in series. Furthermore, because of the similarity, the response of a third-order or any other higher-order system can be approximated by the response of a second-order system plus dead time. This is shown mathematically as follows:

$$\frac{Y(s)}{X(s)} = \frac{K}{\prod_{i=1}^{n} (\tau_i s + 1)} \approx \frac{K e^{-t_o s}}{(\tau_a s + 1)(\tau_b s + 1)} \tag{4-74}$$

Mathematically this approximation is very good.

Up to now the response of processes described by Eq. (4-51) has been shown. The response of processes described by Eq. (4-52) to a step change of unit magnitude in forcing function, $X(s) = 1/s$, is now shown. In general for systems with real and distinct roots and response is given by the following equation:

$$Y(t) = K\left[1 - \sum_{i=1}^{n} \frac{\prod_{j=1}^{m} (\tau_{lg_i} - \tau_{ld_j})\tau_{lg_i}^{n-m-1}}{\prod_{\substack{j=1 \\ j \neq i}}^{n} (\tau_{lg_i} - \tau_{ld_j})} e^{-t/\tau_{lg_i}}\right] \tag{4-75}$$

To obtain a better understanding of the term $(\tau_{ld} s + 1)$, compare the response of the following two processes:

$$Y_1(s) = \frac{1}{s(\tau_{lg_1} s + 1)(\tau_{lg_2} s + 1)(\tau_{lg_3} s + 1)} \tag{4-76}$$

and

$$Y_2(s) = \frac{(\tau_{ld} s + 1)}{s(\tau_{lg_1} s + 1)(\tau_{lg_2} s + 1)(\tau_{lg_3} s + 1)} \tag{4-77}$$

Figure 4-14 compares the responses. The responses for $\tau_{ld} = 0$ corresponds to Eq. (4-76). The effect of the term $(\tau_{ld} s + 1)$ is to "speed up" the response of the process. This is opposite to the effect of the term $1/(\tau_{lg} s + 1)$. In Chapter 3 the term $1/(\tau_{lg} s + 1)$ was referred to as a *first-order lag*. Consequently, the term $(\tau_{ld} s + 1)$ is referred to as a *first-*

order lead. This is why the notation τ_{lg}, indicating a "lag" time constant and τ_{ld}, indicating a "lead" time constant is used. Notice that when τ_{ld} becomes equal to τ_{lg}, the transfer function describing the relationship between $Y(s)$ and $X(s)$ becomes of one order less than it is without a lead term.

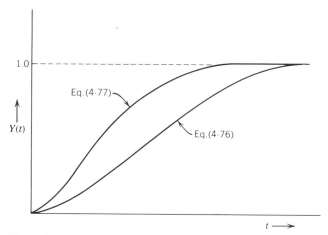

Figure 4-14. Comparison of response of Eqs. (4-76) and (4-77).

An interesting, and important, case is that of the following transfer function:

$$\frac{Y(s)}{X(s)} = \frac{\tau_{ld}s + 1}{\tau_{lg}s + 1} \tag{4-78}$$

This transfer function is called a lead/lag. The expression that describes the response of $Y(t)$ to a unit step-change of magnitude in forcing function, from Eq. (4-75), is

$$Y(t) = 1 + \frac{\tau_{ld} - \tau_{lg}}{\tau_{lg}} e^{-t/\tau_{lg}} \tag{4-79}$$

Figure 4-15 shows the response graphically for the case of $\tau_{lg} = 1$ and different ratios of τ_{ld}/τ_{lg}. It is important to recognize three points about the response of a lead/lag. First and most important, the initial amount of response is dependent on the ratio of τ_{ld}/τ_{lg}. The initial response will be equal to the product of τ_{ld}/τ_{lg} times the magnitude of the step change. Second, the final amount of change in output from the lead/lag is equal to the magnitude of the step change in input. Third, the exponential rate of decay, or increase, in output is only a function of the "lag" time constant, τ_{lg}. Lead/lag systems will become important in the implementation of feedforward control techniques as shown in Chapter 8.

Sinusoidal Function

The Laplace expression for a sine wave forcing function is

$$X(s) = \frac{A\omega}{s^2 + \omega^2}$$

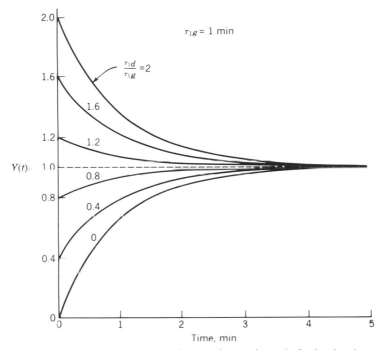

Figure 4-15. Response of lead/lag unit to a unit step change in forcing function.

Thus, the response of a second-order system to this type of forcing function is given by

$$Y(s) = \frac{KA\omega}{(s^2 + \omega^2)(\tau^2 s^2 + 2\tau\xi s + 1)}$$

Inverting back to the time domain yields

$$Y(t) = e^{-\xi t/\tau}\left(C_1 \cos\left(\sqrt{1 - \xi^2}\frac{t}{\tau}\right) + C_2 \sin\left(\sqrt{1 - \xi^2}\frac{t}{\tau}\right)\right)$$

$$+ \frac{KA}{\sqrt{[1 - (\omega\tau)^2]^2 + (2\xi\omega\tau)^2}} \sin(\omega t + \theta)$$

(4-80)

where

$$\theta = -\tan^{-1}\left(\frac{2\xi\omega\tau}{1 - (\omega\tau)^2}\right)$$

(4-81)

C_1, C_2 = constants to be evaluated

As time increases, the term $e^{-\xi t/\tau}$ becomes negligible and the response reaches a steady oscillation, given by

$$Y(t)\big|_{t\to\infty} = \frac{KA}{\sqrt{[1 - (\omega\tau)^2]^2 + (2\xi\omega\tau)^2}} \sin(\omega t + \theta)$$

(4-82)

This response will become important in the study of automatic process control as shown in Chapter 7. This steady oscillation is called the frequency response of the system. The amplitude of this response is equal to the amplitude of the forcing function times the gain of the system and attenuated by the factor

$$\frac{1}{\sqrt{[1 - (\omega\tau)^2]^2 + (2\xi\omega\tau)^2}}$$

As defined in Section 3-7, the amplitude ratio for this system is

$$\frac{K}{\sqrt{[1 - (\omega\tau)^2] + (2\xi\omega\tau)^2}}$$

This amplitude ratio and the phase lag, θ, are both functions of ω, the frequency of the forcing function. Chapter 7 also examines the frequency response of other higher-order systems.

4-5. SUMMARY

In this chapter the development of mathematical models that describe the behavior of higher-order processes have been presented. The processes considered are more complex than the ones considered in Chapter 3; consequently, the models are also more complex. Particularly, second- and third-order transfer functions were developed. The examples used presented cases of interacting and noninteracting processes. The responses of these higher-order systems to a step change and sinusoidal forcing functions were also shown. The differences between these responses and those of first-order systems were also presented and stressed. Probably the most important difference occurs in the response to a step change in forcing function. For first-order systems the initial slope of the response is the steepest one of the complete response curve. For higher-order systems this is not the case; the steepest slope occurs later on the response curve.

The higher-order systems presented in this chapter resulted because they are composed of first-order systems in series. That is, these systems are not intrinsically higher order. Most industrial processes are like the ones shown here. If there are any intrinsically higher-order processes, they are few and far between. The typical examples of intrinsic second-order systems, shown in control textbooks, are measuring elements such as Bourdon-tube pressure gauges and mercury manometers. These measuring elements are part of the total control loop, however, and more often than not their dynamics are not significant compared to the process itself. This is why they have not been considered here. The reader is referred to references 1 and 2 for treatises on this subject.

Since most industrial processes are composed of first-order processes in series, their open-loop dynamic response is overdamped. Once the control loop is closed by installing a controller in the feedback, their response may become underdamped; this subject is studied in Chapter 6.

REFERENCES

1. Close, C. M., and D. K. Frederich, *Modeling and Analysis of Dynamic Systems*, Houghton Mifflin, Boston, 1978.

2. Tyner, M., and F. P. May, *Process Engineering Control*, Ronald Press, New York, 1968.

PROBLEMS

4-1. Consider the process shown in Fig. 4-16. The flow rate of liquid through the tanks, w, is constant at a value of 250 lbm/min. The density of the liquid may be assumed constant at 50 lbm/ft³ at the heat capacity may also be assumed constant at 1.3 Btu/lbm-°F. The volume of each tank is 10 ft³. You may neglect heat losses to the surroundings.

Draw the complete block diagram that shows how changes in inlet temperature, $T_i(t)$, and $q(t)$ affect $T_3(t)$. Give the numerical values and units of each parameter in all transfer functions.

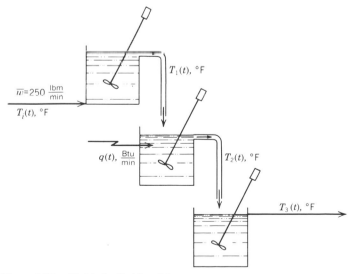

Figure 4-16. Sketch for Problem 4-1.

4-2. Consider the process shown in Fig. 4-17. The following is known about the process:

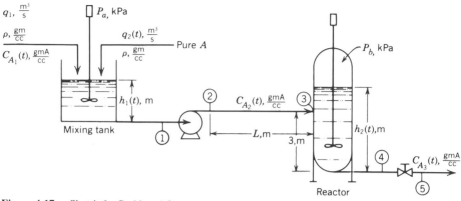

Figure 4-17. Sketch for Problem 4-2.

   a. The density of all streams is approximately equal.
   b. The flow through the constant speed pump is given by

$$q(t) = A(1 + B(p_1(t) - p_2(t))^2), \quad m^3/s$$

   where A and B are constants.
   c. The pipe between points 2 and 3 is rather long with a length of L,m. The flow through this pipe is highly turbulent (plug flow). The diameter of the pipe is D,m. The pressure drop between these two points is fairly constant, Δp, kPa.
   d. We may assume that the energy effects associated with the reaction ($A \rightarrow B$) are negligible and, consequently, the reaction occurs at constant temperature. The rate of reaction is given by

$$r_A(t) = kC_A(t), \quad \frac{kgm}{m^3 - s}$$

   e. The flow through the outlet valve is given by

$$q(t) = C_V vp(t) \sqrt{h_2(t)}$$

Obtain the block diagram that shows the effect of the forcing functions $q_2(t)$, $vp(t)$, and $C_{A_1}(t)$ on the responding variables $h_1(t)$, $h_2(t)$ and $C_{A_3}(t)$.

4-3. Consider the process shown in Fig. 4-18. In this process different streams are mixed. Streams 5, 2, and 7 are solutions of water and component A; stream 1 is pure water. The steady-state values for each stream is given in Table 8-4. Determine the following transfer functions, with numerical values:

$$\frac{X_6(s)}{X_5(s)}, \quad \frac{X_6(s)}{X_2(s)} \quad \text{and} \quad \frac{X_6(s)}{Q_1(s)}$$

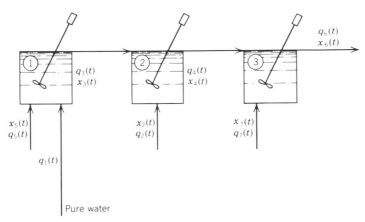

Figure 4-18. Sketch for Problem 4-3.

4-4. Consider the process shown in Fig. 4-19. A gas stream, $q_2(t)$, enters a tank where it is mixed with another stream, $q_1(t)$, which is pure A. From the tank the gas mixture flows into a separator where component A in the gas diffuses out, through a semipermeable membrane, to a pure liquid. The following may be assumed:

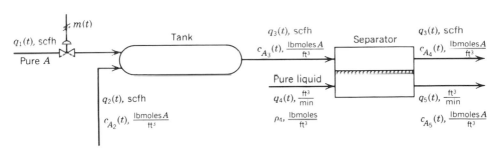

Figure 4-19   Sketch for Problem 4-4.

a. The pressure drop across the valve is constant. The flow of pure A through this valve is given by

$$q_1(t) = k_v vp(t)$$

where $q_1(t)$ is in scfh. The valve position, $vp(t)$ is related to the pneumatic signal, $m(t)$, by the following:

$$vp(t) = \frac{1}{12}(m(t) - 3)$$

b. The output volumetric flow, $q_3(t)$, from the tank is equal to the sum of the input flows. The gas behaves as an incompressible fluid.

c. The gas inside the tank is well mixed.

d. The gas side in the separator is assumed to be well mixed. The liquid side is also assumed to be well mixed.

e. The rate of mass transfer across the semipermeable membrane is given by

$$N_A(t) = A_X K_A [c_{A_4}(t) - c_{A_5}(t)]$$

where

$N_A(t)$ = rate of mass transfer, lb moles A/hr
A_X = cross-sectional area of membrane, ft^2
K_A = overall mass transfer coefficients, ft/hr

f. The amount of component A diffused to the liquid *does not* significantly affect the gas volumetric flow. Therefore, the gas flow out from the separator can be considered equal to the input flow.

g. The amount of component A diffused to the liquid *does* affect the liquid volumetric flow. The density of the liquid stream leaving the separator is given by

$$\rho_5(t) = \rho_4 + k_5 c_{A_5}(t), \quad \text{lb moles/ft}^3$$

Please do the following:

1. Write the mathematical model for the tank.
2. Write the mathematical model for the separator.
3. Draw the block diagram showing how the output variables, $c_{A_4}(t)$ and $c_{A_5}(t)$, are affected by $m(t)$, $q_2(t)$, and $c_{A_2}(t)$. Also obtain the transfer functions.

4-5. Consider the extraction unit shown in Fig. 4-20. The purpose of this unit is to remove component A from a component B–rich phase. The transfer of A to the water medium occurs across a semipermeable membrane. In this process the concentration of A is a function of position along the unit and of time. Thus, the equation that describes this concentration is a partial differential equation (PDE) in length and time. Systems described by PDE's are referred to as "distributed systems"; Chapter 9 presents more of this type of systems. A common way to "get around" this PDE is to divide the unit into sections, or "pools," and assume each "pool" to be well mixed. The dotted lines show the divisions of "pools." Using this method, we find that the differential in length, dL, is approximated by ΔL. The smaller the "pools" the better the approximation.

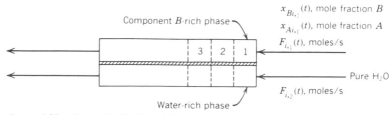

Figure 4-20. Sketch for Problem 4-5.

The mass transfer rate of component A is

$$N_A(t) = Sk_A[x_{An,1}(t) - x^*_{An,2}(t)]$$

where

$N_A(t)$ = moles of A transferred/second

S = surface area, of membrane, across which the transfer takes place, m^2.

k_A = mass transfer coefficient, constant, moles A/m^2-s

$x_{An,1}(t)$ = mole fraction of A in liquid phase 1 (component B–rich phase). The subscript n refers to the ''pool'' number.

$x^*_{An,2}(t)$ = mole fraction of A in liquid phase 2 (water-rich phase) that would be in equilibrium with $x_{An,1}(t)$

Henry's Law helps in relating $x^*_{An,2}(t)$ to the actual concentration of A in liquid phase 2.

$$x^*_{An,2}(t) = H_A x_{An,2}(t)$$

where

H = Henry's law constant

$x_{An,2}(t)$ = mole fraction of A in liquid phase 2

Component B and water are not transferred across the membrane. This process occurs isothermally. Obtain the block diagram that shows how the forcing functions $x_{Ai,1}(t)$, $F_{i,1}(t)$, and $F_{i,2}(t)$ affect the output variables $x_{A2,1}(t)$ and $x_{A2,2}(t)$. That is, do the first two ''pools.''

CHAPTER
5
Basic Components of Control Systems

In Chapter 1 we learned that the four basic components of control systems are the sensors, transmitters, controllers, and final control elements. We also learned that these components perform the three basic operations of every control system: measurement (M), decision (D), and action (A).

This chapter takes a brief look at sensors and transmitters, followed by a more detailed study of control valves and process controllers. Appendix C offers a more in-depth presentation of different types of transmitters, sensors, and control valves. That appendix presents numerous diagrams, schematics, and other figures to aid in the presentation and to expose the reader to as much instrumentation as possible.

5-1. SENSORS AND TRANSMITTERS

Sensors and transmitters perform the measurement (M) operation of the control system. The sensor produces a phenomenon, mechanical, electrical, or the like, related to the process variable that it measures. The transmitter in turn converts this phenomenon into a signal that can be transmitted. The signal, therefore, is related to the process variable.

There are three important terms related to the sensor/transmitter combination. The *range* of the instrument is given by the low and high values of the process variable that is measured. That is, consider a pressure sensor/transmitter that has been calibrated to measure a process pressure between the values of 20 psig and 50 psig. We say that the range of this sensor/transmitter combination is 20–50 psig. The *span* of the instrument is the difference between the high and low values of the range. For the mentioned pressure instrument the span is 30 psi. To summarize, we must specify a low and a high value to define the range of an instrument. That is, two numbers must be given. The span of the instrument is the difference between the two values. Finally, the low value of the range is referred to as the *zero* of the instrument. This value does not have to be zero in order to be called the zero of the instrument. For the above example, the "zero" of the instrument is 20 psig.

Appendix C presents some of the most common industrial sensors: pressure, flow, temperature, and level. That appendix also briefly discusses the working principles of both an electric and a pneumatic transmitter.

Sometimes it is important for system analysis, as will be shown in Chapter 6, to obtain the parameters that describe the sensor/transmitter behavior. The gain term is fairly simple to obtain once the span is known. Consider an electronic pressure sensor/transmitter with a range of 0–200 psig. In Chapter 3 the gain was defined as the change in output, or responding variable, divided by the change in input, or forcing function. In this case the output is the electronic signal (4–20 mA) and the input is the process pressure (0–200 psig). Thus

$$K_T = \frac{20 \text{ mA} - 4 \text{ mA}}{200 \text{ psig} - 0 \text{ psig}} = \frac{16 \text{ mA}}{200 \text{ psi}} = 0.08 \frac{\text{mA}}{\text{psi}}$$

As another example consider a pneumatic temperature sensor/transmitter with a range of 100°F–300°F. The gain is

$$K_T = \frac{15 \text{ psig} - 3 \text{ psig}}{300°F - 100°F} = \frac{12 \text{ psi}}{200°F} = 0.06 \frac{\text{psi}}{°F}$$

That is, we can say that the gain of a sensor/transmitter is the ratio of the span of the output to the span of the input.

The two cases presented show that the gain of the sensor/transmitter is constant over its complete operating range. For most sensors/transmitters this is the case; however, there are some instances, such as a differential pressure sensor used to measure flow, when this is not the case. A differential pressure sensor measures the differential pressure, h, across an orifice. This differential pressure is related to the square of the volumetric flow rate, F. That is

$$F^2 \alpha h$$

The equation that describes the output signal from an electronic differential pressure transmitter when used to measure volumetric flow with a range of $0-F_{\max}$ gpm is

$$M_F = 4 + \frac{16}{(F_{\max})^2} F^2$$

where

M_F = output signal in mA
F = volumetric flow

From this equation the gain of the transmitter is obtained as follows:

$$K_T = \frac{\overline{dM_F}}{dF} = \frac{2(16)}{(F_{\max})^2} \overline{F}$$

with a nominal gain

$$K_T' = \frac{16}{F_{\max}}$$

This expression shows that the gain is not constant but rather a function of flow. The greater the flow the greater the gain. Specifically,

$$\text{At} \left(\frac{\overline{F}}{F_{max}} \right) \quad 0 \quad 0.1 \quad 0.5 \quad 0.75 \quad 1.0$$

$$\left(\frac{K_T}{K_T'} \right) \quad 0 \quad 0.2 \quad 1.0 \quad 1.50 \quad 2.0$$

So the actual gain varies from zero to twice the nominal gain.

This fact results in a nonlinearity in flow control systems. Nowadays most manufacturers offer differential pressure transmitters with built-in square root extractors yielding a linear transmitter. Chapter 8 discusses in more detail the use of square root extractors.

The dynamic response of most sensor/transmitters is much faster than the process. Consequently, their time constants and dead time can often be considered negligible and thus, their transfer function is given by a pure gain. However, when the dynamics must be considered, it is usual practice to represent the transfer function of the instrument by a first-order or second-order system:

$$G(s) = \frac{K_T}{\tau s + 1} \quad \text{or} \quad G(s) = \frac{K_T}{\tau^2 s + 2\tau \xi s + 1}$$

The dynamic parameters are usually obtained empirically using methods similar to the ones presented in Chapters 6 and 7.

5-2. CONTROL VALVES

Control valves are the most common final control elements. They are found in process plants manipulating flows to maintain controlled variables at their set points. In this section an introduction to the most important aspects of control valves as applied to process control is presented.

A control valve acts as a *variable restriction* in a process pipe. By changing its opening it changes the resistance to flow and, thus, the flow itself. Throttling flow is what control valves are all about.

This section presents the subject of control valve action (fail condition), control valve sizing, and their characteristics. Appendix C presents different types of control valves and control valve accessories. The reader is strongly encouraged to read Appendix C along with this section.

Control Valve Action

The first question the engineer must answer when choosing a control valve is: What do I want the valve to do when the energy supply to it fails? The question is concerned with the "fail position" of the valve. The main consideration in answering this question is, or should be, *safety*. If the engineer decides that for safety considerations the valve should close, he must then specify a "fail-closed" (FC) valve. The other possibility is that of a "fail-open" (FO) valve. When the energy supply fails, this valve will move to open its restriction to flow. The great majority of control valves are pneumatically operated and, consequently, the energy supply is air pressure. A fail-closed valve requires energy to open it; therefore, they are also referred to as "air-to-open" (AO) valves. The fail-open valves that require energy to close are also referred to as "air-to-close" (AC) valves.

Let us look at an example to illustrate the choosing of the action of control valves. The example is the process shown in Fig. 5-1. In this process the outlet temperature of a process fluid is controlled by manipulating the steam flow to the heat exchanger. The question is: What do we want the steam valve to do when the air supply to it fails?

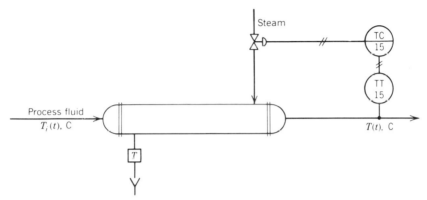

Figure 5-1. Heat exchanger control loop.

As explained above, we want the steam valve to move to the safest position. It seems that the safest condition may be the one that stops the steam flow; that is, we do not want to have steam flowing around in an unsafe operation. This means that a fail-closed valve should be specified. In making this decision we have not taken into consideration the effect of not heating the process fluid by closing the valve. In some cases this may not present any problems; however, in other cases it may have to be considered. As an example, consider the case when the steam is maintaining the temperature of a certain polymer. If the steam valve closes, the temperature will drop and the polymer may solidify in the exchanger. In this case it might be decided that a fail-open valve provides the safest condition.

It is important to note that in this example only the safe condition around the heat exchanger has been taken into consideration. This may not necessarily be the safest overall operation; that is, the engineer needs to look at the complete plant rather than only one piece of equipment. He must look at the effect on the heat exchanger and also on any other equipment from which the steam and process fluid are coming or going. To repeat, he must take into consideration the *complete plant safety*.

Control Valve Sizing

Control valve sizing is the procedure of calculating the valve flow coefficient, C_V. The "C_V method," a well accepted method by all valve manufacturers, was first introduced in 1944 by Masoneilan International, Inc.[2] Once the required C_V is calculated and the type of valve to be used is known, the engineer can obtain, from the manufacturer's catalog, the valve size.

The C_V coefficient is defined as "the number of U.S. gallons per minute of water that flows through a wide open valve with a pressure drop of 1 psi across the valve." For example, a valve with a maximum C_V coefficient of 25 will be able to pass 25 gpm of water when it is wide open with a 1 psi pressure drop.

Even though the C_V method for valve sizing is used by all manufacturers, the equations to calculate C_V differ somewhat from manufacturer to manufacturer. The best policy to follow once a manufacturer is chosen is to use his recommended equations. In this section we present the equations of two manufacturers, Masoneilan[2] and Fisher Controls,[3] to show the differences in their equations and methods. The most pronounced differences occur in the equations to size valves for compressible fluids (gas, vapor, or steam) service. These two manufacturers are by no means the only two valve manufacturers. Table 5-1 gives the names and addresses of some other manufacturers. The table is not inclusive, but it gives the reader names of a number of companies involved in the manufacture of control valves. Masoneilan and Fisher Controls were chosen because their equations and methods are typical of the industry.

Table 5-1   Control Valve Manufacturers

Jamesbury Corporation	640 Lincoln Street Worcester, MA 01605
Jenkins Brothers	101 Merritt Seven Norwalk, CO 06851
Jordan Valve	407 Blade Street Cincinnati, OH 45216
Crane Company	300 Park Avenue New York, NY 10022
DeZurik	250 Riverside Avenue, North Sartell, MN 56377
Fisher Controls Company	P.O. Box 190 Marshalltown, IA 50158
Masoneilan International	63 Nahatan Street Norwood, MA 02062
Honeywell	1100 Virginia Drive Fort Washington, PA 19034
Copes-Vulcan, Inc.	Martin and Rice Avenues Lake City, PA 14623
Valtek	P.O. Box 2200 Sprinville, UT 84663
The Duriron Company, Inc.	1978 Foreman Drive Cookeville, TN 38501
Cashco, Inc.	P.O. Box A Ellsworth, KS 67439
The Foxboro Company	Foxboro, MA 02035

Liquid Service. The basic equation to size a control valve for liquid service is the same for all manufacturers:

$$q = C_V \sqrt{\frac{\Delta P}{G_f}} \qquad (5\text{-}1)$$

or, solving for C_V

$$C_V = q \sqrt{\frac{G_f}{\Delta P}} \qquad (5\text{-}2)$$

where

q = liquid flow in U.S. gpm
ΔP = pressure drop, P_1-P_2, across the valve in psi
P_1 = upstream pressure in psi
P_2 = downstream pressure in psi
G_f = specific gravity of liquid at flowing temperature; water = 1 at 60°F.

Sometimes the units of flow are given in lbm/hr, and for these cases we can write Eqs. (5-1) and (5-2) as follows:

$$W = 500 C_V \sqrt{G_f \Delta P} \qquad (5\text{-}3)$$

and

$$C_V = \frac{W}{500 \sqrt{G_f \Delta P}} \qquad (5\text{-}4)$$

where W = liquid flow in lbm/hr. Eq. (5-3) develops directly from Eq. (5-1).

There are several other considerations, such as viscosity corrections, flashing, and cavitation, in choosing control valves for liquid service. These considerations are presented in Appendix C.

Gas, Vapor, and Steam Service. The most pronounced differences between manufacturers are encountered in their sizing equations for compressible fluids. These differences develop from the way they express, or consider, the phenomenon of *critical flow* in the equations. Critical flow is the condition that exists when the flow is not a function of the square root of the pressure drop across the valve but of only the upstream pressure. This phenomenon occurs after the fluid reaches sonic velocity at the vena contracta. Once the fluid finds itself in the critical flow condition, decreases or increases in the downstream pressure do not affect the flow. Only changes in upstream pressure affect the flow.

Let us now look at the methods used by our two manufacturers to define the condition of critical flow and to size control valves for compressible fluids.

Masoneilan[2] proposes the following set of equations:

Gas Volumetric Flow

$$C_V = \frac{Q \sqrt{GT}}{836 C_f P_1 (y - 0.148 y^3)} \qquad (5\text{-}5)$$

Gas Flow by Weight

$$C_V = \frac{W}{2.8C_f P_1 \sqrt{G_f}\,(y - 0.148y^3)} \qquad (5\text{-}6)$$

Steam

$$C_V = \frac{W(1 + 0.0007T_{SH})}{1.83C_f P_1(y - 0.148y^3)} \qquad (5\text{-}7)$$

where

Q = gas flow rate in scfh. Standard conditions are at 14.7 psia and 60°F.

G = gas specific gravity at 14.7 psia and 60°F (air = 1.0). For ideal gases this is the ratio of the molecular weight of the gas to the molecular weight of air (29).

G_f = gas specific gravity at flowing temperature, $G_f = G\!\left(\dfrac{520}{T}\right)$

T = flowing temperature in °R

C_f = critical flow factor. The numerical value for this factor ranges between 0.6 to 0.95. Fig. C-44 shows this factor for different type valves.

P_1 = upstream pressure in psia

P_2 = downstream pressure in psia

$\Delta P = P_1 - P_2$

W = flow rate in lb/hr

T_{SH} = degrees of superheat in °F

The term y is used to express the critical or subcritical flow condition and is defined as

$$y = \frac{1.63}{C_f} \sqrt{\frac{\Delta P}{P_1}} \qquad (5\text{-}8)$$

with a maximum value of $y = 1.5$. At this value, $y - 0.148y^3 = 1.0$. Therefore, when y reaches a value of 1.5, this indicates a critical flow condition. It is easy to see from the equations that when the term $y - 0.148y^3 = 1.0$, the flow is only a function of the upstream pressure, P_1.

It is important to realize that when the flow is well under critical

$$y - 0.148y^3 \approx y$$

In this case the factor C_f cancels out (there is no need for C_f) and Eq. (5-5) is easily derived from Eq. (5-2). The point is that all of these sizing formulas are derived from the original definition of C_V, Eq. (5-2). The only thing that is special about the gas formulas is the correction factor C_f and the compressibility function $(y - 0.148y^3)$ required to describe the critical flow phenomenom. Similarly, Eq. (5-6) is easily derived from Eq. (5-5).

Fisher Controls defines two new coefficients for the sizing of valves for compressible fluids. The coefficient C_g is related to the flow capacity of the valve and the coefficient C_1, defined as C_g/C_V, provides an indication of the valve's recovery capabilities. This last coefficient, C_1, is heavily dependent on the type of valve, its values usually ranging between 33 and 38. Fisher's equation to size valves for compressible fluids, called the Universal Gas Sizing Equation, is given in two forms:

$$C_g = \frac{Q_{\text{scfh}}}{\sqrt{\dfrac{520}{GT}}\, P_1 \sin\left[\left(\dfrac{59.64}{C_1}\right)\sqrt{\dfrac{\Delta P}{P_1}}\,\right]_{\text{rad}}} \tag{5-9}$$

or

$$C_g = \frac{Q_{\text{scfh}}}{\sqrt{\dfrac{520}{GT}}\, P_1 \sin\left[\left(\dfrac{3417}{C_1}\right)\sqrt{\dfrac{\Delta P}{P_1}}\,\right]_{\text{deg}}} \tag{5-10}$$

The critical flow condition is indicated by the sine term. The argument must be limited to $\pi/2$, in Eq. (5-9), or 90°, in Eq. (5-10). These limiting values indicate critical flow. Values of C_g and C_1 are shown in Fig. C-39c and C-39d.

Equations (5-9) and (5-10) can be derived from Eq. (5-2) for the subcritical flow condition. Well under critical flow the following approximation holds true:

$$\sin\left[\frac{59.64}{C_1}\sqrt{\frac{\Delta P}{P_1}}\,\right]_{\text{rad}} \approx \frac{59.64}{C_1}\sqrt{\frac{\Delta P}{P_1}}$$

The sine term is used to describe the critical flow phenomenon.

It is interesting to note the similarity between the two manufacturers. Both of them use two coefficients to size control valves for compressible fluids. One of the coefficients is related to the valve's flow capacity, C_V for Masoneilan and C_g for Fisher Controls; the other coefficient, C_f for Masoneilan and C_1 for Fisher Controls, is dependent on the type of valve. Masoneilan uses the term $(y - 0.148y^3)$ to indicate critical flow and Fisher Controls uses the sine term. Both terms are empirical and there is no significance to the fact that they are different.

Before concluding this section on control valve sizing it is necessary to mention a few other important points. The sizing of a valve by the calculation of C_V should be done so that when the valve is wide open it can pass more flow than the required flow at normal operating conditions. That is, the valve should be somewhat overdesigned in case more flow is required. Individuals and/or companies have different policies about the overdesign capacity of a valve. In any case, if you decide to overdesign a valve by a factor of 2 times the required flow, then the design flow is given by

$$q_{\text{design}} = 2.0 q_{\text{required}}$$

If a valve is about 3% open when controlling a variable under normal operating conditions, that particular valve is overdesigned. Similarly, if the valve is about 97% open, it is underdesigned. In either case the valve is either almost closed or almost fully opened, less flow or more flow will be difficult to obtain if required.

Related to the capacity of the valve is the term *rangeability*. The rangeability, R, of a valve is defined as the ratio of the maximum controllable flow to the minimum controllable flow:

$$R = \frac{q_{\text{maximum controllable}}}{q_{\text{minimum controllable}}} \qquad (5\text{-}11)$$

The definition of maximum, or minimum, controllable flow is very subjective. Some people like to define a controllable flow between 10% and 90% of valve opening while others define controllable flow between 5% and 95%. There is no set rule, or standard, for this definition. The rangeability of most control valves is limited and usually varies between 20 and 50. It is desirable to have a large rangeability (of the order of 10 or greater) so that the valve can have a significant effect on the flow.

The last two paragraphs have presented the subjects of overdesign and rangeability of control valves. Both characteristics have definite effects on the performance of the control valve while in service, as will be discussed, when "installed valve characteristics" are presented.

Selection of Design Pressure Drop

It is important to recognize that control valves can manipulate flow rates only by producing, or absorbing, a pressure drop in the system. This pressure drop is an economic loss to the process operation since the pressure must usually be supplied by a pump or compressor. Thus, economics dictate the sizing of control valves with low pressure drops. However, this low pressure drop results in larger valve sizes and, consequently, higher initial cost, and in a decrease in range of control. These opposing considerations require some compromise by the engineer in choosing the design pressure drop. Several rules of thumb are in common use to help in the decision. In general, these rules specify the pressure drop to be taken across the valve to be from 20% to 50% of the total dynamic pressure drop of the entire piping system. Another common rule is to specify the design valve pressure drop to be 25% of the total dynamic pressure drop of the entire piping system or 10 psi, whichever is greater. The actual value is dependent on the given situation and on company policy. As expected, the design pressure drop also has an effect on the performance of the valve, as will be shown in the next section.

Example 5-1.   Size a control valve for a gas service with a nominal flow of 25,000 lbm/hr, with an inlet presssure of 250 psia, and a design pressure drop of 100 psi. The gas has a specific gravity of 0.4 at a flowing temperature of 150°F and a molecular weight of 12. The valve to be used is a plug valve.

To use the Masoneilan Eq. (5-6) the C_f factor must be obtained. For the plug valve, from Fig. C-44, $C_f = 0.92$. Then using Eq. (5-8) we get

$$y = \frac{1.63}{C_f} \sqrt{\frac{\Delta P}{P_1}} = \frac{1.63}{0.92} \sqrt{\frac{100}{250}} = 1.12$$

The design flow is

$$W_{\text{design}} = 2W_{\text{nominal}} = 50,000 \ \frac{\text{lbm}}{\text{hr}}$$

and

$$C_V = \frac{W}{2.8 C_f P_1 \sqrt{G_f}(y - 0.148y^3)}$$

$$= \frac{50000}{2.8(0.92)(250)\sqrt{0.35}(1.12 - 0.148(1.12)^3)}$$

$$C_V = 143.3$$

To use the Fisher Controls valve equation, we must determine coefficient C_1 and calculate the flow rate in scfh. For the plug valve, from Fig. C-39d, $C_1 = 35$. The standard volumetric flow is

$$Q_{scfh} = \left(\frac{50000}{12}\right)(379.4) = 1580833.$$

$$\underbrace{\quad\quad}_{\substack{\text{lb moles} \\ \text{hr}}} \cdot \underbrace{\quad\quad}_{\substack{\text{scf} \\ \text{lb mole}}} @ \substack{14.7 \text{ psia} \\ 60°F}$$

Then from Eq. (5-10) we get

$$C_g = \frac{Q_{scfh}}{\sqrt{\dfrac{520}{GT}} P_1 \sin\left[\left(\dfrac{3417}{C_1}\right)\sqrt{\dfrac{\Delta P}{P_1}}\right]_{deg}}$$

$$= \frac{1580833}{\sqrt{\dfrac{520}{0.4(610)}}(250) \sin\left[\left(\dfrac{3417}{35}\right)\sqrt{\dfrac{100}{250}}\right]_{deg}}$$

$$C_g = 4917.5$$

This C_g coefficient can be converted to a C_V equivalent to compare it with the Masoneilan C_V coefficient. Based on the definition of C_1, we obtain

$$C_1 = \frac{C_g}{C_V}$$

then

$$C_V = \frac{C_g}{C_1} = \frac{4917.5}{35} = 140.5$$

Therefore, we can see that both methods give similar results: Masoneilan $C_v = 143.3$ and Fisher Controls $C_v = 140.5$.

Example 5-2. Consider the process shown in Fig. 5-2. In this process a fluid is transferred from a crude tank to a separation tower. The tank is at atmospheric pressure and the tower works at 4 in. Hg vacuum. The operating conditions are the following:

Flow	900 gpm
Temperature	90°F
Specific gravity	0.94
Vapor pressure	13.85 psia
Viscosity	0.29 cp

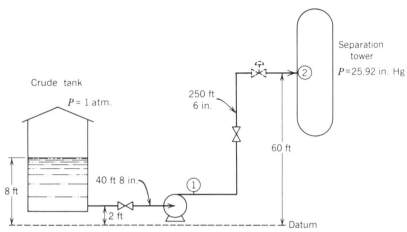

Figure 5-2. Piping system.

The pipe is commercial steel and the efficiency of the pump is 75%.

It is desired to size a valve, shown by dotted lines, between the pump and the separation tower.

To size this valve we must first determine the dynamic pressure drop between point 1, at the pump outlet, and point 2, at the tower inlet. This dynamic pressure drop is due to the frictional losses in the piping system. As shown in the diagram, the piping system consists of 250 ft of 6-in. pipe, two 90° elbows, one block valve, and a sudden expansion due to the entrance to the tower. The dynamic pressure drop in this piping system, ΔP_p, is calculated, using fluid flow principles, to be 6 psi. Knowing this dynamic pressure drop the valve pressure drop, ΔP_v, can then be selected.

If, for example, your company policy is to have a pressure drop across the valve equal to 25% of the total dynamic pressure drop then

$$\frac{\Delta P_v}{\Delta P_p + \Delta P_v} = 0.25 \quad \text{and} \quad \Delta P_v = 2 \text{ psi}$$

Sizing the valve for twice the nominal flow yields

$$C_V = \frac{1800}{\sqrt{\dfrac{2}{0.94}}} = 1234$$

If, on the other hand, the company policy calls for a design pressure drop of 25% of the total dynamic pressure drop or 10 psi, whichever is the greatest, then:

$$\Delta P_v = 10 \text{ psi}$$

Sizing the valve for twice the nominal flow yields

$$C_V = \frac{1800}{\sqrt{\dfrac{10}{0.94}}} = 552$$

Control Valve Flow Characteristics

To help achieve good control the control loop should have "constant personality." This means that the overall process, defined as the combination of sensor/transmitter/process unit/valve, should have its gain, time constants, and dead time as constant as possible. Another way to say that the overall process has a "constant personality" is to say that it is a *linear* system.

As already seen in Chapters 3 and 4, most processes are nonlinear in nature, making the sensor/transmitter/process unit also nonlinear. Since the controller is concerned with the "overall process," it may be that by choosing the correct "control valve personality" the nonlinear characteristics of the sensor/transmitter/process unit combination is offset. If this is done correctly, the sensor/transmitter/process unit/valve combination may end up with a constant gain. The control valve personality is commonly referred to as the "control valve flow characteristic." Therefore, we can say that the purpose of flow characterization is to provide for a relatively constant overall process gain over most of the operating process conditions.

The control valve flow characteristic is defined as the relationship between the flow through the valve and the valve position as the position is varied from 0% to 100%. We may differentiate between "inherent flow characteristic" and "installed flow characteristic." "Inherent flow characteristic" refers to the characteristic observed with constant pressure drop across the valve. "Installed flow characteristic" refers to the characteristic observed when the valve is in service with varying pressure drop and other changes in the system. Let us look first at the inherent flow characteristic.

Fig. 5-3 shows the three most common inherent flow characteristic curves. The shape of the curves is obtained by the contour of the valve plug surface next to the seat of the valve. Fig. 5-4 shows a typical plug for the linear and equal percentage valves.

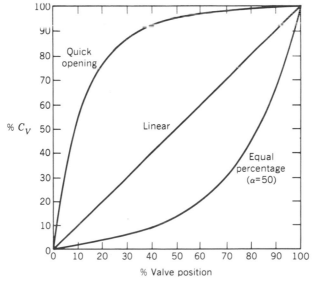

Figure 5-3. Inherent flow characteristic curves.

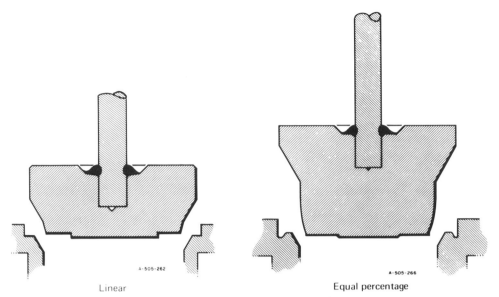

Linear   Equal percentage

Figure 5-4.   Plugs for each type of valve characteristics. (Courtesy of Taylor Instruments.)

The *linear flow characteristic* produces a flow directly proportional to the amount of valve travel, or valve position. At 50% valve travel, the flow through the valve is 50% of its maximum flow.

The *equal percentage flow characteristic* produces a very small flow at the beginning of the valve travel, but as it opens toward its fully open position, the flow increases considerably. The term "equal percentage" arises from the fact that for this valve, for equal increments of valve travel, the flow change with respect to valve travel is a constant percent of the flow rate at the time of change. That is, when increasing the valve position by 1% in going from 20% to 21% valve position, the flow will increase by 1% of its value at the 20% position. If the valve position is increased by 1% in going from 60 to 61%, the flow will increase by 1% of its value at the 60% position.

The *quick opening flow characteristic* produces a large flow at low valve travel. The curve is basically linear for the first part of the travel with a steep slope. It should be pointed out that the quick-opening valve is not good for regulation because it does not affect the flow for most of its travel.

The "matching" of the correct valve characteristic to any process requires a detailed dynamic analysis of the overall process; however, several rules of thumb, based on previous experience, help us in making a decision.[1] Briefly, we can say that valves with the linear flow characteristic are commonly used in liquid level loops and in other processes where the pressure drop across the valve is fairly constant. Valves with the quick-opening flow characteristic are primarily used in on–off services where it is required to have a large flow as soon as the valve starts to open. Finally, valves with the equal percentage flow characteristic are probably the most common ones. They are generally used in services where large variations in pressure drop are expected or in services where a small percentage of the total system pressure drop is to be taken across the valve.

From Fig. 5-3 we can learn important things about the rangeabilities of these three types of valves. Notice that the quick-opening type of valve has provided most of the flow at about 40% open; thus from there on it does not have much control on the flow. This results in a very low rangeability (less than 5 to 1). At the same time, Fig. 5-3 shows that the linear and equal percentage types of valves provide flow control over most of their operating range, resulting in rangeabilities greater than 20 to 1. The reader must remember, however, that these comments refer to the "inherent rangeabilities" since they are based on the inherent characteristics.

When a valve is installed in a piping system, the pressure drop across this valve varies as the flow through the valve varies. In this case the characteristics of the valve also vary and we refer to them, as mentioned above, as "installed characteristics." To help understanding these installed characteristics we consider the piping system shown in Fig. 5-5.

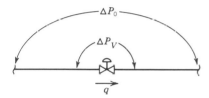

Figure 5-5. Piping system with control valve.

Let

ΔP_o — total dynamic pressure drop (including valve, line, fittings, etc.) in piping system, psi

$\bar{q}$ = design flow rate, gpm

ΔP_V = flow dependent pressure drop across valve, psi

f = fraction of dynamic pressure drop that is taken up by the control valve

$\bar{f}$ = fraction of dynamic pressure drop that is taken up by the control valve at design flow

$C_V|_{vp=1}$ = valve coefficient when valve is wide open

F = factor by which the valve was oversized

ΔP_p = dynamic pressure drop of the piping system, excluding valve, psi

Thus, the design flow through a valve is given by

$$\bar{q} = \frac{C_V|_{vp=1}}{F} \sqrt{\frac{\bar{f} \, \Delta P_o}{G_f}} \tag{5-12}$$

The pressure drop across the valve is given by

$$\Delta P_v = \Delta P_o - \Delta P_p \tag{5-13}$$

The balance of the dynamic pressure drop, which is taken by the piping system, is assumed to follow the relationship from fluid mechanics

$$\Delta P_p = (1 - f) \Delta P_o = K_L G_f q^2 \tag{5-14}$$

where K_L is a constant that has the following value:

$$
K_L = \frac{(1 - \bar{f}) \Delta P_o}{G_f \bar{q}^2}
$$

$$
= \frac{(1 - \bar{f}) \Delta P_o}{G_f} \frac{F^2 G_f}{(C_v|_{vp=1})^2 \bar{f} \, \Delta P_o} \tag{5-15}
$$

$$
K_L = \frac{F^2 (1 - \bar{f})}{\bar{f}(C_v|_{vp=1})^2}
$$

The total dynamic pressure drop for any flow is given by

$$\Delta P_o = \Delta P_V + K_L G_f q^2 \tag{5-16}$$

where

$$
q = C_V \sqrt{\frac{\Delta P_V}{G_f}}
$$

C_v = valve coefficient at any valve position other than $vp = 1$

Then

$$
\Delta P_o = G_f \left(\frac{q}{C_V}\right)^2 + K_L G_f q^2 = G_f (1 + K_L C_V^2) \left(\frac{q}{C_V}\right)^2
$$

from which we obtain

$$
q = \frac{C_V}{\sqrt{1 + K_L C_V^2}} \sqrt{\frac{\Delta P_o}{G_f}} \tag{5-17}
$$

When the valve is fully opened the flow rate is

$$
q_o = \frac{C_V|_{vp=1}}{\sqrt{1 + K_L (C_V|_{vp=1})^2}} \sqrt{\frac{\Delta P_o}{G_f}} \tag{5-18}
$$

Dividing Eq. (5-17) by (5-18) and substituting Eq. (5-15) into the result gives

$$
\frac{q}{q_o} = \frac{C_V}{C_v|_{v=1}} \sqrt{\frac{1 + \dfrac{F^2 (1 - \bar{f})}{\bar{f}}}{1 + \dfrac{F^2 (1 - \bar{f})}{\bar{f}} \left[\dfrac{C_v}{C_v|_{vp=1}}\right]^2}} \tag{5-19}
$$

This equation is valuable because it gives us the flow through a valve when installed in a piping system. We must remember that the equation was derived maintaining the total pressure drop, ΔP_o, constant. However, the pressure drop across the valve, ΔP_V, was allowed to vary. We can relate the C_V coefficient to the valve position for a linear valve, as will be given by Eq. (5-22), by the following relation:

$$C_V = C_V|_{vp=1}(vp)$$

Then, substituting this relation into Eq. (5-19) yields

$$\frac{q}{q_o} = \frac{(C_V|_{vp=1})vp}{C_V|_{vp=1}}\sqrt{\frac{1 + \dfrac{F^2(1-\bar{f})}{\bar{f}}}{1 + \left[\dfrac{F^2(1-\bar{f})}{\bar{f}}\right]\left[\dfrac{(C_V|_{vp=1})vp}{C_V|_{vp=1}}\right]^2}}$$

or

$$\frac{q}{q_o} = vp\sqrt{\frac{1 + \dfrac{F^2(1-\bar{f})}{\bar{f}}}{1 + \left[\dfrac{F^2(1-\bar{f})}{\bar{f}}\right]vp^2}} \qquad (5\text{-}20)$$

For an equal percentage valve the relation between C_V and valve position is given by

$$C_V = (C_V|_{vp=1})\alpha^{vp-1}$$

Substituting this relation into Eq. (5-19) gives

$$\frac{q}{q_o} = \frac{(C_V|_{vp=1})\alpha^{vp-1}}{C_V|_{vp=1}}\sqrt{\frac{1 + \dfrac{F^2(1-\bar{f})}{\bar{f}}}{1 + \left[\dfrac{F^2(1-\bar{f})}{\bar{f}}\right]\left[\dfrac{(C_V|_{vp=1})}{C_V|_{vp=1}}\alpha^{vp-1}\right]^2}}$$

or

$$\frac{q}{q_o} = \alpha^{vp-1}\sqrt{\frac{1 + \dfrac{F^2(1-\bar{f})}{\bar{f}}}{1 + \left[\dfrac{F^2(1-\bar{f})}{\bar{f}}\right](\alpha^{vp-1})^2}} \qquad (5\text{-}21)$$

Using Eqs. (5-20) and (5-21) we can determine the installed characteristics. Let us assume that the valves were sized to take 25% of the total dynamic pressure drop ($\bar{f} = 0.25$) and that the valve was oversized by a factor of 2 ($F = 2$). Under these conditions the installed characteristics at design conditions are shown in Fig. 5-6.

This figure shows that for the case studied ($\bar{f} = 0.25$, $F = 2$, and $\alpha = 50$), the equal percentage valve yields the most linear installed characteristics. The linear valve gives quick-opening installed characteristics with its resulting low rangeability.

The installed characteristics of any valve depends on the inherent characteristics of the valve, the fraction of the total dynamic pressure drop across the valve, $\bar{f}$, and the factor by which the valve is oversized, F.

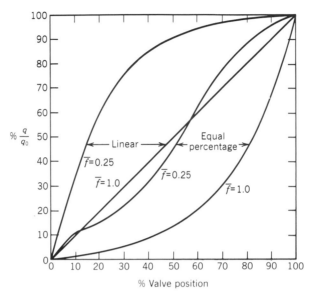

Figure 5-6. Installed valve characteristics.

Control Valve Gain

Figure 5-3 shows the inherent flow characteristic of the three most common types of valves. We define inherent characteristic as the flow characteristic through a valve when the pressure drop across it is maintained constant. Recalling Eq. (5-1) for the flow of liquid through the valve

$$q = C_V \sqrt{\frac{\Delta P}{G_f}}$$

we note that in order for q to change with valve position with a constant pressure drop and specific gravity, the C_V must also change with valve position. We say, therefore, that the C_V coefficient is a function of valve position. The functional relationship between C_V and valve position, vp, for the linear and equal percentage valves are as follows:

Linear Valve

$$C_V = (C_V|_{vp=1}) vp \tag{5-22}$$

Equal Percentage Valve

$$C_V = (C_V|_{vp=1}) \alpha^{vp-1} \tag{5-23}$$

where α = rangeability parameter of the valve.

From these relationships we can calculate the change in flow rate through a valve while maintaining a constant pressure drop. That is, this is the gain of the valve relating the flow and valve position. Consider the flow equation for an equal percentage valve in a liquid service:

$$q = (C_V|_{vp=1}) \alpha^{vp-1} \sqrt{\frac{\Delta P}{G_f}}$$

The gain of this valve is

$$K_V = \frac{\partial q}{\partial vp}\bigg|_{\Delta P} = (C_V|_{vp=1}) \ln(\alpha) \sqrt{\frac{\Delta P}{G_f}} \alpha^{vp-1} \tag{5-24}$$

The flow equation for a linear valve in a liquid service is

$$q = (C_V|_{vp=1}) vp \sqrt{\frac{\Delta P}{G_f}}$$

and the gain for this valve is

$$K_v = \frac{\partial q}{\partial vp}\bigg|_{\Delta P} = (C_V|_{vp=1}) \sqrt{\frac{\Delta P}{G_f}} \tag{5-25}$$

So, as given by the two equations (5-24) and (5-25), the inherent gain (constant pressure drop) for an equal percentage valve varies with valve position while that for a linear valve is constant. This is also easily noted by looking at the inherent flow characteristics shown in Fig. 5-3. The gain is the slope of the flow characteristics curve. The slope of the curve of the linear valve is constant while the one for the equal percentage valve varies.

It is important to realize that the installed gain is different from the inherent gain. Actually, as shown by Fig. 5-6, the installed gain of the equal percentage valve is more constant than that of the linear valve.

An expression for the installed gain can be obtained from Eq. (5-17):

$$K_v = \frac{\overline{dq}}{dvp}$$

$$= \sqrt{\frac{\Delta P_o}{G_f}} \left[\frac{\sqrt{1 + K_L \overline{C}_V^2} - \overline{C}_V(1 + K_L \overline{C}_V^2)^{-1/2} K_L \overline{C}_V}{(1 + K_L \overline{C}_V^2)} \right] \frac{\overline{dC}_V}{dvp}$$

$$= \frac{\overline{q}}{\overline{C}_V} \left(1 - \frac{K_L \overline{C}_V^2}{1 + K_L \overline{C}_V^2} \right) \frac{\overline{dC}_V}{dvp} \tag{5-26}$$

$$= \frac{\overline{q}}{\overline{C}_V(1 + K_L \overline{C}_V^2)} \frac{\overline{dC}_V}{dvp}$$

The term $\overline{dC}_V/dvp$ depends on the type of valve. A bar over a term indicates that the term is evaluated at some known condition. When $K_L = 0$, indicating constant pressure drop across the valve as given by Eq. (5-16), Eq. (5-26) results in Eq. (5-24) or (5-25) depending on the type of valve (this is left to the reader to prove).

It is important to realize that in sizing control valves, as shown earlier, the C_V calculated is the maximum C_V, or $C_V|_{vp=1}$.

Control Valve Summary

This section has presented an introduction to some of the most important considerations of control valves. There are, however, many other considerations that must be taken into account when specifying a control valve. Some of these other considerations, which have not been presented because of lack of time and space, are the sizing of valve actuators, the estimation of noise level, the sizing of valves for two-phase flow and when the compressibility of a gas is important, and the effect of pipe reducers. We hope that with the introduction presented in this section, and that shown in Appendix C, the reader will be motivated to read the many fine references given in this section to understand how to take these considerations into account.

5-3. FEEDBACK CONTROLLERS

This section presents the most important types of industrial controllers. The physical significance of their parameters is stressed to aid in the understanding of how they work. The presentation holds true for pneumatic, electronic, and most microprocessor-based controllers.

Briefly, the controller is the "brain" of the control loop. As mentioned in Chapter 1, the controller is the device that performs the decision (D) making in the control system. To do this, the controller:

1. Compares the process signal from the transmitter, the controlled variable, with the set point, and

2. Sends an appropriate signal to the control valve, or any other final control element, in order to maintain the controlled variable at its set point.

Figure 5-7 shows different types of controllers. Note the different knobs, switches, and buttons that permit adjusting the set point, reading the value of the controlled variable, transferring between manual and automatic modes, and adjusting and reading the output signal from the controller. Most controllers have these options on the front panel to permit ease of operation.

The auto/manual switch is an interesting one. This switch determines the operation of the controller. When the switch is in the auto (automatic) position, the controller decides and outputs the appropriate signal to the final control element to maintain the controlled variable at the set point. When the switch is in the manual position, the controller stops deciding and "freezes" its output. The controller output can then be changed manually by the operator, or engineer, using the manual output dial, thumbwheel, or button. In this mode the controller just provides a convenient (and expensive) way to adjust the final control element. In the auto mode the manual output does not have any meaning; only the set point has influence on the output. In the manual mode the set point does not have any influence on the controller output; only the manual output has influence

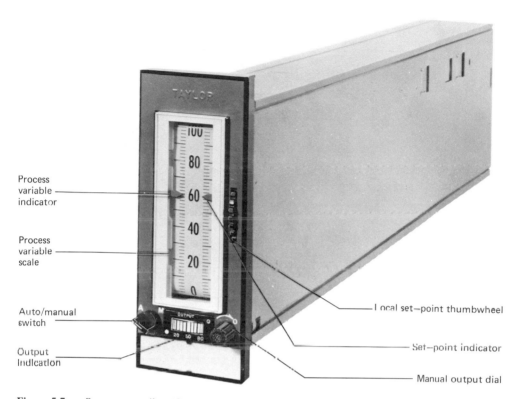

Process variable indicator

Process variable scale

Auto/manual switch

Output indication

Local set—point thumbwheel

Set—point indicator

Manual output dial

Figure 5-7a. Process controller. (Courtesy of Taylor Instruments.)

on the output. If a controller is set in manual, there is not much need for the controller. Only when the controller is in automatic are the benefits of automatic process control obtained.

Figures 5-7b and 5-7d also show the side panel of the controller. Most controllers have similar side panels. One of the options in this side panel is choosing between local and remote set-point operation. If the set point to the controller is to be set by the operating personnel from the front panel, then the local option is chosen. This is usually done by the flip of a knob, or switch. If the set point to the controller is to be set by another device, controller, relay, computer, or the like, then the remote option is chosen. When the knob or switch is set in remote, the front panel set-point knob does not have any effect on the controller set point. In this case the controller expects the set point as a signal from another device. This signal is usually connected through the back panel of the controller. Chapter 8 presents several control schemes that require the controllers to have a remote set point.

Let us now look at the other options in the controllers, including the different types of controllers, while stressing the physical meaning of their parameters.

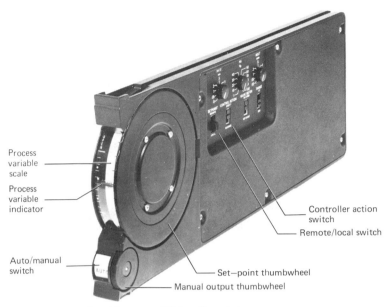

Process
variable
scale

Process
variable
indicator

Controller action
switch

Remote/local switch

Auto/manual
switch

Set—point thumbwheel

Manual output thumbwheel

Figure 5-7b. Process controller. (Courtesy of Fisher Controls.)

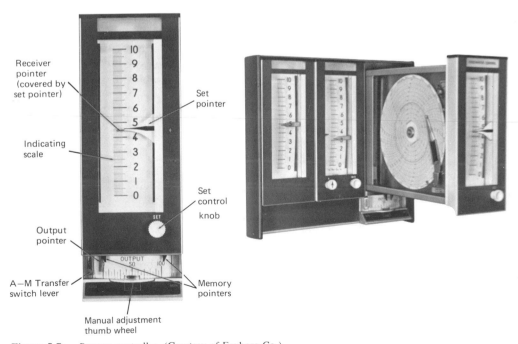

Receiver
pointer
(covered by
set pointer)

Set
pointer

Indicating
scale

Set
control
knob

Output
pointer

A—M Transfer
switch lever

Memory
pointers

Manual adjustment
thumb wheel

Figure 5-7c. Process controller. (Courtesy of Foxboro Co.)

156

Figure 5-7d. Process controller. (Courtesy of Fischer & Porter.)

Actions of Controllers

Consider the heat exchanger control loop shown in Fig. 5-8. If the outlet temperature of the hot fluid moves above the set point, the controller must close the steam valve. Since the valve is an air-to-open (AO) valve, the controller must reduce its output (air pressure or current) signal (see the arrows in the figure). To make this decision the controller must be set to *reverse action*. Some manufacturers refer to this action as *decrease*. That is, upon an *increase* in the input signal to the controller, there is a *decrease* in the output signal from the controller.

Consider now the level control loop shown in Fig. 5-9. If the liquid level moves above its set point, the controller must open the valve to bring the level back to set point. Since the valve is an air-to-open (AO) valve, the controller must increase its output signal (see the arrows in the figure). To make this decision the controller must be set to *direct action*. Some manufacturers refer to this action as *increase*. That is, upon an *increase* in the input signal to the controller, there is an *increase* in the output signal from the controller.

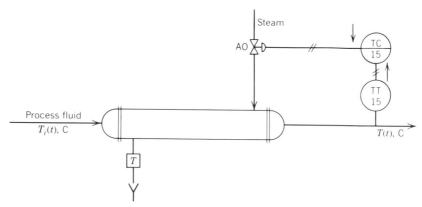

Figure 5-8. Heat exchanger control loop.

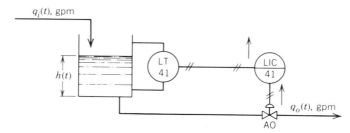

Figure 5-9. Liquid level control loop.

In summary, to determine the action of a controller, the engineer must know:

1. The process requirements for control and,

2. The action of the control valve or other final control element.

Both things *must* be taken into consideration. The reader can ask himself what the correct action of the level controller is if an air-to-close (AC) valve is used or if the level is controlled with the inlet flow instead of the outlet flow. In the first case the control valve action changes while in the second case it is the process requirements for control that change.

The controller action is usually set by a switch on the side panel of pneumatic or electronic controllers, as shown in Fig. 5-7b, or by a configuration bit on most micro-processor-based controllers.

Types of Feedback Controllers

The way feedback controllers make a decision in order to maintain the set point is by calculating the output on the basis of the difference between the controlled variable and the set point. In this section we look at the most common types of controllers by looking at the equations that describe their operation.

Proportional Controller (P).   The proportional controller is the simplest type of controller, with the exception of the on-off controller, which we shall not discuss here. The equation that describes its operation is the following:

$$m(t) = \overline{m} + K_c(r(t) - c(t)) \tag{5-27}$$

or

$$m(t) = \overline{m} + K_c e(t) \tag{5-28}$$

where

$m(t)$ = output from the controller, psig or mA
$r(t)$ = set point, psig or mA
$c(t)$ = controlled variable, psig or mA. This is the signal from the transmitter.
$e(t)$ = error signal, psi or mA. This is the difference between the set point and the controlled variable.
K_c = controller gain, $\dfrac{\text{psi}}{\text{psi}}$ or $\dfrac{\text{mA}}{\text{mA}}$
$\overline{m}$ = bias value, psig or mA. The significance of this value is the output from the controller when the error is zero. This value is usually set, during calibration of the controller, at mid-scale, 9 psig or 12 mA.

Since the input and output ranges are the same (3–15 psig or 4–20 mA), the input and output signals, as well as the set point, are sometimes also expressed in either fraction or percent of range.

It is interesting to note that Eq. (5-27) is written for a reverse-acting controller. If the *controlled variable, c(t), increases* above the set point, $r(t)$, the error becomes negative and, the equation shows, the *output from the controller, m(t), decreases*. The usual way to show a direct-acting controller mathematically is by letting the controller gain, K_c, be negative. We must remember, however, that in industrial controllers there are no negative gains, only positive ones. The reverse/direct switch takes care of this. The negative K_c is used when doing a mathematical analysis of a control system that requires a direct-acting controller.

Equations (5-27) and (5-28) show that the output of the controller is proportional to the error between the set point and the controlled variable. The proportionality is given by the controller gain, K_c. This gain, or controller sensitivity, determines how much the output from the controller changes for a given change in error. This is shown graphically in Fig. 5-10.

Proportional only controllers have the advantage of only one tuning parameter, K_c. However, they suffer a major disadvantage, that of operating the controlled variable with an OFFSET, or "steady-state error." To show this offset graphically, consider the liquid level control loop shown in Fig. 5-9. Assume that the design operating conditions are $\overline{q}_i = \overline{q}_o = 150$ gpm, and $\overline{h} = 6$ ft. Let us also assume that in order for the outlet valve to pass 150 gpm, the air pressure over it must be 9 psig. If the inlet flow, q_i, increases, the response of the system with proportional only controller may look like Fig. 5-11. The controller brings the controlled variable back to a steady value, but this value is not the required set point. The difference between the set point and the steady-state value of

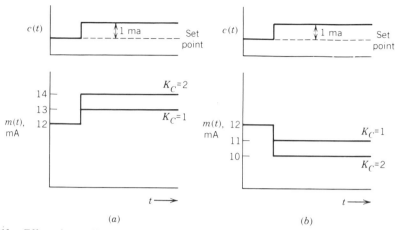

Figure 5-10. Effect of controller gain on output of controller. (*a*) Direct-acting controller. (*b*) Reverse-acting controller.

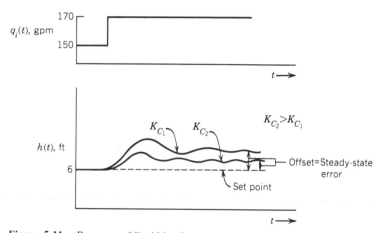

Figure 5-11. Response of liquid level system.

the controlled variable is the offset. Figure 5-11 shows two response curves corresponding to two different values of the tuning parameter K_c. This figure shows that the larger the value of K_c, the smaller the offset but the more oscillatory the response of the process becomes. For most processes, however, there is a maximum value of K_c beyond which the process will go unstable. The calculation of this maximum value of the gain, which we shall call the ultimate gain, K_{cu}, is presented in Chapters 6 and 7.

Let us look at a simple explanation of why offset exists; a more rigorous proof is given in Chapter 6. Consider the same liquid level control system shown in Fig. 5-9 with the operating conditions given previously. Remember that the proportional only controller, direct acting $(-K_c)$, solves the following equation:

$$m(t) = 9 + (-K_c)e(t) \tag{5-29}$$

Assume now that the inlet flow increases up to 170 gpm. When this happens the liquid level increases and the controller will in turn increase its output to open the valve to bring the level back down. In order to reach a steady operation, the outlet flow, q_o, must now be 170 gpm. To pass this new flow the outlet valve must be open more than before when it needed to pass 150 gpm. Since this is an air-to-open valve, let us assume that the new required air pressure over the valve is 10 psig. That is, the output from the controller must be 10 psig. Looking back at Eq. (5-29) we note that the only way for the output of the controller, $m(t)$, to be 10 psig is for the second term of the right-hand side to have a value of $+1$ psig and for this to be so, the error term, $e(t)$, *cannot be zero at steady state. This steady-state error is the offset.* Notice that a negative error means that the controlled variable is greater than the set point. The actual level, in feet, can be calculated from the calibration of the level transmitter.

Two points need to be stressed in this example. First, the magnitude of the error term depends on the value of the controller gain since the total term must have a value of $+1$ psig. Thus:

$-K_c$	$-e(\infty)$ (offset)
1	1
2	0.5
4	0.25

As previously mentioned, the larger the gain, the smaller the offset. The reader must remember that above a certain K_c, most processes go unstable; however, the equation of the controller does not show this. This is proven in Chapter 6.

Second, and as a summary of this example, it seems that all a proportional only controller is trying to do is to reach a steady-state operating condition. The amount of deviation from set point, or offset, depends on the controller gain.

Many controller manufacturers do not use the term gain for the amount of controller sensitivity. They use the term proportional band, PB. The relationship between gain and proportional band is given by

$$PB = \frac{100}{K_c} \tag{5-30}$$

and consequently, the equation that describes the proportional only controller is now written as

$$m(t) = \overline{m} + \frac{100}{PB}(r(t) - c(t)) \tag{5-31}$$

or

$$m(t) = \overline{m} + \frac{100}{PB}e(t) \tag{5-32}$$

The term "100" is used because PB is usually referred to as "percent proportional band."

Equation (5-30) presents a most important fact. A large gain, K_c, is the same as a low, or narrow, proportional band, and a low gain is the same as a large, or wide, proportional band. This means that before we start "fooling around" with the knob in the controller, we must know whether the controller uses gain or proportional band.

Let us offer another definition of proportional band. Proportional band refers to the *error* (expressed in percentage of the range of the controlled variable) *required to move the output of the controller from its lowest to its highest value.* Consider the heat exchanger control loop shown in Fig. 5-8. The temperature transmitter has a range from 100 C to 300 C and the set point of the controller is at 200 C. Figure 5-12 gives a graphical explanation of the definition of PB. This figure shows that a 100% PB means that as the controlled variable varies 100% of its range, the controller output varies 100% of its range. A 50% PB means that as the controlled variable varies by 50% of its range, the controller output varies 100% of its range. Also notice that a proportional only controller with a 200% PB will not move its output the entire range. A 200% PB means very small gain or sensitivity to errors.

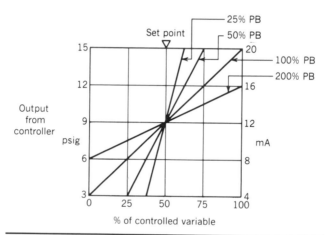

	Controller output		
	3 psig 4 mA	9 psig 12 mA	15 psig 20 mA
PB = 100%	100 C	200 C	300 C
PB = 50%	150 C	200 C	250 C
PB = 25%	175 C	200 C	225 C
PB = 200%	—	200 C	—

Figure 5-12. Definition of proportional band.

To obtain the transfer function for the proportional only controller, Eq. (5-27) can be written as

$$m(t) - \overline{m} = K_c(e(t) - 0)$$

We define the following two deviation variables:

$$M(t) = m(t) - \overline{m} \tag{5-33}$$
$$E(t) = e(t) - 0 \tag{5-34}$$

Then

$$M(t) = K_c E(t)$$

Taking the Laplace transform gives the following transfer function:

$$\frac{M(s)}{E(s)} = K_c \qquad (5\text{-}35)$$

To summarize briefly, proportional only controllers are the simplest controllers with the advantage of only one tuning parameter, K_c or PB. The disadvantage of these controllers is the operation with an offset in the controlled variable. In some processes, such as a surge tank, this may not be of any major consequence. In cases in which the process can be controlled within a band from the set point, proportional only controllers are sufficient. However, in processes in which the control *must* be at the set point, proportional only controllers will not provide satisfactory control.

Proportional-Integral Controller (PI). Most processes cannot be controlled with an offset; that is, they must be controlled at the set point. In these instances an extra amount of intelligence must be added to the proportional only controller to remove the offset. This new intelligence, or new mode of control, is the integral, or reset, action; consequently, the controller becomes a proportional-integral controller (PI). The descriptive equation is as follows:

$$m(t) = \overline{m} + K_c[r(t) - c(t)] + \frac{K_c}{\tau_I}\int[r(t) - c(t)]\,dt \qquad (5\text{-}36)$$

or

$$m(t) = \overline{m} + K_c e(t) + \frac{K_c}{\tau_I}\int e(t)\,dt \qquad (5\text{-}37)$$

where τ_I = integral or reset time, minutes/repeat.

Therefore, the PI controller has two parameters, K_c and τ_I, that must be tuned to obtain satisfactory control.

To understand the physical significance of the reset time, τ_I, consider the hypothetical example shown in Fig. 5-13. τ_I is the time that it takes the controller to repeat the proportional action, and consequently the units are minutes/repeat. The smaller the value of τ_I, the steeper the response curve, which means the faster the controller response becomes. Another way of explaining this is by looking at Eq. (5-37). The smaller the value of τ_I, the larger the term in front of the integral, K_c/τ_I, and, consequently, the more weight is given to the integration, or reset, action.

Also, from Eq. (5-37) we note that as long as the error term is present, the controller will keep changing its output, thereby integrating the error, to remove the error. Remember that integration also means summation.

Let us now recall the liquid level control system used to explain why offset exists. As indicated, when the inlet flow increased to 170 gpm, the outlet flow needed to increase

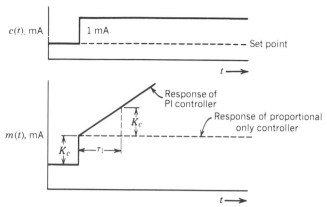

Figure 5-13. Response of proportional-integral (PI) controller (direct action) to a step change in error.

to 170 gpm to reach a final steady-state operating condition. To pass 170 gpm the outlet valve needed a 10 psig air signal, and the only way a proportional only controller could output 10 psig is by maintaining an error term. With a PI controller, as long as the error is present the controller will keep integrating it, thus adding to its output, until the error disappears. When this gets to be the case, the output of the controller is given by

$$m(t) = \overline{m} + \frac{K_c}{\tau_I}\left(\int 0 \, dt\right)$$

The fact that the error is zero does not mean that the integral term is zero. It means that the controller is integrating a function with a value of zero, or better yet, "adding zero" to its output, which keeps it constant. For the liquid level process the integral term

$$\frac{K_c}{\tau_I}\int 0 \, dt$$

will have a value of 1 psig and, therefore, the output of the controller will be 10 psig without any error. This has been a brief explanation of why reset action removes the offset. In Chapter 6 this is proven again from a more rigorous point of view.

Some manufacturers do not use the term reset time, τ_I, for their tuning parameter. They use what is known as reset rate, τ_I^R. The relationship between these two parameters is

$$\tau_I^R = \frac{1}{\tau_I}, \qquad \text{repeats/min} \tag{5-38}$$

Therefore, before tuning the integral parameter, we must know whether the controller uses reset time or reset rate. They are the reciprocal of one another and, consequently, their effects are completely different.

Let us show the equations that describe the operation of PI controllers of some manufacturers. This will reinforce our comment of "you must know who you are fooling around with before you start fooling around."

Foxboro Co.

$$m(t) = \overline{m} + \frac{100}{PB} e(t) + \frac{100}{PB \cdot \tau_I} \int e(t) \, dt \qquad (5\text{-}39)$$

Fisher Controls

$$m(t) = \overline{m} + \frac{100}{PB} e(t) + \frac{100 \tau_I^R}{PB} \int e(t) \, dt \qquad (5\text{-}40)$$

Taylor Co., Honeywell, Inc.

$$m(t) = \overline{m} + K_c e(t) + K_c \tau_I^R \int e(t) \, dt \qquad (5\text{-}41)$$

It is interesting to note that when Honeywell developed their microprocessor-based controller, TDC2000, they changed the PI controller to the one given by Eq. (5-37). On the other hand, when Fisher Controls developed their microprocessor-based control system, PROVOX, they changed their PI equation, Eq. (5-40), to Eq. (5-41). Equation (5-37) is the one proposed as the standard by the Instrument Society of America (ISA) and will be used in this book. The important thing to remember is the relationship between gain and proportional band and between reset time and reset rate.

To obtain the transfer function for the PI controller Eq. (5-37) is written as follows:

$$m(t) - \overline{m} = K_c(e(t) - 0) + \frac{K_c}{\tau_I} \int (e(t) - 0) \, dt$$

Using the same definitions of deviation variables given by Eqs. (5-33) and (5-34), taking the Laplace transform, and rearranging yields

$$\frac{M(s)}{E(s)} = K_c \left(1 + \frac{1}{\tau_I s} \right) \qquad (5\text{-}42)$$

To summarize, proportional-integral controllers have two tuning parameters: the gain or proportional band, and the reset time or reset rate. The advantage of this controller is that the integral, or reset, action removes the offset. Chapter 6 and 7 prove again that this controller removes the offset and how it affects the stability of control loops. Probably close to 75% of all controllers in use are of this type.

Proportional-Integral-Derivative Controller (PID). Sometimes another mode of control is added to the PI controller. This new mode of control is the derivative action, also called the rate action or preact. Its purpose is to *anticipate* where the process is heading by looking at the time rate of change of the error, its derivative. The descriptive equation is the following:

$$m(t) = \overline{m} + K_c e(t) + \frac{K_c}{\tau_I} \int e(t) \, dt + K_c \tau_D \frac{de(t)}{dt} \qquad (5\text{-}43)$$

where τ_D = derivative, or rate, time in minutes.

Therefore, the PID controller has three parameters, K_c or PB, τ_I or τ_I^R, and τ_D, that must be tuned to obtain satisfactory control. Notice that there is only one derivative tuning parameter, τ_D. It has the same units, minutes, for all manufacturers.

As just mentioned, the derivative action gives the controller the capability to anticipate where the process is heading, that is, to "look ahead," by calculating the derivative of the error. The amount of "anticipation" is decided by the value of the tuning parameter, τ_D.

Let us consider the heat exchanger shown in Fig. 5-8 and use it to clarify what is meant by "anticipating where the process is heading." Assume that the inlet process temperature decreases by some amount and the outlet temperature starts to decrease correspondingly as shown in Fig. 5-14. At time t_a the amount of the error is positive and may be small. Consequently, the amount of control correction provided by the proportional and integral mode is small. However, the derivative of this error, the slope of the error curve, is large and positive, making the control correction provided by the derivative mode large. By looking at the derivative of the error, the controller knows that the controlled variable is heading away from the set point rather fast and, consequently, it uses this fact to help in controlling. At time t_b the error is still positive and larger than before. The amount of control correction provided by the proportional and integral modes is also larger than before and is still adding to the output of the controller to further open the steam valve. However, the derivative of the error at this time is negative, signifying that the error is decreasing; that is, the controlled variable has started to come down to set point. Again, using this fact, the derivative mode starts to subtract from the other

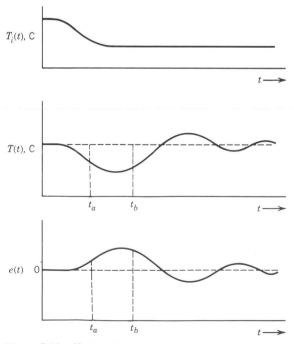

Figure 5-14.   Heat exchanger control.

two modes since it recognizes that the error is decreasing. By doing this it takes longer for the process to return to set point; however, the overshoot and oscillations around set point are reduced.

PID controllers are used in processes with long time constants. Typical examples are temperature and concentration loops. Processes with short time constants (small capacitance) are fast and susceptive to process noise. Typical of these processes are flow loops and loops controlling the pressure of liquid streams. Consider the recording of a flow shown in Fig. 5-15. The application of the derivative mode will only result in the amplification of the noise because the derivative of the fast changing noise is a large value. Long time constant processes (large capacitance) are usually damped and, consequently, are less susceptible to noise. Be aware, however, that you may have a long time constant process, a temperature loop for example, with a noisy transmitter. In this case, before the PID controller is used the transmitter must be fixed.

The transfer function of an "ideal" PID controller is obtained from Eq. (5-43) by rearranging it as follows:

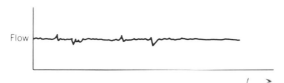

Figure 5-15. Recording of a flow loop.

$$m(t) - \overline{m} = K_c(e(t) - 0) + \frac{K_c}{\tau_I} \int (e(t) - 0)\, dt$$

$$+ K_c\tau_D \frac{d(e(t) - 0)}{dt}$$

Using the same definitions of deviation variables given by Eqs. (5-33) and (5-34), taking the Laplace transform, and rearranging yields:

$$\frac{M(s)}{E(s)} = K_c\left(1 + \frac{1}{\tau_I s} + \tau_D s\right) \qquad (5\text{-}44)$$

This transfer function is called "ideal" because in practice the implementation of the derivative calculation is impossible to obtain. The derivative is then approximated by the use of a lead/lag, resulting in the "actual" transfer function:

$$\frac{M(s)}{E(s)} = K_c\left(1 + \frac{1}{\tau_I s}\right)\left(\frac{\tau_D s + 1}{\alpha\tau_D s + 1}\right) \qquad (5\text{-}45)$$

Typical values of α range between 0.05 and 0.1.

To summarize, PID controllers have three tuning parameters: the gain or proportional band, the reset time or reset rate, and the rate time. The rate, or derivative, time is always in minutes. PID controllers are recommended for long time constant loops which are free of noise. The advantage of the derivative mode is that it provides the capability of "looking where the process is heading." Chapters 6 and 7 show how the use of this controller improves the control and how it affects the stability of control loops.

Proportional-Derivative Controller (PD).

This controller is used in processes where a proportional only controller can be used but some amount of "anticipation" is desired.

The descriptive equation is

$$m(t) = \overline{m} + K_c e(t) + K_c \tau_D \frac{de(t)}{dt} \tag{5-46}$$

and the "ideal" transfer function is

$$\frac{M(s)}{E(s)} = K_c(1 + \tau_D s) \tag{5-47}$$

A disadvantage of the PD controller is the operation with an offset in the controlled variable. Only the integral action can remove the offset. However, a PD controller can stand higher gain, thus resulting in smaller offset, than a proportional only controller on the same loop.

Digital Controllers and Other Comments.

As previously mentioned, Eq. (5-45) is the transfer function of industrial analog controllers; however, the equation of digital controllers is the discrete form of Eq. (5-43). The methods to tune digital controllers are not very different from those to tune analog controllers; this is explained in the following chapter. The reader is referred to reference 4 for a treatise on digital controllers.

A few other comments on controllers are in order before closing this section. Equation (5-43) shows that any time the gain parameter K_c is changed, this affects the integral and derivative actions since τ_I and τ_D are divided or multiplied by it. This means that if you want to change only the gain action and not the amount of reset or anticipation, then the parameters τ_I and τ_D must also be changed to accommodate for the change in K_c. All analog controllers are of this type and are sometimes referred to as "interacting controllers." Most microprocessor-based controllers are also of this type. However, there are some that have avoided this problem by substituting the K_c/τ_I term by the single term K_I and the $K_c \tau_D$ term by K_D. This means that the three tuning parameters are K_c, K_I, and K_D.

The final comment is related to the derivative action. The typical way to change the controller set point is by introducing a change as shown in Fig. 5-16a. When this happens a step change in error is also introduced, as shown in Fig. 5-16b. Since the controller takes the derivative of the error, this derivative produces a sudden change in the controller output, as shown in Fig. 5-16c. This change in controller output is unnecessary and possibly detrimental to the process operation. It has been proposed[4] that a way to get around this problem is by using the negative of the derivative of the controlled variable

$$- \frac{dc(t)}{dt}$$

instead of the derivative of the error. The two derivatives are equal when the set point

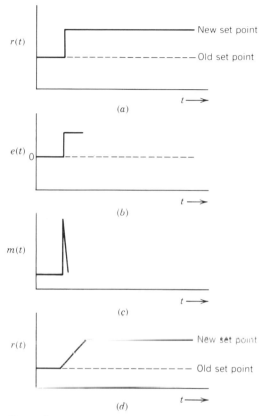

Figure 5-16.   Effect of set-point changes.

remains constant, as can be shown by the following:

$$\frac{de(t)}{dt} = \frac{d[r(t) - c(t)]}{dt} = \frac{dr(t)}{dt} - \frac{dc(t)}{dt}$$

At the moment the set point change is introduced, this "new" derivative does not produce the sudden change. Immediately afterward, the behavior becomes the same as before. This option is offered by some analog and microprocessor-based controllers and is referred to as derivative-on-controlled variable.

Another possible way to avoid this derivative problem can be easily obtained by the use of digital controllers. This option changes the set point in a ramp shape, even though the operator may have changed the set point in a step fashion, as shown in Fig. 5-16d. The slope of the ramp is predetermined by the operating personnel.

Reset Windup

The problem of reset windup is an important and realistic one in process control. It can happen at any time that a controller contains the integral mode of control. The heat exchanger control loop shown in Fig. 5-8 is used to explain this problem.

Suppose that the process inlet temperature drops by a significant amount. This disturbance will drop the outlet process temperature. The controller (PI or PID) will in turn ask the steam valve to open. Since this is an air-to-open valve, the pneumatic signal from the controller will increase until, because of the reset action, the outlet temperature equals the desired set point. But suppose that in the effort of restoring the controlled variable to set point, the controller integrates up to 15 psig. At this point the steam valve is wide open and, therefore, the control loop cannot do any more. Essentially, the process is out of control. This is demonstrated graphically in Fig. 5-17. This figure shows that when the valve is fully open, the controlled variable, the outlet temperature, is still not at set point. Since there is still an error, the controller will try to correct for it by further increasing (integrating the error) its output pressure even though the valve will not open more after 15 psig. The output of the controller can in fact integrate up to the supply pressure, which is usually about 20 psig. At that point the controller cannot increase its output anymore, its output having saturated; this state of the system is also shown in Fig. 5-17. This saturation is due to the integration (reset) action of the controller; as long as an error is present, the controller will continue to change its output. This saturation state is referred to as "reset windup."

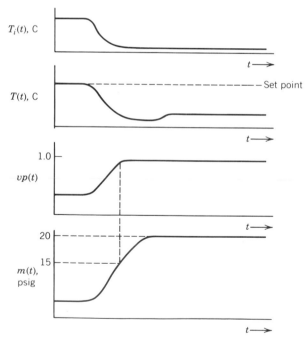

Figure 5-17.   Heat exchanger control loop response.

Suppose now that the inlet temperature goes up again; the outlet process temperature will in turn start to increase, as also shown in Fig. 5-18. This figure shows that the outlet temperature reaches and passes the set point and the valve remains wide open when, in fact, it should be closing. The reason it is not closing is because the controller must

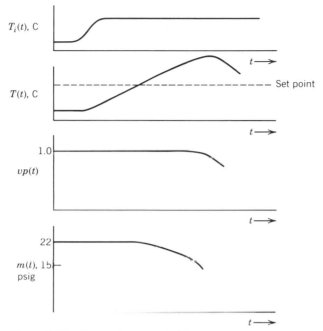

Figure 5-18. Heat exchanger control loop response.

integrate down from 20 psig to 15 psig before the valve starts to close. By the time this happens, the outlet temperature has overshot the set point by a significant amount.

As mentioned earlier, this problem of reset windup can happen any time that integration is present in the controller. It can be avoided if the controller is set in manual as soon as its output reaches 15 psig; this action will stop the integration. The controller can be set back to automatic when the temperature starts to decrease again. The disadvantage of this operation is that it requires the operator's attention. However, most controllers can be bought with "reset windup protection." This protection automatically stops the integration once the controller reaches 15 psig (20 mA) or 3 psig (4 mA). Since this protection may be a special feature in controllers, the engineer must be aware when reset windup is a possibility so that the protection may be specified. Reset windup protection is described in Chapter 6.

Reset windup typically occurs in batch processes, in cascade control, and when a final control element is driven by more than one controller, as in override control schemes. Cascade control and override control are presented in Chapter 8.

Feedback Controller Summary

This section has presented the subject of process controllers. We have seen the purpose of the controllers: to make the decision of how to use the manipulated variable to maintain the controlled variable at set point. We have also looked at the significance of the auto/manual switch and the remote/local switch and how to choose the action, reverse/direct, of the controller. The different types of controllers were also studied, stressing the significance of the tuning parameters gain (K_c) or proportional band (PB), reset time (τ_I) or reset rate (τ_I^R), and rate time (τ_D). Finally, the subject of reset windup was presented and its significance discussed.

We still have not discussed the important subject of obtaining the optimum setting of the tuning parameters. This is called "tuning the controller." In tuning the controller its personality is set so as to match the personality of the process. This is presented in Chapter 6.

5-4. SUMMARY

In this rather extensive chapter, we have looked at some of the hardware necessary to build a control system. A brief look at some terms related to sensors and transmitters and a discussion of the parameters that describe these devices started the chapter. The chapter then continued with a presentation of some important considerations about control valves such as fail action, sizing, and characteristics. The reader is referred to Appendix C for more information on sensors, transmitters, and valves.

A discussion of feedback process controllers followed. The four most common types of controllers were presented. The physical significance of their parameters was explained. The tuning of these parameters is presented in Chapter 6.

We are now ready to use the first five chapters of this book to design process control systems. This is the subject of the next four chapters.

REFERENCES

1. "Control Valve Handbook," Fisher Controls Co., Marshalltown, Iowa.

2. "Masoneilan Handbook for Control Valve Sizing," Masoneilan International, Inc., Norwood, Mass.

3. "Fisher Catalog 10," Fisher Controls Co., Marshalltown, Iowa.

4. C. L. Smith, *Digital Computer Process Control*, International Textbook Co., 1972.

PROBLEMS

5-1. An electronic differential pressure transmitter is used in combination with an orifice to measure flow so that the 4–20 mA signal is proportional to the square of the flow through the orifice. The transmitter is calibrated for a maximum differential pressure of 100 in. of water and the orifice is sized so that the corresponding maximum flow is 750 gpm.

a. Calculate the gain of the transmitter when the flow is 500 gpm. Specify your units.

b. What is the transmitter output signal in percent of its range when the flow is 500 gpm?

5-2.   It is necessary to size a control valve to regulate the flow of 150 psig saturated steam to a heater. The normal steady-state flow is 1000 lbm/hr with an inlet pressure of 150 psig and an outlet pressure of 50 psig. Obtain the required C_V for an overdesign factor of 30%.

5-3.   A control valve for liquid service must be sized. The operating conditions are as follows:

$$\text{Flow} = 52,500 \ \frac{\text{lbm}}{\text{hr}} \ \text{(normal)}, \quad 210,000 \ \frac{\text{lbm}}{\text{hr}} \ \text{(max)}$$

$$
\begin{array}{lll}
P_1 = 229 \text{ psia} & P_2 = 129 \text{ psia} & T = 104°F \\
P_V = 124 \text{ psia} & P_C = 969 \text{ psia} & G_f = 0.92 \\
\mu - 0.2 \text{ cp} & &
\end{array}
$$

Obtain the required C_V.

5-4.   A valve for a gas service with the following operating conditions is needed:

Flow = 55,000 scfh (normal)
$G_f = 1.54$ $T - 40 \text{ C}$ $P_1 - 110 \text{ psig}$
$P_2 = 11 \text{ psig}$

Obtain the required valve size for twice the normal flow.

5-5.   Consider the process shown in Fig. 5-19. Benzene flows through the pipe at a rate of 700 gpm and at 155 C. The dynamic pressure drop between points 1 and 2, at steady-state flow, is 15 psi; this includes the drop across the orifice. At flowing conditions the density of benzene is 45.49 lbm/ft^3 and the viscosity 0.17 cp. Obtain the required valve size for twice the normal flow.

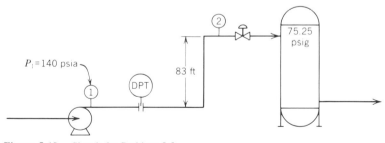

Figure 5-19. Sketch for Problem 5-5.

5-6.   Consider the process shown in Fig. 5-20. Ethylbenzene is pumped at a rate of 1000 gpm and at 445°F. The dynamic pressure drop between points 1 and 2 is 9.36 psi and between points 3 and 4 is 3 psi. At flowing conditions the density of ethylbenzene is 42.05 lbm/ft^3. Using a 25% overdesign factor, obtain the required C_V.

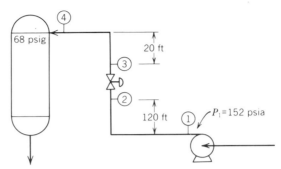

Figure 5-20. Sketch for Problem 5-6.

5-7. A control valve is sized so that at design conditions the flow of liquid (G_f = 0.85) is 420 gpm, the pressure drop across the valve is 3 psi, and the valve is half open. The total dynamic pressure drop in the line, including that across the valve, is 20 psi, and the valve has inherently linear characteristics. The pressure drop in the line may be assumed to be proportional to the square of the flow, but the total dynamic pressure drop is constant.

a. Calculate the valve capacity factor C_V (fully open).
b. Calculate the gain of the valve at design conditions, in gpm per % valve position.
c. Calculate the flow when the valve is fully open.

5-8. Consider the pressure control system shown in Fig. 5-21. The pressure transmitter, PT25, has a range of 0–100 psig. The controller, PIC25, is a proportional only controller, its bias value is set at midscale, and its set point is 10 psig. Obtain the correct action of the controller and the proportional band required so that when the pressure in the tank is 30 psig, the valve will be wide open.

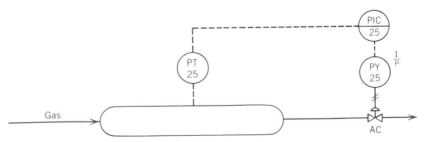

Figure 5-21. Sketch for Problem 5-8.

5-9. Let us change the pressure control system of the previous problem. The new control scheme is shown in Fig. 5-22. This control scheme is called cascade control; its benefits and principles will be explained in Chapter 8. In this scheme, the pressure controller sets the set point of the flow controller. The remote/local switch of the flow controller must be in remote. The pressure transmitter has a range of 0–100 psig and the flow transmitter range is 0–3000 scfh. Both controllers are proportional only. The nominal flow rate through the valve is 1000 scfh, and to give this flow it is open 33%. The control valve has linear characteristics.

a. Obtain the action of the controllers.
b. Choose the bias values for both controllers so that no offset occurs in either controller.
c. Obtain the proportional band setting of the pressure controller so that when the tank pressure reaches 40 psig the set point to the flow controller will be 1700 scfh. The set point of the pressure controller is 10 psig.

5-10. Consider the level loop shown in Fig. 5-9. The steady-state operating conditions are $\bar{q}_i = \bar{q}_v = 150$ gpm and $\bar{h} = 6$ ft. For this steady-state the AO valve requires a 9 psig signal. The level transmitter has a range of 0–20 ft. A proportional only controller, $K_c = 1$, is used in this process. Calculate the offset if the inlet flow increases to 170 gpm and the valve requires 10 psig to pass this flow. Report the offset in psig and in feet.

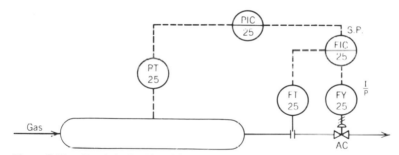

Figure 5-22. Sketch for Problem 5-9.

CHAPTER

6

Design of Single-Loop Feedback Control Systems

In previous chapters we have become familiar with the dynamic characteristics of processes, sensors, transmitters, control valves, and controllers. We have also learned how to write linearized transfer functions for each of these components and to recognize the parameters that are significant to the design of automatic control systems, namely, the steady-state gain, the time constants, and the dead time (transportation lag or time delay). In this chapter we will see how these concepts are put together to design and tune single-loop feedback control systems. We will first analyze a simple feedback control loop and learn how to draw a block diagram for it and determine its characteristic equation. The significance of the characteristic equation will be examined next in terms of how it can be used to determine the stability of the loop. The two methods that we will use to determine the stability of the loop are Routh's method and the direct substitution method. We will then study two methods to tune the feedback controller, that is, to adjust the controller parameters to the characteristics (or personality) of the rest of the components of the loop. These methods are the on-line or closed-loop tuning method, and the step-testing or open-loop method. We will also look at the method of controller synthesis, which, in addition to providing us some simple controller relationships, will give us some insight into the selection of the proportional, integral, and derivative modes for various process transfer functions. Finally we will look into how to prevent the important problem of reset windup, which was discussed in Chapter 5.

The methods that we will study in this chapter are most applicable to the design and tuning of feedback control loops for industrial processes. The more classical design techniques of root-locus and frequency response analysis, which have been traditionally applied to inherently linear systems, will be presented in the Chapter 7.

6-1. THE FEEDBACK CONTROL LOOP

The concept of feedback control, although more than two thousand years old, did not find practical application in industry until James Watt applied it to the control of the speed of his steam engine about two hundred years ago. Since then the number of industrial applications have proliferated to the point that today the great majority of automatic control systems include at least one feedback control loop. None of the advanced control techniques that have been developed in the last fifty years to enhance the performance of feedback control loops has been able to replace it. We will study these advanced techniques in a later chapter.

To review the concept of feedback control, let us again look over the example of the heat exchanger shown in Chapter 1. A sketch of the exchanger is given in Fig. 6-1.

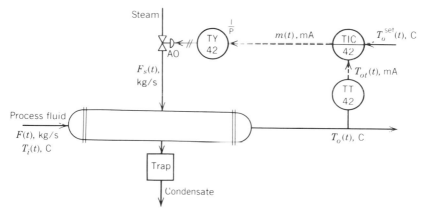

Figure 6-1. Feedback control loop for temperature control of heat exchanger.

The objective is to maintain the outlet temperature of the process fluid $T_o(t)$ at its desired value or set point $T_o^{set}(t)$ in the presence of variations of the process fluid flow $F(t)$ and inlet temperature $T_i(t)$. The steam flow $F_s(t)$ is the variable that can be adjusted to control the outlet temperature, as it determines the amount of energy supplied to the process fluid.

The feedback control scheme works as follows: the outlet temperature or *controlled variable* is measured with a sensor and transmitter (TT42) that generates a signal $T_{ot}(t)$ that is proportional to the temperature. The transmitter signal or *measurement* is sent to the controller (TIC42) where it is compared to the set point. The function of the controller is then to generate an output signal or *manipulated variable*, $m(t)$, on the basis of the *error* or difference between the measurement and the set point. The controller output signal is then connected to the actuator of the steam control valve through a current-to-pressure transducer (I/P). This is because in this example the transmitter and controller generate electric current signals, but the valve actuator must be operated by air pressure. The function of the valve actuator is to position the valve in proportion to the controller output signal (see Appendix C). The steam flow is then a function of the valve position.

The term "feedback" derives from the fact that the controlled variable is measured and this measurement is "fed back" to reposition the steam valve. This causes the signal variations to move around the loop as follows:

Variations in outlet temperature are sensed by the sensor-transmitter and sent to the controller causing the controller output signal to vary. This in turn causes the control valve position and consequently the steam flow to vary. The variations in steam flow cause the outlet temperature to vary, thus completing the loop.

The performance of the control loop can best be analyzed by drawing the block diagram for the entire loop. This is done by drawing the blocks for each component and connecting the output signal from each block to the input of the next block. Let us start with the heat exchanger. As shown in Fig. 6-2, the heat exchanger consists of three blocks, one for each of its three inputs.

$G_T(s)$ is the process transfer function relating the outlet temperature to the inlet temperature, C/C

$G_F(s)$ is the process transfer function relating the outlet temperature to the process flow, C/(kg/s)

$G_s(s)$ is the process transfer function relating the outlet temperature to the steam flow, C/(kg/s)

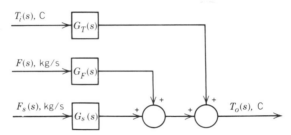

Figure 6-2. Block diagram for heat exchanger of Fig. 6-1.

The complete block diagram for the feedback control loop is shown in Fig. 6-3. The symbols in Fig. 6-3 are as follows:

$E(s)$ is the error signal, mA

$G_c(s)$ is the controller transfer function, mA/mA

$G_v(s)$ is the control valve transfer function, (kg/s)/mA

$H(s)$ is the sensor-transmitter transfer function, mA/C

K_{sp} is the scale factor for the temperature set point, mA/C

It is important at this point to notice the correspondence between the blocks (or groups of blocks) in the block diagram and the components of the control loop as sketched earlier. This comparison is facilitated by matching the symbols used to identify the various signals. It is also important to recall from Chapter 3 that the blocks on the diagram represent linear relationships between the input and output signals, and that the signals are deviations from initial steady-state values and are not absolute variable values.

In order to keep the diagram simple we have included the constant gain of the current-to-pressure transducer (I/P in Fig. 6-1) in the control valve transfer function, $G_v(s)$. The transducer gain is

$$\frac{\Delta P}{\Delta I} = \frac{(15 - 3) \text{ psi}}{(20 - 4) \text{ mA}} = 0.75 \text{ psi/mA}$$

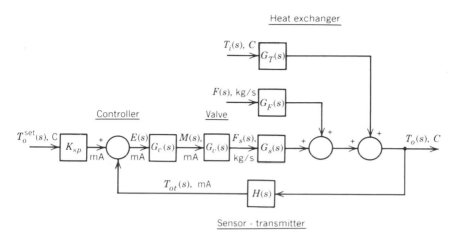

Figure 6-3. Block diagram for heat exchanger temperature control loop. (*Note:* The I/P transducer block has been combined with the control valve block.)

Note that this makes the units of $G_v(s)$ (kg/s)/mA. We have also assumed that the pressure drop across the steam valve is constant.

The term K_{sp} on the set-point signal is included to indicate the conversion of the set-point scale, usually calibrated in the same units as the controlled variable, to the same basis as the transmitter signal, that is, C to mA. When the controller indicates the measurement and the set point on the same scale, K_{sp} is numerically equal to the steady-state gain of the transmitter.

Closed-Loop Transfer Function

We can see by inspection of the closed-loop block diagram (Fig. 6-3) that the loop has one output signal, the controlled variable $T_o(s)$, and three input signals, the set point $T_o^{set}(s)$, and the two disturbances, $T_i(s)$ and $\Gamma(s)$. Since the steam flow is connected to the outlet temperature through the control loop, we might expect that the "closed-loop response" of the system to the various inputs would be different from the response when the loop is "open." Most control loops can be opened by flipping a switch on the controller from the "automatic" to the "manual" position (see Chapter 5). When the controller is in the manual position, its output does not respond to the error signal and it is thus independent of the set point and measurement signals. On the other hand, when in "automatic" the controller output varies when the measurement signal varies.

We can determine the closed-loop transfer function of the loop output with regard to any of its inputs by applying the rules of block diagram algebra learned in Chapter 3 to the diagram of the loop. To review, suppose we want to derive the response of the outlet temperature $T_o(s)$ to the inlet temperature $T_i(s)$. We first write the equations for each block in the diagram, as follows:

Error signal: $$E(s) = K_{sp}T_o^{set}(s) - T_{ot}(s) \tag{6-1}$$

Manipulated variable: $$M(s) = G_c(s)E(s) \tag{6-2}$$

Steam flow: $\qquad\qquad$ $F_s(s) = G_v(s)M(s)$ $\qquad\qquad\qquad$ (6-3)

Outlet temperature: $\qquad$ $T_o(s) = G_s(s)F_s(s) + G_F(s)F(s) + G_T(s)T_i(s)$ $\qquad$ (6-4)

Transmitter signal: $\qquad$ $T_{ot}(s) = H(s)T_o(s)$ $\qquad\qquad\qquad$ (6-5)

Next we assume that the process flow and the set point do not vary, that is, their deviation variables are zero

$$F(s) = 0$$
$$T_o^{set}(s) = 0$$

and eliminate all the intermediate variables by combination of the above equations to obtain a relationship between $T_o(s)$ and $T_i(s)$:

$$T_o(s) = G_s(s)G_v(s)G_c(s)[-H(s)T_o(s)] + G_T(s)T_i(s) \qquad\qquad (6\text{-}6)$$

We can then rearrange this equation into:

$$\frac{T_o(s)}{T_i(s)} = \frac{G_T(s)}{1 + H(s)G_s(s)G_v(s)G_c(s)} \qquad\qquad (6\text{-}7)$$

This is the closed-loop transfer function between the inlet and outlet temperatures. Similarly, if we let $T_i(s) = 0$ and $T_o^{set}(s) = 0$ in Eqs. (6-1) through (6-5), we obtain the closed-loop transfer function between the process flow and the outlet temperature:

$$\frac{T_o(s)}{F(s)} = \frac{G_F(s)}{1 + H(s)G_s(s)G_v(s)G_c(s)} \qquad\qquad (6\text{-}8)$$

Finally, by setting $T_i(s) = 0$ and $F(s) = 0$ and combining Eqs. (6-1) through (6-5), we obtain the closed-loop transfer function between the set point and the outlet temperature:

$$\frac{T_o(s)}{T_o^{set}(s)} = \frac{G_s(s)G_v(s)G_c(s)K_{sp}}{1 + H(s)G_s(s)G_v(s)G_c(s)} \qquad\qquad (6\text{-}9)$$

As we saw in Chapter 3, the denominator is the same for all three inputs while the numerator is different for each input. We recall further that the denominator is one plus the product of the transfer functions of all the blocks that are in the loop itself, while the numerator of each transfer function is the product of the blocks that are in the direct path between the specific input and the output of the loop. These results apply to any block diagram that contains a single loop.

It is enlightening to check the units of the product of the blocks in the loop, as follows:

$$H(s) \cdot G_s(s) \cdot G_v(s) \cdot G_c(s) = \text{dimensionless}$$
$$\left(\frac{mA}{C}\right) \cdot \left(\frac{C}{kg/s}\right) \cdot \left(\frac{kg/s}{mA}\right) \cdot \left(\frac{mA}{mA}\right)$$

This shows that the product of the transfer functions of the blocks in the loop is dimensionless, as it should be. We can also verify that the units of the numerator of each of the closed-loop transfer functions are the units of the output variable divided by the units of the corresponding input variable.

Characteristic Equation of the Loop

As we saw in the preceding discussion, the denominator of the closed-loop transfer function of a feedback control loop is independent of the location of the input to the loop and thus characteristic of the loop. We recall from Chapter 2 that the unforced response of the loop and its stability depend on the eigenvalues or roots of the equation that is obtained when the denominator of the transfer function of the loop is set equal to zero:

$$1 + H(s)G_s(s)G_v(s)G_c(s) = 0 \qquad (6\text{-}10)$$

This is the *characteristic equation* of the loop. Notice that the controller transfer function is very much a part of the characteristic equation of the loop. *This is why the response of the loop can be shaped by tuning the controller.* The other elements that form part of the characteristic equation are the sensor-transmitter, the control valve, and that part of the process that affects the response of the controlled variable to the manipulated variable, that is, $G_s(s)$. On the other hand, the process transfer functions related to the disturbances $[G_T(s)$ and $G_F(s)]$ are not part of the characteristic equation.

To show that the characteristic equation determines the unforced response of the loop, let us derive the response of the closed loop to a change in inlet temperature by inverting the Laplace transform of the output signal, as we learned to do in Chapter 2. Assume that the characteristic equation can be reduced to an nth degree polynomial in the Laplace transform variable s:

$$1 + H(s)G_s(s)G_v(s)G_c(s) = a_n s^n + a_{n-1}s^{n-1} + \ldots + a_0 = 0 \qquad (6\text{-}11)$$

where $a_n, a_{n-1}, \ldots, a_0$ are the polynomial coefficients. With an appropriate computer program (such as the one listed in Appendix D), we can find the n roots of this polynomial, and factor it as follows:

$$a_n s^n + a_{n-1}s^{n-1} + \ldots + a_0 = a_n(s - r_1)(s - r_2) \ldots (s - r_n) = 0 \qquad (6\text{-}12)$$

where $r_1, r_2, \ldots, r_n$ are the eigenvalues or roots of the characteristic equation. These roots can be real numbers or pairs of complex conjugate numbers, and some of them may be repeated, as we saw in Chapter 2.

From Eq. (6-7) we obtain

$$T_o(s) = \frac{G_T(s)}{1 + H(s)G_s(s)G_v(s)G_c(s)}T_i(s) \qquad (6\text{-}13)$$

Next let us substitute Eq. (6-12) for the denominator and assume that other terms will appear because of the input forcing function $T_i(s)$:

$$T_o(s) = \frac{(\text{numerator terms})}{a_n(s - r_1)(s - r_2) \ldots (s - r_n)(\text{input terms})} \qquad (6\text{-}14)$$

We then expand this expression into partial fractions:

$$T_o(s) = \frac{b_1}{s - r_1} + \frac{b_2}{s - r_2} + \ldots + \frac{b_n}{s - r_n} + (\text{input terms}) \qquad (6\text{-}15)$$

where $b_1, b_2, \ldots, b_n$ are the constant coefficients that are determined by the method of partial fractions expansion (see Chapter 2). Inverting this expression with the help of a Laplace transform table (e.g., Table 2-1), we obtain

$$T_o(t) = b_1 e^{r_1 t} + b_2 e^{r_2 t} + \ldots + b_n e^{r_n t} + \text{(input terms)} \qquad (6\text{-}16)$$

$$\underbrace{\phantom{b_1 e^{r_1 t} + b_2 e^{r_2 t} + \ldots + b_n e^{r_n t}}}_{\text{Unforced response}} \qquad \underbrace{\phantom{\text{Forced response}}}_{\text{Forced response}}$$

We have thus shown that each of the terms of the unforced response contains a root of the characteristic equation. We recall that the coefficients $b_1, b_2, \ldots, b_n$ depend on the actual input forcing function and so does the exact response of the loop. However, the speed with which the unforced response terms die out ($r_i < 0$), diverge ($r_i > 0$), or oscillate (r_i is complex) is determined entirely by the roots of the characteristic equation. We will use this concept in the next section to determine the stability of the loop.

The following two examples illustrate the effect of a pure proportional and of a pure integral controller on the closed-loop response of a first-order process. We will see that the pure proportional controller will speed up the first-order response and result in an *offset* or steady-state error, as discussed in Chapter 5. On the other hand, the integral controller produces a second-order response that, as the controller gain increases, changes from overdamped to underdamped. As discussed in Chapter 4, the underdamped response is oscillatory.

Example 6-1.   Proportional Control of a First-Order Process.   The block diagram for a simple process is shown in Fig. 6-4a. The process can be represented by a first-order lag:

$$G(s) = \frac{K}{\tau s + 1}$$

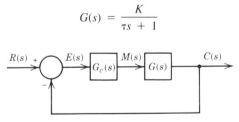

Figure 6-4a.   Block diagram for Example 6-1.

Determine the closed-loop transfer function and the response to a unit step change in set point for a proportional controller:

$$G_c(s) = K_c$$

Solution.   From block diagram algebra we obtain the closed-loop transfer function:

$$\frac{C(s)}{R(s)} = \frac{G(s)G_c(s)}{1 + G(s)G_c(s)}$$

We then substitute the process transfer function and simplify:

$$\frac{C(s)}{R(s)} = \frac{\dfrac{K}{\tau s + 1} G_c(s)}{1 + \dfrac{K}{\tau s + 1} G_c(s)}$$

$$= \frac{K G_c(s)}{1 + \tau s + K G_c(s)}$$

For a proportional controller $G_c(s) = K_c$. Substituting this yields

$$\frac{C(s)}{R(s)} = \frac{KK_c}{1 + KK_c + \tau s} = \frac{KK_c/(1 + KK_c)}{[\tau/(1 + KK_C)]s + 1} = \frac{K'}{\tau's + 1}$$

We can easily see that the steady-state gain

$$K' = KK_c/(1 + KK_c)$$

is always less than unity and that the closed-loop time constant

$$\tau' = \frac{\tau}{1 + KK_c}$$

is always less than the open-loop time constant τ. In other words, the loop always responds faster than the original system but does not quite match the set point at steady state, that is, there will be *offset*. For a unit step change in set point, from Table 2-1

$$R(s) - \frac{1}{s}$$

Substituting in the transfer function and expanding by partial fractions yields

$$C(s) = \frac{K'}{\tau's + 1} \frac{1}{s} = \frac{K'}{s} - \frac{K'\tau'}{\tau's + 1}$$

Inverting results in

$$c(t) - K'(1 - e^{-t/\tau'})$$

$$c(t) = \frac{KK_c}{1 + KK_c}[1 - e^{-(1 + KK_c)t/\tau}]$$

We see from this result that as $t \rightarrow \infty$, $c(t) \rightarrow KK_c/(1 + KK_c)$.

Since the change in set point is 1.0 (unit step change), then the error is given by

$$e(t) = r(t) - c(t) = 1.0 - \frac{KK_c}{1 + KK_c}[1 - e^{-(1 + KK_c)t/\tau}]$$

As $t \rightarrow \infty$

$$e(t) \rightarrow 1 - \frac{KK_c}{1 + KK_c} = \frac{1}{1 + KK_c}$$

Thus, the higher the controller gain, K_c, the closer the controlled variable approaches the set point at steady state, that is, the smaller the offset. This verifies what we discussed in Chapter 5 and is illustrated in Fig. 6-4b.

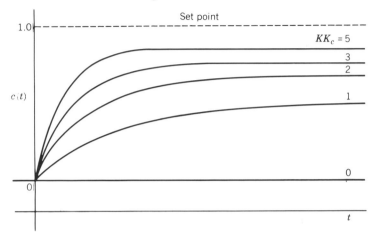

Figure 6-4b.   Loop response to a unit step change in set point for proportional controller in Example 6-1.

Example 6-2.   Pure Integral Control of a First-Order Process.   Determine the closed-loop transfer function and the response to a unit step change in set point for the process of Example 6-1, and a pure integral controller

$$G_c(s) = \frac{K_c}{s}$$

where K_c is the controller gain in min^{-1}.

Solution.   Substitute the integral controller transfer function into the closed-loop transfer function of Example 6-1:

$$\frac{C(s)}{R(s)} = \frac{KK_c/s}{1 + \tau s + KK_c/s} = \frac{KK_c}{\tau s^2 + s + KK_c}$$

By the extension of the final value theorem to transfer functions (see Section 3-3), we substitute $s = 0$ to obtain the steady-state gain:

$$\lim_{s \to 0} \frac{C(s)}{R(s)} = \frac{KK_c}{KK_c} = 1.0$$

This means that, for the integral controller, the controlled variable will always match the set point at steady state, that is, there will not be an offset.

The characteristic equation of the loop is

$$\tau s^2 + s + KK_c = 0$$

The roots of this quadratic equation are

$$s = \frac{-1 \pm \sqrt{1 - 4KK_c\tau}}{2\tau}$$

These roots are real for $0 \leq KK_c\tau \leq 1/4$ and complex conjugates for $KK_c\tau > 1/4$. Next we will determine the response of the loop to a unit step change in set point for various values of KK_c. For a unit step change in set point, from Table 2-1, $R(s) = 1/s$.

Substituting into the transfer function we get

$$C(s) = \frac{KK_c}{s(\tau s^2 + s + KK_c)}$$

Case A.

Two different real roots: $0 \leq KK_c\tau < 1/4$

Let

$$r_1 = \frac{-1 + \sqrt{1 - 4KK_c\tau}}{2\tau}$$

$$r_2 = \frac{-1 - \sqrt{1 - 4KK_c\tau}}{2\tau}$$

By partial fractions expansion

$$C(s) = \frac{1}{s} + \frac{KK_c}{\tau r_1(r_1 - r_2)} \frac{1}{s - r_1} + \frac{KK_c}{\tau r_2(r_2 - r_1)} \frac{1}{s - r_2}$$

Inverting the Laplace transform yields

$$c(t) = 1 + \frac{KK_c}{\tau r_1(r_1 - r_2)} e^{r_1 t} + \frac{KK_c}{\tau r_2(r_2 - r_1)} e^{r_2 t}$$

As both r_1 and r_2 are negative for positive K_c, the exponential terms in this response die down to zero as time increases ($t \rightarrow \infty$). Then the steady-state value of c is 1.0, or equal to the set point. This type of response, known as "overdamped," is illustrated in Fig. 6-5a. As the controller gain K_c is increased, the response becomes faster until it becomes "critically damped" at $KK_c\tau = 1/4$. We will consider this case next.

Case B.

Real repeated roots: $KK_c\tau = 1/4$

Then

$$r_1 = r_2 = -\frac{1}{2\tau}$$

Substitute $KK_c = 1/(4\tau)$ and follow the procedure for partial fractions expansion for repeated roots (Chapter 2):

$$C(s) = \frac{1/(4\tau^2)}{s\left(s + \frac{1}{2\tau}\right)^2} = \frac{1}{s} - \frac{1}{2} \frac{1/\tau}{\left(s + \frac{1}{2\tau}\right)^2} - \frac{1}{\left(s + \frac{1}{2\tau}\right)}$$

Inverting with the help of a Laplace transform table, Table 2-1, results in

$$c(t) = 1 - \left(\frac{1}{2}\frac{t}{\tau} + 1\right)e^{-\frac{t}{2\tau}}$$

This critically damped response is illustrated in Fig. 6-5a. If the controller gain K_c is increased further, an "underdamped" response results. This is the next and final case.

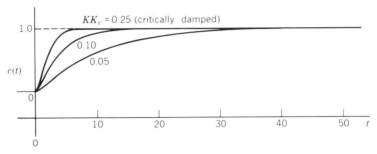

Figure 6-5a. Loop response to a unit step change in set point for integral controller in Example 6-2. Overdamped and critically damped cases ($\tau = 1$).

Case C.

Complex conjugate roots: $KK_c\tau > 1/4$

Let

$$\omega = \frac{1}{2\tau}\sqrt{4KK_c\tau - 1}$$

Then

$$r_1 = -\frac{1}{2\tau} + i\omega$$

$$r_2 = -\frac{1}{2\tau} - i\omega$$

and following the procedure for complex roots (see Chapter 2), the partial fractions expansion yields

$$C(s) = \frac{KK_c}{s(\tau s^2 + s + KK_c)} = \frac{1}{s} - \frac{s + 1/\tau}{\left[\left(s + \frac{1}{2\tau}\right)^2 + \omega^2\right]}$$

Inverting with the help of a Laplace transform table, Table 2-1, gives

$$c(t) = 1 - e^{-\frac{t}{2\tau}}\left(\cos \omega t + \frac{1}{2\omega\tau}\sin \omega t\right)$$

We see from this last result that, as the gain of the loop KK_c is increased, the response oscillates around the set point (1.0) with increasing frequency (ω). However, the amplitude of these oscillations always decays to zero because of the exponential term $e^{-t/2\tau}$. This is illustrated in Fig. 6-5b.

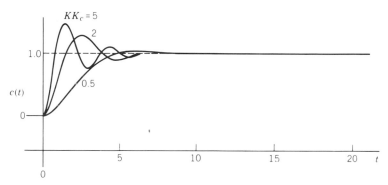

Figure 6-5b. Loop response to a unit step change in set point for integral controller in Example 6-2. Underdamped case ($\tau = 1$).

The control loop in the previous examples, shown in Fig. 6-4a, is a "unity-feedback" loop, that is, a loop that has no elements in the feedback path. This assumes that the process transfer function $G(s)$ includes the gain of the transmitter (it also includes the gain of the valve) so that all of the signals can be assumed to be in percent or fraction of range.

Example 6-2 illustrates the point discussed in Chapter 4 regarding the fact that although most processes are inherently overdamped, their response can be underdamped when forming part of a closed feedback control loop.

Steady-State Closed-Loop Response

We have seen in the preceding example that the final or steady-state value is an important aspect of the closed-loop response. This is because in current industrial process control practice the presence of steady-state error or *offset* is usually unacceptable. We shall learn in this section how to calculate the offset when it is present. To do this we return to the exchanger of Fig. 6-1 and the corresponding block diagram of Fig. 6-3. As we learned earlier, this is a linearized representation of the heat exchanger. Our approach is to obtain the steady-state closed-loop relationships between the output variable and each of the inputs to the loop by application of the final value theorem to the closed-loop transfer function. In Eq. (6-8) the closed-loop transfer function between the outlet temperature and the process fluid flow is given by

$$\frac{T_o(s)}{F(s)} = \frac{G_F(s)}{1 + H(s)G_s(s)G_v(s)G_c(s)} \tag{6-8}$$

We recall that this expression assumes that the deviation variables for the inlet temperature T_i and the set point T_o^{set} are zero when these other inputs remain constant. We also recall from Section 3-3 that the steady-state relationship between the output and the input to a transfer function is obtained by setting $s = 0$ in the transfer function. This follows from the final value theorem of Laplace transforms. Applying this method to Eq. (6-8) we obtain:

$$\frac{\Delta T_o}{\Delta F_s} = \frac{G_F(0)}{1 + H(0)G_s(0)G_v(0)G_c(0)} \; C/(kg/s) \tag{6-17}$$

where

ΔT_o is the steady-state change in outlet temperature, C
ΔF_s is the steady-state change in process fluid flow, kg/s

If we assume, as is usually the case, that the process is stable, then

$G_F(0) = K_F$ the process open-loop gain to a change in process fluid flow, C/(kg/s)
$G_s(0) = K_s$ the process open-loop gain to a change in steam flow, C/(kg/s)

Similarly, for the valve and the sensor-transmitter

$G_v(0) = K_v$ the valve gain, (kg/s)/mA
$H(0) = K_h$ the sensor-transmitter gain, mA/C

Finally, if the controller does not have integral action

$G_c(0) = K_c$ the proportional gain, mA/mA.

Substituting these terms into Eq. (6-17) yields

$$\frac{\Delta T_o}{\Delta F_s} = \frac{K_F}{1 + K_h K_s K_v K_c} \; C/(kg/s) \tag{6-18}$$

Since the change in set point is zero, the steady-state error or offset is given by

$$e = \Delta T_o^{set} - \Delta T_o = -\Delta T_o \; C$$

and, combining this relationship with Eq. (6-18) gives

$$\frac{e}{\Delta F_s} = \frac{-K_F}{1 + K_h K_s K_v K_c} \; C/(kg/s) \tag{6-19}$$

Note that the offset is reduced as the controller gain, K_c, is increased.
     For a proportional-integral-derivative (PID) controller

$$G_c(0) = \lim_{s \to 0} K_c \left(1 + \frac{1}{\tau_I s} + \tau_D s \right) = \infty$$

In which case, by substitution into Eq. (6-17), we can see that the offset is zero for any value of K_c. The same is true for a PI controller ($\tau_D = 0$).

Following an identical procedure for Eq. (6-7), we obtain the steady-state relationship of the outlet temperature to the inlet temperature for constant flow and set point:

$$\frac{\Delta T_o}{\Delta T_i} = \frac{K_T}{1 + K_h K_s K_v K_c} \quad C/C \tag{6-20}$$

where

ΔT_i is the steady-state change in inlet temperature, C

$K_T = G_T(0)$ is the steady-state open-loop gain to a change in inlet temperature, C/C.

Finally, the steady-state relationship to a change in set point at constant process fluid flow and inlet temperature is obtained from Eq. (6-9):

$$\frac{\Delta T_o}{\Delta T_o^{set}} = \frac{K_s K_v K_c K_{sp}}{1 + K_h K_s K_v K_c} \quad C/C \tag{6-21}$$

where ΔT_o^{set} is the steady-state change in set point, C. The offset in this case is given by (Fig. 6-3)

$$e = \Delta T_o^{set} - \Delta T_o C$$

We recall that for the set point and the measurement to be on the same scale, $K_{sp} - K_h$. Combining these relationships with Eq. (6-21) gives

$$\frac{e}{\Delta T_o^{set}} = \frac{1}{1 + K_h K_s K_v K_c} \quad C/C \tag{6-22}$$

Again the offset is smaller the higher the controller gain, and is zero at any controller gain if the controller has integral action.

Example 6-3. For the heat exchanger of Fig. 6-1, calculate the linearized ratios for the steady-state error in outlet temperature to

a. A change in process flow
b. A change in inlet temperature
c. A change in set point

The operating conditions and instrument specifications are

Process fluid flow	$\overline{F} = 12$ kg/s
Inlet temperature	$\overline{T}_i = 50$ C
Set point	$\overline{T}_o^{set} = 90$ C
Heat capacity of fluid	$C_p = 3750$ J/kg C
Latent heat of steam	$\lambda = 2.25 \times 10^6$ J/kg
Capacity of steam valve	$F_{smax} = 1.6$ kg/s
Transmitter range	50 to 150 C

Solution. If we assume that heat losses are negligible, we can write the following steady-state energy balance:

$$FC_p(T_o - T_i) = F_s\lambda$$

and, solving for F_s, the steam flow required to maintain T_o at 90C and the design conditions is

$$F_s = \frac{FC_p(T_o - T_i)}{\lambda} = \frac{(12)(3750)(90 - 50)}{2.25 \times 10^6} = 0.80 \text{ kg/s}$$

The next step is to calculate the steady-state open-loop gains of each of the elements in the loop.

Exchanger

From the steady-state energy balance, solving for T_o yields

$$T_o = T_i + \frac{F_s\lambda}{FC_p}$$

By linearization, as learned in Chapter 2, we obtain

$$K_F = \frac{\partial T_o}{\partial F} = -\frac{\overline{F_s}\lambda}{\overline{F}^2 C_p}$$

$$= -\frac{(0.80)(2.25 \times 10^6)}{(12)^2(3750)} = -3.33 \text{ C/(kg/s)}$$

$$K_T = \frac{\partial T_o}{\partial T_i} = 1 \text{ C/C}$$

$$K_s = \frac{\partial T_o}{\partial F_s} = \frac{\lambda}{\overline{F}C_p} = \frac{2.25 \times 10^6}{(12)(3750)} = 50 \text{ C/(kg/s)}$$

Control Valve

Assuming a linear valve with constant pressure drop and a linear relationship between the valve position, vp, and the controller output, m, we obtain

$$F_s = F_{smax} vp = F_{smax} \frac{m - 4}{20 - 4}$$

By linearization we obtain

$$K_v = \frac{\partial F_s}{\partial m} = \frac{F_{smax}}{16} = \frac{1.6}{16} = 0.10 \text{ (kg/s)/mA}$$

Sensor-Transmitter

The transmitter can be represented by the following linear relationship:

$$T_{ot} = \frac{20 - 4}{150 - 50}(T_o - 50) + 4$$

By linearization we obtain

$$K_h = \frac{\partial T_{ot}}{\partial T_o} = \frac{16}{100} = 0.16 \text{ mA/C}$$

a. Substitute into Eq. (6-18) to get

$$\frac{\Delta T_o}{\Delta F} = \frac{K_F}{1 + K_h K_s K_v K_c} = \frac{-3.33}{1 + (0.16)(50)(0.10)K_c}$$

$$= \frac{-3.33 \text{ C/(kg/s)}}{1 + 0.80 K_c}$$

b. Substitute into Eq. (6-20) to get

$$\frac{\Delta T_o}{\Delta T_i} = \frac{K_T}{1 + K_h K_s K_v K_c} = \frac{1.0}{1 + 0.80 K_c} \text{ C/C}$$

c. Substitute into Eq. (6-21) to get

$$\frac{\Delta T_o}{\Delta T_o^{set}} = \frac{K_s K_v K_c K_{sp}}{1 + K_h K_s K_v K_c} = \frac{0.80 K_c}{1 + 0.80 K_c} \text{ C/C}$$

The results for different values of K_c are

$K_c \dfrac{mA}{mA}$	$\dfrac{\Delta T_o}{\Delta F} \dfrac{C}{kg/s}$	$\dfrac{\Delta T_o}{\Delta T_i} \dfrac{C}{C}$	$\dfrac{\Delta T_o}{\Delta T_o^{set}} \dfrac{C}{C}$
0.5	-2.38	0.714	0.286
1.0	-1.85	0.556	0.444
5.0	-0.67	0.200	0.800
10.0	-0.37	0.111	0.889
20.0	-0.20	0.059	0.941
100.0	-0.04	0.012	0.988

We see that the change in outlet temperature due to disturbances approaches zero as the gain is increased and that the change in outlet temperature due to a set-point change approaches unity.

The results of the preceding example illustrate the point made in Chapter 5 regarding the fact that the offset decreases when the gain of the proportional controller is increased. As pointed out there, the gain of the controller is limited by the stability of the loop. We shall see this in the next section.

The following example illustrates the development of the block diagram for a simple control loop by applying basic process engineering principles.

Example 6-4. Temperature Control of a Continuous Stirred Tank Heater
The stirred tank sketched in Fig. 6-6 is used to heat a process stream so that its premixed components achieve a uniform composition. Temperature control is important because a high temperature tends to decompose the product while a low temperature

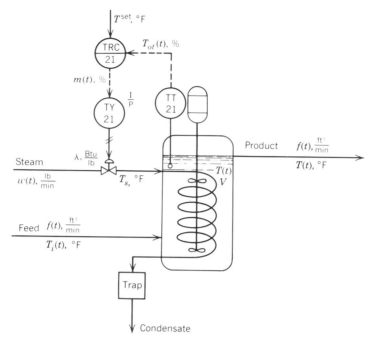

Figure 6-6. Temperature control of continuous stirred tank heater.

results in incomplete mixing. The tank is heated by steam condensing inside a coil. A proportional-integral-derivative (PID) controller is used to control the temperature in the tank by manipulating the steam valve position. It is desired to derive the complete block diagram and the characteristic equation of the loop from the following design data.

Process

The feed has a density ρ of 68.0 lb/ft^3 and a heat capacity C_p of 0.80 Btu/lb°F. The volume V of liquid in the reactor is maintained constant at 120 ft^3. The coil consists of 205 ft of 4-in. schedule 40 steel pipe, weighing 10.8 lb/ft with a heat capacity of 0.12 Btu/lb°F and an outside diameter of 4.500 in. The overall heat transfer coefficient U, based on the outside area of the coil, has been estimated as 2.1 Btu/min ft^2°F. The steam available is saturated at a pressure of 30 psia; it can be assumed that its latent heat of condensation λ is constant at 966 Btu/lb.

Design Conditions

The feed flow f at design conditions is 15 ft^3/min and its temperature T_i is 100°F. The contents of the tank must be maintained at a temperature T of 150°F. Possible disturbances are changes in feed rate and temperature.

Temperature Sensor and Transmitter

The temperature sensor has a calibrated range of 100 to 200°F and a time constant τ_t of 0.75 min.

Control Valve

The control valve is to be designed for 100% overcapacity and pressure drop variations can be neglected. The valve is an equal percentage valve with a rangeability parameter α of 50. The actuator has a time constant τ_v of 0.20 min.

Solution Our approach will be to derive the equations that describe the dynamic behavior of the tank, the control valve, the sensor-transmitter, and the controller. Then we will linearize them and Laplace transform them to obtain the block diagram of the loop.

Process

An energy balance on the liquid in the tank, assuming negligible heat losses, perfect mixing, and constant volume and physical properties, results in the following equation:

$$V\rho C_p \frac{dT(t)}{dt} = f(t)\rho C_p T_i(t) + UA[T_s(t) - T(t)] - f(t)\rho C_p T(t)$$

1 eqn., 2 unks. (T, T_s)

where

A is the heat transfer area, ft^2
$T_s(t)$ is the condensing steam temperature, °F

and the other symbols have been defined in the statement of the problem. For the liquid contents of the tank, the C_v in the accumulation term has been approximated with C_p.

An energy balance on the coil, assuming that the coil metal is essentially at the same temperature as the condensing steam, results in

$$C_M \frac{dT_s(t)}{dt} = w(t)\lambda - UA[T_s(t) - T(t)]$$

2 eqns., 3 unks. (w)

where

$w(t)$ is the steam rate, lb/min
C_M is the heat capacitance of the coil metal, Btu/°F

Since the steam rate is the output of the control valve and an input to the process, our process model is complete.

Control Valve

The equation for an equal-percentage valve with constant upstream pressure and constant pressure drop can be written as

$$w(t) = W_{max}\alpha^{vp(t)-1}$$

3 eqns., 4 unks. (vp)

where

W_{max} is the maximum flow through the valve, lb/min
α is the equal-percentage rangeability parameter
$vp(t)$ is the valve position in a scale of 0 to 1

The variation in pressure drop across the valve with steam condensing temperature (and pressure) has been neglected in this simple example. The valve actuator can be modeled by a first-order lag:

$$VP(s) = \frac{1/100}{\tau_v s + 1} M(s) \qquad\qquad \text{4 eqns., 5 unks. } (M)$$

where $M(s)$ is the controller output signal in percent.

Sensor-Transmitter (TT21)

The sensor-transmitter can be represented by a first-order lag:

$$\frac{T_{ot}(s)}{T(s)} = \frac{K_t}{\tau_t s + 1} \qquad\qquad \text{5 eqns., 6 unks. } (T_{ot})$$

where

$T_{ot}(s)$ is the Laplace transform of the transmitter output signal, %.

Feedback Controller (TRC21)

The transfer function of the PID controller is

$$G_c(s) = K_c\left(1 + \frac{1}{\tau_I s} + \tau_D s\right) = \frac{M(s)}{R(s) - T_{ot}(s)}$$

$$\text{6 eqns., 6 unks.}$$

where K_c is the controller gain, τ_I the integral time, and τ_D the derivative time.

   This completes the derivation of the equation for the temperature control loop. Our next step will be to linearize the model equations and Laplace transform them to obtain the block diagram of the loop.

Linearization and Laplace Transformation

By the methods learned in Section 2-3, we obtain the linearized tank model equations in terms of deviation variables

$$V\rho C_p \frac{dT(t)}{dt} = \bar{f}\rho C_p T_i(t) + \rho C_p(\bar{T}_i - \bar{T})F(t)$$

$$+ UA T_s(t) - (UA + \bar{f}\rho C_p)T(t)$$

$$C_M \frac{dT_s(t)}{dt} = \lambda W(t) - UA T_s(t) + UA T(t)$$

where $T(t)$, $T_s(t)$, $F(t)$, $T_i(t)$, and $W(t)$ are the deviation variables.

Taking the Laplace transform of these equations and rearranging as learned in Chapter 2 yields

$$T(s) = \frac{K_F}{\tau s + 1} F(s) + \frac{K_i}{\tau s + 1} T_i(s) + \frac{K_s}{\tau s + 1} T_s(s)$$

$$T_s(s) = \frac{1}{\tau_c s + 1} T(s) + \frac{K_W}{\tau_c s + 1} W(s)$$

where

$$\tau = \frac{V\rho C_p}{UA + \bar{f}\rho C_p} \qquad \tau_c = \frac{C_M}{UA}$$

$$K_F = \frac{\rho C_p(\bar{T}_i - \bar{T})}{UA + \bar{f}\rho C_p} \qquad K_i = \frac{\bar{f}\rho C_p}{UA + \bar{f}\rho C_p}$$

$$K_s = \frac{UA}{UA + \bar{f}\rho C_p} \qquad K_W = \frac{\lambda}{UA}$$

Linearization of the valve equation results in

$$W(t) = W_{max}(\ln \alpha)\alpha^{\overline{vp}-1} VP(t)$$

$$= \overline{W}(\ln \alpha) VP(t)$$

where $VP(t)$ is the valve-position deviation variable.

Taking the Laplace transform of this equation results in

$$W(s) = \overline{W}(\ln \alpha) VP(s)$$

Combining this equation with the transfer function of the actuator we can eliminate $VP(s)$:

$$\frac{W(s)}{M(s)} = \frac{K_v}{\tau_v s + 1}$$

where

$$K_v = \frac{\overline{W}(\ln \alpha)}{100}.$$

From what we learned in Chapter 5, we can determine that the gain of the transmitter is

$$K_t = \frac{100 - 0}{200 - 100} = 1.0 \ \%/°F$$

The complete block diagram for the loop is given in Fig. 6-7a. All of the transfer functions in the diagram have been derived above.

Using the rules for block diagram manipulation learned in Chapter 3, we can simplify the diagram as shown in Fig. 6-7b. The transfer functions shown in the diagram are

$$G_F(s) = \frac{K_F(\tau_c s + 1)}{(\tau s + 1)(\tau_c s + 1) - K_s}$$

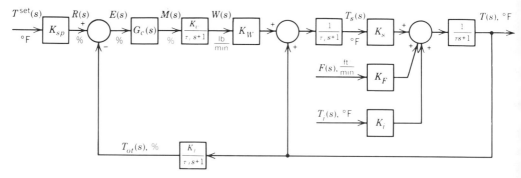

Figure 6-7a. Block diagram of temperature control loop.

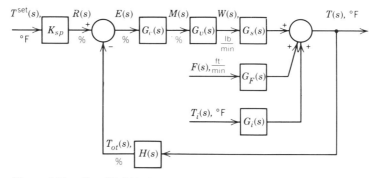

Figure 6-7b. Simplified block diagram of temperature control loop.

$$G_i(s) = \frac{K_i(\tau_c s + 1)}{(\tau s + 1)(\tau_c s + 1) - K_s}$$

$$G_s(s) = \frac{K_W K_s}{(\tau s + 1)(\tau_c s + 1) - K_s}$$

The closed-loop transfer functions to each of the inputs are

$$\frac{T(s)}{T^{\text{set}}(s)} = \frac{K_{sp} G_c(s) G_v(s) G_s(s)}{1 + H(s) G_c(s) G_v(s) G_s(s)}$$

$$\frac{T(s)}{F(s)} = \frac{G_F(s)}{1 + H(s) G_c(s) G_v(s) G_s(s)}$$

$$\frac{T(s)}{T_i(s)} = \frac{G_i(s)}{1 + H(s) G_c(s) G_v(s) G_s(s)}$$

where

$$G_v(s) = \frac{K_v}{\tau_v s + 1}$$

$$H(s) = \frac{K_t}{\tau_t s + 1}$$

The characteristic equation of the loop is

$$1 + \frac{K_t}{\tau_t s + 1} K_c \left(1 + \frac{1}{\tau_I s} + \tau_D s\right) \frac{K_v}{\tau_v s + 1} \frac{K_W K_s}{(\tau_s + 1)(\tau_c s + 1) - K_s} = 0$$

The numerical values are obtained next:

$$K_{sp} = K_t = 1.0\%/°F \qquad \tau_I = 0.75 \text{ min} \qquad \tau_v = 0.20 \text{ min}$$

From the description of the coil, we have

$$A = (205 \text{ ft}) \pi \left(\frac{4.500}{12} \text{ ft}\right) = 241.5 \text{ ft}^2$$

$$C_M = (205 \text{ ft})\left(10.8 \frac{\text{lb}}{\text{ft}}\right)\left(0.12 \frac{\text{Btu}}{\text{lb°F}}\right) = 265.7 \text{ Btu/°F}$$

$$\tau = \frac{(120)(68.0)(0.80)}{(2.1)(241.5) + (15)(68)(0.80)} = 4.93 \text{ min}$$

$$\tau_c = \frac{265.7}{(2.1)(241.5)} - 0.524 \text{ min}$$

$$K_F = \frac{(68)(0.80)(100 - 150)}{(2.1)(241.5) + (15)(68)(0.80)} = -2.06°F/(\text{ft}^3/\text{min})$$

$$K_i = \frac{(15)(68)(0.80)}{(2.1)(241.5) + (15)(68)(0.80)} = 0.617°F/°F$$

$$K_s = \frac{(2.1)(241.5)}{(2.1)(241.5) + (15)(68)(0.80)} = 0.383°F/°F$$

$$K_W = \frac{966}{(2.1)(241.5)} = 1.905°F/(\text{lb/min})$$

To size the control valve we make use of the fact that the design conditions are at steady state:

$$\bar{f}\rho C_p \bar{T}_i + UA(\bar{T}_s - \bar{T}) - \bar{f}\rho C_p \bar{T} = 0$$

$$\bar{W}\lambda - UA(\bar{T}_s - \bar{T}) = 0$$

$$\bar{T}_s = \frac{(15)(68)(0.80)(150 - 100)}{(2.1)(241.5)} + 150 = 230°F$$

$$\bar{W} = \frac{(2.1)(241.5)(230 - 150)}{966} = 42.2 \text{ lb/min}$$

$$K_v = \frac{(42.2)(\ln 50)}{100} = 1.652 \text{ lbm/min-}\%$$

and

$$W_{max} = 2\bar{W} = 84.4 \text{ lb/min}$$

with these numbers the characteristic equation is

$$s(0.75s + 1)(0.20s + 1)[(4.93s + 1)(0.524s + 1) - 0.383]$$
$$+ (1.0)K_c\left(s + \frac{1}{\tau_I} + \tau_D s^2\right)(1.652)(1.905)(0.383) = 0$$

$$0.387s^5 + 3.272s^4 + 7.859s^3 + (6.043 + 1.205K_c\tau_D)s^2$$
$$+ (0.617 + 1.205K_c)s + 1.205K_c/\tau_I = 0$$

This example illustrates how the basic principles of process engineering can be put to work in analyzing simple feedback control loops. From the characteristic equation we can study the stability of the loop, and from the closed-loop transfer functions we can calculate the response of the closed loop to various input forcing functions for different values of the controller tuning parameters K_c, τ_I, and τ_D.

6-2. STABILITY OF THE CONTROL LOOP

As defined in Chapter 2, a system is stable if its output remains bound for a bound input. Most industrial processes are open-loop stable, that is, they are stable when not a part of a feedback control loop. This is equivalent to saying that most processes are self-regulating, that is, the output will move from one steady state to another when driven by changes in its input signals. A typical example of an open-loop unstable process is an exothermic stirred-tank reactor. This type of reactor sometimes exhibits an unstable operating point where increasing temperature produces an increase in reaction rate with the consequent increase in the rate of heat liberation. This in turn causes a further increase in temperature.

Even for open-loop stable processes, stability becomes a consideration when the process becomes a part of a feedback control loop. This is because the signal variations may reinforce each other as they travel around the loop causing the output—and all the other signals in the loop—to become unbound. As noted in Chapter 1, the behavior of a feedback control loop is essentially oscillatory—"trial and error." Under some circumstances the oscillations may increase in magnitude, resulting in an unstable process. The easiest illustration of an unstable feedback loop is the controller whose direction of action is the opposite of what it should be. For example, in the heat exchanger sketched in the preceding section, if the controller output were to increase with increasing temperature (direct-acting controller), the loop would be unstable because the opening of the steam valve would cause a further increase in temperature. What is needed in this case is a reverse-acting controller that decreases its output when the temperature increases so as to close the steam valve and bring the temperature back down. However, even for a controller with the proper action, the system may become unstable because of the lags in the loop. This usually happens as the loop gain is increased. The controller gain at which the loop reaches the threshold of instability is therefore of utmost importance in the design of a feedback control loop. This maximum gain is known as the *ultimate gain*.

In this section we will determine a criterion for the stability of dynamic systems and study two methods to calculate the ultimate gain: Routh's test and direct substitution. Then we will study the effect of various loop parameters on its stability.

Criterion of Stability

We have seen earlier that the response of a control loop to a given input can be represented [Eq. (6-16)] by

$$c(t) = b_1 e^{r_1 t} + b_2 e^{r_2 t} + \ldots + b_n e^{r_n t} + \text{(input terms)} \tag{6-23}$$

where

$c(t)$ is the loop output or controlled variable
$r_1, r_2, \ldots, r_n$ are the eigenvalues or roots of the characteristic equation of the loop

Assuming that the input terms remain bound as time increases, the stability of the loop requires that the unforced response terms also remain bound as time increases. This depends only on the roots of the characteristic equation and can be expressed as follows:

> For real roots: If $r < 0$ then $e^{rt} \to 0$ as $t \to \infty$
> For complex roots: $r = \sigma + i\omega$ $e^{rt} = e^{\sigma t}(\cos \omega t + i \sin \omega t)$
>   If $\sigma < 0$ then $e^{\sigma t}(\cos \omega t + i \sin \omega t) \to 0$ as $t \to \infty$

In other words, the real part of the complex roots and the real roots must be negative in order for the corresponding terms in the response to decay to zero. This result is not affected by repeated roots, as this only introduces a polynomial of time into the solution (see Chapter 2) that cannot overcome the effect of the decaying exponential term. Notice that, if any root of the characteristic equation is a positive real number, or a complex number with a positive real part, that term on the response [Eq. (6-23)] will be unbound and the entire response will be unbound even though all the other terms may decay to zero. This brings us to the following statement of the criterion for the stability of a control loop:

> In order for a feedback control loop to be stable, all of the roots of its characteristic equation must be either negative real numbers or complex numbers with negative real parts.

If we now define the complex s plane as a two-dimensional graph with the horizontal axis for the real parts of the roots and the vertical axis for the imaginary parts, we can enunciate the following graphical statement of the criterion of stability (see Fig. 6-8):

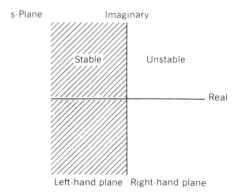

Figure 6-8. The s-plane showing regions of stability and instability for the location of the roots of the characteristic equation.

In order for a feedback control loop to be stable, all of the roots of its characteristic equation must fall on the left-hand half of the s plane, also known as the "left-hand plane."

We must point out that both of these statements of the stability criterion in the Laplace domain apply in general to any physical system, not just feedback control loops. In each case the characteristic equation is obtained by setting the denominator of the linearized transfer function of the system equal to zero.

Having enunciated the criterion of stability, let us turn our attention to the determination of the stability of a control loop.

Routh's Test

Routh's test is a procedure to determine how many of the roots of a polynomial have positive real parts without having to actually find the roots by iterative techniques. Since the stability of a system requires that none of the roots of its characteristic equation have positive real parts, Routh's test is most useful to determine stability.

With today's availability of computer and calculator programs to solve for the roots of polynomials, Routh's test would not be useful if the problem were merely to find out whether a feedback loop is stable once all of the parameters of the loop have been specified. However, the more relevant problem is to find the limits on a given loop parameter—usually the controller gain—for which the loop is stable. Routh's test is most useful for solving this problem.

The mechanics of Routh's test can be presented as follows: Given the nth degree polynomial,

$$a_n s^n + a_{n-1} s^{n-1} + \ldots + a_1 s + a_0 = 0 \tag{6-24}$$

where $a_n, a_{n-1}, \ldots, a_1, a_0$ are the coefficients of the polynomial, determine how many roots have positive real parts.

To perform the test we must first prepare the following array:

Row 1	a_n	a_{n-2}	a_{n-4}	$\ldots$	a_1	0
Row 2	a_{n-1}	a_{n-3}	a_{n-5}	$\ldots$	a_0	0
Row 3	b_1	b_2	b_3	$\ldots$	0	0
Row 4	c_1	c_2	c_3	$\ldots$	0	0
$\vdots$	$\vdots$	$\vdots$	$\vdots$	$\vdots$	$\vdots$	$\vdots$
Row n	d_1	d_2	0	$\ldots$	0	0
Row $n+1$	e_1	0	0	$\ldots$	0	0

where rows 3 through $n+1$ are calculated by

$$b_1 = \frac{a_{n-1}a_{n-2} - a_n a_{n-3}}{a_{n-1}} \qquad b_2 = \frac{a_{n-1}a_{n-4} - a_n a_{n-5}}{a_{n-1}} \qquad \text{etc.}$$

$$c_1 = \frac{b_1 a_{n-3} - a_{n-1}b_2}{b_1} \qquad c_2 = \frac{b_1 a_{n-5} - a_{n-1}b_3}{b_1} \qquad \text{etc.}$$

and so on. The process is continued until all new terms are zero. Once the array is completed, the number of roots of the polynomial that have positive real parts can be determined by counting the number of *changes* of sign in the extreme left-hand column of the array. In other words, for the polynomial to have all of its roots on the left half of the *s*-plane, all of the terms on the left-hand column of the array must be of the same sign.

In order to illustrate the use of Routh's test, let us apply it to the determination of the ultimate gain of the temperature controller for the exchanger discussed in the previous section.

Example 6-5.   Ultimate Gain of Temperature Controller by Routh's Test
Let us assume that the transfer functions for the various elements of the temperature control loop of Fig. 6-3 are as follows:

Exchanger

The exchanger response to the steam flow has a gain of 50 C/(kg/s) and a time constant of 30 s:

$$G_s(s) = \frac{50}{30\ s\ +\ 1}\ \mathrm{C/(kg/s)}$$

Sensor-Transmitter

The sensor-transmitter has a calibrated range of 50 to 150 C and a time constant of 10 s.

$$\mathrm{Gain} = \frac{100\%}{(150\ -\ 50)\mathrm{C}} = 1.0\ \%/\mathrm{C}$$

$$H(s) = \frac{1.0}{10\ s\ +\ 1}\ \%/\mathrm{C}$$

Note: In this and other examples we will use "percent of range" (%) as the units of the transmitter and controller signals. For electronic signals, 100% = 16 mA, and for penumatic signals, 100% = 12 psi.

Control Valve

The control valve has a maximum capacity of 1.6 kg/s of steam, linear characteristics, and a time constant of 3 s.

$$\mathrm{Gain} = \frac{1.6\ \mathrm{(kg/s)}}{100\%} = 0.016\ \mathrm{(kg/s)}/\%$$

$$G_v(s) = \frac{0.016}{3\ s\ +\ 1}\ \mathrm{(kg/s)}/\%$$

(we have assumed a constant pressure drop across the valve.)

Controller

The controller is proportional only.

$$G_c(s) = K_c \ \%/\%$$

The problem is then to determine the ultimate controller gain, that is, the value of K_c at which the loop becomes marginally stable.

Solution. The characteristic equation is given by Eq. (6-10):

$$1 + H(s)G_s(s)G_v(s)G_c(s) = 0$$

or

$$1 + \frac{1}{10s + 1} \cdot \frac{50}{30s + 1} \cdot \frac{0.016}{3s + 1} \cdot K_c = 0$$

We must now rearrange this equation into polynomial form:

$$(10s + 1)(30s + 1)(3s + 1) + 0.80K_c = 0$$

$$900s^3 + 420s^2 + 43s + 1 + 0.80K_c = 0$$

The next step is to prepare Routh's array:

Row 1	900	43	0
Row 2	420	$1 + 0.80K_c$	0
Row 3	b_1	0	0
Row 4	$1 + 0.80K_c$	0	0

where

$$b_1 = \frac{(420)(43) - 900(1 + 0.80K_c)}{420} = \frac{17160 - 720K_c}{420}$$

In order for the control loop to be stable, all of the terms in the left-hand column must be of the same sign, in this case positive. This requires that

$$b_1 \geqq 0 \quad \text{or} \quad 17160 - 720K_c \geqq 0 \quad K_c \leqq 23.8$$

$$1 + 0.80K_c \geqq 0 \quad \text{or} \quad 0.80K_c \geqq -1 \quad K_c \geqq -1.25$$

In this case the lower limit on K_c is negative. This is meaningless because a negative gain means that the controller has the wrong action (opens the steam valve on increasing temperature). The upper limit on the controller gain is the ultimate gain that we seek:

$$K_{cu} = 23.8 \ \%/\%$$

This tells us that in tuning the proportional controller for this loop, we must not exceed the gain of 23.8 or reduce the proportional band below $100/23.8 = 4.2\%$.

We saw in the preceding section that the offset or steady-state error inherent in proportional controllers can be reduced by increasing the controller gain. We see here that stability imposes a limit on how high that gain can be. It is of interest to study how the other parameters of the loop affect the ultimate gain.

Effect of Loop Parameters on the Ultimate Gain

Let us assume that the calibrated range of the temperature sensor-transmitter is reduced to 75–125 C. The new transmitter gain and transfer function are

$$\text{Gain} = \frac{100\%}{(125 - 75)\text{C}} = 2.0 \ \%/\text{C}$$

$$H(s) = \frac{2.0}{10s + 1} \ \%/\text{C}$$

If we repeat the Routh test with this new gain we obtain an ultimate gain of

$$K_{cu} = 11.9 \ \%/\% \quad (\text{PB} = 8.4\%)$$

This is exactly half the original ultimate gain, showing that the ultimate loop gain remains the same. The loop gain is defined as the product of the gains of all of the blocks in the loop:

$$K_l = K_h K_s K_v K_c \tag{6-25}$$

where

K_L = loop gain (dimensionless)
K_h = sensor-transmitter gain, %/C
K_s = process gain, C/(kg/s)
K_v = control valve gain, (kg/s)/%
K_c = controller gain, %/%

For the two cases considered so far, the ultimate loop gains are

$$K_{Lu} = (1.0)(50)(0.016)(23.8) = 19.04$$

$$K_{Lu} = (2.0)(50)(0.016)(11.9) = 19.04$$

Similarly, if we were to double the size of the control valve and thus its gain, the ultimate controller gain would be reduced to half its original value.

Next let us assume that a faster sensor-transmitter with a time constant of 5 s is installed in this service replacing the 10-s instrument. The new transfer function is given by

$$H(s) = \frac{1.0}{5s + 1} \ \%/\text{C}$$

We repeat the Routh test with this new transfer function:

$$1 + \frac{1.0}{5s + 1} \cdot \frac{50}{30s + 1} \cdot \frac{0.016}{3s + 1} \cdot K_c = 0$$

$$450s^3 + 255s^2 + 38s + 1 + 0.80K_c = 0$$

Row 1	450	38	0
Row 2	255	$1 + 0.80K_c$	0
Row 3	b_1	0	0
Row 4	$1 + 0.80K_c$	0	0

where

$$b_1 = \frac{(255)(38) - (450)(1 + 0.80K_c)}{255} = \frac{9240 - 360K_c}{255}$$

And the ultimate controller gain is

$$K_{cu} = \frac{9240}{360} = 25.7 \ \%/\%$$

The reduction of the time constant of the sensor has resulted in a slight increase in the ultimate gain. This is because we have reduced the measurement lag on the control loop. A similar result would be obtained if the time constant of the control valve were to be reduced. However, the increase in the ultimate gain would be even less because the valve is not as slow as the sensor-transmitter. You are invited to verify this.

Finally let us consider a case in which a change in exchanger design results in a shorter time constant for the process, namely from 30 to 20 s. The new transfer function is

$$G_s(s) = \frac{50}{20s + 1} \ \text{C/(kg/s)}$$

Again we repeat the procedure for the Routh test:

$$1 + \frac{1.0}{10s + 1} \cdot \frac{50}{20s + 1} \cdot \frac{0.016}{3s + 1} \cdot K_c = 0$$

$$600s^3 + 290s^2 + 33s + 1 + 0.80K_c = 0$$

Row 1	600	33	0
Row 2	290	$1 + 0.80K_c$	0
Row 3	b_1	0	0
Row 4	$1 + 0.80K_c$	0	0

where

$$b_1 = \frac{(290)(33) - 600(1 + 0.80K_c)}{290} = \frac{8970 - 480K_c}{290}$$

The ultimate gain is then

$$K_{cu} = \frac{8970}{480} = 18.7 \ \%/\%$$

Surprisingly, the ultimate gain is reduced by a reduction in the process time constant. This is opposite to the effect of reducing the time constant of the sensor-transmitter. The reason is that when the longest or dominant time constant is reduced, the relative effect of the other lags in the loop becomes more pronounced. In other words, in terms of the ultimate gain, reducing the longest time constant is equivalent to proportionately increasing the other time constants in the loop. What cannot be shown by the Routh test is that the loop with the shorter time constant responds faster than the original one. This can be illustrated by the method of direct substitution.

Direct Substitution Method

The method of direct substitution is based on the fact that if the roots of the characteristic equation vary continuously with the loop parameters, at the point at which the loop becomes unstable at least one and usually two of the roots must lie on the imaginary axis of the complex plane, that is, there must be pure imaginary roots. Another way of looking at it is that in order for the roots to move from the left-hand plane to the right-hand plane they must cross the imaginary axis. At this point the loop is said to be "marginally stable" and the corresponding term on the loop output is, in the Laplace domain

$$C(s) = \frac{b_1 s + b_2}{s^2 + \omega_u^2} + \text{(other terms)} \tag{6-26}$$

or, inverting, this term, from Table 2-1, is a sine wave in the time domain:

$$c(t) = b_1' \sin(\omega_u T + \theta) + \text{(other terms)} \tag{6-27}$$

where

ω_u = frequency of sine wave
θ = phase angle of sine wave
b_1' = amplitude of sine wave (constant)

This means that at the point of marginal stability the characteristic equation must have a pair of pure imaginary roots at

$$r_{1,2} = \pm i\omega_u$$

The frequency ω_u with which the loop will oscillate is the *ultimate frequency*. Just before reaching this point of marginal instability, the system oscillates with a decaying amplitude, while just beyond this point, the amplitude of the oscillations increases with time. At the point of marginal stability the amplitude of the oscillations remains constant with time. This is shown in Fig. 6-9, where the relationship between the ultimate period, T_u, and the ultimate frequency, ω_u, in rad/s, is given by

$$\omega_u = \frac{2\pi}{T_u} \tag{6-28}$$

The method of direct substitution consists of substituting $s = i\omega_u$ in the characteristic equation. This results in a complex equation that can be converted into two simultaneous equations:

Real part = 0

Imaginary part = 0

From these we can solve for two unknowns: one is the ultimate frequency ω_u, the other is any of the parameters of the loop, usually the ultimate gain. Let us illustrate this by working the temperature control example by the direct substitution method.

Example 6-6. Determine the ultimate gain and frequency of the temperature controller of Example 6-5 by direct substitution.

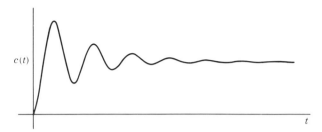

Figure 6-9a. Response of stable system.

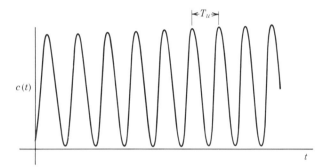

Figure 6-9b. Response of marginally stable system with ultimate period T_u.

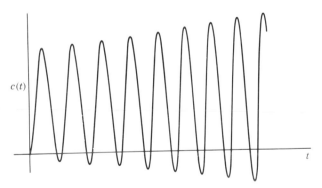

Figure 6-9c. Response of unstable system.

Solution. From Example 6-5 we obtain the characteristic equation for the base case in polynomial form:

$$900s^3 + 420s^2 + 43s + 1 + 0.80K_c = 0$$

Next we substitute $s = i\omega_u$ at $K_c = K_{cu}$:

$$900i^3\omega_u^3 + 420i^2\omega_u^2 + 43i\omega_u + 1 + 0.80K_{cu} = 0$$

Then we substitute $i^2 = -1$ and separate the real and imaginary parts:

$$(-420\omega_u^2 + 1 + 0.80K_{cu}) + i(-900\omega_u^3 + 43\omega_u) = 0 + i0$$

From this complex equation we obtain the following two, as both the real and imaginary parts must be zero:

$$-420\omega_u^2 + 1 + 0.80K_{cu} = 0$$
$$-900\omega_u^3 + 43\omega_u = 0$$

The solution of this set has the following possibilities:

For $\omega_u = 0$ $K_{cu} - 1.25\ \%/\%$

For $\omega_u - 0.2186$ rad/s $K_{cu} = 23.8\ \%/\%$

The first solution corresponds to the monotonic instability caused by having the wrong action on the controller. In this case the system does not oscillate but moves monotonically in one direction or the other. The crossing of the imaginary axis occurs at the origin ($s = 0$).

The ultimate gain for the second solution is identical to the one obtained by Routh's test, but this time we obtain the additional information that this gain will cause the loop to oscillate with a frequency of 0.2186 rad/s (0.0348 hertz) or a period of

$$T_u = \frac{2\pi}{0.2186} = 28.7\ \text{s}$$

The results of the direct substitution method for the other cases considered before are tabulated below.

	K_{cu}	ω_u, rad/s	T_u, s
1. Base case	23.8	0.2186	28.7
2. $H(s) = \dfrac{2.0}{10s + 1}$	11.9	0.2186	28.7
3. $H(s) = \dfrac{1.0}{5s + 1}$	25.7	0.2906	21.6
4. $G_s(s) = \dfrac{50}{20s + 1}$	18.7	0.2345	26.8

We see that the ultimate gains are the same as were obtained by Routh's test. However, the results of the direct substitution method also tell us that the loop can oscillate significantly faster when the time constant of the sensor-transmitter is reduced from 10 to 5 s. It also tells us that the loop oscillates slightly faster when the exchanger time constant is reduced from 30 to 20 s, in spite of the significant reduction in ultimate gain. We can also observe that changing the gains of the blocks on the loop has no effect on the frequency of oscillation.

Effect of Dead Time

We have seen how Routh's test and direct substitution allow us to study the effect of various loop parameters on the stability of the feedback control loop. Unfortunately both of these methods fail when any of the blocks on the loop contains a dead-time (transportation lag or time delay) term. This is because the dead time introduces an exponential function of the Laplace transform variable in the characteristic equation. This means that this equation is no longer a polynomial and the methods we have learned in this section no longer apply. An increase in dead time tends to reduce the ultimate loop gain very rapidly. This effect is similar to the effect of increasing the nondominant time constants of the loop in that it is relative to the magnitude of the dominant time constant. We will be able to study the effect of dead time on loop stability when we study the method of frequency response in the next chapter.

We must point out that the exchanger we have used in this chapter is a distributed-parameter system, that is, the temperature of the process fluid is distributed throughout the exchanger. The transfer functions for such systems usually contain at least one dead-time term, which, for simplicity, we have ignored.

An estimate of the ultimate gain and frequency of a loop with dead time may sometimes be obtained by using an approximation to the dead-time transfer function. A popular approximation is the first-order Padé approximation[9] given by

$$e^{-t_0 s} \doteq \frac{1 - \frac{1}{2} t_0 s}{1 + \frac{1}{2} t_0 s} \tag{6-29}$$

where t_0 is the dead time. More accurate higher-order approximations are also available, but they are too complex to be practical. The following example illustrates the use of the Padé approximation with the direct substitution method.

Example 6-7   Ultimate Gain and Frequency of First-Order Plus Dead-time Process.
The process transfer function of the loop of Fig. 6-4a is given by

$$G(s) = \frac{Ke^{-t_0 s}}{\tau s + 1}$$

where

   K is the gain
   t_0 is the dead time
   τ is the time constant

Determine the ultimate gain and frequency of the loop as a function of the process parameters if the controller is a proportional controller:

$$G_c(s) = K_c$$

Solution.   From Example 6-1, the characteristic equation of the loop is

$$1 + G_c(s) \, G(s) = 0$$

or for the transfer functions considered here

$$1 + \frac{KK_c e^{-t_0 s}}{\tau s + 1} = 0$$

Substitute the first-order Padé approximation, Eq. (6-29) to get

$$1 + \frac{KK_c(1 - \frac{1}{2}t_0 s)}{(\tau s + 1)(1 + \frac{1}{2}t_0 s)} = 0$$

Clear the fraction:

$$\frac{1}{2}t_0 \tau s^2 + (\tau + \frac{1}{2}t_0 - \frac{1}{2}KK_c t_0)s + 1 + KK_c = 0$$

By the direct substitution method, $s = i\omega_u$ yields

$$\frac{1}{2}t_0 \tau i^2 \omega_u^2 + (\tau + \frac{1}{2}t_0 - \frac{1}{2}KK_c t_0) i\omega_u + 1 + KK_c = 0$$

$$(-\frac{1}{2}t_0 \tau \omega_u^2 + 1 + KK_c) + i(\tau + \frac{1}{2}t_0 - \frac{1}{2}KK_c t_0)\omega_u = 0$$

After rearrangement, the solution is

$$(KK_c)_u = 1 + 2\left(\frac{\tau}{t_0}\right)$$

$$\omega_u = \frac{2}{t_0}\sqrt{\frac{t_0}{\tau} + 1}$$

These formulas show that the ultimate loop gain goes to infinity—with no stability limit—as the dead time approaches zero, which agrees with the results of Example 6-1. However, any finite amount of dead time imposes a stability limit on the loop gain. The ultimate frequency increases with decreasing dead time and becomes very small as the dead time increases. This means that dead time slows down the response of the loop.

From the results of direct-substitution analysis in the preceding examples we can summarize the following general effects of various loop parameters:

1. Stability imposes a limit on the overall loop gain, so that an increase in the gain of the control valve, the transmitter, or the process, results in a decrease in the ultimate controller gain.

2. An increase in dead time or in any of the nondominant (smaller) time constants of the loop results in a reduction of the ultimate loop gain and in the ultimate frequency.

3. An increase in the dominant (longest) time constant of the loop results in an increase in the ultimate loop gain and a decrease in the ultimate frequency of the loop.

We will next look at the important task of tuning feedback controllers.

6-3. TUNING OF FEEDBACK CONTROLLERS

Tuning is the procedure of adjusting the feedback controller parameters to obtain a specified closed-loop response. The tuning of a feedback control loop is analogous to the tuning of an automobile engine or of a television set. In each of these cases the difficulty of the problem increases with the number of parameters that must be adjusted. For example, tuning a simple proportional only or integral only controller is similar to adjusting the volume of a television set. As only one parameter or "knob" needs to be adjusted, the procedure consists of moving it in one direction or the other until the desired response (or volume) is obtained. The next degree of difficulty is the tuning of a two-mode or proportional-integral (PI) controller, which is similar to adjusting the contrast and brightness on a black-and-white television set. Since two parameters, the gain and the reset time, must be adjusted, the tuning procedure is significantly more complicated than when only one parameter needs to be adjusted. Finally, the tuning of three-mode or proportional-integral-derivative (PID) controllers represents the next higher degree of difficulty since three parameters, the gain, the reset time, and the derivative time, must be adjusted. This is analogous to the adjusting of the intensities of the green, red, and blue beams on a color television set.

Although we have drawn an analogy between the tuning of a television set and that of a feedback control loop, we do not want to give the impression that the two tasks have the same degree of difficulty. The main difference lies in the speed of response of the television set versus that of a process loop. On the television set we get almost immediate feedback on the effect of our tuning adjustments; on the other hand, although some process loops do have relatively fast responses, for many process loops we may have to wait several minutes and maybe even hours to observe the response that results from our tuning adjustments. This makes the tuning of feedback controllers by trial and error a tedious and time-consuming task. Yet this happens to be the most common method used by control and instrument engineers in industry. A number of tuning procedures and formulas have been introduced to tune controllers for various response criteria. We will study some of these in this section because each of them gives us some insight into the tuning procedure. However, we must keep in mind that no one procedure will give better results than any other for all process control situations.

The values of the tuning parameters depend on the desired closed-loop response and on the dynamic characteristics or personality of the other elements of the control loop, particularly the process. We saw earlier that if the process is nonlinear, as is usually the case, its characteristics will change from one operating point to the next. This means that a particular set of tuning parameters can produce the desired response at only one operating point, given that standard feedback controllers are basically linear devices. For operation in a range of operating conditions, a compromise must be reached in arriving at an acceptable set of tuning parameters, as the response will be sluggish at one end of the range and oscillatory at the other. With this in mind, let us look at some of the procedures that have been proposed to tune industrial controllers.

Quarter Decay Ratio Response by Ultimate Gain

This pioneer method, also known as the closed-loop or on-line tuning method, was proposed by Ziegler and Nichols[1] in 1942. It consists of two steps, as do all the other tuning methods:

Step 1. The determination of the dynamic characteristics or personality of the control loop.

Step 2. The estimation of the controller tuning parameters that produce a desired response for the dynamic characteristics determined in the first step—in other words, the matching of the personality of the controller to that of the other elements in the loop.

In this method the parameters by which the dynamic characteristics of the process are represented are the *ultimate gain* of a proportional controller, and the *ultimate period* of oscillation. These parameters, introduced in the preceding section, could be determined by the direct substitution method if the transfer functions of all of the components of the loop were known quantitatively. As this is not usually the case, we must often experimentally determine the ultimate gain and period from the actual system by the following procedure:

1. Switch off the integral and derivative actions of the feedback controller so as to have a proportional controller. In some models the integral action cannot be switched off, but can be detuned only by setting the integral time to its maximum value or, equivalently, the integral rate to its minimum value.

2. With the controller in *automatic* (i.e., the loop closed), increase the proportional gain (or reduce the proportional band) until the loop oscillates with constant amplitude. Record the value of the gain that produces sustained oscillations as K_{cu}, the ultimate gain. This step must be carried out in discrete gain increments, bumping the system by applying a small set point change at each gain setting. The increments in gain should be smaller as the ultimate gain is approached.

3. From a time recording of the controlled variable, the period of oscillation is measured and recorded as T_u, the ultimate period, as shown in Fig. 6-10.

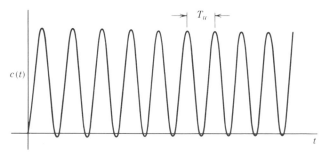

Figure 6-10. Response of the loop with the controller gain set equal to the ultimate gain K_{cu}. T_u is the ultimate period.

For the desired response of the closed loop Ziegler and Nichols[1] specified a decay ratio of one-fourth. The decay ratio is the ratio of the amplitudes of two successive oscillations. It should be independent of the input to the system and depend only on the roots of the characteristic equation for the loop. Typical quarter decay ratio responses for a disturbance input and a set point change are shown in Fig. 6-11.

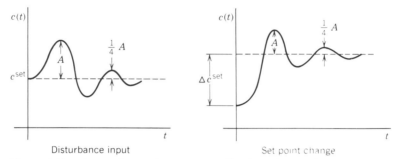

Disturbance input   Set point change

Figure 6-11. Quarter-decay ratio responses to disturbance input and to set point change.

Once the ultimate gain and period are determined, they are used in the formulas of Table 6-1 for calculating the controller tuning parameters that produce quarter decay ratio responses.

Notice that the introduction of integral action forces a reduction of 10% in the gain of the PI controller as compared to the proportional controller gain. Derivative action, on the other hand, allows an increase in both the proportional gain and the integral rate (a decrease in integral time) of the PID controller as compared to the PI controller. This

Table 6-1 Quarter Decay Tuning Formulas

Controller type		Proportional gain K_c	Integral time τ_I	Derivative time τ_D
Proportional only	P	$K_{cu}/2$	—	—
Proportional-integral	PI	$K_{cu}/2.2$	$T_u/1.2$	—
Proportional-integral-derivative	PID	$K_{cu}/1.7$	$T_u/2$	$T_u/8$

is because the integral action introduces a lag in the operation of the feedback controller while the derivative action introduces an advance or lead. This will be discussed in more detail in Chapter 7.

The quarter decay ratio response is not desirable for step changes in set point because it causes a 50% overshoot ($A/\Delta c^{set} = 0.5$). This is because the maximum deviation from the new set point in each direction is one-half the preceding maximum deviation in the opposite direction (see Fig. 6-11). However, the quarter decay ratio response is very desirable for disturbance inputs because it prevents a large initial deviation from the set point without being too oscillatory. The major difficulty with the quarter decay ratio response is that the set of tuning parameters necessary to obtain it is not unique, except for the case of the proportional controller. In the case of PI controllers we can easily verify that for each value of the integral time we could find a value of the gain that produces a quarter decay ratio response and vice versa. The same is true for the PID controller. The settings proposed by Ziegler and Nichols are ball park figures that produce fast response for most industrial loops.

Example 6-8. Given the characteristic equation of the continuous stirred tank heater derived in Example 6-4, determine the quarter decay ratio tuning parameters for the PID controller by the ultimate gain method. Also calculate the roots of the characteristic equation for the controller tuned with these parameters.

Solution. In Example 6-4 we derived the following characteristic equation for the heater:

$$0.387s^5 + 3.272s^4 + 7.859s^3 + (6.043 + 1.205\, K_c\, \tau_D)s^2$$
$$+ (0.617 + 1.205K_c)s + 1.205\, K_c/\tau_I = 0$$

We first use the direct substitution method to calculate the ultimate gain and period of oscillation for a proportional controller. With $\tau_D = 0$, $1/\tau_I = 0$, the characteristic equation reduces to:

$$0.387s^4 + 3.272s^3 + 7.859s^2 + 6.043s + 0.617 + 1.205K_c = 0$$

Next we substitute $s = i\omega_u$ and $K_c = K_{cu}$ to obtain the following equation set after simplification:

$$-3.272\omega_u^3 + 6.043\omega_u = 0$$
$$0.387\omega_u^4 - 7.859\omega_u^2 + 0.617 + 1.205K_{cu} = 0$$

From these we can obtain the ultimate frequency and gain:

$$\omega_u = \sqrt{\frac{6.043}{3.272}} = 1.359 \text{ rad/min}$$

$$K_{cu} = \frac{1}{1.205}(-0.387\omega_u^4 + 7.859\omega_u^2 - 0.617) = 10.44 \ \%/\%.$$

The ultimate period is $T_u = \dfrac{2\pi}{1.359} = 4.62$ min. According to Table 6-1, the tuning parameters for quarter decay ratio response of a PID controller are

$$K_c = K_{cu}/1.7 = 6.14 \ \%/\%$$
$$\tau_I = T_u/2 \quad = 2.31 \text{ min.}$$
$$\tau_D = T_u/8 \quad = 0.58 \text{ min.}$$

With these tuning parameters the characteristic equation is

$$0.387s^5 + 3.272s^4 + 7.858s^3 + 10.34s^2 + 8.017s + 3.20 = 0$$

Using the program in Appendix D, the roots of this characteristic equation are found to be

$$-0.42 \pm i0.99, \qquad -1.03 \pm i0.49, \qquad -5.55$$

The unit step response of the closed loop has the following form:

$$T(t) = b_0 u(t) + b_1 e^{-0.42t} \sin(0.99t + \theta_1)$$
$$+ b_2 e^{-1.03t} \sin(0.49t + \theta_2)$$
$$+ b_3 e^{-5.55t}$$

where the parameters b_0, b_1, θ_1, b_2, θ_2, and b_3 must be evaluated by partial fractions expansion for the particular input (set point, inlet flow, or inlet temperature) under consideration. The technique of partial fractions expansion was discussed in Chapter 2.

Process Characterization

The Ziegler-Nichols on-line tuning method we have just introduced is the only one that characterizes the process by the ultimate gain and the ultimate period. Most of the other controller tuning methods characterize the process by a simple first- or second-order model with dead time. In order to better understand the assumptions involved in such characterization, let us consider the block diagram of a feedback control loop given in Fig. 6-12a. The symbols shown in the block diagram are

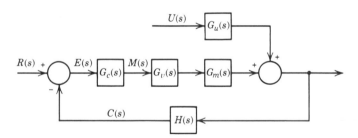

Figure 6-12a.   Block diagram of typical feedback control loop.

$R(s)$	the Laplace transform of the set-point signal
$M(s)$	the Laplace transform of the controller-output signal
$C(s)$	the Laplace transform of the transmitter-output signal
$E(s)$	the Laplace transform of the error signal
$U(s)$	the Laplace transform of the disturbance signal
$G_c(s)$	the controller transfer function
$G_v(s)$	the transfer function of the control valve (of final control element)
$G_m(s)$	the process transfer function between the controlled and the manipulated variable
$G_u(s)$	the process transfer function between the controlled variable and the disturbance
$H(s)$	the transfer function of the sensor-transmitter

Using the simple block diagram algebra manipulations learned in Chapter 3, we can draw the equivalent block diagram of Fig. 6-12b. In this diagram there are only two blocks in the control loop, one for the controller and the other for the rest of the components of the loop. The advantage of this simplified representation is that it highlights the two signals in the loop that can be usually observed and recorded: the controller output $M(s)$ and the transmitter signal $C(s)$. For most loops no signal or variable can be observed except these two. Therefore, the lumping of the transfer functions of the control valve, the process, and the sensor-transmitter into a single block is not just a convenience, but a practical necessity. Let us call this combination of the transfer functions $G(s)$:

$$G(s) = G_v(s)\ G_m(s)H(s) \tag{6-30}$$

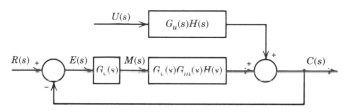

Figure 6-12b. Equivalent simplified block diagram in which all of the field instruments and the process have been lumped into single blocks.

It is precisely this combined transfer function that is approximated by low-order models for the purpose of characterizing the dynamic response of the process. The point is that the characterized "process" includes the dynamic behavior of the control valve and of the sensor/transmitter. The two most common models used to characterize the process are the following.

First-Order Plus Dead-Time (FOPDT) Model

$$G(s) = \frac{Ke^{-t_0 s}}{\tau s + 1} \tag{6-31}$$

Second-Order Plus Dead-Time (SOPDT) Model

$$G(s) = \frac{Ke^{-t_0 s}}{(\tau_1 s + 1)(\tau_2 s + 1)} \tag{6-32}$$

$$G(s) = \frac{Ke^{-t_0 s}}{\tau^2 s + 2\xi\tau s + 1} \tag{6-33}$$

for underdamped processes ($\xi < 1$), where

$$K = \text{the process steady-state gain}$$
$$t_0 = \text{the effective process dead time}$$
$$\tau, \tau_1, \tau_2 = \text{the effective process time constants}$$
$$\xi = \text{the effective process damping ratio}$$

Of these, the FOPDT model is the one on which most controller tuning formulas are based. This model characterizes the process by three parameters: the gain K, the dead time t_0, and the time constant τ. The question then is, How can these parameters be determined for a given loop? The answer is that some dynamic test must be performed on the actual system or on a computer simulation of the loop. The simplest test that can be performed is a step test.

Process Step Testing

The step test procedure is carried out as follows:

1. With the controller on ''manual'' (that is, the loop opened) a step change in the controller output signal $m(t)$ is applied to the process. The magnitude of the change should be large enough for the consequent change in the transmitter signal to be measurable, but not so large that the response will be distorted by the process nonlinearities.

2. The response of the transmitter output signal $c(t)$ is recorded on a strip chart recorder or equivalent device, making sure that the resolution is adequate in both the amplitude and the time scale. The resulting plot of $c(t)$ versus time must cover the entire test period from the introduction of the step test until the system reaches a new steady state. Typically a step test lasts between a few minutes and several hours, depending upon the speed of response of the process.

It is of course imperative that no disturbances enter the system while the step test is performed. A typical test plot, also known as a *process reaction curve*, is sketched in Fig. 6-13. As we saw in Chapter 4, the S-shaped response is characteristic of second- and higher-order processes with or without dead time. The next step is to match the process reaction curve to a simple process model in order to determine the model parameters. Let us do this for the first-order plus dead-time (FOPDT) model.

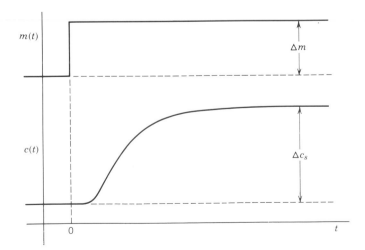

Figure 6-13. Process reaction curve or open-loop step response.

In the absence of disturbances, and for the conditions of the test, the block diagram of Fig. 6-12b can be redrawn as in Fig. 6-14. The response of the transmitter output signal is given by

$$C(s) = G(s) M(s)$$

For a step change in controller output of magnitude Δm and a FOPDT model, Eq. (6-31), we have

$$C(s) = \frac{Ke^{-t_0 s}}{\tau s + 1} \cdot \frac{\Delta m}{s} \tag{6-34}$$

Expanding this expression by partial fractions, we obtain

$$C(s) = K\Delta m \, e^{-t_0 s} \left[\frac{1}{s} - \frac{\tau}{\tau s + 1} \right] \tag{6-35}$$

Inverting with the help of a Laplace transform table (Table 2-1) and applying the real translation theorem of Laplace transforms (see Chapter 2), we get

$$\Delta c(t) = K\Delta m \, u(t - t_0) \, [1 - e^{-(t-t_0)/\tau}] \tag{6-36}$$

where the unit step function $u(t - t_0)$ is included to explicitly indicate that

$$\Delta c(t) = 0 \qquad \text{for } t \leq t_0$$

The term Δc is the perturbation or change of the transmitter output from its initial value:

$$\Delta c(t) = c(t) - c(0) \tag{6-37}$$

A graph of Eq. (6-36) is shown in Fig. 6-15. In this figure the term Δc_s is the steady-state change in $c(t)$. From Eq. (6-36) we find

$$\Delta c_s = \lim_{t \to \infty} \Delta c(t) = K\Delta m \tag{6-38}$$

From this equation, and realizing that the model response must match the process reaction curve at steady state, we can calculate the steady-state gain of the process, which is one of the model parameters:

$$K = \frac{\Delta c_s}{\Delta m} \tag{6-39}$$

This result was also shown in Chapter 3.

The determination of the dead time t_0 and time constant τ can be done by at least three methods, each of which results in different values.

Fit 1. This method makes use of the line that is tangent to the process reaction curve at the point of maximum rate of change. For the FOPDT model this happens at $t = t_0$, as is evident from inspecting the model response of Fig. 6-15. From Eq. (6-36), this initial (maximum) rate of change is found to be

$$\frac{d(\Delta c)}{dt} \bigg|_{t_0} = K\Delta m \left[\frac{1}{\tau} \right] = \frac{\Delta c_s}{\tau} \tag{6-40}$$

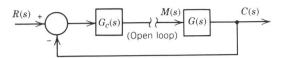

Figure 6-14.   Block diagram for open-loop step test.

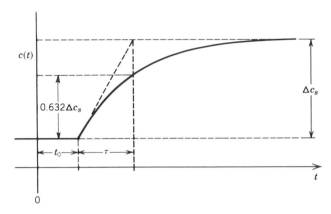

Figure 6-15.   Step response of a first-order plus dead-time process showing the graphical definition of the dead time t_0 and the time constant τ.

From Fig. 6-15 we see that this result tells us that the line of maximum rate of change crosses the initial value line at $t = t_0$ and the final value line at $t = t_0 + \tau$. This finding suggests the construction for determining t_0 and τ shown in Fig. 6-16a The line is drawn tangent to the actual process reaction curve at the point of maximum rate of change. The model response using these values of t_0 and τ is shown by the dashed line in the figure. Evidently, the model response obtained with this fit does not match the actual response very well.

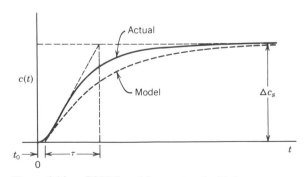

Figure 6-16a.   FOPDT model parameters by Fit 1.

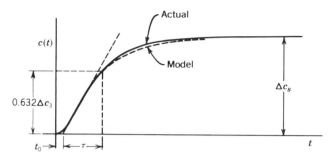

Figure 6-16b. FOPDT model parameters by Fit 2.

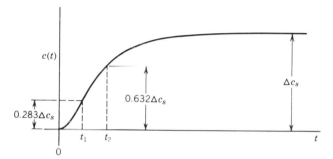

Figure 6-16c. FOPDT model parameters by Fit 3.

Fit 2. In this fit t_0 is determined in the same manner as in fit 1, but the value of τ is the one that forces the model response to coincide with the actual response at $t = t_0 + \tau$. According to Eq. (6-36) this point is

$$\Delta c(t_0 + \tau) = K\Delta m \,[1 - e^{-1}] = 0.632 \,\Delta c_s \qquad (6\text{-}41)$$

The comparison of the model and actual responses for fit 2 are shown to be much closer than for fit 1 in Fig. 6-16b. The value of the time constant obtained by fit 2 is usually less than that obtained by fit 1.

Fit 3. The least precise step in the determination of t_0 and τ by the previous two methods is the drawing of the line tangent to the process reaction curve at the point of maximum rate of change. Even for fit 2, for which the value of $(t_0 + \tau)$ is independent of the tangent line, the estimated values of both t_0 and τ depend on the line. To eliminate this dependence on the tangent line, Dr. Cecil L. Smith[2] proposes that the values of t_0 and τ be selected such that the model and actual responses coincide at two points in the region of high rate of change. The two points recommended are $(t_0 + 1/3\tau)$ and $(t_0 + \tau)$. To locate these points we make use of Eq. (6-36):

$$\Delta c(t_0 + \tau) = K\Delta m \,[1 - e^{-1}] = 0.632 \,\Delta c_s \qquad (6\text{-}42)$$

$$\Delta c\left(t_0 + \frac{1}{3}\tau\right) = K\Delta m \,[1 - e^{-1/3}] = 0.283 \,\Delta c_s$$

These two points are labeled t_2 and t_1, respectively, in Figure 6-16c. The values of t_0 and τ can then be easily obtained by the simple solution of the following set of equations:

$$t_0 + \tau = t_2 \tag{6-43}$$

$$t_0 + \frac{1}{3}\tau = t_1$$

Which reduces to

$$\tau = \frac{3}{2}(t_2 - t_1) \tag{6-44}$$

$$t_0 = t_2 - \tau$$

where

t_1 = time at which $\Delta c = 0.283\ \Delta c_s$
t_2 = time at which $\Delta c = 0.632\ \Delta c_s$

Our past experience has demonstrated that results obtained by this method are more reproducible than those obtained by the other two. We therefore recommend this method for the estimation of t_0 and τ from the process reaction curve. However, we must always keep in mind that some correlations for the controller tuning parameters are based on different FOPDT model fits.

Various methods have been proposed in the literature to estimate the parameters of a second-order plus dead-time (SOPDT) model to the process reaction curve. Our experience is that the precision of these methods is very low. The reason is that the step test does not provide enough information to extract the additional parameter—time constant or damping ratio—that is required by the SOPDT. In other words, the increased complexity of the model demands a more sophisticated dynamic test. Pulse testing is an adequate method to obtain second and higher-order model parameters. It will be presented in Section 7-3.

Since most of the controller tuning formulas that we are about to introduce are based on FOPDT model parameters, we may find ourselves in some situation in which we have the parameters of a high-order model and need to estimate the equivalent first-order model parameters. Although there is no general procedure to do this, the following rule of thumb might provide a rough estimate for a first approximation:

If one of the time constants of the high-order model is much longer than the others, the effective time constant of the first-order model can be estimated to be equal to the longest time constant. The effective dead time of the first-order model can then be approximated by the sum of all of the smaller time constants plus the dead time of the high-order model.

Example 6-9. Estimate the FOPDT parameters for the temperature control loop of the exchanger of Example 6-5. The combined transfer function for the control valve, exchanger, and sensor-transmitter for that example is given by

$$G(s) = \frac{1}{10s + 1} \cdot \frac{50}{30s + 1} + \frac{0.016}{3s + 1}$$

Solution Assuming that the 30 s time constant is much longer that the other two, we can roughly approximate:

$$\tau = 30 \text{ s}$$

$$t_0 = 10 + 3 = 13 \text{ s}$$

and the gain is of course the same, that is, $K = 0.80$. The resulting FOPDT model transfer function is then

$$G(s) = \frac{0.80 \, e^{-13s}}{30 \, s + 1} \qquad \text{(Model A)}$$

Let us next compare this rough approximation with the experimentally determined FOPDT parameters from the process reaction curve. Figure 6-17 shows the process reaction curve for the three first-order lags in series that we have assumed represent the heat exchanger, control valve, and sensor-transmitter. The response of Fig. 6-17 was obtained by simulating the three first-order lags on a desk-top analog computer, applying a 5% step change on the controller output signal, and recording the output of the sensor-transmitter versus time. From this result we can calculate the FOPDT parameters using the three fits presented above.

Process Gain

$$K = \frac{\Delta c}{\Delta m} = \frac{4C}{5\%} \cdot \frac{100\%}{(150-50)C} = 0.80 \frac{\%}{\%}$$

Fit 1

$$t_0 = 7.2 \text{ s} \qquad t_s = 61.5 \text{ s} \qquad \text{(see Fig. 6-17)}$$

$$\tau = 61.5 - 7.2 = 54.3 \text{ s}$$

$$G(s) = \frac{0.80 \, e^{-7.2s}}{54.3 \, s + 1} \qquad \text{(Model B)}$$

Fit 2

$$t_0 = 7.2 \text{ s}$$

$$\text{At } \Delta c(t) = 0.632 \, (4 \text{ C}) = 2.53 \text{ C} \qquad t_2 = 45.0 \text{ s}$$

$$\tau = 45.0 - 7.2 = 37.8 \text{ s}$$

$$G(s) = \frac{0.80 \, e^{-7.2s}}{37.8 \, s + 1} \qquad \text{(Model C)}$$

Fit 3

$$\text{At } \Delta c(t) = 0.283 \, (4 \text{ C}) = 1.13 \text{ C} \qquad t_1 = 22.5 \text{ s}$$

$$\tau = \frac{3}{2} (t_2 - t_1) = \frac{3}{2} (45.0 - 22.5) = 33.8 \text{ s}$$

$$t_0 = 45.0 - 33.8 = 11.2 \text{ s}$$

$$G(s) = \frac{0.80 \, e^{-11.2s}}{33.8 \, s + 1} \qquad \text{(Model D)}$$

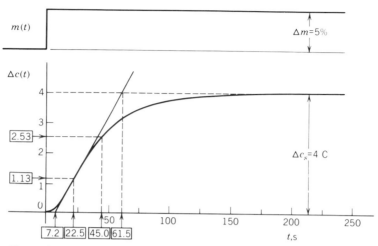

Figure 6-17. Process reaction curve for Example 6-9.

As we shall see in the following sections, an important parameter in terms of tuning is the ratio of the dead time to the time constant. The values for the four FOPDT model approximations are as follows:

Model	A(rough)	B(Fit 1)	C(Fit 2)	D(Fit 3)
t_0,s	13.0	7.2	7.2	11.2
τ,s	30.0	54.3	37.8	33.8
t_0/τ	0.433	0.133	0.190	0.331

We see that the ratio t_0/τ is the most sensitive parameter, varying by a factor of slightly over 3:1. Although fits 2 and 3 provide the closest approximations to the actual step response, we must keep in mind that some tuning correlations are based on specific fits.

Example 6-10.   Given a second-order process

$$G(s) = \frac{C(s)}{M(s} = \frac{K}{(\tau_1 s + 1)(\tau_2 s + 1)} \qquad \tau_1 \geq \tau_2$$

determine the parameters of a first-order plus dead-time (FOPDT) model

$$G'(s) = \frac{K'e^{-t_0 s}}{\tau' s + 1}$$

using fit 3, as a function of the ratio of τ_2/τ_1.

Solution.   We first obtain the unit step response of the actual process:

$$M(s) = \frac{1}{s}$$

$$C(s) = \frac{K}{(\tau_1 s + 1)(\tau_2 s + 1)} \frac{1}{s}$$

By partial fractions expansion, for the case $\tau_1 > \tau_2$:

$$C(s) = \frac{K}{s} - \frac{K\tau_1}{\tau_1 - \tau_2} \frac{1}{s + 1/\tau_1} + \frac{K\tau_2}{\tau_1 - \tau_2} \frac{1}{s + 1/\tau_2}$$

Inverting, with the help of a Laplace transform table (Table 2-1), we obtain

$$\Delta c(t) = K \left[1 - \frac{\tau_1}{\tau_1 - \tau_2} e^{-t/\tau_1} + \frac{\tau_2}{\tau_1 - \tau_2} e^{-t/\tau_2} \right]$$

As

$$t \to \infty \qquad \Delta c(t) \to K$$

and since

$$\Delta m = 1 \qquad K' = \frac{\Delta c}{\Delta m} = K$$

For fit 3, at $t_1 = t_0' + \tau'/3$

$$\Delta c = (1 - e^{-1/3})K = K \left[1 - \frac{\tau_1}{\tau_1 - \tau_2} e^{-t_1/\tau_1} + \frac{\tau_2}{\tau_1 - \tau_2} e^{-t_1/\tau_2} \right]$$

and at $t_2 = t_0' + \tau'$

$$\Delta c = (1 - e^{-1})K = K \left[1 - \frac{\tau_1}{\tau_1 - \tau_2} e^{-t_2/\tau_1} + \frac{\tau_2}{\tau_1 - \tau_2} e^{-t_2/\tau_2} \right]$$

or

$$e^{-1/3} = \frac{\tau_1}{\tau_1 - \tau_2} \left[e^{-t_1/\tau_1} - \frac{t_2}{\tau_1} e^{-t_1/\tau_2} \right] \tag{A}$$

$$e^{-1} = \frac{\tau_1}{\tau_1 - \tau_2} \left[e^{-t_2/\tau_1} - \frac{\tau_2}{\tau_1} e^{-t_2/\tau_2} \right] \tag{B}$$

For the case $\tau_1 = \tau_2$, by partial fractions expansion

$$C(s) = \frac{K}{(\tau_1 s + 1)^2 s} = \frac{K}{s} - \frac{K}{\tau_1} \frac{1}{\left(s + \frac{1}{\tau_1} \right)^2} - K \frac{1}{s + \frac{1}{\tau_1}}$$

Inverting, with the help of a Laplace transforms table (Table 2-1), we obtain

$$\Delta c(t) = K \left[1 - \frac{t}{\tau_1} e^{-t/\tau_1} - e^{-t/\tau_1} \right]$$

resulting in

$$e^{-1/3} = \left[\frac{t_1}{\tau_1} + 1\right] e^{-t_1/\tau_1} \tag{C}$$

$$e^{-1} = \left[\frac{t_2}{\tau_1} + 1\right] e^{-t_2/\tau_1} \tag{D}$$

From Eqs. (A) and (B) or (C) and (D) we must solve for t_1 and t_2 by trial and error. Then, from Eq. (6-44)

$$\tau' = \frac{3}{2}(t_2 - t_1)$$

$$t_0' = t_2 - \tau'$$

This problem was solved by Martin[3] using a computer program and the results are plotted in Fig. 6-18. As can be seen from this figure the maximum effective dead time takes place when the two time constants are equal:

For $\tau_1 = \tau_2$ $t_0' = 0.505\tau_1$ $\tau' = 1.641\,\tau_1$

For $\tau_2 \ll \tau_1$ $t'_0 \to \tau_2$ $\tau' \to \tau_1$

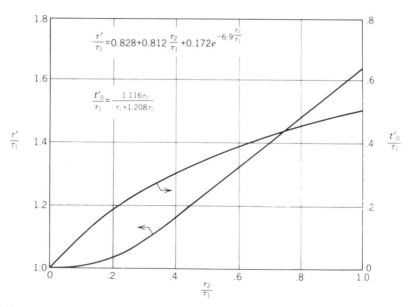

Figure 6-18. FOPDT model dead time and time constant for Fit 3 approximation to overdamped second-order system. (Reproduced by permission from Reference 3.)

This is the basis for the rule of thumb presented earlier. We can use Fig. 6-18 to refine this rule of thumb for systems represented by three or more first-order lags in series. For instance, for the heat exchanger example, we can refine the rough approximation model as follows:

Assume

$$\tau_1 = 30 \text{ s} \qquad \tau_2 \approx 10 + 3 = 13 \text{ s}$$

Then

$$\frac{\tau_2}{\tau_1} = \frac{13}{30} = 0.433$$

From Fig. 6-18

$$t_0' = 0.33\tau_1 = 9.9 \text{ s}$$

$$\tau' = 1.2\tau_1 = 36 \text{ s}$$

These values are closer to the values obtained by fit 3 in Example 6-9 than those obtained by the rough approximation (Model A). The reason is that in this case τ_2 is not that much smaller than τ_1.

Quarter Decay Ratio Response

In addition to their on-line tuning formulas, Ziegler and Nichols[1] proposed a set of formulas based on the parameters of a first-order model fit to the process reaction curve. These formulas are given in Table 6-2. Although the parameters they used were not precisely the gain, time constant, and dead time, their formulas can be modified and expressed in terms of these parameters. They used fit 1 to determine the model parameters.

Table 6-2 Tuning Formulas for Quarter Decay Ratio Response

Controller Type		Proportional Gain K_c	Integral Time τ_I	Derivative Time τ_D
Proportional only	P	$\dfrac{1}{K}\left(\dfrac{t_0}{\tau}\right)^{1}$	—	—
Proportional-integral	PI	$\dfrac{0.9}{K}\left(\dfrac{t_0}{\tau}\right)^{-1}$	$3.33\, t_0$	—
Proportional-integral-derivative	PID	$\dfrac{1.2}{K}\left(\dfrac{t_0}{\tau}\right)^{-1}$	$2.0\, t_0$	$\dfrac{1}{2} t_0$

As we can see in Table 6-2, the relative magnitudes of the gain, integral time, and derivative time between the P, PI, and PID controllers are the same as for the on-line tuning formulas that are based on the ultimate gain and period (Table 6-1). The formulas for the gain show that the loop gain, KK_c, is inversely proportional to the ratio of the effective dead time to the effective time constant.

In using these formulas we must keep in mind that they are empirical and apply only to a limited range of dead time to time constant ratios. This means that they should not be extrapolated outside a range of t_0/τ of around 0.10 to 1.0.

As was pointed out on the discussion of on-line tuning, the difficulty of the quarter decay ratio performance specification for PI and PID controllers is that there is an infinite set of values of the controller parameters that can produce it. The formulas given are just one such set.

Example 6-11. Compare the values of the tuning parameters for the temperature control of the exchanger of Example 6-5 using the quarter decay ratio on-line tuning and the FOPDT parameters estimated in Example 6-8. In earlier examples we determined the following results for the exchanger temperature control loop:

By direct substitution method: $K_{cu} = 23.8 \%/\%$.
  (see Example 6-6) $T_u = 28.7$ s
By fit 1 approximation: $K = 0.80 \%/\%$
  (see Example 6-9) $t_0 = 7.2$ s
                                         $\tau = 54.3$ s

Solution The following are the tuning parameters for quarter decay ratio:

On-line Tuning (Table 6-1)

Proportional Only

$$K_c = \frac{1}{2}(23.8) = 11.9 \ \%/\%$$

Proportional Integral

$$K_c = \frac{23.8}{2.2} = 10.8\%/\%$$

$$\tau_I = \frac{28.7}{1.2} = 23.9 \text{ s } (0.40 \text{ min})$$

Proportional-Integral-Derivative

$$K_c = \frac{23.8}{1.7} = 14 \ \%/\%$$

$$\tau_I = \frac{28.7}{2.0} = 14.3 \text{ s } (0.24 \text{ min})$$

$$\tau_D = \frac{28.7}{8} = 3.6 \text{ s } (0.06 \text{ min})$$

Process Reaction Curve (Table 6-2)

$$K_c = \frac{1}{0.80}\left(\frac{7.2}{54.3}\right)^{-1} = 9.4 \ \%/\%$$

$$K_c = \frac{0.9}{0.80}\left(\frac{7.2}{54.3}\right)^{-1} = 8.5 \ \%/\%$$

$$\tau_I = 3.33 (7.2) = 24.0 \text{ s } (0.40 \text{ min})$$

$$K_c = \frac{1.2}{0.80}\left(\frac{7.2}{54.3}\right)^{-1} = 11.3 \ \%/\%$$

$$\tau_I = 2.0(7.2) = 14.4 \text{ s } (0.24 \text{ min})$$

$$\tau_D = 0.5(7.2) = 3.6 \text{ s } (0.06 \text{ min})$$

The agreement is evident. Notice however that this agreement depends on using the fit 1 model parameters, which happens to be what Ziegler and Nichols used.

Tuning for Minimum Error Integral Criteria

Because of the nonuniqueness of the quarter decay ratio tuning parameters, a substantial research project was conducted at Louisiana State University under Professors Paul W. Murrill and Cecil L. Smith to develop tuning relationships that were unique. They used the first-order plus dead-time (FOPDT) model parameters to characterize the process. Their specification of the closed-loop response is basically a minimum error or deviation of the controlled variable from its set point. Since the error is a function of time for the duration of the response, the sum of the error at each instant of time must be minimized. This is by definition the integral of the error with time, or the shaded area in the responses illustrated in Fig. 6-19. Since the tuning relationships are intended to minimize the integral of the error, they are referred to as *minimum error integral tuning*. However, the integral of the error cannot be minimized directly because a very large negative error would be the minimum. In order to prevent negative values of the performance function, the following formulations of the integral can be proposed:

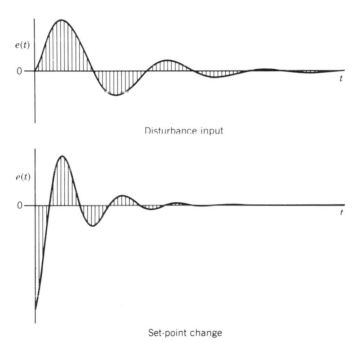

Figure 6-19. Definition of error integrals for disturbance and for set-point changes.

Integral of the Absolute Value of the Error (IAE)

$$\text{IAE} = \int_0^\infty |e(t)|dt \qquad (6\text{-}45)$$

Integral of the Square of the Error (ISE)

$$\text{ISE} = \int_0^\infty e^2(t)\, dt \qquad (6\text{-}46)$$

The integrals extend from the occurrence of the disturbance or set-point change ($t = 0$) to a very long time thereafter ($t = \infty$). This is because the ending of the responses cannot be fixed beforehand. The only problem with this definition of the integral is that it becomes undetermined when the error is not forced to zero. This happens only when the controller does not have integral action because of offset or steady-state error. In this case, the error in the definition is replaced with the difference between the controlled variable and its final steady-state value.

The difference between the IAE and ISE criteria is that the ISE puts more weight on large errors, which usually occur at the beginning of the response, and less weight on the smaller errors, which happen toward the end of the response. In trying to reduce the initial error, the minimum ISE criterion results in high controller gains and very oscillatory responses (i.e., a high decay ratio) with the error oscillating around zero for a relatively long time. This phenomenon suggests that the performance criteria should contain a penalty for the time elapsed from the start of the response. The following error integrals contain such a penalty by including a weight for the elapsed time.

Integral of the Time-Weighted Absolute Value of the Error (ITAE)

$$ITAE = \int_0^\infty t \, |e(t)| \, dt \tag{6-47}$$

Integral of the Time-Weighted Square of the Error (ITSE)

$$ITSE = \int_0^\infty t \, e^2(t) \, dt \tag{6-48}$$

Equations (6-45) through (6-48) constitute the four basic error integrals that can be minimized for a given loop by adjusting the controller parameters. Unfortunately, the optimum set of parameter values is not only a function of which of the four integral definitions is selected, but also of the type of input, that is, disturbance or set point, and of its shape, for example, step change, ramp, and so on. In terms of the shape of the input, the step change is usually selected because it is the most disruptive that can occur in practice, while in terms of the input type, we must select either set point or disturbance input for tuning, according to which one is expected to affect the loop more often. When set-point inputs are more important, the purpose of the controller is to have the controlled variable track the set-point signal and the controller is referred to as a "servo regulator." When the purpose of the controller is to maintain the controlled variable at a constant set point in the presence of disturbance inputs, the controller is called a "regulator." The optimum tuning parameters in terms of minimum error integral are different for each case. Most process controllers are considered to be regulators, except for the slave controllers in cascade control schemes, which are servo regulators. We will study cascade control in Section 8-3.

When the controller is tuned for optimum response to a disturbance input, an additional decision must be made regarding the process transfer function to the particular disturbance. This is complicated by the fact that the controller response cannot be optimum for each disturbance if there are more than one major disturbance signals entering the loop. Since the process transfer function is different for each disturbance and for the controller output signal, the optimum tuning parameters are functions of the relative speed of response of the controlled variable to the disturbance input. The slower the response to the disturbance input, the tighter the controller can be tuned, that is, the higher the controller gain can be. At the other extreme, if the controlled variable were to respond instantaneously to the disturbance, the controller tuning would be the least tight it can be, which would be the equivalent to the tuning for set-point changes. This is evident from examining the block diagram for the case when the response to the disturbance input is instantaneous, which is shown in Fig. 6-20a. As we can see, the disturbance enters at the same point in the loop as the set point, making the responses to step changes in disturbance and set point identical to each other, except for the sign. This is illustrated in Fig. 6-20b.

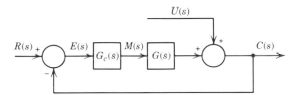

Figure 6-20a   Block diagram for instantaneous response to disturbance input.

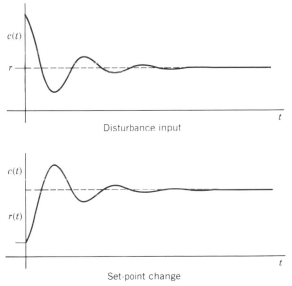

Disturbance input

Set-point change

Figure 6-20b. Instantaneous response to disturbance is identical to the response to a set-point change except for sign.

Lopez et al.[5] developed tuning formulas for minimum error integral criteria based on the assumption that the process transfer function to disturbance inputs is identical to the transfer functions to the controller output signal. The loop block diagram for this case is shown in Fig. 6-21. The tuning formulas are given in Table 6-3.

These formulas indicate the same trend as the quarter decay ratio formulas, except that the integral time depends to some extent on the effective process time constant and

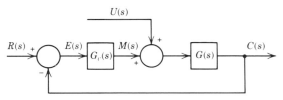

Figure 6-21. Block diagram for identical process response to manipulated variable and to disturbance.

Table 6-3. Minimum Error Integral Tuning Formulas for Disturbance Inputs

Process Model: $G(s) = \dfrac{K\,e^{-t_0 s}}{\tau s + 1}$

Proportional (P) Controller: $G_c(s) = K_c$

Error Integral	ISE	IAE	ITAE
$K_c = \dfrac{a}{K}\left(\dfrac{t_0}{\tau}\right)^b$	$a = \quad 1.411$	0.902	0.490
	$b = -0.917$	-0.985	-1.084

Proportional-Integral (PI) Controller:

$$G_c(s) = K_c\left(1 + \frac{1}{\tau_I s}\right)$$

Error Integral	ISE	IAE	ITAE
$K_c = \dfrac{a_1}{K}\left(\dfrac{t_0}{\tau}\right)^{b_1}$	$a_1 = \quad 1.305$	0.984	0.859
	$b_1 = -0.959$	-0.986	-0.977
$\tau_I = \dfrac{\tau}{a_2}\left(\dfrac{t_0}{\tau}\right)^{b_2}$	$a_2 = \quad 0.492$	0.608	0.674
	$b_2 = \quad 0.739$	0.707	0.680

Proportional-Integral-Derivative (PID) Controller:

$$G_c(s) = K_c\left(1 + \frac{1}{\tau_I s} + \tau_D s\right)$$

Error Integral	ISE	IAE	ITAE
$K_c = \dfrac{a_1}{K}\left(\dfrac{t_0}{\tau}\right)^{b_1}$	$a_1 = \quad 1.495$	1.435	1.357
	$b_1 = -0.945$	-0.921	-0.947
$\tau_I = \dfrac{\tau}{a_2}\left(\dfrac{t_0}{\tau}\right)^{b_2}$	$a_2 = \quad 1.101$	0.878	0.842
	$b_2 = \quad 0.771$	0.749	0.738
$\tau_D = a_3\tau\left(\dfrac{t_0}{\tau}\right)^{b_3}$	$a_3 = \quad 0.560$	0.482	0.381
	$b_3 = \quad 1.006$	1.137	0.995

less on the process dead time. We must again keep in mind that these formulas are empirical and should not be extrapolated beyond a range of (t_0/τ) of between 0.1 and 1.0. (This is the range of values used by Lopez in his correlations.) As is the case for the quarter decay ratio tuning formulas, these formulas predict that both the proportional and integral actions go to infinity as the process approaches a first-order process without dead time. This behavior is typical of tuning formulas for disturbance inputs.

The set-point tuning formulas given in Table 6-4 were developed by Rovira et al.,[6] who considered that the minimum ISE criterion was unacceptable because of its highly oscillatory nature. They also omitted relationships for pure proportional controllers on the assumption that the minimum error integral criteria is not appropriate for those applications for which a pure proportional controller is indicated, for example, flow averaging by proprortional level control. These formulas are also empirical and should not be extrapolated beyond the range of (t_0/τ) between 0.1 and 1.0. They predict that for a single capacitance process without dead time, the integral time approaches the time

Table 6-4. Minimum Error Integral Tuning Formulas for Set-Point Changes

Process Model: $G(s) = \dfrac{Ke^{-t_0 s}}{\tau s + 1}$

Proportional-Integral (PI) Controller:

$$G_c(s) = K_c\left(1 + \frac{1}{\tau_I s}\right)$$

Error Integral:	IAE	ITAE
$K_c = \dfrac{a_1}{K}\left(\dfrac{t_0}{\tau}\right)^{b_1}$	$a_1 = 0.758$	0.586
	$b_1 = -0.861$	-0.916
	$a_2 = 1.02$	1.03
$\tau_I = \dfrac{\tau}{a_2 + b_2(t_0/\tau)}$	$b_2 = -0.323$	-0.165

Proportional-Integral-Derivative (PID) Controller:

$$G_c(s) = K_c\left(1 + \frac{1}{\tau_I s} + \tau_D s\right)$$

Error Integral:	IAE	ITAE
$K_c = \dfrac{a_1}{K}\left(\dfrac{t_0}{\tau}\right)^{b_1}$	$a_1 = 1.086$	0.965
	$b_1 = -0.869$	-0.855
	$a_2 = 0.740$	0.796
$\tau_I = \dfrac{\tau}{a_2 + b_2(t_0/\tau)}$	$b_2 = -0.130$	-0.147
	$a_3 = 0.348$	0.308
$\tau_D = a_3\tau\left(\dfrac{t_0}{\tau}\right)^{b_3}$	$b_3 = 0.914$	0.9292

constant of the process while the proportional gain goes to infinity and the derivative time to zero. These trends are typical of set-point tuning formulas.

Example 6-12. Compare the tuning parameters that result form the various error integral criteria for disturbance inputs for the heat exchanger temperature controller using the FOPDT model transfer function of Example 6-9. Consider (a) a P controller, (b) a PI controller, and (c) a PID controller.

Solution. The FOPDT model parameters from Example 6-9 are, for fit 3 are

$$K = 0.80 \ \%/\%; \qquad \tau = 33.8 \ s; \qquad t_0 = 11.2 \ s$$

The minimum error integral tuning parameters for disturbance inputs can be calculated using the formulas from Table 6-3:

a. P Controller

$$\text{ISE: } K_c = \frac{1.411}{K} \left(\frac{t_0}{\tau} \right)^{-0.917}$$

$$= \frac{1.411}{0.80} \left(\frac{11.2}{33.8} \right)^{-0.917} = 4.9 \ \%/\%$$

$$\text{IAE: } K_c = \frac{0.902}{K} \left(\frac{t_0}{\tau} \right)^{-0.985} = 3.3 \ \%/\%$$

$$\text{ITAE: } K_c = \frac{0.490}{K} \left(\frac{t_0}{\tau} \right)^{-1.084} = 2.0 \ \%/\%$$

b. PI Controller

A similar application of the formulas from Table 6-3 results in the following parameters.

Criteria	ISE	IAE	ITAE
K_c, %/%	4.7	3.7	3.2
τ_I, s	30.3	25.5	23.7
(min)	(0.51)	(0.42)	(0.39)

c. PID Controller

Criteria	ISE	IAE	ITAE
K_c, %/%	5.3	5.0	4.8
τ_I, s	13.1	16.8	17.8
(min)	(0.22)	(0.28)	(0.30)
τ_D, s	6.2	4.6	4.3
(min)	(0.104)	(0.077)	(0.072)

The first conclusion we can draw from comparing these tuning parameters is that all of these formulas result in values of the same order of magnitude or "ball park." The second is that the ISE performance criterion usually results in the tightest control parameters (highest gain and shortest integral time), while ITAE usually results in the least tight.

Example 6-13. Compare the responses to unit step changes in disturbance and in set point obtained when a PI controller is tuned for minimum IAE for disturbance inputs with those obtained for the same criteria but for set point changes. The loop can be represented by the block diagram of Fig. 6-21 and the process is modeled by the following FOPDT model:

$$G(s) = \frac{1.0 \, e^{-0.5s}}{s + 1} \, \%/\%$$

where the time parameters are in minutes.

Solution. The FOPDT parameters are

$$K = 1.0 \, \%/\%; \qquad \tau = 1.0 \text{ min}; \qquad t_0 = 0.5 \text{ min}$$

We can now calculate the tuning parameters for a PI controller using the formulas from Tables 6-3 and 6-4:

IAE-Disturbance Criterion (Table 6-3)

$$K_c = \frac{0.984}{K} \left(\frac{t_0}{\tau}\right)^{-0.986}$$

$$= \frac{0.984}{1.0} (0.5)^{-0.986} = 1.95 \, \%/\%$$

$$\tau_I = \frac{\tau}{0.608} \left(\frac{t_0}{\tau}\right)^{0.707}$$

$$= \frac{1.0 \text{ min}}{0.608} (0.5)^{0.707} = 1.01 \text{ min}$$

IAE–Set-Point Criterion (Table 6-4)

$$K_c = \frac{0.758}{K} \left(\frac{t_0}{\tau}\right)^{0.861}$$

$$= \frac{0.758}{1.0} (0.50)^{-0.861} = 1.38 \, \%/\%$$

$$\tau_I = \frac{\tau}{1.02 - 0.323 (t_0/\tau)}$$

$$= \frac{1.0 \text{ min}}{1.02 - 0.323(0.50)} = 1.16 \text{ min}$$

In order to calculate the responses we must solve for the output variable in the block diagram of Fig. 6-21.

Disturbance Input

$$\frac{C(s)}{U(s)} = \frac{G(s)}{1 + G(s) \, G_c(s)}; \qquad U(s) = \frac{1}{s}$$

Set-Point Input

$$\frac{C(s)}{R(s)} = \frac{G(s)\,G_c(s)}{1 + G(s)\,G_c(s)}; \qquad R(s) = \frac{1}{s}$$

However, the presence of the dead-time term in the FOPDT transfer function $[G(s)]$ makes it impractical to invert the Laplace transform by partial fractions expansion. A more practical approach is to solve the loop equations on a digital computer. The differential equations for this loop are as follows.

FOPDT process

$$\tau\,\frac{dc(t)}{dt} + c(t) = K\,[m(t - t_0) + U(t - t_0)]$$

PID controller

$$m(t) = m(0) + K_c\left[e(t) + \frac{1}{\tau_I}\int e(t)\,dt + \tau_D\,\frac{de(t)}{dt}\right]$$

$$e(t) = r(t) - c(t)$$

Disturbance input

$$U(t) = u(t) \qquad \text{(unit step)}$$

Set-point input

$$r(t) = u(t) \qquad \text{(unit step)}$$

Initial conditions

$$c(0) = 0$$
$$m(0) = 0$$

These equations were solved by Rovira[6] using a computer program that generated the response plots shown in Fig. 6-22. The plot in Fig. 6-22a is for a unit step change in disturbance; it shows that the disturbance tuning parameters result in a slightly smaller initial deviation and a faster return to the set point than the set-point tuning parameters. Fig. 6-22b is for a unit step change in set point and shows that the set point tuning parameters result in a significantly smaller overshoot, less oscillatory behavior, and a shorter settling time than the disturbance tuning parameters. As would be expected, each set of tuning parameters performs best for the input for which it is designed. The responses obtained are a direct result of the higher gain and shorter reset time obtained with disturbance tuning.

Tuning Sampled-Data Controllers

Today the trend in industry is toward the implementation of control functions using microprocessors (distributed controllers), minicomputers, and regular digital computers. A common characteristic of these installations is that the control calculations are performed

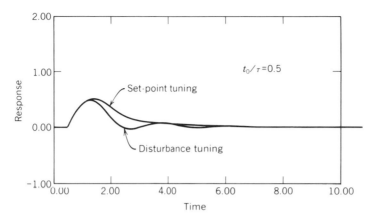

Figure 6-22a. Response to a distrubance change. Disturbance vs. set-point tuning, PI controller, minimum IAE criteria. (Reproduced by permission from Reference 6.)

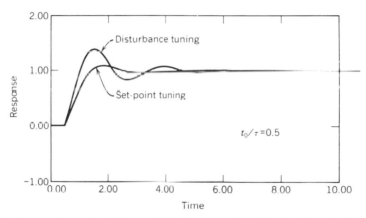

Figure 6-22b. Response to a set-point change. Disturbance vs. set-point tuning, PI controller, minimum IAE criteria. (Reproduced by permission from reference 6.)

at regular intervals of time T, the *sample time*. This is in contrast to analog (electronic and pneumatic) instruments that perform their functions continuously with time. Sampling is also characteristic of some analyzers, such as on-line gas chromatographs.

The discrete mode of the operation characteristic of computers requires that at each sampling instant the transmitter signals be sampled, the value of the manipulated variable be calculated, and the controller output signal be updated. The output signals are then held constant for a full sample time until the next update. This is illustrated in Fig. 6-23. As might be expected, this sampling and holding operation has an effect on the performance of the controller and thus on its tuning parameters.

The sampling time of computer controllers varies from about 1/3 second to several minutes, depending on the application. A good rule of thumb is that the sample time should be from one-tenth to one-twentieth of the effective process time constant. When

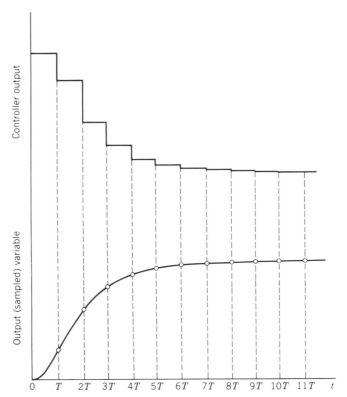

Figure 6-23.   The output of a sampled-data (computer) controller is held constant during each sampling period T.

the sample time is of this order of magnitude, its effect can be taken into consideration in the tuning formulas by adding one-half the sample time to the process dead time and then using this corrected dead time in the tuning formulas for continuous controllers (Tables 6-2, 6-3, and 6-4). This method, proposed by Moore et al.[10], says that the dead time to use in the tuning formulas is

$$t_{0c} = t_0 + \frac{1}{2}T \tag{6-49}$$

where

    t_{0c} is the corrected dead time
    t_0 is the dead time of the process
    T is the sample time

Notice that the on-line tuning method inherently incorporates the effect of sampling when the ultimate gain and period are determined for the loop with the sampled-data controller in automatic.

    Tuning formulas that are specific for sampled-data controllers have been developed by Chiu et al.[11] and reproduced by Corripio.[12]

Summary

In this section we have presented two methods for measuring the dynamic characteristics of the process in a feedback control loop: the ultimate gain method, and the step test or process reaction curve. We have also presented one set of tuning formulas for the ultimate gain method and three sets of formulas for the first-order plus dead-time model parameters. We have seen that for a given process all four sets of tuning formulas result in controller parameters that are in the same "ball park." These tuning parameters are just starting values that must be adjusted in the field so that the true "personality" of the specific process can be matched by the controller. We must reiterate a point made at the beginning of this section. As discussed in previous chapters, most processes are nonlinear and their dynamic characteristics (e.g., ultimate gain and frequency, FOPDT model parameters) vary from one operating point to another. It follows that the controller parameters arrived at by the tuning procedure are at best a compromise between slow behavior at one end of the operating range and oscillatory behavior at the other. In short, tuning is not an exact science. However, we must also keep in mind that the tuning formulas offer us insight into how the various controller parameters depend on such process parameters as the gain, the time constant, and the dead time.

6-4. SYNTHESIS OF FEEDBACK CONTROLLERS

In the preceding section we have taken the approach of tuning a feedback controller by adjusting the parameters of the proportional-integral-derivative (PID) control structure. In this section we will take a different approach to controller design, that of controller synthesis:

Given the transfer functions of the components of a feedback loop, synthesize the controller required to produce a specified closed-loop response.

Although we get no assurances that the controller resulting from our synthesis procedure can be built in practice, we stand to gain some insight into the selection of the various controller modes and their tuning.

Development of the Controller Synthesis Formula

Let us consider the simplified block diagram of Fig. 6-24 in which the transfer functions of all the loop components other than the controller have been lumped into a single block, $G(s)$. From block diagram algebra the transfer function for the closed loop is

$$\frac{C(s)}{R(s)} = \frac{G_c(s)G(s)}{1 + G_c(s)G(s)} \tag{6-50}$$

Then, from this expression, we can solve for the controller transfer function:

$$G_c(s) = \frac{1}{G(s)} \cdot \frac{C(s)/R(s)}{1 - [C(s)/R(s)]} \tag{6-51}$$

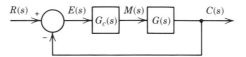

Figure 6-24.   Simplified block diagram for controller synthesis.

This is the *controller synthesis formula*. It gives us the controller transfer function $G_c(s)$ from the process transfer function $G(s)$ and the specified closed-loop response $C(s)/R(s)$. In order to illustrate how this formula is used, consider the specification of perfect control, that is, $C(s) = R(s)$ or $C(s)/R(s) = 1$. The resulting controller is

$$G_c(s) = \frac{1}{G(s)} \cdot \frac{1}{1-1} = \frac{1}{G(s)} \cdot \frac{1}{0} \tag{6-52}$$

This says that in order to force the output equal to the set point at all times, the controller gain must be infinite. In other words, perfect control cannot be achieved with feedback control. This is because any feedback corrective action must be based on an error.

The controller synthesis formula, Eq. (6-51), results in different controllers for different combinations of closed-loop response specifications and process transfer functions. Let us look at each of these elements in turn.

Specification of the Closed-Loop Response

The simplest achievable closed-loop response is a first-order lag response. In the absence of process dead time, this response is the one shown in Fig. 6-25 and results from the closed-loop transfer function:

$$\frac{C(s)}{R(s)} = \frac{1}{\tau_c s + 1} \tag{6-53}$$

where τ_c is the time constant of the closed-loop response and, being adjustable, becomes the single *tuning parameter* for the synthesized controller; the shorter τ_c the tighter controller tuning.

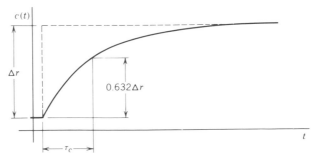

Figure 6-25.   First-order lag specification for closed-loop response of synthesized controller.

Note: This response was originally proposed by Dahlin,[7] who defined the tuning parameter as the reciprocal of the closed-loop time constant, $\lambda = 1/\tau_c$. In this book we will use τ_c.

Substituting Eq. (6-53) into Eq. (6-51) we obtain

$$G_c(s) = \frac{1}{G(s)} \cdot \frac{\dfrac{1}{\tau_c s + 1}}{1 - \dfrac{1}{\tau_c s + 1}} = \frac{1}{G(s)} \cdot \frac{1}{\tau_c s + 1 - 1}$$

or

$$G_c(s) = \frac{1}{G(s)} \cdot \frac{1}{\tau_c s} \tag{6-54}$$

We can see that this controller has integral action, which results from the specification of unity gain in the closed-loop transfer function, Eq. (6-53). This assures the absence of offset.

Although second- and higher-order closed-loop responses could be specified, it is seldom necessary to do so. However, when the process contains dead time, the closed-loop response must also contain a dead-time term with the dead time equal to the process dead time. We will look at this shortly, but first let us see how controller synthesis can guide us in the selection of controller modes for various process transfer functions.

Controller Modes and Tuning Parameters

Controller synthesis allows us to establish a relationship between the process transfer function and the modes of a PID controller. This is so because for simple transfer functions without dead time, the synthesized controller can be expressed in terms of the proportional, integral, and derivative modes. Controller synthesis also provides us with relationships for the controller tuning parameters in terms of the closed-loop time constant, τ_c, and the parameters of the process transfer function. In what follows we will derive these relationships by substituting process transfer functions of increasing complexity into Eq. (6-54).

Instantaneous Process Response: $G(s) = K$.

$$G_c(s) = \frac{1}{K\tau_c} \cdot \frac{1}{s} \tag{6-55}$$

where K is the process gain.

This is a *pure integral* controller, which is indicated for very fast processes such as flow controllers, steam turbine governors, and the control of outlet temperatures from reformer furnaces.

First-Order Process: $G(s) = K/(\tau s + 1)$.

$$G_c(s) = \frac{\tau s + 1}{K} \cdot \frac{1}{\tau_c s}$$

$$= \frac{\tau}{K\tau_c}\left(1 + \frac{1}{\tau s}\right) \tag{6-56}$$

where τ is the process time constant.

This is a proportional-integral (PI) controller with tuning parameters

$$K_c = \frac{\tau}{K\tau_c} \qquad \tau_I = \tau \tag{6-57}$$

or, in words, the integral time is set equal to the process time constant and the proportional gain is adjustable or tunable. Notice that if the process time constant τ is known, tuning is reduced to the adjustment of a single parameter, the controller gain. This is because the tuning parameter τ_c affects only the controller gain.

Second-Order Process: $G(s) = K/[(\tau_1 s + 1)(\tau_2 s + 1)]$.

$$G_c(s) = \frac{(\tau_1 s + 1)(\tau_2 s + 1)}{K} \cdot \frac{1}{\tau_c s}$$

$$= \frac{\tau_1}{K\tau_c}\left(1 + \frac{1}{\tau_1 s}\right)(\tau_2 s + 1) \tag{6-58}$$

where

τ_1 is the longer or dominant process time constant
τ_2 is the shorter process time constant

Equation (6-58) matches the transfer function of the industrial PID controller discussed in Chapter 5, provided we ignore the noise filter term $(\alpha\tau_D s + 1)$:

$$G_c(s) = K_c\left(1 + \frac{1}{\tau_I s}\right)\left(\frac{\tau_D s + 1}{\alpha\tau_D s + 1}\right) \tag{6-59}$$

The tuning parameters are then

$$K_c = \frac{\tau_1}{K\tau_c} \qquad \tau_I = \tau_1 \qquad \tau_D = \tau_2 \tag{6-60}$$

Again the tuning procedure is reduced to the adjustment of the process gain with the integral time set equal to the longer time constant and the derivative time set to the shorter time constant. This arbitrary choice is the result of experience that indicates that the derivative time should always be smaller than the integral time. In industrial practice, PID controllers are commonly used for temperature control loops so that the derivative action can compensate for the sensor lag. We have arrived at this same result by controller synthesis.

We can easily see that a third-order process would demand a second derivative term in series with the first and with its time constant set to the third longest process time constant, and so on. A reason why this idea has not caught on in practice is that the controller would be very complex and expensive. Besides, the values of the third and subsequent process time constants are very difficult to determine in practice. The common practice has been to approximate high-order processes with low-order plus dead-time models. Let us next synthesize the controller for such an approximation of the process transfer function.

First-Order plus Dead-Time Process: $G(s) = (Ke^{-t_0s})/(\tau s + 1)$.

$$G_c(s) = \frac{\tau s + 1}{Ke^{-t_0s}} \cdot \frac{1}{\tau_c s}$$

$$= \frac{\tau}{K\tau_c}\left(1 + \frac{1}{\tau s}\right)e^{t_0s}$$

(6-61)

where t_0 is the process dead time.

We note immediately that this is an *unrealizable* controller because it requires knowledge of the future, that is, a negative dead time. This is even more obvious when the specified and the best possible closed-loop responses are graphically compared, as in Fig. 6-26. It is evident from this comparison that the specified response must be delayed by one process dead time:

$$\frac{C(s)}{R(s)} = \frac{e^{-t_0s}}{\tau_c s + 1}$$

(6-62)

This results in the following synthesized controller transfer function:

$$G_c(s) = \frac{\tau s + 1}{Ke^{-t_0s}}\frac{e^{-t_0s}}{\tau_c s + 1 - e^{-t_0s}}$$

or

$$G_c(s) = \frac{\tau s + 1}{K} \cdot \frac{1}{\tau_c s + 1 - e^{-t_0s}}$$

(6-63)

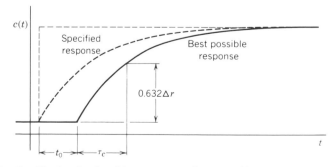

Figure 6-26. Specification for closed-loop response of system with dead time t_0.

Although this controller is now realizable in principle, its implementation is far from common practice. This is mostly because the original PID controllers were implemented with analog components and the term $e^{-t_0 s}$ cannot be implemented in practice with analog devices. Modern implementation of PID controllers on microprocessors and digital computers makes it possible to implement the dead-time term. When this is done the term is called a "predictor" or "dead-time compensation" term.

In order to convert the algorithm of Eq. (6-63) to the standard PI form, we approximate the exponential term by a Taylor series expansion:

$$e^{-t_0 s} = 1 - t_0 s + \frac{1}{2!}(t_0 s)^2 - \frac{1}{3!}(t_0 s)^3 + \dots \tag{6-64}$$

Dropping all but the first two terms, a first-order approximation results:

$$e^{-t_0 s} \doteq 1 - t_0 s \tag{6-65}$$

Substituting this expression into Eq. (6-63) and simplifying yields

$$G_c(s) = \frac{\tau s + 1}{K} \cdot \frac{1}{(\tau_c + t_0)s}$$

$$= \frac{\tau}{K(\tau_c + t_0)}\left(1 + \frac{1}{\tau s}\right) \tag{6-66}$$

This is a PI controller tuned as follows:

$$K_c = \frac{\tau}{K(\tau_c + t_0)} \qquad \tau_I = \tau \tag{6-67}$$

The first-order Taylor expansion approximation is valid as long as the dead time is small compared with the closed-loop response speed. In other words, a PI controller without dead-time compensation is a good approximation of the synthesized controller as long as the process dead time is small compared to its time constant. This is usually the case when the process does not have true dead time and the model dead time accounts mostly for the high order of the process.

The most important insight that we can extract from the tuning relationships of Eq. (6-67) is that an increase in dead time results in a reduction of the controller gain for a given closed-loop time constant specification. If we compare Eqs. (6-57) and (6-67), we see that the presence of dead time imposes a limit on the controller gain. In other words, for the first-order process without dead time, Eq. (6-57), the gain can be increased without limit to obtain faster and faster responses ($\tau_c \to 0$). However, for the process with an effective dead time, from Eq. (6-67), we have the following limit on the controller gain:

$$K_{c_{max}} = \lim_{\tau_c \to 0} \frac{\tau}{K(\tau_c + t_0)} = \frac{\tau}{K t_0} \tag{6-68}$$

The closed-loop response will deviate from the specified first-order response as the controller gain is increased. That is, increasing the gain will eventually result in overshoot and even instability of the closed-loop response. This is because the error of the first-order Taylor expansion approximation increases with the speed of response, as s increases with speed. (Recall that s, the Laplace transform variable, has units of reciprocal time or frequency. Thus higher speeds of response, or frequencies, correspond to a high magnitude of s.)

Derivative Mode for Dead-Time Processes

If a second-order plus dead-time model is used to approximate a high-order process, a procedure similar to the one followed for the first-order plus dead-time model results in a synthesized controller, which is the equivalent of an industrial PID controller, Eq. (6-59). This derivation is left to the reader as an exercise. However, because of the difficulty of determining the parameters of a second-order model, it is more attractive to derive tuning formulas for a PID controller that are based on the parameters of a first-order plus dead-time model. This can be done by using a different approximation to the dead-time term in Eq. (6-63). The first-order Padé approximation to the exponential, presented earlier, is given by

$$e^{-t_0 s} \doteq \frac{1 - \dfrac{t_0}{2}s}{1 + \dfrac{t_0}{2}s} \tag{6-69}$$

A long-hand division of this expression results in the following infinite series:

$$\frac{1 - \dfrac{t_0}{2}s}{1 + \dfrac{t_0}{2}s} = 1 - t_0 s + \frac{1}{2}(t_0 s)^2 - \frac{1}{4}(t_0 s)^3 + \ldots \tag{6-70}$$

Notice that, by comparison with Eq. (6-64), this expression matches the first three terms of the Taylor series expansion of the exponential, and thus it is a bit more accurate than the approximation given by Eq. (6-65). This means that Eq. (6-69) is closer to the true exponential expression for processes with higher dead time to time constant ratios.

Substituting Eq. (6-69) into Eq. (6-63) and simplifying, we obtain the following synthesized controller:

$$G_c(s) = \frac{\tau}{K(\tau_c + t_0)}\left(1 + \frac{1}{\tau s}\right)\left(\frac{1 + \dfrac{t_0}{2}s}{1 + \tau' s}\right) \tag{6-71}$$

where

$$\tau' = \frac{\tau_c t_0}{2(\tau_c + t_0)}$$

This is equivalent to an industrial PID controller, Eq. (6-59) with tuning parameters:

$$K_c = \frac{\tau}{K(\tau_c + t_0)} \qquad \tau_I = \tau \qquad \tau_D = \frac{t_0}{2} \tag{6-72}$$

Although a lag term is present in the transfer function of the industrial controller to prevent high-frequency noise amplification, the time constant τ' is usually fixed and much shorter than τ_D. In order to interpret the meaning of the term $(1 + \tau's)$, we first note that for small dead time $(t_0 \ll \tau_c)$

$$\tau' = \frac{t_0}{2} \tag{6-73}$$

Substitution of this equation into Eq. (6-71) results in exactly the same PI controller given by Eq. (6-66). This confirms our earlier conclusion that the PI controller is indicated when the dead time is short. For long dead time and tight control $(\tau_c \to 0)$, the value of τ' becomes

$$\tau' = \frac{\tau_c}{2} \to 0 \tag{6-74}$$

Therefore, for long dead time, the tighter the control the closer the synthesized algorithm, Eq. (6-71), is to the industrial PID controller with the tuning parameters of Eq. (6-72).

It is interesting to note that the derivative time of Eq. (6-72) is exactly the same as the value from the Ziegler-Nichols quarter decay ratio formulas (see Table 6-2). However, the proportional gain for quarter decay is 20% higher than the maximum synthesis gain $(\tau_c = 0)$ and the integral time of the synthesis formula is related to the model time constant whereas that of the quarter decay ratio formula is related to the model dead time.

Table 6-5 summarizes the selection of controller modes and tuning parameters that result from the synthesis procedure for Dahlin's response. The fact that the controller gain is a function of the tuning parameter τ_c is both an advantage and a disadvantage of the tuning formulas derived by the synthesis procedure. It is an advantage in that it allows the engineer to achieve a specified response by adjusting a single parameter, the gain, regardless of the controller modes involved. The tunable gain is a disadvantage, however, because the formulas do not provide a "ball-park" value for it. The following guidelines are given in order to remedy this situation.

Minimum IAE. For *disturbance inputs*, $\tau_c = 0$ approximately minimizes the IAE when t_0/τ is in the range 0.1 to 0.5 for PI controllers $(\tau_D = 0)$ and 0.1 to 1.5 for PID controllers. For *set-point changes*, the following formulas result in approximately minimum IAE when t_0/τ is in the range 0.1 to 1.5:

PI controller $(\tau_D = 0)$

$$\tau_c = \frac{2}{3} t_0 \tag{6-75a}$$

PID controller

$$\tau_c = \frac{1}{5} t_0 \tag{6-75b}$$

These formulas are to be used with the last entry of Table 6-5.

Table 6-5 Controller Modes and Tuning Formulas for Dahlin Synthesis

Process	Controller	Tuning Parameters
$G(s) = K$	I	$K_c = \dfrac{1}{K\tau_c}$ tunable
$G(s) = \dfrac{K}{\tau s + 1}$	PI	$K_c = \dfrac{\tau}{K\tau_c}$ tunable $\tau_I = \tau$
$G(s) = \dfrac{K}{(\tau_1 s + 1)(\tau_2 s + 1)}$ $\tau_1 > \tau_2$	PID	$K_c = \dfrac{\tau_1}{K\tau_c}$ tunable $\tau_I = \tau_1$ $\tau_D = \tau_2$
$G(s) = \dfrac{Ke^{-t_0 s}}{\tau s + 1}$	PID[a]	$K_c = \dfrac{\tau}{K(t_0 + \tau_c)}$ tunable $\tau_I = \tau$ $\tau_D = \dfrac{t_0}{2}$

[a]This last set of formulas applies to both PID and PI ($\tau_D = 0$) controllers. PID is recommended when t_0 is greater than $\tau/4$.

Caution: The PID tuning parameters on this table are for the analog PID controllers [Eq. (5-45)]. For use with microprocessor-based PID controllers [Eq. (5-44)], the tuning parameters on this table must be converted by the following formulas:

$$K_c' = K_c\left(1 + \frac{\tau_D}{\tau_I}\right)$$

$$\tau_I' = \tau_I + \tau_D$$

$$\tau_D' = \tau_D\frac{\tau_I}{\tau_I + \tau_D}$$

5% Overshoot. For set-point inputs a response having an overshoot of 5% of the change in set point is highly desirable. For this type of response Martin et al.[8] recommend that τ_c be set equal to the effective dead time of the FOPDT model. This results in the following formula for the controller gain that produces 5% overshoot on *set-point changes*:

$$K_c = \frac{0.5}{K}\left(\frac{\tau}{t_0}\right) \tag{6-76}$$

Comparison of this formula with the one in Table 6-2 shows that this is about 40% of the PID gain required for quarter decay ratio (50% overshoot).

One interesting point about the controller synthesis method is that if controllers had been designed this way from the start, the evolution of controller modes would have followed the pattern I, PI, PID. This pattern follows from a consideration of the simplest process model to the more complex. Contrast this to the actual evolution of industrial controllers: P, PI, PID, that is, from the simplest controller to the more complex.

An important insight that we can gain from the controller synthesis procedure is that the main effect of the proportional mode, when added to the basic integral mode, is to compensate for the longest or dominant process lag, while that of the derivative mode is to compensate for the second-longest lag or for the effective process dead time. The entire synthesis procedure is based on the assumption that the primary closed-loop response specification is the elimination of offset or steady-state error. This is what makes the integral action the basic controller mode.

Example 6-14. Determine the tuning parameters for the heat exchanger of Example 6-5 using the formulas derived by the controller synthesis method. Compare these results with those obtained from the minimum IAE tuning formulas for set-point inputs.

Solution. The FOPDT parameters obtained for the heat exchanger by fit 3 in Example 6-9 are

$$K = 0.80 \ \%/\% \qquad \tau = 33.8 \ s \qquad t_0 = 11.2 \ s$$

We obtain the transfer function of the synthesized controller by plugging these values into Eq. (6-71):

$$G_c(s) = \frac{33.8}{0.80(\tau_c + 11.2)} \left(1 + \frac{1}{33.8s}\right) \left\{ \frac{1 + 5.6s}{1 + \frac{5.6\tau_c}{\tau_c + 11.2}s} \right\}$$

As the dead time in this case is greater than one-fourth the time constant, a PID controller is indicated. The minimum IAE gain for disturbance input is obtained with $\tau_c = 0$:

$$K_c = \frac{33.8}{(0.80)(11.2)} = 3.8 \ \%/\%$$

For minimum IAE set-point input, from Eq. (6-75b) we get

$$\tau_c = \frac{1}{5}(11.2) = 2.24 \ s$$

$$K_c = \frac{33.8}{(0.80)(2.24 + 11.2)} = 3.1 \ \%/\%$$

For 5% overshoot on set-point input, from Eq. (6.76) we obtain

$$K_c = \frac{(0.5)(33.8)}{(0.8)(11.2)} = 1.9 \ \%/\%$$

The integral and derivative times are

$$\tau_I = \tau = 33.8 \ s \qquad (0.56 \ min)$$

$$\tau_D = \frac{t_0}{2} = 5.6 \ s \qquad (0.093 \ min)$$

For comparison, the minimum IAE parameters for set-point inputs are calculated by the PID formulas in Table 6-4:

$$K_c = \frac{1.086}{0.80} \left(\frac{11.2}{33.8}\right)^{-0.869} = 3.5 \ \%/\%$$

$$\tau_I = \frac{33.8}{0.740 - 0.130(11.2/33.8)} = 48.5 \text{ s} \qquad (0.81 \text{ min})$$

$$\tau_D = 0.348(33.8) \left(\frac{11.2}{33.8}\right)^{0.914} = 4.3 \text{ s} \qquad (0.071 \text{ min})$$

These two sets of parameters are in the same ball park.

Example 6-15. A second-order plus dead-time process has the following transfer function:

$$G(s) = \frac{1.0 \, e^{-0.26s}}{s^2 + 4s + 1}$$

Compare the responses to a step change in set point of a PI controller tuned by (a) Ziegler-Nichols quarter-decay ratio, (b) minimum IAE for set-point changes, (c) controller synthesis with the gain adjusted for a 5% overshoot. Use Fig. 6-18 to obtain the first-order plus dead-time (FOPDT) model parameters.

Solution. The first step is to approximate the second-order transfer function by a FOPDT model. We start by factoring the denominator into two time constants:

$$(\tau_1 s + 1)(\tau_2 s + 1) = \tau_1 \tau_2 s^2 + (\tau_1 + \tau_2)s + 1$$

$$= s^2 + 4s + 1$$

$$\tau_1 \tau_2 = 1 \qquad \tau_1 + \tau_2 = 4$$

$$\tau_1 + \frac{1}{\tau_1} - 4 = 0 \qquad \text{or} \qquad \tau_1^2 - 4\tau_1 + 1 = 0$$

$$\tau_1 = \frac{4 + \sqrt{16 - 4}}{2} = 3.73 \text{ min}$$

$$\tau_2 = 4 - 3.73 = 0.27 \text{ min}$$

The transfer function can then be written as

$$G(s) = \frac{1.0 \, e^{-0.26s}}{(3.73s + 1)(0.27s + 1)}$$

The second step is to approximate the second-order lag with a FOPDT model, using Fig. 6-18:

$$\frac{\tau_2}{\tau_1} = \frac{0.27}{3.73} = 0.072 \qquad \tau' = 1.0\tau_1 = 3.73 \text{ min}$$
$$t'_0 = 0.072\tau_1 = 0.27 \text{ min}$$

Note that the second time constant is small enough for the simple rule of thumb given in section 6-3 to apply, that is, $\tau' = \tau_1$, $t'_0 = \tau_2$. The effective dead time of the FOPDT model must be added to the actual-process dead time to obtain the total dead time:

$$t_0 = 0.26 + t_0' = 0.53 \text{ min}$$

The FOPDT parameters are then

$$K = 1.0 \qquad \tau = 3.73 \text{ min} \qquad t_0 = 0.53 \text{ min}$$

The third step is to calculate the tuning parameters from the formulas specified. For a PI controller:

a. *Quarter Decay Ratio (from Table 6-2)*

$$K_c = \frac{0.90}{1.0}\left(\frac{0.53}{3.73}\right)^{-1} = 6.3 \text{ %/% } (16\% \text{ P.B.})$$

$$\tau_I = 3.33 t_0 = 3.33(0.53) = 1.76 \text{ min}$$

b. *Minimum IAE for Set-Point Changes (from Table 6-4)*

$$K_c = \frac{0.758}{1.0}\left(\frac{0.53}{3.73}\right)^{-0.861} = 4.1 \text{ %/% } (24.6\% \text{ P.B.})$$

$$\tau_I = \frac{3.73}{1.02 - 0.323(0.53/3.73)} = 3.83 \text{ min}$$

c. *Controller Synthesis Tuned for 5% Overshoot, from Eq. (6-67)*

$$\tau_I = \tau = 3.73 \text{ min}$$

For 5% overshoot, from Eq. (6-76) we get

$$\tau_c = t_0 = 0.53 \text{ min}$$

$$K_c = \frac{3.73}{2(0.53)} = 3.57 \text{ %/% } (28.4\% \text{ PB})$$

The final step is to compare the responses to a unit step in set point using each of the three sets of tuning parameters. Martin et al.[8] published the solution to this problem, which they obtained using an analog computer to simulate the following set of equations (see Fig. 6-22 for the corresponding block diagram):

$$\frac{d^2c(t)}{dt^2} + 4\frac{dc(t)}{dt} + c(t) = m(t - 0.26) \qquad c(0) = 0$$

$$m(t) = K_c\left[e(t) + \frac{1}{\tau_I}\int e(t)\,dt\right] \qquad m(0) = 0$$

$$e(t) = r(t) - c(t) \qquad r(0) = 0$$

$$r(t) = 1 \qquad \text{for } t > 0$$

A standard Padé approximation was used to simulate the dead time.

The resulting responses are shown in Fig. 6-27. Comparison of the responses shows that the controller synthesis formulas for 5% overshoot yield a response that is very close to the minimum IAE set point response. These responses are superior to the quarter decay ratio response in terms of stability and settling time for changes in set point.

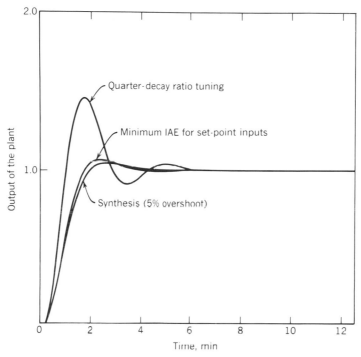

Figure 6-27. Closed loop response of second-order plant with PI controller to set point change. Model parameters: $\tau - 3.7$ min, $t_0 = 0.53$ min, $K - 1$. (Reproduced by permission from Reference 8, © ISA, 1976)

Summary

In this section we have presented the technique of feedback controller synthesis. From the resulting synthesized controllers we have gained insight into the functions of the proportional, integral, and derivative modes. We have also derived a set of tuning relationships for PID controllers.

6-5. PREVENTION OF RESET WINDUP

We have seen in the preceding sections that integral or reset action is necessary to eliminate offset or steady-state error in feedback controllers. As discussed in Chapter 5, one of the penalties that we must pay for this convenience is "reset windup" or excessive overshoot of the controlled variable as the controller output signal returns to its normal range after a period of saturation. It is because of reset windup that controllers must be switched to "manual" during process startups and shutdowns, as it is under these conditions that controllers saturate most often. The controller is said to be saturated when its output signal is at or outside the limits of operation of the control valve or final control element. When this happens the feedback control loop is interrupted and the controlled variable deviates from the set point, as would be expected. Because of the integral action, a large deviation in the opposite direction may eventually be required in order to drive the controller output back into its normal operating range. Reset windup is this inability of the controller to recover quickly from a saturation condition.

In order to review the concept of reset windup, let us consider the startup of the steam-heated tank sketched in Fig. 6-28a. A proportional-integral (PI) feedback controller (TIC) is used to control the temperature in the tank by adjusting the steam control valve. The instrumentation is pneumatic with a normal range of 3 to 15 psig and a supply pressure of 20 psig. If the controller is left in automatic during startup, its output is driven to its maximum value, the supply pressure of 20 psig, by the integral action. This is because the temperature remains below its set point for a long period of time. A time recording of the startup procedure is shown in Fig. 6-28b. The steam valve is initially fully open while the controller output signal is at the supply pressure of 20 psig. Although the controller is saturated, it keeps the steam valve fully open. This is the correct strategy to heat the tank contents up to the set point in minimum time. The windup problem begins to show when the temperature of the tank reaches its set point. At that instant the controller output is

$$m(t) = \overline{m} + K_c e(t) + \frac{K_c}{\tau_I} \int e(t)\, dt$$

$$= 20 \text{ psig}$$

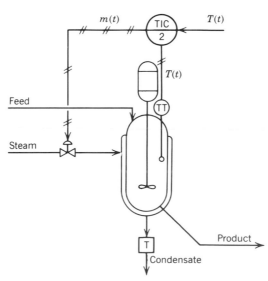

Figure 6-28a.   Steam-heated tank.

This is because the integral action has driven the controller output to the supply pressure. (Note that this is equivalent to resetting the bias value $\overline{m}$ to 20 psig and integral term to zero.) Since the steam control valve will not begin to close until the controller output reaches 15 psig, a large negative error will be required to cause the 5 psi decrease in

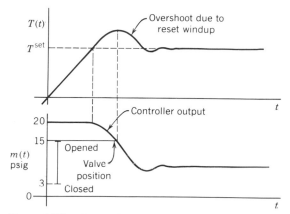

Figure 6-28b. Reset windup on start-up of steam-heated tank.

controller output. For example, by proportional action alone, assuming 25% proportional band ($K_c = 4$), the minimum error required to start closing the steam valve would be

$$m = 20 + K_c e \text{ psig}$$

$$e = \frac{m - 20}{K_c} = \frac{15 - 20}{4} = -1.25 \text{ psig } (-10.4\% \text{ of range})$$

This is a significant error and means that the temperature will continue to rise above the set point while the steam valve remains fully open. The controller would not appear to be responding to the rising temperature and will be said to be "wound up."

The integral action will start to reduce the controller output as soon as the error goes negative, so that the error will peak at a value lower than the one estimated above (-10.4%). However, had it not been for the integral action in the first place, the bias value $\overline{m}$ would not have been driven to 20 psig. As we learned in Chapter 5, the formula for a proportional controller is

$$m(t) = \overline{m} + K_c e(t)$$

If the bias value $\overline{m}$ is, say, 9 psig, the proportional controller would start closing the steam valve *before* the temperature reaches the set point. The integral action is thus the cause of the large temperature overshoot shown in Fig. 6-28b. Since this effect is highly undesirable, how can it be prevented?

One way to prevent the large overshoot caused by reset windup is to keep the controller on manual until the temperature reaches the set point, and then switch it to automatic. In this case the steam valve is kept fully open by manually setting the controller output to 15 psig. This assures that the control valve will start to close as soon as the controller is switched to automatic. A second alternative is to install a limiter on the controller output to keep it from going beyond the operating range of the control valve, that is, above 15 psig or below 3 psig. But will this work? In order to answer this question let us break down the block diagram of the controller into its parts. The transfer function of the PI controller is given by

$$M(s) = K_c\left[1 + \frac{1}{\tau_I s}\right]E(s)$$

$$= K_c E(s) + M_I(s) \tag{6-77}$$

where

$$M_I(s) = \frac{1}{\tau_I s} K_c E(s)$$

A straightforward implementation of this transfer function is shown in the block diagram of Fig. 6-29a. The diagram illustrates why the placing of a limiter on the controller output does not prevent the windup problem: the output of the integral action, $M_I(s)$, will still be driven beyond the controller output limits and cause windup. In other words, in order to prevent windup, the output of the integral action must somehow be limited. In penumatic and electronic analog controllers this limiting is accomplished in a very ingenious way: First, from the definition of $M_I(s)$ we get

$$\tau_I s M_I(s) = K_c E(s) \tag{6-78}$$

Solving for $K_c E(s)$ from Eq. (6-77), we get

$$K_c E(s) = M(s) - M_I(s) \tag{6-79}$$

Combining Eqs. (6-78) and (6-79) and rearranging yields

$$M_I(s) = \left[\frac{1}{\tau_I s + 1}\right]M(s) \tag{6-80}$$

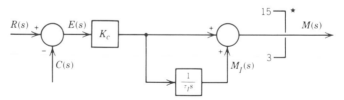

Figure 6-29a.   Straightforward implementation of PI controller is difficult to limit.

This implementation of the integral action is represented in the block diagram of Fig. 6-29b. We can see from this diagram that if we place the limiter as shown, $M_I(s)$ will be automatically limited. This is because $M_I(s)$ is always lagging $M(s)$ with a gain of 1.0 and an adjustable time constant τ_I; thus it can never get outside the range within which $M(s)$ is limited. In other words, if $M(s)$ reaches one of its limits, $M_I(s)$ will approach that limit, say, 15 psig; then the moment the error turns negative the controller output becomes

$$m(t) = 15 + K_c e(t) < 15 \text{ psig} \qquad \text{as} \qquad e(t) < 0$$

That is, the controller output will come off the limit, closing the control valve, the instant the controlled variable crosses the set point!

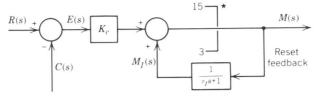

*Limiter on absolute signal

Figure 6-29b. Reset-feedback implementation of PI controller is easy to limit.

Note that at steady state the error must be zero, because

$$M = M_I = M_I + K_c e \qquad \therefore e = 0$$

Thus there can be no offset.

The limiter shown in Fig. 6-29b is sometimes known as a "batch switch" because batch processes encounter windup situations often enough to justify the extra expense of the limiter. Today, with feedback controllers being implemented with microprocessors, the limiter is a standard controller feature.

The "reset feedback" structure of Fig. 6-29b has the advantage of providing for the elimination of reset windup in override and cascade control systems in a very clean and straightforward manner. We will mention this in the sections that cover these topics.

6-6. SUMMARY

Feedback control is the basic strategy of industrial process control. In this chapter we have presented methods to determine the linearized response of a feedback control loop and its stability limits. We have also presented several techniques to tune feedback controllers and discussed the problem of reset windup and its prevention in simple loops.

We have so far presented two methods of analyzing control loop stability, the Routh test and direct substitution, and one method for measuring process dynamics, step testing. In the following chapter we will look at two classical methods for analyzing control loop responses, root locus and frequency response. We will also present a more powerful process identification method, pulse testing.

REFERENCES

1. Ziegler, J. G., and Nichols, N. B., "Optimum Settings for Automatic Controllers," *Transactions ASME*, Vol. 64, Nov. 1942, p. 759.

2. Smith, Cecil L., *Digital Computer Process Control*, Intext Educational Publishers, Scranton, Pa., 1972.

3. Martin, Jacob Jr., Ph.D. dissertation, Department of Chemical Engineering, Louisiana State University, Baton Rouge, 1975.

4. Murrill, Paul W., *Automatic Control of Processes*, International Textbook Company, Scranton, Pa., 1967.

5. Lopez, A. M., P. W. Murrill, and C. L. Smith, "Controller Tuning Relationships Based on Integral Performance Criteria," *Instrumentation Technology*, Vol. 14, No. 11, Nov. 1967, p. 57.

6. Rovira, Alberto A., Ph.D. dissertation, Department of Chemical Engineering, Louisiana State University, Baton Rouge, 1981.

7. Dahlin, E. B., "Designing and Tuning Digital Controllers," *Instruments and Control Systems*, Vol. 41, No. 6, June 1968, p. 77.

8. Martin, Jacob Jr., A. B. Corripio, and C. L. Smith, "How to Select Controller Modes and Tuning Parameters from Simple Process Models," *ISA Transactions*, Vol. 15, No. 4, 1976, pp. 314–319.

9. Carlson, A., G. Hannauer, T. Carey, and P. J. Holsberg, *Handbook of Analog Computation*, 2nd ed., Electronic Associates, Inc., Princeton, N.J., 1967, p. 226.

10. Moore, C. F., C. L. Smith, and P. W. Murrill, "Simplifying Digital Control Dynamics for Controller Tuning and Hardware Lag Effects," *Instrument Practice*, Vol. 23, No. 1, Jan. 1969, p. 45.

11. Chiu, K. C., A. B. Corripio, and C. L. Smith, "Digital Control Algorithms. Part III. Tuning PI and PID Controllers," *Instruments and Control Systems*, Vol. 46, No. 12, Dec. 1973, pp. 41–43.

12. Corripio, A. B., "Digital Control Techniques," in Edgar, T. F., Ed., *Process Control*, AIChEMI, Series A, Vol. 3, American Institute of Chemical Engineers, New York, 1982, p. 69.

PROBLEMS

6-1. A feedback control loop is represented by the block diagram of Fig. 6-4a. The process can be represented by two lags in series:

$$G(s) = \frac{K}{(\tau_1 s + 1)(\tau_2 s + 1)}$$

where the process gain is $K = 0.50 \ \%/\%$ and the time constants are

$$\tau_1 = 1 \ \text{min}$$
$$\tau_2 = 0.5 \ \text{min}$$

The controller is a proportional controller: $G_c(s) = K_c$.

a. Write the closed-loop transfer function and the characteristic equation of the loop.

b. For what values of the controller gain is the loop response to a step change in set point overdamped, critically damped, and underdamped? Can the loop be made unstable?

c. Determine the response of the closed loop to a step change in set point for $K_c = 0.16$, 0.25, and 0.50.

6-2. Do Problem 6-1 for a process transfer function of

$$G(s) = \frac{6(1 - s)}{(s + 1)(0.5s + 1)} \frac{\%}{\%}$$

Transfer functions such as this are typical of processes consisting of two lags in parallel with opposite action. The controller is a proportional controller as in Problem 6-1.

6-3. A feedback control loop is represented by the block diagram of Fig. 6-4a. The process can be represented by a first-order lag and the controller is proportional-integral (PI):

$$G(s) = \frac{K}{\tau s + 1}$$

$$G_c(s) = K_c \left(1 + \frac{1}{\tau_I s} \right)$$

Without loss of generality you can set the process time constant $\tau = 1$, and the process gain $K = 1$.

a. Write the closed-loop transfer function and the characteristic equation of the loop.
b. Is there an ultimate gain for this loop?
c. Determine the response of the closed loop to a step change in set point for $\tau_I = \tau$ as the controller gain varies from zero to infinity.

6-4. Given the feedback control loop of Problem 6-1 and a pure integral controller

$$G_c(s) = \frac{K_c}{s}$$

a. By Routh's test determine the ultimate controller gain.
b. Recalculate the ultimate controller gain for $\tau_2 = 0.10$ and for $\tau_2 = 2$. Are your results what you expected?
c. By the direct substitution method, check the ultimate gains calculated in parts (a) and (b) and determine the ultimate frequency of oscillation of the loop.

6-5. For the feedback control loop of Problem 6-1 and a proportional-integral controller

$$G_c(s) = K_c \left(1 + \frac{1}{\tau_I s} \right)$$

a. Determine the ultimate loop gain KK_{cu} and the ultimate frequency of oscillation as functions of the integral time τ_I.
b. Determine the closed-loop response to a step change in set point with the controller tuned for minimum IAE. To determine the first-order plus dead-time (FOPDT) model parameters use Fig. 6-18.
c. Repeat part (b) with the controller tuned for 5% overshoot using the controller synthesis formulas, Eqs. (6-67) and (6-76).

6-6.   For a feedback control loop represented by the block diagram of Fig. 6-4a, use Routh's test to determine the ultimate gain for a proportional controller and each of the following process transfer functions:

a. $G(s) = \dfrac{1}{(s + 1)^3}$

b. $G(s) = \dfrac{1}{(s + 1)^5}$

c. $G(s) = \dfrac{1}{(3s + 1)(2s + 1)(s + 1)}$

d. $G(s) = \dfrac{(0.5s + 1)}{(3s + 1)(2s + 1)(s + 1)}$

e. $G(s) = \dfrac{1}{(3s + 1)(0.2s + 1)(0.1s + 1)}$

f. $G(s) = \dfrac{e^{-0.3s}}{3s + 1}$

Notice that part (f) is a first-order plus dead-time approximation of part (e) using the rule of thumb given in Section 6-3.

6-7.   Do Problem 6-6 using the direct substitution method to determine the ultimate controller gain and the ultimate frequency of oscillation of the loop.

6-8.   A feedback control loop is represented by the block diagram of Fig. 6-21. The process transfer function is given by

$$G(s) = \frac{K}{(\tau_1 s + 1)(\tau_2 s + 1)(\tau_3 s + 1)}$$

where the process gain is $K = 2.5$ %/% and the time constants are

$$\tau_1 = 5 \text{ min} \qquad \tau_2 = 0.8 \text{ min} \qquad \tau_3 = 0.2 \text{ min}$$

Determine the controller tuning parameters for quarter decay response by the ultimate gain method for

a.   A proportional (P) controller
b.   A proportional-integral (PI) controller
c.   A proportional-integral-derivative (PID) controller

6-9.   Using the tuning parameters determined for the loop of Problem 6-8, find the response of the closed loop to a unit step change in disturbance, $U(s) = 1/s$.

Note: The student may work this problem by inverting the Laplace transform or using one of the computer simulation programs listed in Chapter 9. The Laplace transform solution will require the use of the polynomial root-solving method of Chapter 2 or a computer program such as the one listed in Appendix D.

6-10. Given the feedback control loop of Fig. 6-21 and the following process transfer function

$$G(s) = \frac{Ke^{-t_0 s}}{(\tau_1 s + 1)(\tau_2 s + 1)}$$

where the process gain, time constants, and dead time are

$$K = 1.25\%/\% \qquad \tau_1 = 1 \text{ min} \qquad \tau_2 = 0.6 \text{ min} \qquad t_0 = 0.20 \text{ min}$$

calculate the first-order plus dead-time (FOPDT) model parameters by use of Fig. 6-18. Then use these parameters to compare the tuning parameters for a proportional-integral (PI) controller using the following formulas:
   a.   Quarter decay ratio response
   b.   Minimum IAE for disturbance inputs
   c.   Minimum IAE for set point inputs
   d.   Controller synthesis for 5% overshoot on a set-point change

6-11. Do Problem 6-10 for a proportional-integral-derivative (PID) controller.

6-12. Do Problem 6-10 for a sampled-data (computer) controller with a sample time T = 0.10 min.

6-13. For the control loop of Problem 6-10, derive the tuning formulas for an industrial PID controller using the Dahlin synthesis procedure. Consider two cases:
   a.   No dead time, $t_0 = 0$
   b.   Dead time
   Check your answers with the entries in Table 6-5

6-14. Use the computer program listed in Example 9-5 to obtain the responses to step changes in set point and in disturbance of the control loop in Problems 6-10 and 6-11. Use the controller tuning parameters determined there. Can you improve on the control performance by trial-and error adjustment of the tuning parameters? (The program prints the integral of the absolute error (IAE) that you may use as your measure of control performance.)

   Note: In the problems that follow the student is required to use concepts learned in Chapters 1 through 6 of this book.

6-15. Consider the vacuum filter shown in Fig. 6-30. This process is part of a waste treatment plant. The sludge enters the filter at about 5% solids. In the vacuum filter the sludge is dewatered to about 25% solids. The filterability of the sludge in the rotating filter depends on the pH of the sludge entering the filter. One way to control the moisture of the sludge to the incinerator is by the addition of chemicals (ferric chloride) to the sludge feed to maintain the necessary pH. Figure 6-30 shows a control scheme sometimes used. The moisture transmitter has a range 60 to 95%.

   The following data have been obtained from a step test on the output of the controller (MIC70) of +2 mA:

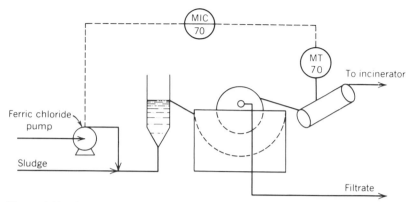

Figure 6-30. Vacuum filter for Problem 6-15.

Time, min	Moisture, %	Time, min	Moisture, %
0	75.0	10.5	70.9
1	75.0	11.5	70.3
1.5	75.0	13.5	69.3
2.5	75.0	15.5	68.6
3.5	74.9	17.5	68.0
4.5	74.6	19.5	67.6
5.5	74.3	21.5	67.4
6.5	73.6	25.5	67.1
7.5	73.0	29.5	67.0
8.5	72.3	33.5	67.0
9.5	71.6		

When the input moisture to the filter was changed by 2% the following data were obtained:

Time, min	Moisture,%	Time, min	Moisture, %
0	75	11	75.9
1	75	12	76.1
2	75	13	76.2
3	75	14	76.3
4	75.0	15	76.4
5	75.0	17	76.6
6	75.1	19	76.7
7	75.3	21	76.8
8	75.4	25	76.9
9	75.6	29	77.0
10	75.7	33	77.0

a. Draw a block diagram for the moisture control loop. Include the possible disturbances.

b. Approximate the transfer functions by first-order plus dead-time models and determine the parameters of the models. Redraw the block diagram showing the transfer function for each block using fit 3.

c. Give an idea of the controllability of the output moisture. Indicate the action of the controller.

d. Obtain the gain of a proportional controller for minimum IAE response. Calculate the offset for a 1% change in inlet moisture.

e. Obtain the ultimate gain and ultimate period for this control loop. The dead time term can be approximated by a first-order Padé approximation as given by Eq. (6-29).

f. Tune a PI controller for quarter decay ratio response.

6-16. Consider the absorber shown in Fig. 6-31. There is a gas flow entering the absorber with a composition of 90 mole % air and 10 mole % ammonia (NH_3). Before this gas is vented to the atmosphere, it is necessary to remove most of the NH_3 from it. This will be done by absorbing it with water. The NH_3 concentration in the exit vapor stream cannot be above 200 ppm. The absorber has been designed so that the outlet NH_3 concentration in the vapor is 50 ppm. During the design

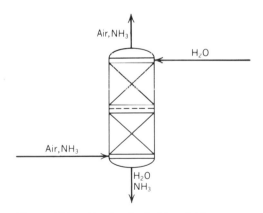

Figure 6-31. Absorber for Problem 6-16.

stage several dynamic simulations were done. From these simulations the following data were obtained:

Response to a step change in water flow to the absorber

Time, s	Water flow, gpm	Outlet NH_3 concentration, ppm
0	250	50
0	250	50
0	200	50
10	200	50
20	200	50
30	200	50.12
40	200	50.30
50	200	50.60
60	200	50.77
70	200	50.90
80	200	51.05
90	200	51.20
100	200	51.26
110	200	51.35
120	200	51.48
130	200	51.55
140	200	51.63
160	200	51.70
180	200	51.76
200	200	51.77
220	200	51.77

a. Design a control loop for maintaining the outlet NH_3 concentration at a set point of 50 ppm. Draw the instrument diagram for the loop. There are some instruments in the stock room that you can use for this purpose. There is an electronic concentration sensor and transmitter calibrated for 0–200 ppm. This instrument has a negligible time lag. There is also an air-actuated valve that, at full opening and for the 10 psi pressure drop that is available, will pass 500 gpm. The time constant of the valve actuator is 5 s. You may need more instrumentation to complete the design, so go ahead and use anything you need. Specify the action of the control valve and controller.

b. Draw a block diagram for the closed loop and obtain the transfer function for each block. Approximate the response of the absorber with a first-order plus dead-time model using fit 3. Obtain the ultimate gain and ultimate period for the loop. You may approximate the dead time in the system with a first-order Padé approximation, Eq. (6-29).

c. Tune a proportional only controller for quarter decay ratio response and obtain the offset when the set point is changed to 60 ppm.

d. Repeat part (c) using a PID controller.

6-17. Consider the furnace shown in Fig. 6-32, used to heat the supply air to a catalyst regenerator. The temperature transmitter is calibrated for 300–500°F. The following response data were obtained for a step change of +5% in the output of the controller:

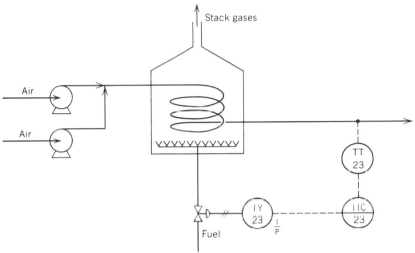

Figure 6-32. Furnace for Problem 6-17.

Time, min	$T(t)$, °F	Time, min	$T(t)$, °F
0	425	5.5	436.6
0.5	425	6.0	437.6
1.0	425	7.0	439.4
2.0	425	8.0	440.7
2.5	426.4	9.0	441.7
3.0	428.5	10.0	442.5
3.5	430.6	11.0	443.0
4.0	432.4	12.0	443.5
4.5	434.0	14.0	444.1
5.0	435.3	16.0	444.5
		19.0	445.0

a. Specify the action of the controller
b. Draw the complete block diagram specifying the units of each signal to/from each block. Identify each block.
c. Fit the process data by a first-order plus dead-time using fit 3. Draw the block diagram showing the transfer function for each block.

    d.   Tune the controller, proportional only, for quarter decay ratio response, and obtain the offset for a step change in set point of $+5°F$.

    e.   Tune a PI controller by the controller synthesis method for a 5% overshoot.

6-18. A storage tank, shown in Figure 6-33, supplies a gas with a molecular weight of 50 to two processes. The first process receives a normal flow of 500 scfm and operates at a pressure of 30 psig, while the second process operates at a pressure of 15 psig. A process operating at 90 psig supplies gas to the storage tank at a rate of 1500 scfm. The tank has a capacity of 550,000 ft^3 and operates a 45 psig and 350°F. You may assume that the pressure transmitter responds instantaneously with a calibrated range of 0–100 psig.

    a.   Size all three valves with an overcapacity factor of 50%. For all valves you can use the factor $C_f = 0.9$ (Masoneilan).

    b.   Draw the complete block diagram for the system. You can consider as disturbances $P_1(t)$, $P_3(t)$, $P_4(t)$, $vp_3(t)$, $vp_4(t)$ and the set point to the controller. The time constant of the pressure control valve is 5 s.

    c.   Can the feedback loop go unstable? If yes, what is its ultimate gain?

    d.   Using a proportional only controller with a gain of 50%/%, what is the offset observed for a set point change of $+5$ psi?

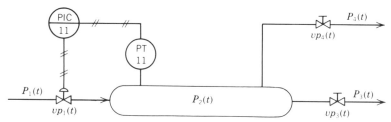

Figure 6-33. Storage tank for Problem 6-18.

6-19. Consider the chemical reactor system shown in Fig. 6-34. The exothermic catalytic reaction $A + E \rightarrow C + D$ takes place inside the reactor tubes. The reactor is cooled by an oil flowing through the shell of the reactor. As the oil flows out of the reactor, it goes to a boiler where it is cooled by producing low-pressure steam. The temperature in the reactor is controlled by manipulating the bypass flow around the boiler. The following process conditions are known:

Reactor design temperature at point of measurement: 275°F
Oil flow the pump can deliver: 400 gpm (constant)
Temperature control valve: pressure drop at design conditions: 15 psi
Flow at design conditions: 200 gpm
Range of the temperature transmitter: 100–400°F
Density of the oil: 55 lbm/ft^3
Open loop testing: A 5% decrease in the valve position results in a temperature change of $-3°F$ after a very long time.

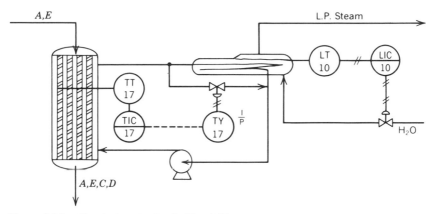

Figure 6-34. Chemical reactor for Problem 6-19.

Closed loop testing: At a controller gain of 7 mA/mA, the temperature loop starts to oscillate with a constant amplitude and a period of 15 min.

a. Size the temperature control valve for 100% overcapacity. What are your recommended valve and controller actions for the temperature loop?
b. If the pressure drop across the boiler tubes varies with the square of the flow and the valve is equal percentage with a rangeability parameter of 50, what is the flow through the valve when fully open? What is the valve position at design conditions?
c. Draw a general block diagram for the temperature loop.
d. Calculate the process gain, at design conditions, including the control valve and the temperature transmitter.
e. Calculate the tuning parameters for a PID controller for quarter-decay response. Report them as proportional band, repeats/minute, and minutes.
f. Tune a proportional controller for quarter decay ratio response and calculate the offset for a step change in set point of $-10°F$.

6-20. Consider the typical control system for the double-effect evaporator shown in Fig. 6-35. These evaporator systems are characterized by slow dynamics. The concentration out of the last effect is controlled by controlling the boiling point rise (BPR). That is, by keeping the BPR at a certain value the outlet concentration is maintained at the desired value. The higher concentration of the solute, the higher the BPR. The BPR is controlled by manipulating the steam to the first effect.

Step test data are available. Figure 6-36 shows the response of the BPR for a change of 2.5 lbm/ft^3 in density of the solution entering the first effect. Figure 6-37 shows the response of the BPR for a change of $+2$ psig from the controller output.

The BPR transmitter has a range of 20–30°F and a time of constant of 5 s.

a. Draw a complete block diagram with the transfer function of each block.

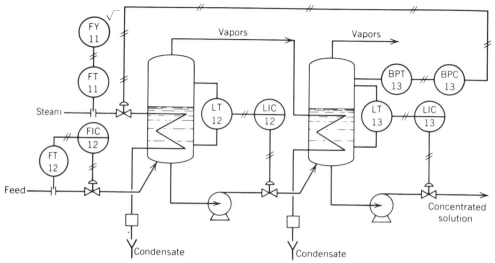

Figure 6-35. Evaporator for Problem 6-20.

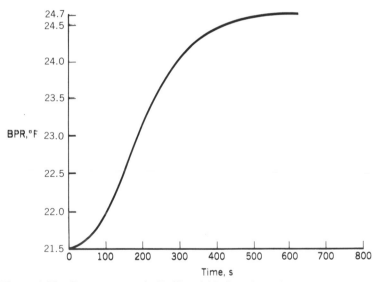

Figure 6-36. Response curve for Problem 6-20. Step change in density of the feed.

b. Based on the data given, what is the fail action of the control valve? What is the correct controller action?

c. Tune a proportional only controller for quarter decay ratio response and calculate the offset when the density of the entering solution changes by $+2$ lbm/ft^3.

d. Tune a PI controller for 5% overshoot using the controller synthesis method.

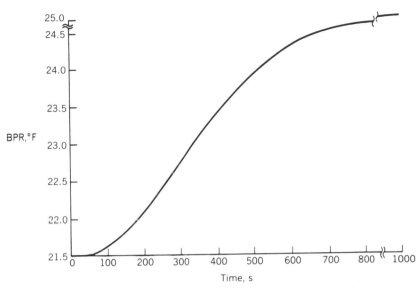

Figure 6-37. Response curve for Problem 6-20. Step change in BPR-controller output.

6-21. The temperature in a continuous stirred tank exothermic chemical reactor is controlled by manipulating the cooling water rate through a coil, as shown in Fig.
6-38.

The following are the process design conditions:

Reactor temperature: 210°F
Cooling water rate: 350 gal/min
Pressure drop across the coil: 10 psi
Range of temperature transmitter: 180–230°F
Valve trim: Equal percentage with rangeability parameter of 50; air-to-close valve

The following testing has been performed on the system:

Open-loop sensitivity: A 10 gal/min increase in water rate results in a temper-
                        ature drop of 3°F after a long time.
Closed-loop testing: At a controller gain of 5 the temperature oscillates with
                        constant amplitude and a period of 12 min.

a. Determine the valve coefficient C_V for an overcapacity factor of 2.
b. Draw the block diagram for the control loop and determine the total process
    gain, including the transmitter and the control valve.
c. Calculate the PID controller tuning parameters for quarter decay ratio re-
    sponse. Report them as proportional band, repeats/minute, and derivative
    minutes. What is the required controller action?

6-22. Consider the process shown in Fig. 6-39 for drying phosphate pebbles. A table

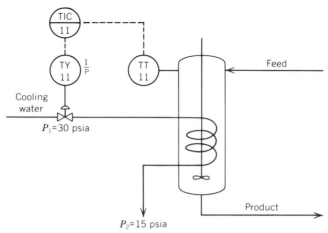

Figure 6-38.   Reactor for Problem 6-21.

feeder transports the pebble–water slurry into the bed of the dryer. In this bed the pebbles are dried by direct contact with hot combustion gases. From the dryer the pebbles are conveyed to a silo for storage. It is *most* important to control the moisture of the pebbles leaving the dryer. If the pebbles are too dry, they may fracture into fine dust resulting in possible loss of material. If too wet, they may form large chunks, or clinkers, in the silo.

It is proposed to control the moisture of the exiting pebbles by the speed of the table feeder, as shown in Fig. 6-39. The speed of the feeder is directly proportional to its input signal. The moisture of the inlet pebbles is usually about 14% and is reduced to 3% in the dryer. The transmitter has a range of 1–6% moisture. An important disturbance to this process is the moisture of the inlet pebbles.

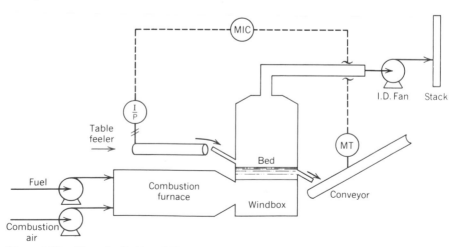

Figure 6-39.   Dryer for Problem 6-22.

a. Draw a complete block diagram of the control loop showing all units. Include the disturbances.
b. Figure 6-40 shows the response of the outlet moisture to an increase of 1 mA in controller output, while Fig. 6-41 shows the response of the outlet moisture to an increase of 2% in inlet moisture. Approximate each process curve by a first-order plus dead-time model. Redraw the block diagram showing the transfer functions of these approximate models. Use fit 2.
c. Determine the tuning of a PID controller for minimum ISE response on disturbance inputs. Report the controller gain as proportional band. What is the correct controller action?
d. Determine the ultimate gain and ultimate period for the loop using a first-order Padé approximation for the dead time.
e. If the inlet moisture of the pebbles drops by 2%, what is the new steady state value of the outlet moisture? Assume that the controller is proportional only tuned for quarter decay response from the information determined in part (d).
f. What is the controller output (in mA) required to avoid offset for the disturbance of part (e)?

Note: In the problems that follow, the student is required to model the process from fundamental principles. These problems are better suited for term projects than for regular homework assignments.

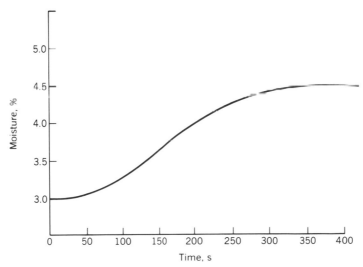

Figure 6-40. Response curve for Problem 6-22. Step change in controller output.

6-23. Consider the electric heater shown in Fig. 6-42. Two liquid streams with variable mass rates $F_A(t)$ and $F_B(t)$ come together in a tee and pass through the heater where they are thoroughly mixed and heated to temperature $T(t)$. The outlet

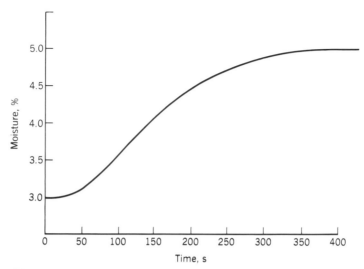

Figure 6-41. Response curve for Problem 6-22. Step change in inlet moisture.

temperature is controlled by manipulating the current through an electric coil. The outlet mass fraction of component B is also controlled by manipulating the inlet flow of stream B. The following information is known:

1. The pressure drop across the valves can be assumed constant so, the flow through the valves is given by

$$F_A(t) = K_{v1} \, vp_1(t) \qquad F_B(t) = K_{v2} \, vp_2(t)$$

These two streams are each pure in components A and B, respectively.

2. The easiest way to control the outlet composition, the mass fraction of B, is to measure the electrical conductivity of the outlet stream. The conductivity of this stream is inversely proportional to the mass fraction, x_B, that is

$$\text{Conductivity} = \frac{\alpha}{x_B}$$

where α is a constant, mho-mass fraction/m.
The conductivity transmitter has a range of C_L to C_H mho/m.

3. The heater is always full of liquid. The total outlet flow is equal to the inlet flow at all times.

4. The heat capacity and density are constant.

5. You may assume that the heat transferred, $\dot{q}$, is directly proportional to the output of the controller. For 4 mA output, $\dot{q} = 0$; for 20 mA, $\dot{q} = \dot{q}_{max}$.

6. The heater is well insulated.

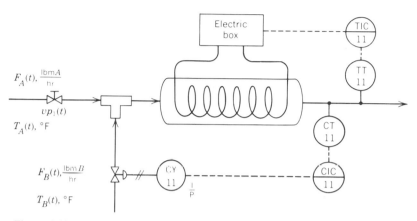

Figure 6-42. Electric heater for Problem 6-23.

7. The load disturbances to this system are $vp_1(t)$, $T_A(t)$, and $T_B(t)$.

Do the following:

a. Obtain the action of the conductivity controller assuming that the control valve is air-to-open. Obtain the action of the temperature controller.
b. Derive, from basic principles, the set of equations that describe the composition (conductivity) control loop.
c. Linearize the equations from part (b) and draw the complete block diagram for the conductivity loop. Show the transfer function of each block.
d. Derive, from basic principles, the set of equations that describe the temperature control loop.
e. Linearize the equations from part (d) and draw the complete block diagram for the temperature loop. Show the transfer function of each block.
f. Write the characteristic equation for each of the control loops. Can either loop be made unstable by increasing the controller gain? Discuss briefly.

6-24. Consider the system shown in Fig. 6-43. In each of the two tanks the reaction $A \rightarrow E$ takes place. The rate of reaction is given by

$$r(t) = kc_A(t), \quad \frac{\text{lb moles}}{\text{gal-min}}$$

where

k is the reaction rate coefficient, $\min^{-1}$

$c_A(t)$ is the concentration, $\dfrac{\text{lb moles}}{\text{gal}}$

The disturbances to this process are $q_i(t)$ and $c_{Ai}(t)$. The concentration out of the second reactor is controlled by manipulating a stream of pure A to the first reactor.

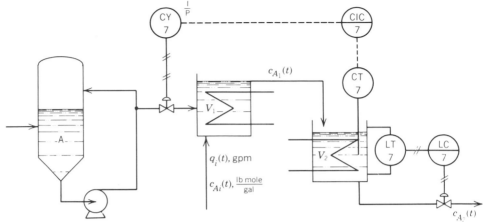

Figure 6-43. Reactors in series for Problem 6-24.

The density of this stream is ρ_A in lb moles/gal.

The level in the first reactor is constant because the outlet stream is by overflow. You can assume that the level in the second reactor is also constant (perfect level control) and thus ignore the level control loop.

The temperature in each reactor can be assumed constant.

DATA

Reactor volumes, gal: $V_1 = 500$ $V_2 = 500$

Reaction rate coefficients, min^{-1}: $k_1 = 0.25$ $k_2 = 0.50$

Properties of stream A: $\rho_A = 2.0$ lb mole/gal $MW_A = 25$

Design conditions: $\bar{c}_{Ai} = 0.8$ lb mole/gal $\bar{q}_i = 50$ gal/min

$\bar{q}_A = 50$ gal/min

Control valve: $\Delta P = 10$ psi (constant). Linear characteristics.

Concentration transmitter has a range of 0.05–0.5 lb mole/gal. The dynamics of this transmitter can be represented by a second-order lag with time constants of 0.5 min and 0.25 min.

a. Size the control valve to handle twice the nominal flow rate.

b. Derive, from basic principles, the set of equations that describe the composition control loop. State all the assumptions made in the derivation.

c. Linearize the equations from part (b) and draw the complete block diagram of the composition control loop. Show the numerical values and units of all gains and time constants, except for the controller.

d. Obtain the closed-loop transfer functions:

$$\frac{C_{A_2(s)}}{C_{A_2(s)}^{set}} \; ; \qquad \frac{C_{A_2(s)}}{Q_i(s)} \; ; \qquad \frac{C_{A_2(s)}}{C_{A_i(s)}}$$

e. Determine the ultimate gain and frequency of the control loop and the parameters of a PID controller tuned for quarter decay-ratio response.

6-25. Consider the process shown in Fig. 6-44. In the first tank, streams $q_1(t)$ and $q_2(t)$
are being mixed and heated. The heating medium flows at such a high rate that
its temperature change from inlet to exit is not significant. Thus, the heat transfer
rate can be described by $UA[T_{c1}(t) - T_3(t)]$. It can also be assumed that the outlet
flow is equal to the inlet flow (total) and that the densities and heat capacities of
all streams are not strong functions of temperature or composition.

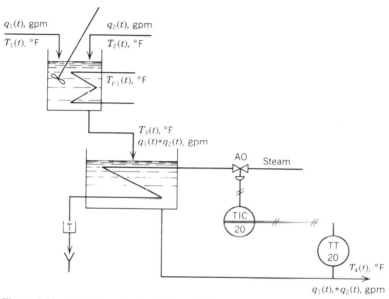

Figure 6-44. Tanks in series for Problem 6-25.

The outlet flow from the first tank flows into the second tank where it is
again heated, this time by condensing steam. The energy gained by the process
fluid is equal to the energy lost by the condensing steam. Thus, the rate of heat
transfer can be described by $w_s(t)h_{fg}$, where $w_s(t)$ is the mass flow rate of steam
and h_{fg} is the latent heat of vaporization of the steam. Assuming that the pressure
drop across the steam valve is constant, the flow through this valve can be
described as $w_s(t) = K_v vp(t)$. The time constant of this valve is τ_v.

The temperature transmitter has a range $T_L - T_H$, and a time constant τ_T.

Assuming that the heat losses from both tanks are negligible and that the
important disturbances are $T_1(t)$, $T_2(t)$, $T_{c1}(t)$, and $T_4^{set}(t)$, obtain the complete
block diagram of the temperature control loop and its characteristic equation.
Show the transfer function of each block.

6-26. Consider the process shown in Fig. 6-45. The process fluid entering the tank is
an oil with a density of 53 lbm/ft^3, a heat capacity of 0.45 Btu/lbm-°F, and an
inlet temperature of 70°F. This oil is to be heated up to 200°F by steam supplied
at 115 psig. The pressure in the tank, above the oil level, is maintained at 40 psia
by a blanket of inert gas, N_2.

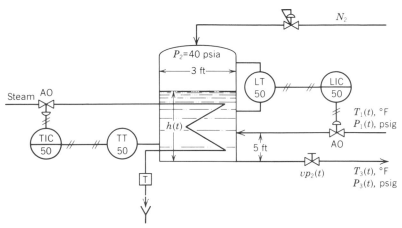

Figure 6-45. Oil heater for Problem 6-26.

You may assume that the tank is well insulated, that the physical properties of the oil are not strong functions of temperature, that the liquid is well mixed, and that the level covers the heating coil. The following data are also known:

$\overline{P}_1$ = 45 psig; $\overline{P}_3$ = 15 psig
h, steam side = 1500 Btu/hr-ft²-°F
h, oil side = 150 Btu/hr-ft²-°F
Heating surface area = 127.5 ft²
Heating coil: ½ in. O.D., 20 BWG tubes, 974 linear ft., and 0.035 in. wall
         thickness.
         Density of tube = 500 lbm/ft³
         C_P of tube = 0.12 Btu/lbm-°F
Tank diameter : 3ft
Level transmitter: Range: 7–10 ft
                             Time constant: 0.01 min
Temperature transmitter: Range: 100–300°F
                             Time constant: 0.5 min

a. Size valves 1 and 2 with an overcapacity factor of 50%. The nominal oil flow rate is 100 gpm. Size valve 3, the steam valve, with an overcapacity factor of 50%. The pressure drop across this valve can be assumed to be constant.
b. Obtain the complete block diagram for the level control loop. Use a proportional only controller.
c. Obtain the complete block diagram and the characteristic equation of the temperature control loop. Use a PID controller. Show the numerical values of all the gains and time constants in the transfer functions.

6-27. Consider the process shown in Fig. 6-46. The following information is available.

1. The density of the liquid is constant.

2. The two outlet valves remain at constant opening and their downstream pressure is also constant.

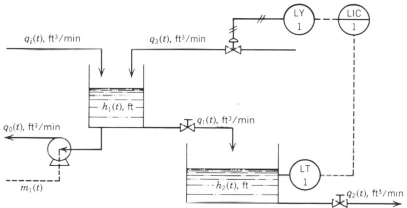

Figure 6-46. Level controller for Problem 6-27.

3. The pump flow is given by

$$q_0(t) = K_p[m_1(t) - 4]$$

4. The variable speed pump has a time constant, relating the flow to the input signal, $m_1(t)$, of τ_p s.

5. The control valve is linear and has a time constant, relating the flow to the pneumatic signal, of τ_v s. The pressure drop across the control valve is constant.

6. The diameters of the tanks are D_1 and D_2.

7. The valve coefficients are C_{v_1}, C_{v_2}, and C_{v_3}.

8. The level transmitter has a range Δh and a negligible time constant.

a. Obtain the block diagram, with the corresponding transfer functions, for this control system. The disturbances are $q_i(t)$ and $m_1(t)$.

b. Write the characteristic equation of the level control loop and determine its ultimate gain and period as functions of the system parameters.

6-28. Consider the process shown in Fig. 6-47. In this process a waste gas is enriched with natural gas to be used as fuel in a small furnace. The enriched waste gas must have a certain heating value to be used as fuel. The control strategy calls for measuring the heating value of the gas leaving the process and manipulating the natural gas flow (using a variable-speed fan) to maintain the heating value set point.

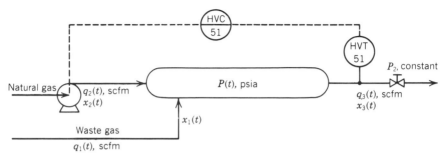

Figure 6-47.   Sketch for Problem 6-28.

The waste gas is composed of methane (CH_4) and some low heating value combustibles. The natural gas composition can be considered constant and is composed mainly of methane and some small amount of other hydrocarbons. The heating value of the enriched waste gas is related to the mole fraction of methane by the following relation:

$$hv(t) = c + d\, x_3(t)$$

where

$hv(t)$ is the heating value
x_3 is the mole fraction of methane
c,d are constants

The variable-speed fan is such that at full speed its flow is q_{2max}. It can be assumed that the relationship between the flow and the input signal to the fan driver is linear. This driver has a time constant τ_F.

The outlet valve has a constant opening. A Masoneilan valve equation can be used to describe the flow through this valve. A proportional-integral controller is used to control the heating value. The sensor-transmitter has a time constant of τ_T min.

The specific gravity of the enriched gas is related to the mole fraction of methane by

$$G(t) = a + b\, x_3(t)$$

where a,b are constants.

a. Determine the complete block diagram for this control system. Show all transfer functions. The possible disturbances are $q_1(t)$, $x_1(t)$, and $x_2(t)$.
b. Write the characteristic equation of the feedback control loop.
c. Determine the ultimate gain and period of oscillation of the loop (if they exist).

CHAPTER
7
Classical Single-Loop Feedback Control Design

The previous chapter started the study of the stability and design of control systems by presenting two techniques: Routh test and direct substitution. This chapter continues this study, presenting two additional techniques: root locus and frequency response. The significance and use of these techniques are discussed from a practical point of view. Finally, the chapter ends with a presentation of the frequency response technique for process identification.

Before the root locus and frequency response techniques can be learned, some new terms that will become important in their study must be defined. Consider the general closed-loop block diagram shown in Fig. 7-1. As learned in Chapter 6 the closed-loop transfer functions are

$$\frac{C(s)}{R(s)} = \frac{G_c(s)G_v(s)G_{p_1}(s)}{1 + H(s)G_c(s)G_v(s)G_{p_1}(s)} \tag{7-1}$$

and

$$\frac{C(s)}{L(s)} = \frac{G_{p_2}(s)}{1 + H(s)G_c(s)G_v(s)G_{p_1}(s)} \tag{7-2}$$

with the characteristic equation

$$1 + H(s)G_c(s)G_v(s)G_{p_1}(s) = 0 \tag{7-3}$$

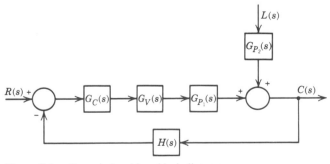

Figure 7-1.   General closed-loop block diagram.

The *open-loop transfer function* (OLTF) is defined as the product of all the transfer functions in the control loop.

$$\text{OLTF} = H(s)G_c(s)G_v(s)G_{p_1}(s) \qquad (7\text{-}4)$$

Therefore, the characteristic equation can also be written as

$$1 + \text{OLTF} = 0 \qquad (7\text{-}5)$$

Now suppose that the individual transfer functions are known and that the OLTF looks as follows:

$$\text{OLTF} = \frac{K_c K_v K_{p_1} K_T (1 + \tau_D s)}{(\tau_T s + 1)(\tau_1 s + 1)(\tau_2 s + 1)}$$

or

$$\text{OLTF} = \frac{K(1 + \tau_D s)}{(\tau_T s + 1)(\tau_1 s + 1)(\tau_2 s + 1)}$$

where $K = K_c K_v K_{p_1} K_T$.

The *poles* are defined as the roots of the denominator of the OLTF. In this case the poles are $-1/\tau_T$, $-1/\tau_1$, and $-1/\tau_2$. The *zeros* are defined as the roots of the numerator of the OLTF, in this case $-1/\tau_D$.

These definitions are generalized by writing the OLTF as follows:

$$\text{OLTF} = \frac{K \prod\limits_{i=1}^{m} (\tau_i s + 1)}{\prod\limits_{j=1}^{n} (\tau_j s + 1)}, \qquad n > m$$

or

$$\text{OLTF} = \frac{K' \prod\limits_{i=1}^{m} \left(s + \dfrac{1}{\tau_i} \right)}{\prod\limits_{j=1}^{n} \left(s + \dfrac{1}{\tau_j} \right)}, \qquad n > m \qquad (7\text{-}6)$$

where

$$K' = \frac{K \prod\limits_{i=1}^{m} \tau_i}{\prod\limits_{j=1}^{n} \tau_j}$$

From Eq. (7-6) we see that the poles are immediately recognized as equal to $-1/\tau_j$ for $j = 1$ to n. Similarly, the zeros are given by $-1/\tau_i$ for $i = 1$ to m. These definitions will be frequently used in the study of root locus and frequency response techniques.

7-1. ROOT LOCUS TECHNIQUE

Root locus is a graphical technique that consists of graphing the roots of the characteristic equation, that is, the eigenvalues, as a gain or any other control loop parameter changes. The resulting graph allows us to see at a glance whether a root of the characteristic equation crosses the imaginary axis from the left- to the right-hand side of the s plane. This would indicate the possibility of instability of the control loop.

Several examples are presented on how to draw the root locus. Then, based on these examples, general rules for plotting are given. These examples also help to understand the effects of the different parameters of the control loop on its stability. These effects were presented in Chapter 6; therefore, the following should also serve as a review for the reader.

Examples

Example 7-1. Consider the block diagram of a certain control loop shown in Fig. 7-2. The characteristic equation for this system is

$$1 + \frac{K_c}{(3s + 1)(s + 1)} = 0 \tag{7-7}$$

and

$$OLTF = \frac{K_c}{(3s + 1)(s + 1)}$$

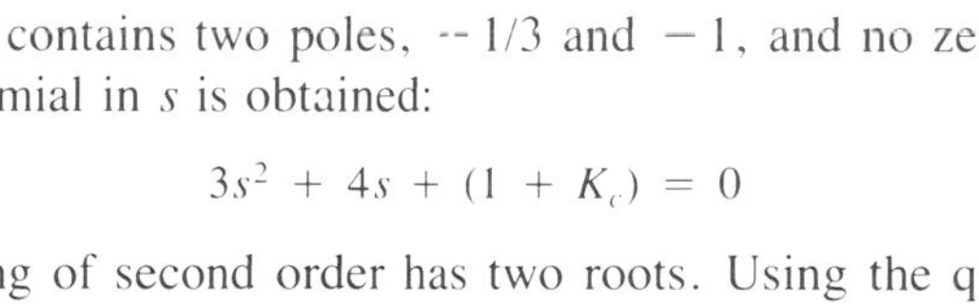

Figure 7-2. Block diagram of control loop—Example 7-1.

Note that this OLTF contains two poles, $-1/3$ and -1, and no zeros. From Eq. (7-7) the following polynomial in s is obtained:

$$3s^2 + 4s + (1 + K_c) = 0 \tag{7-8}$$

This polynomial being of second order has two roots. Using the quadratic equation to solve for the roots, the following expression is developed:

$$r_1, r_2 = \frac{-4 \pm \sqrt{16 - 12(1 + K_c)}}{6}$$

$$r_1, r_2 = -\frac{2}{3} \pm \frac{1}{3}\sqrt{1 - 3K_c} \tag{7-9}$$

Equation 7-9 shows that the roots of the characteristic equation depend on the value of K_c. This is the same thing as saying that the stability of the control loop depends on the tuning of the feedback controller! Of course, this was also shown to be the case in Chapter 6. By giving values to K_c, the loci of the roots of the characteristic equation can be determined. The graph of the roots, or root locus, is shown in Fig. 7-3. Several things can be learned by examining this root locus diagram:

1. The most important point is that this particular control loop will never go unstable no matter how high the value of K_c is set. As the value of K_c increases, the loop response becomes more oscillatory, or underdamped, but never unstable. The underdamped response can be recognized because the roots of the characteristic equation move away from the real axis as K_c increases. The fact that a control loop with a pure second-order (or first-order) characteristic equation does not go unstable was also shown in Chapter 6 using the Routh test and direct substitution methods.

2. When $K_c = 0$, the root loci originates from the OLTF poles: $-1/3$ and -1.

3. The number of root loci or branches is equal to the number of OLTF poles, $n = 2$.

4. As K_c increases, the root loci approach infinity.

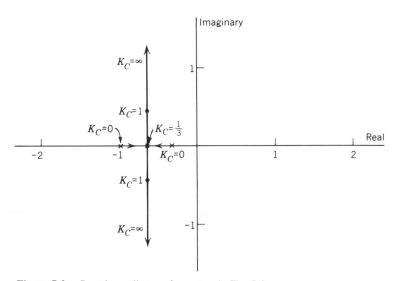

Figure 7-3.   Root locus diagram for system in Fig. 7-2.

Example 7-2.   Suppose now that the sensor-transmitter combination of the previous example has a time constant of 0.5 time units. The block diagram is shown in Fig. 7-4. The new characteristic equation and open-loop transfer function are

Characteristic equation

$$1 + \frac{K_c}{(3s + 1)(s + 1)(0.5s + 1)} = 0$$

or

$$1.5s^3 + 5s^2 + 4.5s + (1 + K_c) = 0$$

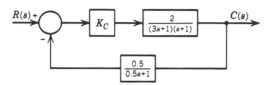

Figure 7-4. Block diagram of control loop—Example 7-2.

$$\text{OLTF} = \frac{K_c}{(3s + 1)(s + 1)(0.5s + 1)}$$

$$\text{with poles:} \quad -\frac{1}{3}, \ -1, \ -2; \qquad n = 3$$

$$\text{zeros:} \quad \text{none}; \qquad m = 0$$

In this case the characteristic equation is a third-order polynomial; thus, the calculation of its roots is not straightforward. Newton's method of Chapter 2 or the computer program in Appendix D could be used. However, as we will see, there is an easier method to plot the root locus without the need to calculate any roots.

Figure 7-5 shows the root locus diagram. Again, several things can be learned by a simple glance at this diagram.

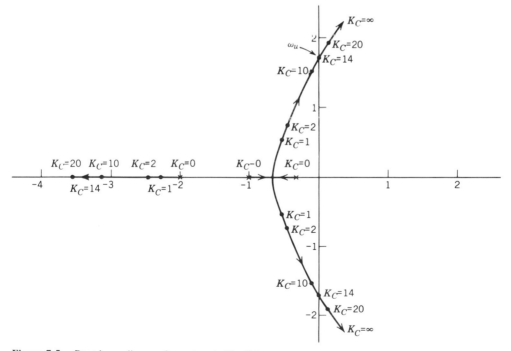

Figure 7-5. Root locus diagram for system in Fig. 7-4.

1. The most important thing is that this control system *can* go unstable. At some value of K_c, in this case $K_c = 14$, the root loci will cross the imaginary axis. For values of K_c greater than 14, the roots of the characteristic equation will be on the right-hand side of the s plane, yielding an unstable control system. The value of K_c at which the root locus crosses the imaginary axis, yielding a conditionally stable system, is called the ultimate gain, K_{cu}, as learned in Chapter 6. The ultimate frequency, ω_u, is given by the coordinate where the branches cross the imaginary axis. Any loop with a characteristic equation of third or higher order can go unstable; pure, first- or second-order systems will not go unstable, as shown in Example 7-1. Any system with dead time *can* also go unstable, as will be shown shortly in this chapter.

2. The root loci again originate, with $K_c = 0$, at the OLTF poles $-1/3$, -1, -2.

3. Also, again, the number of root loci is equal to the number of poles of the OLTF, $n = 3$.

4. Finally, the root loci again approach infinity as K_c increases.

Example 7-3.   Suppose a proportional-derivative controller is now used in the original control loop of Example 7-1. Fig. 7-6 shows the block diagram. The new characteristic equation and open-loop transfer function are as follows:

Characteristic Equation

$$1 + \frac{K_c(1 + 0.2s)}{(3s + 1)(s + 1)} = 0$$

or

$$3s^2 + (4 + 0.2K_c)s + (1 + K_c) = 0$$

$$\text{OLTF} = \frac{K_c(1 + 0.2s)}{(3s + 1)(s + 1)}$$

$$\text{with poles:} \quad -\frac{1}{3}, \ -1; \qquad n = 2$$

$$\text{zeros:} \quad -5; \qquad m = 1$$

Since the characteristic equation is of second order, its roots are determined by the quadratic equation

$$r_1, r_2 = \frac{-(4 + 0.2K_c) \pm \sqrt{4 - 10.4K_c + 0.04K_c^2}}{6}$$

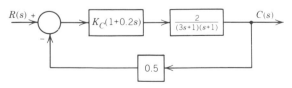

Figure 7-6.   Block diagram of control loop—Example 7-3.

By giving values to K_c, the root locus for this system can be obtained as shown on Fig. 7-7.

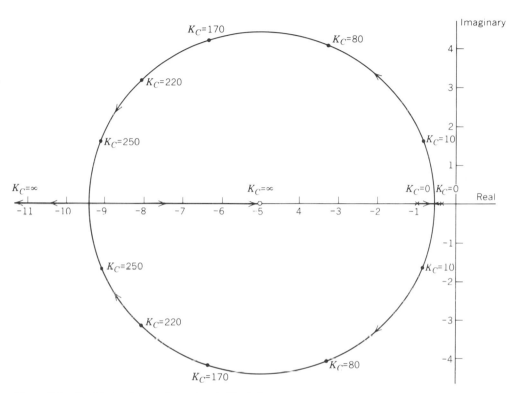

Figure 7-7.   Root locus diagram for system in Fig. 7-6.

As with the other examples, several things can be learned from this diagram:

1. This control loop will never go unstable. Furthermore, as K_c increases the root loci move away from the imaginary axis and the control loop becomes more stable. The derivative action in effect adds a "lead" term to the control loop. The addition of any lead term "adds" stability to control loops. The addition of a lag term "removes" stability from control systems, as shown in Example 7-2.

2. The root loci originate at the OLTF poles: $-1/3$ and -1. This is similar to the previous examples.

3. The number of root loci is equal to the number of OLTF poles, $n = 2$. This is also the case for the previous examples.

4. As K_c increases, one of the root loci approaches the OLTF zero, -5, and the other root loci approaches minus infinity.

Rules for Plotting Root Locus Diagrams

The previous examples have shown the development of root locus diagrams. As long as the characteristic equation is of second order, it is fairly simple to develop a diagram. For higher-order systems it becomes quite tedious to obtain the roots of the characteristic equation as one of the parameters changes.

Several rules have been developed to aid the engineer in the plotting of root locus diagrams. To use these rules, the characteristic equation and the open-loop transfer function are both written in the following form:

Characteristic Equation

$$1 + \frac{K' \prod_{i=1}^{m} (s - z_i)}{\prod_{j=1}^{n} (s - p_j)} = 0 \tag{7-10}$$

$$\text{OLTF} = \frac{K' \prod_{i=1}^{m} (s - z_i)}{\prod_{j=1}^{n} (s - p_j)} \tag{7-11}$$

where

$$z_i = -\frac{1}{\tau_i} = \text{zeros}$$

$$p_j = -\frac{1}{\tau_j} = \text{poles}$$

$$K' = \frac{K \prod_{i=1}^{m} \left(-\frac{1}{z_i}\right)}{\prod_{j=1}^{n} \left(-\frac{1}{p_j}\right)}$$

K = total loop gain. The multiplication of all the gains in the loop.

The rules to be presented here were developed from the fact that the root locus must satisfy what is called the magnitude and angle conditions. To explain what these conditions are, consider again the characteristic equation given by Eq. (7-10). This equation can also be written as follows:

$$\frac{K' \prod_{i=1}^{m} (s - z_i)}{\prod_{j=1}^{n} (s - p_j)} = -1$$

Since the equation is complex in nature, it can be separated into two parts: magnitude and phase angle. Carrying out the multiplication and division in polar form, as presented in Section 2-3, yields

$$\frac{K' \prod_{i=1}^{m} |s - z_i|}{\prod_{j=1}^{n} |s - p_j|} = 1 \tag{7-12}$$

and

$$\sum_{i=1}^{m} \angle (s - z_i) - \sum_{j=1}^{n} \angle (s - p_j) = -\pi \pm 2\pi k \tag{7-13}$$

where k is a positive integer with values $k = 0, 1, 2, \ldots, n - m - 1$.

Eq. (7-12) is called the *magnitude condition* and Eq (7-13) is called the *angle condition*. The roots, or eigenvalues, of the characteristic equation must satisfy both of these criteria.

The angle condition is used to locate the root loci on the s plane. The magnitude condition is then used to calculate the K' value that provides the root at a specific point in the root locus diagram.

Let us show by a simple example how to use the angle and magnitude conditions to search for a root. Consider Fig. 7 8, which depicts a system having two OLTF poles, shown as x, and one OLTF zero, shown as o. Choose a value of s, say s_1, and try to determine whether it indeed is a root and therefore a part of the root locus. The first step is to check the angle condition. To do this the line segments joining point s_1 with each pole and with each zero are formed, as shown in Fig. 7-8. The angles between each of these lines segments and the real axis are then measured. If the angle condition is satisfied, then point s_1 is part of the root locus. If the angle condition is not satisfied, another point in the s plane is chosen and tried—a trial-and-error procedure. This enhances the im-

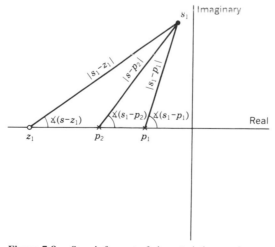

Figure 7-8.   Search for root of characteristic equation.

portance of the use of the rules to be presented. Once a point in the s plane has been identified as part of the root locus, then the magnitude condition is used to calculate the value of K' that corresponds to the root. This calculation is shown later in an example.

The preceding paragraph briefly explained how the magnitude and angle conditions serve as the basis to draw the root locus. The rules developed from these conditions can be used to qualitatively sketch the root locus. Certain points can then be determined, such as the ultimate gain and frequency, as well as the gain required to obtain a given damping ratio. It must be realized that the root loci will always be symmetrical with respect to the real axis. This is a consequence of the fact that the roots of the characteristic equation are either real or complex conjugates. Here are the rules:

Rule 1. On the real axis the locus exists at a point where an odd number of poles and zeros is found to the right of the point.

Rule 2. The root loci will always originate, for a total loop gain $= 0$, at the poles of the OLTF. Repeated poles originate repeated loci. That is, a qth-order pole originates q loci or branches.

Rule 3. The number of loci or branches is equal to the number of poles of the OLTF, n.

Rule 4. As the total gain of the loop increases, the loci or branches will approach either the zeros of the OLTF or infinity. The number of loci that approach infinity is given by $n - m$. Repeated zeros attract repeated loci. That is, a qth-order zero attracts q loci or branches.

Rule 5. Those loci that approach infinity will do so along straight line asymptotes. All asymptotes must pass through the "center of gravity" of the poles and zeros of the OLTF. The location of this center of gravity (CG) is computed as follows:

$$\text{CG} = \frac{\sum_{j=1}^{n} p_j - \sum_{i=1}^{m} z_i}{n - m} \tag{7-14}$$

These asymptotes make the following angles with the positive real axis:

$$\phi = \frac{180° + (360°)k}{n - m} \tag{7-15}$$

where $k = 0, 1, \ldots, n - m - 1$

Rule 6. The points on the real axis where the loci meet and leave, or enter from the complex region of the s plane, are called "breakaway points." These breakaway points are determined, most often by trial and error, from the solution to the equation

$$\sum_{i=1}^{m} \frac{1}{s - z_i} = \sum_{j=1}^{n} \frac{1}{s - p_j} \tag{7-16}$$

The loci always leave or enter the real axis at the breakaway points at angles of $\pm 90°$.

When a locus leaves from a complex conjugate pole, p_k, the angle of departure relative to the real axis is found from

$$\text{angle of departure} = 180° + \sum_{i=1}^{m} \measuredangle\ (p_k - z_i) - \sum_{\substack{j=1 \\ j \neq k}}^{n} \measuredangle\ (p_k - p_j)$$

[The symbol $\measuredangle\ (p_k - z_i)$ signifies the angle, relative to the real axis, between pole p_k and zero z_i.]

When a locus arrives at a complex conjugate zero, z_k, the angle of arrival relative to the real axis is found from

$$\text{angle of arrival} = -180° + \sum_{j=1}^{n} \measuredangle\ (z_k - p_j) - \sum_{\substack{i=1 \\ i \neq k}}^{m} \measuredangle\ (z_k - z_i)$$

Let us now present the use of these rules for plotting root locus diagrams.

Example 7-4. Consider the heat-exchanger temperature control loop presented in Chapter 6. The block diagram, Fig. 6-3, has been redrawn in Fig. 7-9 showing each transfer function.

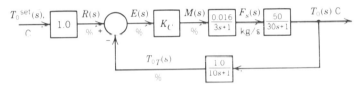

Figure 7-9. Block diagram of heat exchanger temperature control loop—P controller.

Characteristic equation

$$1 + \frac{0.8 K_c}{(10s + 1)(30s + 1)(3s + 1)} = 0$$

$$\text{OLTF} = \frac{0.8 K_c}{(10s + 1)(30s + 1)(3s + 1)}$$

As indicated by Eq. (7-11) the OLTF can also be written as follows:

$$\text{OLTF} = \frac{K'}{\left(s + \dfrac{1}{10}\right)\left(s + \dfrac{1}{30}\right)\left(s + \dfrac{1}{3}\right)}$$

$$\text{with poles:} \quad -\frac{1}{10}, \quad -\frac{1}{30}, \quad -\frac{1}{3}; \qquad n = 3$$

$$\text{zeros:} \quad \text{none;} \qquad m = 0$$

where

$$K' = \frac{0.8 K_c}{(10)(30)(3)} = 0.000888 K_c$$

Fig. 7-10 shows the locations of the poles (×) in the s plane.

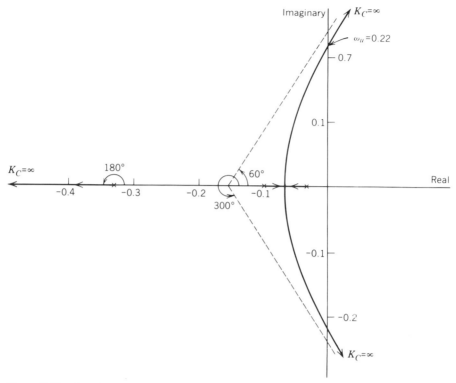

Figure 7-10. Root locus diagram of heat exchanger control loop.

● Rule 1 indicates that the negative real axis between poles $-1/30$ and $-1/10$ and from pole $-1/3$ to $-\infty$ is part of the root locus.

● From Rule 2 it is known that the root loci will originate at the poles of the OLTF: $-1/10$, $-1/30$, and $-1/3$.

● Since there are three poles, $n = 3$, Rule 3 indicates that there are three loci or branches.

● Since there are no zeros, $m = 0$, Rule 4 indicates that all loci will approach infinity as K_c increases.

● Rule 5 allows us to obtain the center of gravity through which the asymptotes must pass and to obtain their angles with the positive real axis. Since there are three branches that approach infinity, there must also be three asymptotes. From Eq. (7-14) we obtain

$$CG = \frac{-\dfrac{1}{10} - \dfrac{1}{30} - \dfrac{1}{3}}{3} = -0.155$$

and from Eq. (7-15) we find

$$\phi = \frac{180° + 360°(0)}{3}, \quad \frac{180° + 360°(1)}{3}, \quad \frac{180° + 360°(2)}{3}$$

$$\phi = 60°, 180°, 300°$$

These asymptotes and angles are shown in Fig. 7-10. One of the asymptotes lies on the real axis, $\phi = 180°$, and moves from the center of gravity to minus infinity. The other two asymptotes move away from the real axis into the complex region of the s plane. These asymptotes cross the imaginary axis, indicating the possibility of instability since the loci will approach infinity along these asymptotes.

• Using Rule 6 the breakaway points can be calculated. Applying Eq. (7-16) we get

$$\frac{1}{s + \dfrac{1}{30}} + \frac{1}{s + \dfrac{1}{10}} + \frac{1}{s + \dfrac{1}{3}} = 0$$

from where the two possibilities are -0.247 and -0.063. The only valid breakaway point is -0.063 since it is the one that lies in the region of the real axis where two loci move toward each other.

Before the final root locus diagram is drawn, it is convenient to know where the loci cross the imaginary axis. This provides one more point to draw the root loci through and enhances the accuracy of the diagram. This point is the ultimate frequency, ω_u, and is easily found by applying the direct substitution method learned in Chapter 6. The application of this method to the present problem yields $\omega_u = \pm 0.22$. The controller gain that produces this state of conditional stability can also be calculated using the direct substitution method. This value is $K_{cu} = 24.0$. Fig. 7-10 shows the complete root locus diagram.

This example has shown that it is fairly simple to draw the root locus diagram and that it is not necessary to find any root of the system. The loci between the breakaway point and the crossover frequency have been drawn by hand. This is usually good enough for most process control work. For convenience a drawing instrument called the spirule can be used. The spirule is principally used by electrical engineers.

The diagram shown in Fig. 7-10 helps to illustrate another use of root locus. Suppose it is desired to tune the feedback controller so that the closed-loop control system response is oscillatory with a damping ratio of 0.707, $\xi = 0.707$.

In Chapter 4 we defined damping ratio as a parameter of a second-order system. In using this performance specification, ξ, we assume that either the process is of second order or that there are two time constants that are much longer than the others. These two time constants will "dominate" the dynamics of the process and will provide the roots that are farthest to the right (closest to the imaginary axis). These two roots are referred to as the *dominant roots*. Tuning a feedback controller for the above specification means that the two dominant roots of the characteristic equation must satisfy the equation

$$\tau^2 s_1^2 + 2\tau \xi s_1 + 1 = 0$$

with $\xi = 0.707$. These roots are

$$s_1, s_1^* = -\frac{\xi}{\tau} \pm \frac{1}{\tau} \sqrt{1 - \xi^2}\, i$$

These roots are shown graphically in Fig. 7-11. From this figure the following can be determined:

$$\theta = \cos^{-1} \xi$$

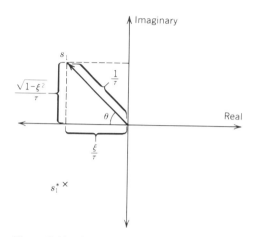

Figure 7-11.   Roots s_1 and s_1^* of second-order system.

Then, for

$$\xi = 0.707$$
$$\theta = \cos^{-1} 0.707 = 45°$$

Figure 7-12 shows how to find the roots, s and s^*, of the system with this damping factor. In this case the roots are roughly located at $-0.06 \pm 0.06i$. The gain of the controller that yields this closed-loop behavior must now be calculated; the magnitude criterion, Eq. (7-12), is used. To do so the distance between the root s and each pole and zero must be measured. This measurement is simply done with a ruler (using the same magnitude scale as the axis) or by the use of the Pythagorean theorem. The latter method is preferred because it minimizes measurement errors. For this system the magnitude criterion is

$$\frac{K'}{|s - p_1||s - p_2||s - p_3|} = 1$$

Since the first pole occurs at $p_1 = -0.033 + 0i$, the distance $|s - p_1|$ is calculated by the Pythagorean theorem to be

$$|s - p_1| = \sqrt{(0.060 - 0.033)^2 + (0.06 - 0)^2} = 0.066$$

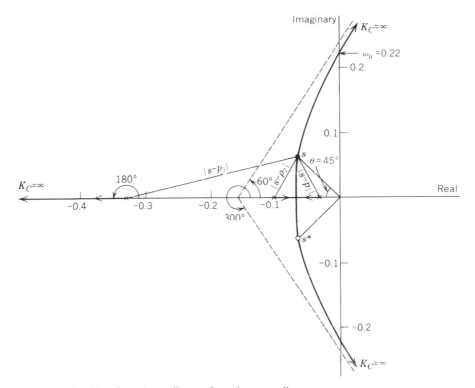

Figure 7-12. Use of root locus diagram for tuning controller.

similarly

$$|s - p_2| = \sqrt{(0.060 - 0.1)^2 + (0.06 - 0)^2} = 0.072$$
$$|s - p_3| = \sqrt{(0.060 - 0.33)^2 + (0.06 - 0)^2} = 0.276$$

then

$$\frac{K'}{(0.066)(0.072)(0.276)} = 1$$

$$K' = 0.00131$$

and since

$$K' = 0.000888 \, K_c$$

the controller gain is

$$K_c = 1.475$$

This controller gain will yield an oscillatory response of the control loop with a damping factor of 0.707.

One more detailed example is now presented.

Example 7-5. Suppose it is desired to use a PI controller to control the heat exchanger of Example 7-4. Plot the root locus diagram for this new control system, using a reset time of 1 min. What is the effect of adding reset action to the controller?

Figure 7-13 shows the block diagram. Note that the reset time has been written in seconds (60 s) instead of minutes (1 min) to maintain consistent units.

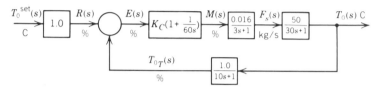

Figure 7-13. Block diagram of heat exchanger temperature control loop—PI controller.

Characteristic equation

$$1 + \frac{0.8 K_c \left(1 + \dfrac{1}{60s}\right)}{(10s + 1)(30s + 1)(3s + 1)} = 0$$

$$\text{OLTF} = \frac{0.8 K_c \left(1 + \dfrac{1}{60s}\right)}{(10s + 1)(30s + 1)(3s + 1)}$$

or

$$\text{OLTF} = \frac{K' \left(s + \dfrac{1}{60}\right)}{s \left(s + \dfrac{1}{10}\right)\left(s + \dfrac{1}{30}\right)\left(s + \dfrac{1}{3}\right)}$$

where

$$K' = \frac{0.8 K_c}{(10)(30)(3)} = 0.000888 K_c$$

$$\text{with poles: } 0, \ -\frac{1}{10}, \ -\frac{1}{30}, \ -\frac{1}{3}; \quad n = 4$$

$$\text{zeros: } -\frac{1}{60}; \quad m = 1$$

Figure 7-14 shows the locations of the poles (x) and zeros (o) in the s plane.

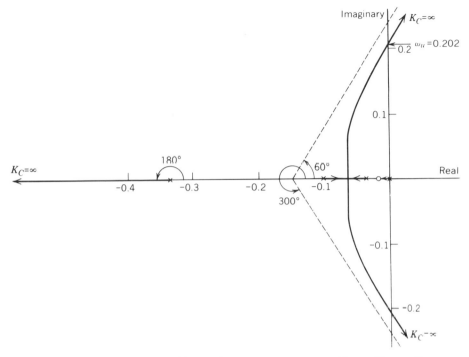

Figure 7-14. Root locus diagram of heat exchanger temperature control loop—PI controller.

- Rule 1 indicates that the negative real axis between the pole at 0 and the zero at $-1/60$ will be part of the root locus. This is also the case between the poles at $-1/30$ and $-1/10$ and from the pole at $-1/3$ to $-\infty$.

- Rule 2 indicates that the root loci will originate at 0, $-1/10$, $-1/30$, and $-1/3$, the poles of the OLTF.

- Since there are four poles, $n = 4$, Rule 3 indicates that there are four branches or loci.

- Since there is one zero, $m = 1$, Rule 4 indicates that one of the branches will terminate at this zero. In this case, this is the branch that originates at the pole equal to zero. The other three branches, $n - m = 3$, will approach infinity as K_c increases.

- Using Rule 5 the center of gravity can be determined as well as the angles the asymptotes make with the positive real axis. Since there are three branches that approach infinity, there must be three asymptotes. According to Eq. (7-14)

$$CG = \frac{-\dfrac{1}{10} - \dfrac{1}{30} - \dfrac{1}{3} + \dfrac{1}{60}}{3} = -0.15$$

and from Eq. (7-13) we get

$$\phi = \frac{180° + 360°(0)}{3}, \quad \frac{180° + 360°(1)}{3}, \quad \frac{180° + 360°(2)}{3}$$

$$\phi = 60°, \ 180°, \ 300°$$

These asymptotes and angles are shown in Fig. 7-14. One of the asymptotes lies on the real axis and moves from the center of gravity to minus infinity. The other two asymptotes move away from the real axis into the complex region of the s plane crossing the imaginary axis and thus indicating the possibility of instability.

- Rule 6 provides the breakaway points. Applying Eq. (7-16) we obtain

$$\frac{1}{s + \dfrac{1}{60}} = \frac{1}{s + 0} + \frac{1}{s + \dfrac{1}{10}} + \frac{1}{s + \dfrac{1}{30}} + \frac{1}{s + \dfrac{1}{3}}$$

The following four possibilities are obtained: $-0.0137 \pm 0.0149i$, -0.0609, -0.245. The only valid breakaway point is at -0.0609 since it is the only point that lies in the region of the real axis where two loci move toward each other.

Applying the direct substitution method to the characteristic equation of this system, the crossover or ultimate frequency is determined to be $\omega_u = 0.202$. The controller gain that yields this condition is $K_{cu} = 20.12$. Fig. 7-14 shows the final root locus diagram.

Comparing Figs. 7-10 and 7-14 we see that the addition of the reset action to the proportional only controller does not significantly change the shape of the root locus. The most important effect is the decrease in the ultimate gain and ultimate frequency. It can be said that the reset action tends to make the control loop more unstable since it cannot tolerate as much controller gain before it goes unstable. What about the effect of making τ_I larger or smaller?

Root Locus Summary

This section has presented the root locus technique for process control analysis and design. The development of the root locus diagram has been shown to be quite simple without the need for sophisticated mathematics. Probably the major advantage of the method is its graphical nature. On the other hand, its major disadvantage is that it cannot be applied to processes with dead time. In this respect it is similar to the other techniques already learned: Routh test and direct substitution. The examples used have also shown the effect of derivative and reset actions on the stability of the control loop.

7-2. FREQUENCY RESPONSE TECHNIQUES

Frequency response techniques are one of the most popular techniques for the analysis and design of control systems of linear systems. This section presents what is meant by frequency response and how to use these techniques to analyze and synthesize control systems.

To begin, consider the general block diagram shown in Fig. 7-15. The control loop has been opened before the valve and after the transmitter. A variable frequency generator provides the input to the valve, $x(t) = X_o \sin \omega t$, and a recorder records the output signal from the transmitter and the input signal to the valve. Figure 7-16 shows the two recordings. After the transients have died out the transmitter output reaches a sinusoidal response, $y(t) = Y_o \sin (\omega t + \theta)$. This experiment is referred to as "sinusoidal testing."

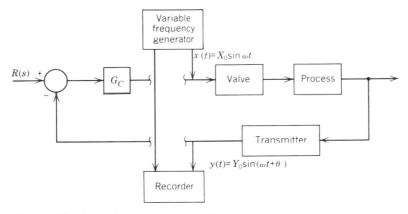

Figure 7-15. Block diagram showing variable-frequency generator and recorder.

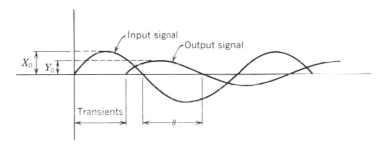

Figure 7-16. Recordings from sinusoidal testing.

Let us now "perform" the same experiment using the transfer functions that describe the process. Assume the following simple transfer function:

$$G(s) = \frac{Y(s)}{X(s)} = \frac{K}{\tau s + 1} \tag{7-17}$$

This transfer function describes the valve, process, and transmitter combination. The input signal to the valve is

$$x(t) = X_o \sin \omega t$$

or

$$X(s) = \frac{X_o \omega}{s^2 + \omega^2} \tag{7-18}$$

Therefore

$$Y(s) = \frac{KX_o \omega}{(\tau s + 1)(s^2 + \omega^2)}$$

The time domain expression can be obtained using the techniques learned in Chapter 2:

$$y(t) = \frac{KX_o \omega\tau}{1 + \omega^2\tau^2}e^{-t/\tau} + \frac{KX_o}{1 + \omega^2\tau^2}[-\omega\tau \cos \omega t + \sin \omega t]$$

Finally, using the identity

$$A \cos at + B \sin at = r \sin (at + \theta)$$

where

$$r = \sqrt{A^2 + B^2}$$

and

$$\theta = \tan^{-1}\left(\frac{A}{B}\right)$$

the $y(t)$ expression can be changed to

$$y(t) = \frac{KX_o \omega\tau}{1 + \omega^2\tau^2}e^{-t/\tau} + \frac{KX_o}{\sqrt{1 + \omega^2\tau^2}} \sin (\omega t + \theta) \tag{7-19}$$

with

$$\theta = \tan^{-1}(-\omega\tau) = -\tan^{-1}(\omega\tau) \tag{7-20}$$

Equation (7-19) shows that as time increases, the exponential term goes to zero; this is the transient term that dies out. When this happens the output expression becomes

$$Y(t)\big|_{t \text{ very large}} = \frac{KX_o}{\sqrt{1 + \omega^2\tau^2}} \sin (\omega t + \theta) \tag{7-21}$$

which constitutes the sinusoidal behavior of the output signal. The amplitude of this output is

$$Y_o = \frac{KX_o}{\sqrt{1 + \omega^2\tau^2}}$$

The minus sign in Eq. (7-20) indicates that the output signal "lags" the input signal by the amount θ calculated from the equation. All of this information is shown graphically in Fig. 7-16.

A word of advice is necessary here. Care should be taken when calculating the sine term in Eq. (7-21). The term ω is in radians/time, the term ωt is in radians. Thus, for the operation $(\omega t + \theta)$ to be in the correct units, θ must be in radians. If degrees are to be used, then the term must be written as

$$\left(\frac{180}{\pi}\omega t + \theta\right)$$

In short, be aware and careful with the units.

Some terms often used in frequency response studies are now defined.

Amplitude ratio (AR)   is defined as the ratio of the amplitude of the output signal to the amplitude of the input signal. That is

$$AR = \frac{Y_o}{X_o}$$

Magnitude ratio (MR)   is defined as the amplitude ratio divided by the steady-state gain.

$$MR - \frac{AR}{K}$$

Phase angle (θ)   is the amount, in degrees or radians, by which the output signal lags or leads the input signal. When θ is positive, it is a lead angle; when θ is negative, it is a lag angle.

For the first-order transfer function shown above:

$$AR = \frac{K}{\sqrt{1 + \omega^2\tau^2}}; \qquad MR = \frac{1}{\sqrt{1 + \omega^2\tau^2}}; \qquad \theta = \tan^{-1}(-\omega\tau)$$

Notice that all three terms are functions of the input frequency. Certainly, different processes have different AR (MR) and θ dependence on ω.

Frequency response is essentially the study of how the AR (MR) and θ of different components or systems behave as the input frequency changes. The next paragraphs show that frequency response is a powerful technique to analyze and synthesize control systems. The development of the frequency response of process systems is covered first, followed by its use for analysis and synthesis.

There are in general two different ways to generate the frequency response.

1.   Experimental methods.   This is essentially the experiment, shown above, with the variable frequency generator and the recorder. The idea is to run the experiment at different frequencies, so that a table of AR vs. ω and θ vs. ω be obtained. These experimental methods are covered later in this chapter; they also provide a way to identify process systems.

2.   Transforming the open-loop transfer function after a sinusoidal disturbance.   This method consists of using the open-loop transfer function to obtain the response of the system to a sinusoidal input. From the response the amplitude and phase angle of the output can then be determined. It is essentially the mathematical manipulations shown above that resulted in Eqs. (7-20) and (7-21).

Fortunately, operational mathematics provide a very simple method to determine AR (MR) and θ. The necessary mathematics have already been presented in Chapter 2. This section now develops the general method to determine AR (MR) and θ. Consider

$$\frac{Y(s)}{X(s)} = G(s)$$

for $x(t) = X_o \sin \omega t$. From Table 2-1 we obtain

$$X(s) = \frac{X_o \omega}{s^2 + \omega^2}$$

Then

$$Y(s) = G(s) \frac{X_o \omega}{s^2 + \omega^2}$$

Expansion by partial fractions yields

$$Y(s) = \frac{A}{s + i\omega} + \frac{B}{s - i\omega} + [\text{terms for the poles of } G(s)] \qquad (7\text{-}22)$$

To obtain A we use

$$A = \lim_{s \to -i\omega} \left[\frac{(s + i\omega)X_o\omega G(s)}{(s^2 + \omega^2)} \right] = \frac{G(-i\omega)X_o\omega}{-2i\omega}$$

As shown in Chapter 2 any complex number can be represented by a magnitude and argument; then

$$A = \frac{X_o|G(i\omega)|e^{-i\angle G(i\omega)}}{-2i} = \frac{X_o|G(i\omega)|e^{-i\theta}}{-2i}$$

To obtain B we use

$$B = \lim_{s \to i\omega} \left[\frac{(s - i\omega)X_o\omega G(s)}{(s^2 + \omega^2)} \right] = \frac{X_o|G(i\omega)|e^{i\theta}}{2i}$$

Then, substituting the expressions for A and B into Eq. (7-22) yields

$$Y(s) = \frac{X_o|G(i\omega)|}{2i} \left[\frac{-e^{-i\theta}}{s + i\omega} + \frac{e^{i\theta}}{s - i\omega} \right] + [\text{terms of } G(s)]$$

Inverting back into the time domain, we get

$$y(t) = \frac{X_o|G(i\omega)|}{2i} [-e^{-i\theta}e^{-i\omega t} + e^{i\theta}e^{i\omega t}] + [\text{transient terms}]$$

After a very long time the transient terms die out; then

$$y(t) = X_o|G(i\omega)| \frac{e^{i(\omega t + \theta)} - e^{-i(\omega t + \theta)}}{2i}$$

$$y(t) = X_o|G(i\omega)| \sin(\omega t + \theta) \qquad (7\text{-}23)$$

This is the response for $G(s)$ after the transients die out.

The amplitude ratio is

$$AR = \frac{Y_o}{X_o} = \frac{X_o|G(i\omega)|}{X_o} = |G(i\omega)| \tag{7-24}$$

and the phase angle is

$$\theta = \angle G(i\omega) \tag{7-25}$$

So in order to obtain the AR and θ, one substitutes $i\omega$ for s in the transfer function and then calculates the magnitude and argument of the resulting complex number. The magnitude is equal to the amplitude ratio (AR) and the argument equals the phase angle (θ).

This, indeed, simplifies tremendously the required calculations.

Let us apply these results to the first-order system used earlier:

$$G(s) = \frac{K}{\tau s + 1}$$

Now substitute $i\omega$ for s

$$G(i\omega) = \frac{K}{i\omega\tau + 1}$$

resulting in a complex number expression. This expression is composed of the ratio of two terms: the numerator, a real number, and the denominator, a complex number. The equation can also be written as follows:

$$G(i\omega) = \frac{G_1}{G_2} = \frac{K}{i\omega\tau + 1} \tag{7-26}$$

As shown, the AR is equal to the magnitude of this complex number:

$$AR = |G(i\omega)| = \frac{|G_1|}{|G_2|} = \frac{K}{\sqrt{\omega^2\tau^2 + 1}} \tag{7-27}$$

which is the same AR as previously obtained.

The θ is equal to the argument of the complex number:

$$\theta = \angle G(i\omega) = \angle G_1 - \angle G_2 = 0 - \tan^{-1}(\omega\tau) = -\tan^{-1}(\omega\tau) \tag{7-28}$$

which is also the same θ as previously obtained.

Let us now look at several other examples.

Example 7-6. Consider the following second-order system:

$$G(s) = \frac{K}{\tau^2 s^2 + 2\tau\xi s + 1}$$

Determine the expressions for AR and θ.

The first step is to substitute $i\omega$ for s.

$$G(i\omega) = \frac{K}{-\omega^2\tau^2 + i2\tau\xi\omega + 1} = \frac{K}{(1 - \omega^2\tau^2) + i2\tau\xi\omega}$$

Again, a complex number expression results that is a ratio of two other numbers.

$$G(i\omega) = \frac{G_1}{G_2} = \frac{K}{(1 - \omega^2\tau^2) + i2\tau\xi\omega}$$

The amplitude ratio is

$$AR = |G(i\omega)| = \frac{|G_1|}{|G_2|} = \frac{K}{\sqrt{(1 - \omega^2\tau^2)^2 + (2\tau\xi\omega)^2}} \qquad (7\text{-}29)$$

and the phase angle is

$$\theta = \measuredangle G(i\omega) = \measuredangle G_1 - \measuredangle G_2$$

$$= 0 - \tan^{-1}\left(\frac{2\tau\xi\omega}{1 - \omega^2\tau^2}\right)$$

$$\theta = -\tan^{-1}\left(\frac{2\tau\xi\omega}{1 - \omega^2\tau^2}\right) \qquad (7\text{-}30)$$

Example 7-7. Consider the following transfer function:

$$G_c(s) = K(1 + \tau s) \qquad (7\text{-}31)$$

This is the transfer function for a gain times a first-order lead. Determine the expressions for AR and θ.

Substituting $i\omega$ for s results in the following complex number expression:

$$G(i\omega) = K(1 + i\omega\tau)$$

which can also be thought of as being formed by two other numbers:

$$G(i\omega) = G_1G_2 = K(1 + i\omega\tau)$$

The amplitude ratio is

$$AR = |G(i\omega)| = |G_1||G_2| = K\sqrt{1 + \omega^2\tau^2} \qquad (7\text{-}32)$$

and the phase angle is

$$\theta = \measuredangle G(i\omega) = \measuredangle G_1 + \measuredangle G_2 = 0 + \tan^{-1}(\omega\tau)$$

$$\theta = \tan^{-1}(\omega\tau) \qquad (7\text{-}33)$$

The phase angles of systems described by Eqs. (7-17) and (7-31) can be compared. Systems described by Eq. (7-17), referred to in Chapter 3 as first-order lags, provide negative phase angles, as shown by Eq. (7-28). Systems described by Eq. (7-31), referred to in Chapter 3 as first-order leads, provide positive phase angles, as shown by Eq. (7-33). This fact will become important in the study of process control stability by frequency response techniques.

Example 7-8. Determine the expressions for AR and θ for a dead time:

$$G(s) = e^{-t_0 s}$$

Substituting $i\omega$ for s yields

$$G(i\omega) = e^{-it_0\omega}$$

Since this expression is already in polar form, using the principles learned in Chapter 2, we obtain

$$G(i\omega) = |G(i\omega)| e^{iG(i\omega)} = e^{-it_0\omega}$$

which means that

$$AR = |G(i\omega)| = 1 \qquad (7\text{-}34)$$

and

$$\theta = \angle G(i\omega) = -t_0\omega \qquad (7\text{-}35a)$$

Recall our discussion of the units of θ. As written in Eq. (7-35a), the unit of θ is radians. If it is desired to obtain θ in degrees, then

$$\theta = \left(\frac{180°}{\pi}\right)(-t_0\omega) \qquad (7\text{-}35b)$$

It is interesting, and *most important*, to notice that θ gets increasingly negative as ω increases. The rate at which θ drops depends on t_0. The larger t_0, the faster θ drops. This fact will become important in the analysis of process control systems. The dead time does not affect the amplitude ratio or magnitude ratio.

Example 7-9. Determine the expressions for AR and θ for an integrator:

$$G(s) = \frac{1}{s}$$

Substituting $i\omega$ for s yields

$$G(i\omega) = \frac{1}{i\omega}$$

which can be thought of as being formed by two complex numbers:

$$G(i\omega) = \frac{G_1}{G_2} = \frac{1}{i\omega}$$

The amplitude ratio is

$$AR = |G(i\omega)| = \frac{|G_1|}{|G_2|} = \frac{1}{\omega} \qquad (7\text{-}36)$$

and the phase angle is

$$\theta = \angle G(i\omega) = \angle G_1 - \angle G_2$$

$$= 0 - \tan^{-1}\left(\frac{\omega}{0}\right) = -\tan^{-1}(\infty) \qquad (7\text{-}37)$$

$$\theta = -90°$$

So for an integrator the amplitude ratio decreases as the frequency increases while the phase angle remains constant at $-90°$. That is, the integrator provides a constant phase lag.

At this point we can generalize the expressions for AR and θ. Consider the following general OLTF:

$$\text{OLTF}(s) = \frac{K \prod_{i=1}^{m} (\tau_i s + 1) e^{-t_0 s}}{s^k \prod_{j=1}^{n} (\tau_j s + 1)}, \qquad (n + k) > m \qquad (7\text{-}38)$$

Then substituting s by $i\omega$ we get

$$\text{OLTF}(i\omega) = \frac{K \prod_{i=1}^{m} (i\tau_1 \omega + 1) e^{-i\omega t_0}}{(i\omega)^k \prod_{j=1}^{n} (i\tau_j \omega + 1)}$$

and finally we arrive at

$$\text{AR} = \frac{K \prod_{i=1}^{m} |i\tau_i \omega + 1|}{|(i\omega)^k| \prod_{j=1}^{n} |i\tau_j \omega + 1|}$$

or

$$\text{AR} = \frac{K \prod_{i=1}^{m} [(\tau_i \omega)^2 + 1]^{1/2}}{\omega^k \prod_{j=i}^{n} [(\tau_j \omega)^2 + 1]^{1/2}} \qquad (7\text{-}39)$$

and

$$\theta = \sum_{i=1}^{m} \tan^{-1}(\tau_i \omega) - t_0 \omega \left(\frac{180°}{\pi}\right) - \sum_{j=1}^{n} \tan^{-1}(\tau_j \omega) - k(90°) \qquad (7\text{-}40)$$

So far expressions for AR and θ as a function of ω have been developed. There are several ways for graphical representations of these expressions. The three most common ways are the Bode plots, the Nyquist plot, and the Nichols chart. Bode plots are presented in detail in the following section.

Bode Plots

The Bode plot is the most common of the graphical representations of AR (MR) and θ functions. This plot consists of two graphs: (1) log AR (or log MR) vs. log ω and (2) θ vs. log ω. Sometimes, instead of plotting log AR, the quantity 20 log AR, which is referred to as decibels, is plotted. This term is used extensively in the electrical engineering field and sometimes also in the process control field; this book plots log AR. The Bode plot of some of the most common process transfer functions are now presented.

Gain Element. A pure gain element has the transfer function

$$G(s) = K$$

Substituting $i\omega$ for s gives

$$G(i\omega) = K$$

and using the mathematics previously presented yields

$$AR = |G(i\omega)| = K$$

and

$$\theta = \tan^{-1} G(i\omega) = 0°$$

Figs. 7-17 and 7-18a show the Bode plot for this element. Fig. 7-17 plots log AR while Fig. 7-18a plots log MR. Notice that in both cases log-log and semi-log graph papers have been used.

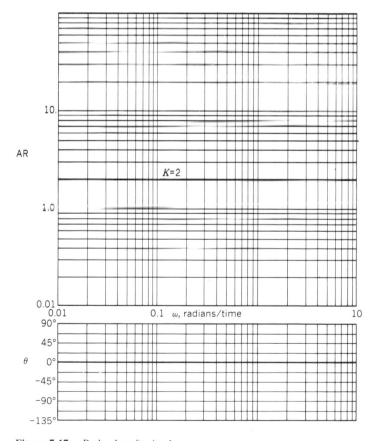

Figure 7-17. Bode plot of gain element.

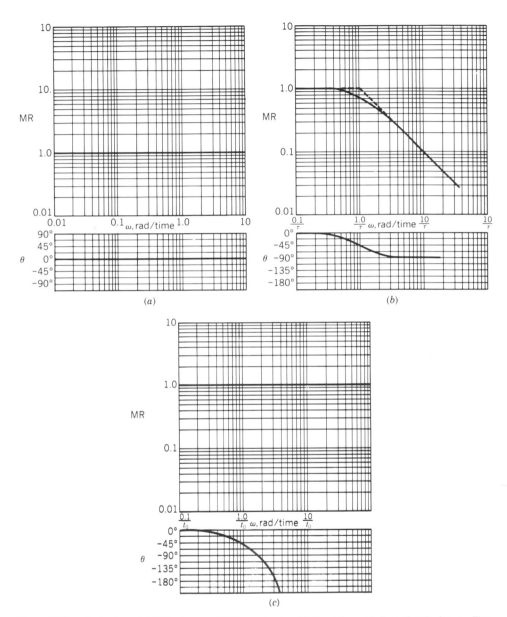

Figure 7-18. Bode plots. (*a*) Gain element. (*b*) First-order lag. (*c*) Dead time. (*d*) Second-order lag. (*e*) First-order lead. (*f*) Integrator.

First-Order Lag. For a first-order lag the AR and θ are given by Eqs. (7-27) and (7-28), respectively.

$$AR = \frac{K}{\sqrt{\omega^2 \tau^2 + 1}} \qquad (7\text{-}27)$$

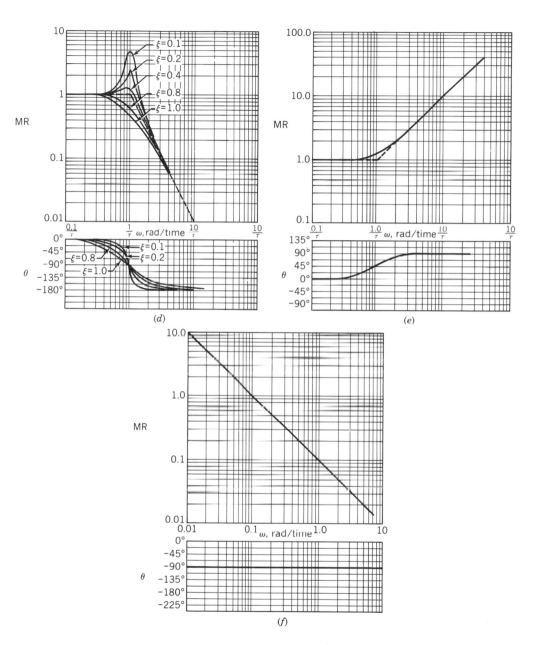

Figure 7-18 (*Continued*)

or

$$MR = \frac{1}{\sqrt{\omega^2\tau^2 + 1}}$$

and

$$\theta = -\tan^{-1}(\omega\tau) \tag{7-28}$$

Fig. 7-18b shows the Bode plot for this system.

The magnitude ratio plot of Fig. 7-18b shows two dotted lines. These lines are asymptotes to the frequency response of the system at low and high frequencies. These asymptotes are very important in the study of frequency response. As seen from the figure, they do not deviate much from the actual frequency response. Therefore, quite often frequency response analysis is done with the asymptotes since they are easier to draw and not much error is involved in their use. Let us now see how these asymptotes are developed.

From the magnitude ratio equation as $\omega \to 0$, $MR \to 1$, which results in the horizontal asymptote.

Before the high-frequencies asymptote is developed, the magnitude ratio equation is written in log form:

$$\log MR = -1/2 \log (\tau^2\omega^2 + 1)$$

now as $\omega \to \infty$

$$\log MR \to -1/2 \log \tau^2\omega^2 = -\log \tau\omega$$
$$\log MR \to -\log \tau - \log \omega \tag{7-41}$$

which is the expression of a straight line in a log-log graph of MR vs. ω. This straight line has a slope of -1. The location of this line in the graph must now be determined. The simplest way to do this is to find where the high-frequency asymptote meets the low-frequency asymptote. It is known that as $\omega \to 0$, $MR \to 1$ so,

$$\log MR = 0$$

Equating this equation to Eq. (7-41) yields

$$\omega = \frac{1}{\tau} \tag{7-42}$$

This is the frequency, referred to as the *corner frequency* (ω_c) or *breakpoint frequency*, at which both asymptotes meet, as shown in Fig. 7-18b. It is also at this frequency at which the maximum error between the frequency response and the asymptotes exists. The actual magnitude ratio is

$$MR = \frac{1}{\sqrt{\omega^2\tau^2 + 1}} = \frac{1}{\sqrt{2}} = 0.707$$

and not $MR = 1$ as given by the asymptotes.

Before concluding with the Bode plot of this system, it is worthwhile to go over the θ at low and high frequencies. At low frequencies

$$\omega \to 0$$
$$\theta \to -\tan^{-1}(\omega\tau) = -\tan^{-1}(0)$$
$$\theta \to 0°$$

At high frequencies

$$\omega \to \infty$$
$$\theta \to -\tan^{-1}(\infty)$$
$$\theta \to -90°$$

These values of phase angle, $0°$ and $-90°$, are the asymptotes for the phase angle plot. At the corner frequency

$$\omega_c = \frac{1}{\tau}, \qquad \theta = -\tan^{-1}(1) = -45°$$

To summarize, the following are the important characteristics of the Bode plot of a first-order lag:

1. AR (MR) graph. The low-frequency asymptote has a slope of 0 while the high-frequency asymptote has a slope of -1. The corner frequency, where these two asymptotes meet, occurs at $1/\tau$.

2. Phase angle graph. At low frequencies the phase angle approaches $0°$ while at high frequencies it approaches $-90°$. At the corner frequency, the phase angle is $-45°$.

Second-Order Lag. As shown in Example 7-6, the AR and θ expressions for a second-order lag are given by Eqs. (7-29) and (7-30), respectively:

$$AR = \frac{K}{\sqrt{(1 - \omega^2\tau^2)^2 + (2\tau\xi\omega)^2}} \tag{7-29}$$

and

$$\theta = -\tan^{-1}\left(\frac{2\tau\xi\omega}{1 - \omega^2\tau^2}\right) \tag{7-30}$$

From the AR expression the expression for MR is obtained:

$$MR = \frac{1}{\sqrt{(1 - \omega^2\tau^2)^2 + (2\tau\xi\omega)^2}}$$

Giving values to ω for a given τ and ξ, the frequency response is determined as shown in Fig. 7-18d.

The asymptotes are obtained in a similar manner as for the first-order lag. At low frequencies

$$\omega \to 0$$
$$MR \to 1$$

and

$$\theta \rightarrow 0°$$

At high frequencies

$$\omega \rightarrow \infty$$

$$\log MR \rightarrow -1/2 \log [(1 - \omega^2\tau^2)^2 + (2 \tau\xi\omega)^2]$$

$$\log MR \rightarrow -1/2 \log (\omega^2\tau^2)^2 = -2 \log \tau - 2 \log \omega$$

which is the expression of a straight line with a slope of -2. The phase angle at these high frequencies approaches

$$\theta \rightarrow -180°$$

To find the corner frequency, ω_c, at which the two asymptotes meet, the same procedure as for the first-order lag is followed and yields

$$\omega_c = \frac{1}{\tau}$$

Notice from Fig. 7-18d that the transition of the frequency response from low to high frequencies depends on ξ.

At the corner frequency

$$\omega_c = \frac{1}{\tau}$$

$$\theta = -\tan^{-1}(\infty) = -90°$$

To summarize, the following are the important characteristics of the Bode plot of a second-order lag:

1.   AR (MR) graph.   The slope of the low-frequency asymptote is 0 while that of the high-frequency asymptote is -2. The corner frequency, ω_c, occurs at $1/\tau$. The transition of the AR from low to high frequency depends on the value of ξ.

2.   Phase angle graph.   At low frequencies the phase angle approaches $0°$ while at high frequencies it approaches $-180°$. At the corner frequency, the phase angle is $-90°$.

Dead Time.   As shown in Example 7-8, the AR and θ expressions for a dead time are given by Eqs. (7-34) and (7-35), respectively:

$$AR = MR = 1 \tag{7-34}$$

and

$$\theta = -\omega t_0 \tag{7-35a}$$

or

$$\theta = \left(\frac{180°}{\pi}\right)(-\omega t_0) \tag{7-35b}$$

The Bode plot is shown in Fig. 7-18c. Notice that as the frequency increases, the phase angle becomes more negative. The larger the value of the dead time, the faster the phase angle drops (becomes increasingly negative). The phase angle plot does not asymptotically approach any final value.

First-Order Lead.   As shown in Example 7-7 the AR and θ expressions for a first order lead are given by Eqs. (7-32) and (7-33), respectively:

$$AR = K\sqrt{1 + \omega^2\tau^2} \tag{7-32}$$

or

$$MR = \sqrt{1 + \omega^2\tau^2}$$

and

$$\theta = \tan^{-1}(\omega\tau) \tag{7-33}$$

The Bode plot is shown in Fig. 7-18e. Notice that the low-frequency asymptote has a slope of 0 while the high-frequency asymptote has a slope of $+1$. At low frequencies the phase angle approaches $0°$ while at high frequencies it approaches $+90°$. At the corner frequency, the phase angle is $+45°$. Thus, a first-order lead provides "phase lead."

Integrator.   As shown in Example 7-9, the AR and θ expressions for an integrator are given by Eqs. (7-36) and (7-37), respectively:

$$AR = MR = \frac{1}{\omega} \tag{7-36}$$

and

$$\theta = 90° \tag{7-37}$$

The Bode plot is shown in Fig. 7-18f. Notice that the MR graph consists of a straight line with slope of -1. This is easily shown by taking the log of Eq. (7-36):

$$\log MR = -\log \omega$$

This equation also shows that $MR = 1$ at $\omega = 1$ radian/time.

Development of Bode Plots of Complex System.   Most complex transfer functions of process systems are formed by the product of simpler components. The Bode plot of these complex transfer functions can be obtained by adding the Bode plots of the simpler components. Most of these simpler components have been presented in the preceding sections. Consider the following:

$$G(s) = \frac{K(\tau_1 s + 1)\,e^{-t_0 s}}{(\tau_2 s + 1)(\tau_3 s + 1)} \tag{7-43}$$

This particular transfer function can be considered as being composed of the following five simpler transfer functions:

$$G(s) = G_1(s)G_2(s)G_3(s)G_4(s)G_5(s)$$

where

$$G_1(s) = K$$

$$G_2(s) = (\tau_1 s + 1)$$

$$G_3(s) = e^{-t_0 s}$$

$$G_4(s) = \frac{1}{\tau_2 s + 1}$$

$$G_5(s) = \frac{1}{\tau_3 s + 1}$$

The development of the Bode plot for Eq. (7-43) is very simple. For AR and MR we have

$$\begin{aligned} \text{AR} &= |G(i\omega)| \\ &= |G_1(i\omega)|\,|G_2(i\omega)|\,|G_3(i\omega)|\,|G_4(i\omega)|\,|G_5(i\omega)| \end{aligned} \tag{7-44}$$

$$\begin{aligned} \text{MR} &= \frac{|G(i\omega)|}{K} \\ &= |G_2(i\omega)|\,|G_3(i\omega)|\,|G_4(i\omega)|\,|G_5(i\omega)| \end{aligned} \tag{7-45}$$

or since the logarithms are usually plotted, we see that

$$\begin{aligned} \log \text{AR} &= \log |G_1(i\omega)| + \log |G_2(i\omega)| + \log |G_3(i\omega)| \\ &\quad + \log |G_4(i\omega)| + \log |G_5(i\omega)| \end{aligned} \tag{7-46}$$

$$\begin{aligned} \log \text{MR} &= \log |G_2(i\omega)| + \log |G_3(i\omega)| \\ &\quad + \log |G_4(i\omega)| + \log |G_5(i\omega)| \end{aligned} \tag{7-47}$$

and for the phase angle we have

$$\angle\, G(i\omega) = \angle\, G_1(i\omega) + \angle\, G_2(i\omega) + \angle\, G_3(i\omega) + \angle\, G_4(i\omega) + \angle\, G_5(i\omega) \tag{7-48}$$

Equations (7-47) and (7-48) show that to obtain the composite Bode plot, the individual Bode plots are added. To obtain the composite asymptote, the individual asymptotes are added.

Example 7-10.   Consider the following transfer function:

$$G(s) = \frac{K(s + 1)e^{-s}}{s(2s + 1)(3s + 1)}$$

Using the principles learned, we see that

$$\text{MR} = \frac{\sqrt{\omega^2 + 1}}{\omega \sqrt{4\omega^2 + 1}\,\sqrt{9\omega^2 + 1}}$$

or

$$\log MR = \frac{1}{2} \log (\omega^2 + 1) - \log(\omega) - \frac{1}{2} \log (4\omega^2 + 1) - \frac{1}{2} \log (9\omega^2 + 1)$$

and

$$\theta = \tan^{-1}(\omega) - \left(\frac{180°}{\pi}\right) \omega - 90° - \tan^{-1}(2\omega) - \tan^{-1}(3\omega)$$

From these last two equations the Bode plot is developed as shown in Fig. 7-19. Notice that the composite asymptote is obtained by adding the individual asymptotes. At low frequencies, $\omega < 0.33$, the slope is -1, because of the integrator term. At $\omega = 0.33$, one of the first-order lags starts to contribute to the graph and thus, the slope changes to -2 at this frequency. At $\omega = 0.5$, the other first-order lag starts to contribute, changing the slope of the asymptote to -3. Finally, at $\omega = 1$ the first-order lead enters, with a slope of $+1$, and the slope of the asymptote changes back to -2. Similarly, the composite phase angle plot is obtained by algebraically adding the individual angles.

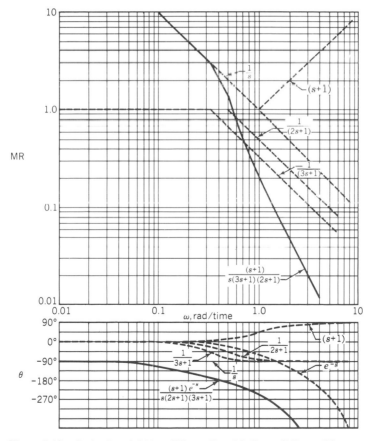

Figure 7-19. Bode plot of $G(s) = [K(s + 1)e^{-s}]/[s(2s + 1)(3s + 1)]$.

One final comment can be made about the slopes of the low-frequency and high-frequency asymptotes (initial and final slopes) and angles of Bode plots. Consider a general transfer function such as

$$G(s) = \frac{K(a_m s^m + a_{m-1} s^{m-1} + \ldots + 1)}{s^k(b_n s^n + b_{n-1} s^{n-1} + \ldots + 1)}; \qquad (n + k) > m \qquad (7\text{-}49)$$

The slope of the low-frequency asymptote is given by

$$\text{slope of AR (MR)} \Big|_{\omega \to 0} \to (-1)k$$

and the angle is given by

$$\theta \Big|_{\omega \to 0} \to (-90°)k$$

The slope of the high-frequency asymptote is given by

$$\text{slope of AR(MR)} \Big|_{\omega \to \infty} \to (n + k - m)(-1)$$

and the angle is given by

$$\theta \Big|_{\omega \to \infty} \to (n + k - m)(-90°)$$

Most systems follow these slopes and angles; these systems are called *minimal phase systems*. There are three exceptions, however, and these are called *nonminimal phases systems*. The three exceptions are:

1. Systems with dead time: $G(s) = e^{-t_0 s}$
2. Systems that show inverse response (positive zeros): $G(s) = (1 - \tau_1 s)/(1 + \tau_2 s)$
3. Systems that are open-loop unstable (positive poles), such as some exothermic chemical reactors: $G(s) = 1/(1 - \tau s)$

In each of these cases the magnitude ratio plot is not changed, but the phase angle plot is, always making the phase lag more negative than would be predicted. The dead time term was presented earlier; the Bode plot of the other two systems is the subject of one of the problems at the end of the chapter.

The expression for the slope of the high-frequency asymptote also serves to show why transfer functions must have more lags than leads. If $(n + k - m) < 0$ then the final slope is positive and noise of very high frequency is amplified with infinite gain.

Frequency Response Stability Criterion.   The frequency response stability criterion is now developed by means of an example to enhance understanding of the significance of the criterion.

Consider the heat exchanger temperature control loop presented in Chapter 6 and used in Example 7-4. The heat exchanger is shown again, for convenience, in Fig. 7-20 and the block diagram in Fig. 7-21. The open-loop transfer function is

$$OLTF = \frac{0.8 \, K_c}{(10s + 1)(30s + 1)(3s + 1)} \tag{7-50}$$

The MR and θ expressions are

$$MR = \frac{AR}{0.8K_c}$$

$$= \frac{1}{\sqrt{(10\omega)^2 + 1} \, \sqrt{(30\omega)^2 + 1} \, \sqrt{(3\omega)^2 + 1}} \tag{7-51}$$

$$\theta = -\tan^{-1}(10\omega) - \tan^{-1}(30\omega) - \tan^{-1}(3\omega) \tag{7-52}$$

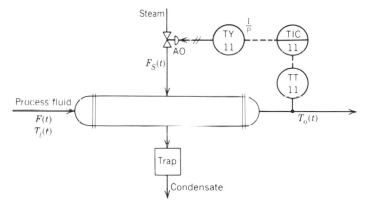

Figure 7-20. Feedback control loop for temperature control of heat exchanger.

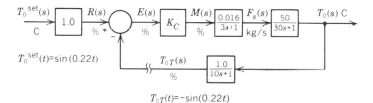

$$T_0T(t) = -\sin(0.22t)$$

Figure 7-21. Block diagram of heat exchanger temperature control loop—P controller.

The Bode plot is shown in Fig. 7-22. From this figure the frequency at which $\theta = -180°$ (or $-\pi$ radians) is read to be 0.22 rad/s. At this frequency

$$\frac{AR}{0.8K_c} = 0.052$$

The controller gain that yields AR = 1 is

$$K_c = \frac{AR}{0.8(0.052)} = \frac{1}{0.8(0.052)} = 24.0$$

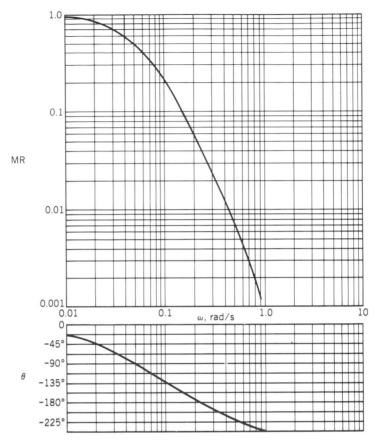

Figure 7-22. Bode plot of heat exchanger temperature control loop—P controller.

These calculations are highly significant. The value $K_c = 24.0$ is the gain of the controller that yields AR $= 1$. Remember that AR is defined as the ratio of the amplitude of the output signal to the amplitude of the input signal, Y_o/X_o. This means that if the input set point to the temperature controller is varied as follows

$$T_o^{\text{set}}(t) = \sin(0.22t)$$

then the output signal from the transmitter, after the transients disappear, will vary as

$$T_{oT}(t) = \sin(0.22t - \pi) = -\sin(0.22t)$$

Notice that the feedback signal is disconnected from the controller, as shown in the block diagram of Fig. 7-21, and that the frequency of the set point oscillation is 0.22 rad/s. This is the frequency at which $\theta = -180° = -\pi$ radians and AR $= 1$.

Suppose now that at some time, $t = 0$, the set point oscillation is stopped, $T_o^{\text{set}}(t) = 0$, and the transmitter signal is connected to the controller. *The error signal, $E(s)$, inside the controller remains unchanged and the oscillations are sustained.* If nothing changes in the control loop, the oscillations remain indefinitely.

If at some time the controller gain is slightly increased to 25.0, the amplitude ratio becomes 1.04:

$$AR = 0.052 \ (0.8) \ K_c$$

$$AR = 0.052 \ (0.8)(25) = 1.04$$

This means that as the signal goes through the control loop, it is amplified. After the first time, the output signal from the transmitter is $-1.04 \sin (0.22t)$. After the second time, it is $-(1.04)^2 \sin (0.22t)$ and so on. If this is not stopped the outlet temperature will increase continuously, yielding an unstable control loop.

On the other hand, if the controller gain is slightly decreased to 23.0, the amplitude ratio becomes 0.957:

$$AR - 0.052(0.8)23 - 0.957$$

This means that as the signal goes through the control loop, it decreases in amplitude. After the first time, the output signal from the transmitter is $-0.957 \sin (0.22t)$. After the second, time it is $-(0.957)^2 \sin (0.22t)$ and so on. This results in a stable control loop.

In summary, the stability criterion based on frequency response can be stated as follows:

In order for a control system to be stable the amplitude ratio must be less than unity when the phase angle is $-180°$ ($-\pi$ radians).

If $AR < 1$ at $\theta = -180°$ the system is stable; if, however, $AR > 1$ at $\theta = -180°$ the system is unstable. This was shown to be the case in the above example.

The controller gain that provides the condition of $AR = 1$ at $\theta = -180°$ is the ultimate gain, K_{cu}. In the above example, $K_{cu} = 24$. The frequency at which this condition happens is the ultimate frequency, ω_u. From this frequency the ultimate period can be calculated as shown in Chapter 6.

Before proceeding with more examples, it is important to point out that the ultimate frequency and ultimate gain can be obtained directly from the MR and θ equations, Eqs. (7-51) and (7-52) for this example, without the need for the Bode plot. The Bode plot was developed from these equations. Using these equations saves developing the plot. Several years ago when handheld calculators were not available (remember slide rules?), it was probably easier to draw the Bode plot, using the high- and low-frequency asymptotes. Today the use of these calculators makes the determination of ω_u and K_{cu} a rather easy procedure. The determination of ω_u requires a small amount of trial and error using the θ equation. Once ω_u has been determined, the equation for AR is used to calculate K_{cu}. This complete procedure is usually faster and yields more accurate results than drawing and using the Bode plot. The use of this plot is still very useful, however, because it shows at a glance how AR and θ vary as the frequency varies.

Several examples are now presented to provide more practice with this powerful technique.

Example 7-11. Consider the same heat exchanger, Fig. 7-20, used above to explain the frequency response stability criterion. Suppose now that for some reason the outlet temperature cannot be measured at the exit of the exchanger but rather farther down the pipe, as shown in Fig. 7-23. The effect of this new sensor location is the addition of some dead time, say two seconds, to the control loop. Fig. 7-24 shows the block diagram with the new transfer function.

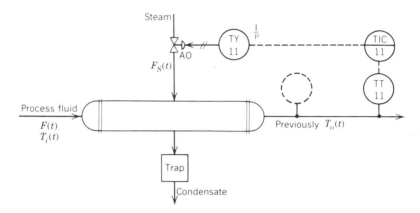

Figure 7-23. Feedback control loop for temperature control of heat exchanger.

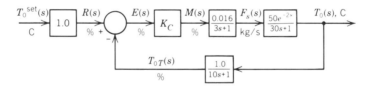

Figure 7-24. Block diagram of heat exchanger temperature control loop—P controller.

The new OLTF is

$$OLTF = \frac{0.8 \, K_c e^{-2s}}{(10s + 1)(30s + 1)(3s + 1)}$$

with

$$MR = \frac{AR}{0.8K_c} = \frac{1}{\sqrt{(10\omega)^2 + 1} \, \sqrt{(30\omega)^2 + 1} \, \sqrt{(3\omega)^2 + 1}}$$

and

$$\theta = -2\omega - \tan^{-1}(10\omega) - \tan^{-1}(30\omega) - \tan^{-1}(3\omega)$$

As explained above, these last two equations can be used to determine the ultimate frequency and ultimate gain. Performing these calculations yields for $\theta = \pi$ rad

$$\omega_u = 0.16 \text{ rad/s}$$

and

$$K_{cu} = 12.8$$

The Bode plot is shown in Fig. 7-25.

The results of Example 7-11 show the effect of dead time on the stability (and consequently also the controllability) of the control loop. The ultimate gain and ultimate period for the heat exchanger without dead time were previously found to be

$$K_{cu} = 24 \quad \text{and} \quad \omega_u = 0.22 \text{ rad/s}$$

when the dead time of 2 seconds was added in Example 7-11 the results were

$$K_{cu} = 12.8 \quad \text{and} \quad \omega_u = 0.16 \text{ rad/s}$$

Thus, it is easier for the process with dead time to go unstable. The difference in ω_u also indicates that a process with dead time is slower than one without dead time.

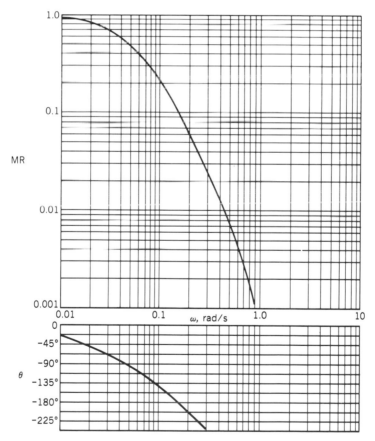

Figure 7-25. Bode plot of heat exchanger temperature control loop with dead time—P controller.

As mentioned in previous chapters, dead time is the worst thing that can happen to any control loop. Example 7-11 proves this point. The dead time term "adds phase lag" to the control loop and consequently the phase angle crosses the $-180°$ value at a lower frequency. The longer the dead time, the lower the ultimate frequency and ultimate gain.

Example 7-5 showed that the addition of reset mode to a proportional controller decreases the ultimate frequency and ultimate gain. This can be explained, from a frequency response point of view, by saying that the addition of reset mode "adds phase lag" to the control loop. A proportional only controller has a phase angle of 0°, as shown in Fig. 7-17. Consider now a proportional-integral controller:

$$G(s) = K_c \left(1 + \frac{1}{\tau_I s} \right) = K_c \left(\frac{\tau_I s + 1}{\tau_I s} \right)$$

The transfer function is composed of a lead term, $\tau_I s + 1$, and an integrator term, $1/\tau_I s$. At low frequencies

$$\omega << \frac{1}{\tau_I}$$

the lead term does not affect the phase angle but the integrator contributes $-90°$ thus, adding phase lag. At higher frequencies

$$\omega >> \frac{1}{\tau_I}$$

the lead term cancels the integrator term with a resulting 0° phase angle. Figures 7-18*f* and 7-18*e* show the Bode plot of an integrator and a first-order lead, respectively.

The reader must remember, however, that the reset mode in a controller is the only action that can remove the offset. On the other hand, it is more difficult to tune the controller because it has two terms, K_c and τ_I, to set and, as explained in the previous paragraph, it is easier for the process to go unstable. It seems that the second law of thermodynamics also applies to process control: You cannot get something for nothing.

The following example demonstrates the effect of derivative action on the stability of a control loop.

Example 7-12. Consider the same heat exchanger control loop, no dead time, with a proportional derivative controller. Suppose the rate time is 0.25 min (15 s). The equation for a "real" PD controller is, as shown in Chapter 5 is

$$G_c(s) = K_c \left(\frac{1 + \tau_D s}{1 + \alpha \tau_D s} \right)$$

or for this example, using $\alpha = 0.1$

$$G_c(s) = K_c \left(\frac{1 + 15s}{1 + 1.5s} \right)$$

The OLTF is then

$$\text{OLTF} = \frac{0.8K_c(1 + 15s)}{(10s + 1)(30s + 1)(3s + 1)(1 + 1.5s)}$$

with

$$\text{MR} = \frac{\text{AR}}{0.8K_c}$$

$$= \frac{\sqrt{(15\omega)^2 + 1}}{\sqrt{(10\omega)^2 + 1}\sqrt{(30\omega)^2 + 1}\sqrt{(3\omega)^2 + 1}\sqrt{(1.5\omega)^2 + 1}}$$

and

$$\theta = \tan^{-1}(15\omega) - \tan^{-1}(10\omega) - \tan^{-1}(30\omega) - \tan^{-1}(3\omega) - \tan^{-1}(1.5\omega)$$

The Bode plot for this system is shown in Fig. 7-26. If we compare this Bode plot

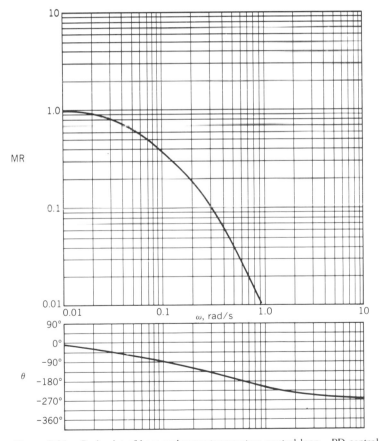

Figure 7-26. Bode plot of heat exchanger temperature control loop—PD controller.

to that shown in Fig. 7-22, we see that the phase angle plot has been "moved up"; the derivative action "adds phase lead." In this system the ultimate gain and period are found to be

$$K_{cu} = 33.05 \quad \text{and} \quad \omega_u = 0.53 \text{ rad/s}$$

Thus, these results show that the derivative action makes the control loop more stable and faster.

The examples presented in this section have demonstrated the use of frequency response, and in particular Bode plots, for analysis of control loops. These examples have also shown the effect of different terms, dead time, and derivative action, on the stability of control loops.

Based on the frequency response stability criterion, any control loop with a pure (no dead time) first- or second-order open-loop transfer function will never go unstable. This is becuase its phase angle will never go below $-180°$. Once a dead time is added (no matter how small), the system can go unstable because the phase angle will always cross the $-180°$ value.

Controller Performance Specifications.   Chapter 6 presented several ways to tune controllers to obtain a desired loop performance. The methods presented were the Ziegler-Nichols (quarter decay ratio), the error-integral criteria (IAE, ISE, and ITAE) and the controller synthesis. Frequency response techniques provide still other methods (based on different performance specifications) to tune controllers. There are three such methods.

The first method is the same Ziegler-Nichols. As shown in this section, frequency response provides a convenient, accurate procedure by which to obtain the ultimate gain and ultimate frequency of a control loop. Once these terms have been determined, the equations presented in Chapter 6 can be used to tune the controller.

The *gain margin* (GM) is a typical control loop specification associated wtih the frequency response technique. The gain margin represents the factor by which the total loop gain must increase to make the system unstable. The controller gain that yields a desired gain margin is calculated as follows:

$$K_c = \frac{K_{cu}}{GM} \tag{7-53}$$

Typical specification is for GM > 1.5.

Phase margin (PM) is another typical specification commonly associated with the frequency response technique. The phase margin is the difference between $-180°$ and the phase angle at the frequency for which the amplitude ratio (AR) is unity. That is

$$PM = 180° + \theta \Big|_{AR=1} \tag{7-54}$$

PM represents the additional amount of phase lag required to make the system unstable. Typical specification is for PM > 45°.

Example 7-13 shows the use of these last two specifications to tune a controller.

Example 7-13. Consider the heat exchanger of Example 7-11. Tune the proportional controller for (a) GM = 1.5 and (b) PM = 45°.
    a. In Example 7-11 the ultimate gain of the controller was determined to be

$$K_{cu} = 12.8$$

In this case the overall loop gain is

$$K_{loop_u} = 0.8 \, K_{cu} = 10.24$$

To obtain a GM specification of 1.5 the controller gain is then set to

$$K_c \Big|_{GM=1.5} = \frac{K_{cu}}{1.5} = 8.53$$

The overall loop gain is then

$$K_{loop} = 0.8 \, K_c = 6.83$$

b. In Example 7-11 expressions for MR and θ were given as

$$MR = \frac{AR}{0.8 \, K_c} = \frac{1}{\sqrt{(10\omega)^2 + 1} \, \sqrt{(30\omega)^2 + 1} \, \sqrt{(3\omega)^2 + 1}}$$

and

$$\theta = -(2\omega)\left(\frac{180°}{\pi}\right) - \tan^{-1}(10\omega) - \tan^{-1}(30\omega) - \tan^{-1}(3\omega)$$

Based on the definition of phase margin given above, for a PM = 45°, $\theta = -135°$. Using the equation for θ, or the Bode plot of Fig. 7-25, the frequency for this phase angle can be determined as

$$\omega \Big|_{PM=45°} = 0.087 \text{ rad/s}$$

Then, substituting into the equation for the magnitude ratio, we get

$$\frac{AR}{0.8K_c} = 0.261$$

$$K_c \Big|_{PM=45°} = \frac{AR}{0.8(0.261)} = \frac{1}{0.8(0.261)}$$

$$K_c \Big|_{PM=45°} = 4.78$$

Example 7-13 shows how to obtain the tuning of the feedback controller for a certain GM and PM. In part (a) the controller was tuned to yield a control loop with a GM of 1.5. This means that the overall loop gain must increase (because of process nonlinearities, or for any other reason) by a factor of 1.5 before instability is reached. In choosing the value of *GM*, the engineer must understand the process to decide how much the process gain can change over the operating range. Based on this understanding a realistic GM value can then be used. The larger the GM used, the greater the "safety factor" designed into the control loop. However, the larger this safety factor (GM) is, the smaller the controller gain that results and therefore the less sensitive the controller is to errors. Thus, the selection of a realistic GM is important.

In part (b) of the example, the controller was tuned to yield a PM of 45°. This means that 45° of phase lag must be added to the control loop before it goes unstable. Changes in phase angle of the control loop are due mainly to changes in its dynamic terms (time constants and dead time), because of process nonlinearities.

Gain margin and phase margin are two different performance criteria. The choice of one of them as the criterion for a particular loop depends on the process being controlled. If due to the process nonlinearities and characteristics the gain is expected to change more than the dynamic terms, then the GM may be the indicated criterion. If, on the other hand, the dynamic terms are expected to change more than the gain, then the PM may be the indicated criterion.

Example 7-13 demonstrated how the gain of a proportional only controller is calculated to yield the desired performance specification. If a PI or PID controller is used, the reset time and rate time must first be set before K_c can be calculated. This means that more than one set of tuning parameters yields the desired performance. It is up to the engineer to choose what he or she considers the "best."

In this section we have learned the meaning of gain margin and phase margin, as well as how to tune feedback controllers based on these performance specifications. In the process industries, however, the performance specifications of Chapter 6 are preferred.

Polar Plots

The polar plot is another way to graph the frequency response of control systems. It has the advantage of being only one graph as opposed to two graphs as with Bode plots. The polar plot is a plot of the complex-valued function $G(i\omega)$ as ω goes from 0 to ∞. For every value of ω there will be a vector in the complex plane. The end of this vector will generate a locus as ω changes. The vector has its base at the origin and has length equal to the amplitude ratio of the $G(i\omega)$ function; its angle with the positive real axis is the phase angle. This section presents the fundamentals of polar plots and how to graph them. The polar plots of some of the most common process components are first presented.

First-Order Lag. The amplitude ratio and the phase angle of a first-order lag is given by Eqs. (7-27) and (7-28), respectively:

$$AR = \frac{K}{\sqrt{(\omega\tau)^2 + 1}} \tag{7-27}$$

$$\theta = -\tan^{-1}(\omega\tau) \tag{7-28}$$

For $\omega = 0$, AR $= K$ and $\theta = 0°$. For $\omega = \dfrac{1}{\tau}$, AR $= 0.707K$ and $\theta = -45°$. For $\omega = \infty$, AR $= 0$ and $\theta = -90°$.

The polar plot for this system is shown in Fig. 7-27. The solid curve represents the amplitude ratio and phase angle as the frequency goes from 0 to ∞. Each point in the curve represents a different ω. The length of the vector from the origin to a point on the curve is equal to the amplitude ratio at that ω. The angle that the vector makes with the real axis is equal to the phase angle. Figure 7-27 shows two vectors. The first one represents AR and θ for $\omega = 0$. The second vector represents AR and θ for $\omega = 1/\tau$. Notice that as AR approaches zero, θ approaches $-90°$; this is what Eqs. (7-27) and (7-28) indicate. The dotted curve is the plot of AR and θ as ω goes from $-\infty$ to 0.

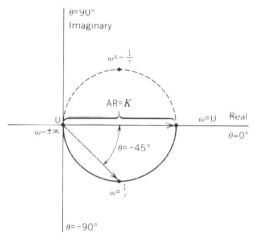

Figure 7-27. Polar plot of first-order lag.

Second-Order Lag. The AR and θ equations for a second-order lag are given by Eqs. (7-29) and (7-30), respectively.

$$AR = \frac{K}{\sqrt{(1 - \omega^2\tau^2)^2 + (2\tau\xi\omega)^2}} \tag{7-29}$$

and

$$\theta = -\tan^{-1}\left(\frac{2\tau\xi\omega}{1 - \omega^2\tau^2}\right) \tag{7-30}$$

For $\omega = 0$, AR $= K$ and $\theta = 0°$. For $\omega = 1/\tau$, AR $= K/2\xi$ and $\theta = -90°$. For $\omega = \infty$, AR $= 0$ and $\theta = -180°$.

Figure 7-28 shows the polar plot for this system. In this system AR approaches zero from the negative real axis because θ approaches $-180°$.

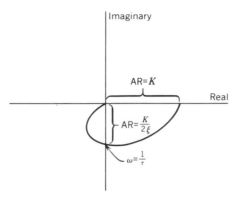

Figure 7-28.   Polar plot of second-order system.

Dead Time.   The AR and θ expressions for a pure dead-time system are given by Eqs. (7-34) and (7-35).

$$AR = 1 \tag{7-34}$$

and

$$\theta = -t_o\omega\left(\frac{180°}{\pi}\right) \tag{7-35}$$

These equations indicate that the vector will always have a magnitude of unity and that as ω increases the vector will start rotating. The resulting polar plot, shown in Fig. 7-29, is a unit circle.

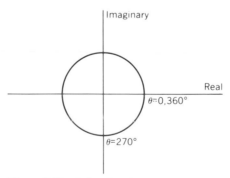

Figure 7-29.   Polar plot of pure dead time.

Conformal Mapping.   Some examples of polar plots have been shown. However, before continuing with this subject, it is important to talk about conformal mapping since the polar plots lean heavily on this theory. A brief introduction to conformal mapping is presented so that a better understanding of polar plots may be obtained.

Consider the general transfer function $G(s)$. As already learned, the variable s is the independent variable, which can be real, imaginary, or complex; that is, in general, $s =$

$\sigma + i\omega$. This variable s can be graphed in the s plane, as shown in Fig. 7-30a. Substituting the value of s into the transfer function $G(s)$, the value for this function at the given s can be obtained. This value of $G(s)$, which can also be real, imaginary, or complex, $G(s) = \delta + i\gamma$, can be graphed in the $G(s)$ plane, as shown in Fig. 7-30b For every point in the s plane there is a corresponding point in the $G(s)$ plane. It is said that the function G "maps" the s plane onto the $G(s)$ plane. The function G maps not only points but also paths or regions.

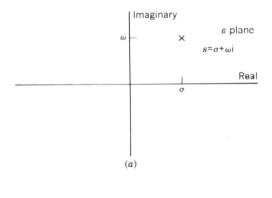

(a)

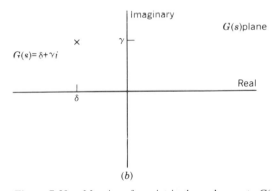

(b)

Figure 7-30. Mapping of a point in the s plane onto $G(s)$ plane.

The word "conformal" is used because, in mapping, the $G(s)$ plane "conforms" with the s plane. To explain what is meant by this, suppose that a path from the s plane must be mapped onto the $G(s)$ plane. Further, if the path in the s plane has a sharp turn, the mapped path in the $G(s)$ plane will also have a sharp turn. That is, the $G(s)$ plane "conforms" with the s plane.

To be a bit more specific about this conformal mapping, consider the following transfer function:

$$G(s) = \frac{10}{(2s + 1)(4s + 1)}$$

Figure 7-31 shows the mapping of a region in the s plane, given by points 1-2-3-4-1, onto the $G(s)$ plane, given by points $1'$-$2'$-$3'$-$4'$-$1'$. Table 7-1 shows the mathematical manipulations to produce the mapping.

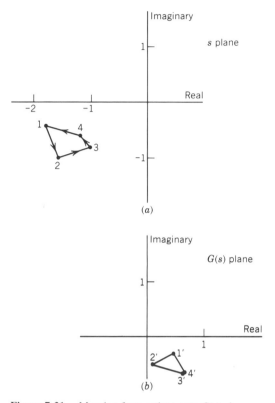

Figure 7-31. Mapping from s plane onto $G(s)$ plane.

This has been a brief explanation of conformal mapping. As mentioned earlier, polar plots are based on this theory. Conformal mapping will also become important when the use of polar plots to study process control stability is presented.

The Nyquist Stability Criterion[1]. The polar plots are most useful in the stability analysis of process control loops. This section presents the Nyquist stability criterion, which makes use of these plots. No proof of the theorem is given; however, the reader is strongly encouraged to read the original paper. Since this criterion is based on the use of polar plots, these are also often referred to as Nyquist plots.

The Nyquist criterion may be stated as follows:

A closed loop control system is stable if the region R, (consisting of the entire right half of the s plane including the imaginary axis), when mapped onto the $G(s)$ plane,

Table 7-1

Point	Coordinate	$2s + 1$	$4s + 1$	$(2s + 1)(4s + 1)$	$G(s)$	Point in $G(s)$ Plane
1	$-1.8 - 0.4i$	$-2.6 - 0.8i$	$-6.2 - 1.6i$	$14.84 + 9.12i$	$\dfrac{10}{(14.84 + 9.12i)} = 0.49 - 0.3i$	$1'$
2	$-1.6i$	$-2.2 - 2i$	$-5.4 - 4i$	$3.88 + 19.6i$	$\dfrac{10}{3.88 + 19.6i} = 0.10 - 0.49i$	$2'$
3	$-1 - 0.8i$	$-1 - 1.6i$	$-3 - 3.2i$	$3.12 + 8i$	$\dfrac{10}{8.12 + 8i} = 0.62 - 0.61i$	$3'$
4	$-1.2 - 0.6i$	$-1.4 - 1.2i$	$-3.8 - 2.4i$	$3.2 + 7.92i$	$\dfrac{10}{8.2 - 7.92i} = 0.63 - 0.61i$	$4'$

the open-loop transfer function plane results in region R', which does not include the point $(-1,0)$.

The following example demonstrates the application of this criterion.

Example 7-14.   Consider the heat exchanger control loop presented in Chapter 6 and used in example 7-4. The closed-loop block diagram for this control loop is shown in Fig. 7-9. The open-loop transfer function is

$$\text{OLTF} = \frac{0.8\,K_c}{(10s + 1)(30s + 1)(3s + 1)}$$

The Nyquist criterion requires us to map the entire right-hand plane (RHP) of the s plane, as shown in Fig. 7-32, onto the $G(s)$ plane. To best show this the mapping is divided into three steps.

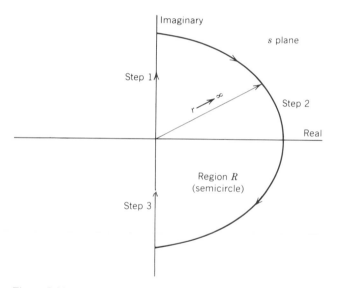

Figure 7-32.   Region R of s plane.

Step 1.

Frequency ω goes from 0 to ∞ on the positive imaginary axis, thus, $s = i\omega$. Substituting this expression for s into the $G(s)$ expression yields

$$G(s) = \frac{0.8\,K_c}{(i10\omega + 1)(i30\omega + 1)(i3\omega + 1)}$$

The plot starts, $\omega = 0$, on the positive real axis at the value $0.8K_c$ and terminates, $\omega \rightarrow \infty$, at the origin (AR = 0) with a phase angle of $-270°$, as shown in Fig. 7-33. The frequency at which the locus crosses the $-180°$ angle can readily be found by solving

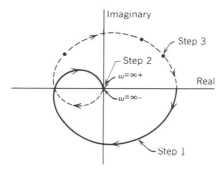

Figure 7-33. Polar plot of $G(s) = (0.8K_c)/[10s + 1)(30s + 1)(3s + 1)]$.

Eq. (7-52) for ω. Once this value of ω is obtained, the value of AR is calculated by using Eq. (7-51).

Step 2.

In this step the frequency moves from $\omega = \infty$ to $\omega = -\infty$ along the path shown in Fig. 7-32. Along this path, $s = re^{i\sigma}$, with σ going from 90° to 0° to $-90°$. Substituting this expression for s into the $G(s)$ expression gives

$$G(s) = \frac{0.8 K_c}{(10re^{i\sigma} + 1)(30re^{i\sigma} + 1)(3re^{i\sigma} + 1)}$$

Since $r \to \infty$ then the $+1$ term in each of the parentheses can be neglected and

$$\lim_{r \to \infty} G(i\omega) = \lim_{r \to \infty} \left[\frac{0.8 K_c}{(10re^{i\sigma})(30re^{i\sigma})(3re^{i\sigma})} \right]$$

$$\lim_{r \to \infty} G(i\omega) = \lim_{r \to \infty} \left[\frac{0.8 K_c}{900r^3 e^{i3\sigma}} \right] = 0$$

which says that the semicircle with $r \to \infty$ maps in the $G(s)$ plane as a point at the origin.

Step 3.

In this step the frequency moves from $-\infty$ to 0 along the negative imaginary axis; thus $s = -i\omega$. Again, substituting this expression for s into the $G(s)$ expression yields

$$G(i\omega) = \frac{0.8K_c}{(-i10\omega + 1)(-i30\omega + 1)(-i3\omega + 1)}$$

The plot starts at the origin, $\omega = -\infty$, and terminates, $\omega = 0$, on the positive real axis at the value of $0.8K_c$ with a phase angle of 0°. The path is shown in Fig. 7-33.

It is interesting to note that Step 3 is just the "mirror image" of Step 1. This is understandable if one realizes that $G(s)$ is a complex conjugate function, which means that the map is symmetical around the abscissa axis in the $G(s)$ plane.

Step 1 explained how to obtain the value of AR for a phase angle of $-180°$. This AR is the distance from the origin at which the path crosses the negative real axis. If this cross point is less than -1, then the system is stable [the mapped region does not include point $(-1,0)$]. If the cross point is greater than -1, then the system is unstable [the mapped region includes point $(-1,0)$]. As K_c increases, AR also increases, resulting in a less stable system. This is the same statement as the frequency response stability criterion.

This section has presented a brief introduction to polar plots and the Nyquist stability criterion. The reader has no doubt noticed the equivalence between the Nyquist and the frequency response stability criteria.

Nichols Plots

The Nichols plot is still another way to graphically represent the frequency response of systems. Essentially, it is a plot of the amplitude ratio (or magnitude ratio) versus phase angle. Fig. 7-34 shows this type of plot for some typical systems. In these plots, frequency is the parameter along the curve.

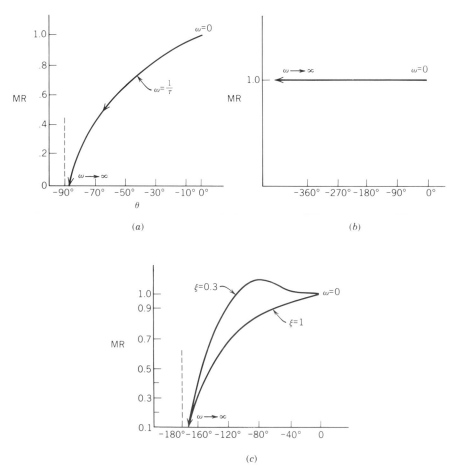

Figure 7-34. Nichols plots. (*a*) First-order lag. (*b*) Dead time. (*c*) Second-order lag.

Frequency Response Summary

This presentation on frequency response has been rather extensive and hopefully has shown the power of the technique. Equations (7-38) and (7-39) are general equations to determine AR and θ. The use of these equations allows one to perform the analysis and design of control systems. Different ways to graph the frequency response of systems were shown.

7-3. PULSE TESTING

One of the most practical and interesting applications of frequency response is the use of pulse testing for determining the transfer functions of existing processes, instruments, and other control devices. A number of industrial applications of pulse testing are described by Hougen[2]. In this section we will describe the technique and derive the basic formulas required for its application.

In Chapter 6 we studied the method of step testing for determining the parameters of a first-order plus dead-time model of the process. The advantages of step testing are simplicity and minimal computation requirements. Its major disadvantage is that it is limited by accuracy to first-order plus dead-time models.

The technique of sinusoidal testing described in Section 7-2 can in principle determine the transfer function of a process of any order. Although it is used extensively to determine the transfer functions of sensors, transmitters, and control valve actuators, sinusoidal testing can seldom be used to test actual processes. The reason is that most processes are too slow for sinusoidal testing. Consider a process in which the longest time constant is one minute. The breakpoint frequency on the Bode plot of such a process is, from Eq. (7-42), at

$$\omega_c = \frac{1}{\tau} = 1.0 \text{ rad/min}$$

In order to locate the low-frequency asymptote of the process, we must perform at least two tests at lower frequencies than the breakpoint frequency, say 0.5 and 0.25 rad/min. The second of these sinusoidal signals has a period of

$$T = \frac{2\pi}{\omega} = \frac{6.28 \text{ rad}}{0.25 \text{ rad/min}} = 25.1 \text{ min!}$$

It is not only difficult to find a sine-wave generator capable of consistently generating such a slow signal, but the test would take at least a couple of hours, given that a minimum of four or five cycles are necessary to complete the test. Furthermore, such a test would give us only one point on the Bode plot. If our process time constant were 10 minutes, a sine wave with a period of over 4 hours would have to be applied to the process.

Pulse testing produces a complete Bode plot of the process from a single test lasting considerably less than the test described in the preceding paragraph. Given that we cannot get something for nothing, we must pay for the savings in testing effort with additional computational effort.

Performing the Pulse Test

The diagram for the pulse test is identical to the one given for the sinusoidal test in Fig. 7-15, except that instead of the sinusoidal signal and response of Fig. 7-16, the input signal is a pulse such as the one shown in Fig. 7-35. Notice that the duration of the response, T_F, is longer than that of the pulse T_D. The three parameters to be selected in order to perform the test are the shape of the pulse, its amplitude, and its duration.

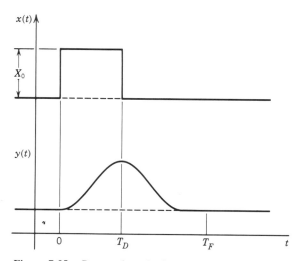

Figure 7-35.   Rectangular pulse input and output response.

Although the rectangular pulse of Fig. 7-35 is the easiest to generate and to analyze, other pulse shapes, such as those shown in Fig. 7-36, can be used. The rectangular pulse is predictably the most popular, followed by the rectangular doublet of Fig. 7-36b. The only requirement for the pulse shape is that it return to the initial steady-state value.

As in the case of step and sinusoidal testing, the amplitude of the pulse, X_0, must be large enough for the measurements of the response to be accurate, but not so large that the response gets outside the range within which the linear transfer function is a valid approximation of the process response. This requirement usually necessitates a very sensitive recorder or an on-line digital computer for recording the response.

The duration T_D of the pulse depends entirely on the time constants of the process being tested. The pulse should not be of such short duration that the process does not have time to react or so long that the response has time to reach steady state before the pulse is completed. Such a long pulse not only represents a waste of test time but also results in a reduction of the highest frequency for which the test results are useful, as we shall see shortly.

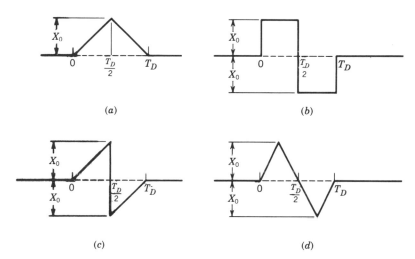

Figure 7-36. Common pulse shapes for process dynamic testing. (a) Triangular pulse. (b) Rectangular doublet. (c) Ramp. (d) Triangular doublet.

Derivation of the Working Equation

The frequency response of the tested process is determined by calculating the complex transfer function $G(i\omega)$, as a function of frequency, from the process response to the pulse input. In order to do this we must make use of the definition of the *Fourier transform of a signal*:

$$F(i\omega) = \int_{-\infty}^{\infty} f(t)\, e^{-i\omega t}\, dt \tag{7-55}$$

By comparison of Eq. (7-55) with Eq. (2-1) we see that, except for the lower limit on the integral, the Fourier transform can be obtained by substituting $s = i\omega$ in the definition of the Laplace transform. Since the signals of interest in process control are deviations from the initial steady-state value and thus zero for negative time, the lower limit on the integral of Eq. (7-55) can be changed to zero. The Fourier transform was developed before the Laplace transform as an extension of the Fourier series to nonperiodic signals.

By definition of the process transfer function (see Chapter 3)

$$G(s) = \frac{Y(s)}{X(s)} \tag{7-56}$$

where

Y(s) is the Laplace transform of the process response (as a deviation from its initial steady-state value)

X(s) is the Laplace transform of the pulse

Substitute $s = i\omega$ to find

$$G(i\omega) = \frac{Y(i\omega)}{X(i\omega)} \tag{7-57}$$

and apply Eq. (7-55) to both signals to obtain

$$G(i\omega) = \frac{\int_0^\infty y(t) e^{-i\omega t} \, dt}{\int_0^\infty x(t) e^{-i\omega t} \, dt} \tag{7-58}$$

Equation (7-58) is our working equation for calculating the frequency response of the process being tested. From the response $y(t)$ to a single pulse $x(t)$ we can calculate, for each value of frequency ω of interest, the integrals in the numerator and the denominator of Eq. (7-58). The result of the calculation is a single complex number $G(i\omega)$. The magnitude of this number is then, from Eq. (7-24), the amplitude ratio, and its argument is, from Eq. (7-25), the phase angle, at frequency ω. By repeating the calculations for a number of values of ω, the entire Bode plot can be generated from the results of a single test.

The Fourier transform integral of the process output singal $y(t)$ must be calculated numerically, as we will see shortly. Although the integral of the pulse $x(t)$ can also be calculated numerically, an analytical formula can be derived from Eq. (7-55). This is illustrated in the following example.

Example 7-15. Derive the Fourier transform of a rectangular pulse of amplitude X_0 and duration T_D (see Fig. 7-35).

Solution. From Eq. (7-55) we know that

$$X(i\omega) = \int_0^\infty x(t) e^{-i\omega t} \, dt$$

Since the pulse is zero at all times except between 0 and T_D, we can say that

$$X(i\omega) = \int_0^{T_D} X_0 e^{-i\omega t} \, dt = -\frac{X_0}{i\omega} [e^{-i\omega t}]_0^{T_D}$$

$$= \frac{X_0}{i\omega} [1 - e^{-i\omega T_D}]$$

$$= \frac{X_0}{\omega} [\sin \omega T_D - i(1 - \cos \omega T_D)]$$

where we have made use of the identity

$$e^{-i\omega T_D} = \cos \omega T_D - i \sin \omega T_D$$

and $1/i = -i$. The magnitude and arguments of $X(i\omega)$ are

$$|X(i\omega)| = \frac{X_0}{\omega} \sqrt{\sin^2 \omega T_D + (1 - \cos \omega T_D)^2}$$

$$= \frac{X_0}{\omega} \sqrt{2(1 - \cos \omega T_D)}$$

$$\angle X(i\omega) = \tan^{-1}\left(-\frac{1 - \cos \omega T_D}{\sin \omega T_D}\right)$$

At $\omega = 0$, $X(0) = X_0 T_D$.

The magnitude of the pulse is a maximum at $\omega = 0$ and then drops to zero as the frequency increases to infinity. The magnitude is also zero at values of the frequency ω that are multiples of $2\pi/T_D$ radians/time. These values occur more frequently the longer the pulse duration, T_D.

The preceding example shows that the maximum magnitude of the Fourier transform of the rectangular pulse is proportional to the area of the pulse, $X_0 T_D$. Since the Fourier transform of the pulse appears in the denominator of Eq. (7-57), we want to avoid values of ω at which the transform is zero. This imposes an upper limit on the range of frequencies for which the frequency response can be calculated from the pulse test. For the rectangular pulse, the upper limit on the frequency is $2\pi/T_D$, which, as mentioned earlier, decreases with the duration of the pulse, T_D.

Numerical Evaluation of the Fourier Transform Integral

The evaluation of the integral in the numerator of Eq. (7-58) is normally carried out numerically using a digital computer. An efficient and accurate technique for accomplishing the numerical integration is based on the trapezoidal rule. The calculations must be performed using complex-number arithmetic (see Section 2-4) because of the complex exponential term. Most modern computer programming languages such as FORTRAN are capable of performing complex-number operations.

In order to perform the numerical integration of the function $y(t)$, we must first divide it into N time increments of Δt as shown in Fig. 7-37. We shall assume that the increments are of uniform duration of Δt, although this is not necessary in practice. The trapezoidal rule for numerical integration consists of approximating the function by a straight line within each interval:

$$Y(i\omega) = \int_0^\infty y(t) e^{-i\omega t} \, dt$$

$$\doteq \sum_{k=0}^{N-1} \int_0^{\Delta t} \left[y_k + \frac{t}{\Delta t}(y_{k+1} - y_k) \right] e^{-\omega(t + k\Delta t)} \, dt \qquad (7\text{-}59)$$

where

y_k is the response at time $k\Delta t$

$N = T_F/\Delta t$ is the number of increments.

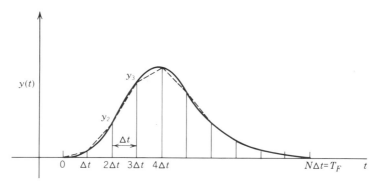

Figure 7-37. Discretization of response pulse for numerical calculation of Fourier transform integral.

In writing Eq. (7-59) we have assumed that the response is zero for negative time and returns to zero after N increments. We can do this because y is the deviation from the initial steady-state value. Integrating Eq. (7-59) by parts and simplifying, we obtain

$$Y(i\omega) = \sum_{k=0}^{N-1} \frac{e^{-i\omega k \Delta t}}{\omega^2 \Delta t} \{y_k(1 - e^{-i\omega \Delta t} - i\omega \Delta t)$$
$$+ y_{k+1}[(1 + i\omega \Delta t)e^{-i\omega \Delta t} - 1]\} \tag{7-60}$$

This formula may be simplified further by making use of the fact that y_0 and y_N are zero:

$$Y(i\omega) = \frac{(2 - e^{i\omega \Delta t} - e^{-i\omega \Delta t})}{\omega^2 \Delta t} \sum_{k=1}^{N-1} y_k e^{-i\omega k \Delta t} \tag{7-61}$$

The following working formula results from the appropriate trigonometric identities:

$$Y(i\omega) = \frac{4 \sin^2 \left(\dfrac{\omega \Delta t}{2}\right)}{\omega^2 \Delta t} \sum_{k=1}^{N-1} y_k e^{-i\omega k \Delta t} \tag{7-62}$$

This formula can be programmed directly in FORTRAN or in another high-level computer language. Equation (7-62) is very efficient because the time-consuming evaluation of the sine function need be performed only once for each frequency of interest.

When the Fourier transforms of both the output response and the input pulse are calculated numerically, substitution of Eq. (7-62) for both the numerator and denominator of Eq. (7-57) results in the following formula for the process transfer function:

$$G(i\omega) = \frac{\displaystyle\sum_{k=1}^{N-1} y_k e^{-i\omega k \Delta t}}{\displaystyle\sum_{j=1}^{M-1} x_j e^{-i\omega j \Delta t}} \tag{7-63}$$

where M is the number of increments used to integrate the pulse ($M = T_D/\Delta t$). Equation (7-63) assumes that the same integration interval is used for the numerator as for the denominator. When the Fourier transform of the pulse can be calculated from an analytical formula such as the one derived in Example 7-15, it is usually more efficient to do so than to use Eq. (7-63). In such a case the Fourier transform of the output response is calculated from Eq. (7-62) and the transfer function is calculated from Eq. (7-57). These calculations must be repeated at a number of values of the frequency ω in order to generate the Bode plot of the process. Since we want the points on the Bode plot to be equally spaced in the logarithmic frequency axis, each frequency selected must be a constant factor of the previous one:

$$\omega_i = \beta \omega_{i-1} \qquad \text{for} \quad i = 1, 2, \ldots, N_\omega \qquad (7\text{-}64)$$

where

$$\beta = \left(\frac{\omega_{max}}{\omega_{min}}\right)^{1/N_\omega}$$

ω_{max} is the upper limit of the frequency range on the Bode plot, rad/time
ω_{min} is the lower limit of the frequency range on the Bode plot, rad/time
N_ω is the number of increments into which the frequency range is divided

The procedure just described allows us to obtain the complete Bode plot from a single pulse test provided that both the input pulse and the output response return to their initial values. Since the variables $y(t)$ and $x(t)$ are the *deviations* from the initial steady-state conditions, they are initially zero and, in order for $y(t)$ to return to zero, the final steady-state conditions must be the same as the initial ones. This latter requirement is not met when the process contains an integration element. We will look at this case next.

Processes with Integration. The response of a process with integration to a pulse input is shown in Fig. 7-38. The final steady-state value of the output deviation variable is proportional to the integral of the input pulse:

$$y_\infty = K_I \int_0^{T_D} x(t) \, dt \qquad (7\text{-}65)$$

Luyben[3] proposes that the process be postulated as consisting of a pure integrator with gain K_I in parallel with a fictitious process with transfer function $G_A(s)$, so that the actual process transfer function is

$$G(s) = G_A(s) + \frac{K_I}{s} \qquad (7\text{-}66)$$

The output of the integrator is given by

$$y_I(t) = \begin{cases} K_I \int_0^t x(t) \, dt & 0 \le t < T_D \\ y_\infty & t \ge T_D \end{cases} \qquad (7\text{-}67)$$

The output of the fictitious process is then given by

$$y_A(t) = y(t) - y_I(t) \qquad (7\text{-}68)$$

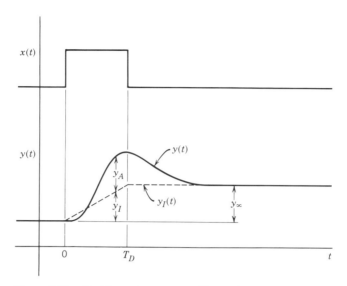

Figure 7-38.   Handling pulse response of integrating processes.

The signal $y_A(t)$ is zero at both the initial and final times, as illustrated in Fig. 7-38. We can calculate $y_A(t)$ from Eqs. (7-67) and (7-68) and then use the result in Eqs. (7-62) and (7-57) to calculate $G_A(i\omega)$. Then, from Eq. (7-66), the process transfer function is

$$G(i\omega) = G_A(i\omega) + \frac{K_I}{i\omega} \qquad (7\text{-}69)$$

The gain K_I of the integrator is calculated from Eq. (7-65).

In this section we have outlined the pulse testing method for experimentally determining the transfer function of a process. A computer program for generating Bode plot data from pulse response data is given by Luyben[3]. A number of successful industrial applications of pulse testing are presented by Hougen[2].

7-4. SUMMARY

In this chapter we have presented two techniques for the analysis and design of feedback control systems: root locus and frequency response. Both of these are excellent tools for the graphical representation of the response of control systems. Of the two, frequency response can handle the presence of dead time directly, without approximation. Frequency response also serves as a basis for the quantitative determination of process dynamic parameters through the pulse testing method. As we saw in this chapter, the entire frequency response of a process or instrument can be determined from the results of a single pulse test.

Having studied the design and analysis of feedback control systems, we will next look into other important control techniques that are commonly used in industry. This is the subject of our next chapter.

REFERENCES

1. Nyquist, H., "Regeneration Theory," *Bell System Technical Journal*, Vol. 11 (1932), pp. 126–147.
2. Hougen, Joel O., "Experiences and Experiments with Process Dynamics," *CEP Monograph Services*, Vol. 60, No. 4, AIChE, New York, 1964.
3. Luyben, William L., *Process Modeling, Simulation, and Control for Chemical Engineers*, McGraw-Hill, New York, 1973, Section 9-3.

PROBLEMS

7-1. Draw the root locus diagram for each of the following open-loop transfer functions (use a Padé approximation for the dead time term):

    a. $G(s) = \dfrac{K}{(s + 1)(2s + 1)(10s + 1)}$

    b. $G(s) = \dfrac{K(3s + 1)}{(s + 1)(2s + 1)(10s + 1)}$

    c. $G(s) = \dfrac{Ke^{-s}}{(s + 1)(2s + 1)(10s + 1)}$

    d. $G(s) = \dfrac{K(3s + 1)e^{-s}}{(s + 1)(2s + 1)(10s + 1)}$

7-2. Draw the root locus diagram for each of the following open-loop transfer functions:

    a. $G(s) = \dfrac{K}{s(s + 1)(4s + 1)}$

    b. $G(s) = \dfrac{K(3s + 1)}{(2s + 1)}$

    c. $G(s) = K\left(1 + \dfrac{1}{\tau_I s}\right)\left(\dfrac{\tau_D s + 1}{\alpha\tau_D s + 1}\right)$

7-3. Consider the following transfer function of a certain process:

$$G(s)H(s) = \dfrac{1.5}{(s + 1)(5s + 1)(10s + 1)}$$

    a.   Tune a proportional only controller for a gain margin of 2.

    b.   Determine the damping factor, ξ, of the oscillatory response of the control loop using the tuning parameter obtained in part (a).

7-4. Draw the root locus diagram for the following two open-loop transfer functions.

    a.   System with inverse response:

$$G(s) = \dfrac{K(1 - 0.25s)}{(2s + 1)(s + 1)}$$

b. Open-loop unstable system:

$$G(s) = \frac{K}{(\tau_1 s + 1)(1 - \tau_2 s)}$$

for two cases:

$$\tau_1 = 2, \qquad \tau_2 = 1$$

and

$$\tau_1 = 1, \qquad \tau_2 = 1$$

7-5. Consider the pressure control system shown in Fig. 7-39. The pressure in the tank can be described by

$$\frac{P(s)}{F(s)} = \frac{0.4}{(0.15s + 1)(0.8s + 1)} \text{ psi/scfm}$$

The valve can be described by the following transfer function:

$$\frac{F(s)}{M(s)} = \frac{5}{0.1s + 1} \text{ scfm/psi}$$

The pressure transmitter has a range of 0 psig to 30 psig. The dynamics of the transmitter are negligible.

a. Draw the block diagram for this system including all the transfer functions.
b. Sketch the root locus diagram.
c. Determine the gain of the controller at the breakaway point.
d. Determine the ultimate gain and ultimate period.
e. Calculate the tuning of a P controller so as to obtain a damping factor of 0.707.
f. Explain graphically how the addition of reset action to the controller affects the stability of the control loop.
g. Explain graphically how the addition of rate action to the PI controller of part (f) affects the stability of the control loop.

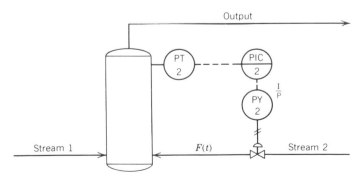

Figure 7-39. Sketch for Problem 7-5.

7-6. Repeat Problem 7-3 for the following transfer function:

$$G(s)H(s) = \frac{0.8e^{-s}}{10s + 1}$$

7-7. For the process shown in Problem 6-20, sketch the root locus diagram. Use a first-order Padé approximation, as given by Eq. (6-29), to approximate the dead-time term.

7-8. Draw the asymptotes of the Bode plot magnitude ratio (or amplitude ratio) and roughly sketch the phase angle plot for the transfer functions given in Problem 7-1.

7-9. Repeat Problem 7-8 for the transfer functions given in Problem 7-2.

7-10. Given the following transfer functions:

$$G(s) = \frac{2s + 1}{s + 1} \quad \text{and} \quad G(s) = \frac{1 + s}{(2s + 1)(s + 1)}$$

   a. Sketch the asymptotes of the magnitude ratio part of the Bode plot, marking the breakpoint frequencies.
   b. Indicate the phase lag (or lead) at high frequencies ($\omega \to \infty$).

7-11. Sketch the Bode plot of the transfer functions given in Problem 7-4.

7-12. On Fig. 7-40 the Bode plot of an open-loop system is shown. Obtain the transfer function for this system What controller gain can be tolerated if a gain margin of 2 is desired? What is the phase margin with a controller gain of 0.6?

7-13. Consider the vacuum filter process shown in Fig. 6-15. For this process using the data given in Problem 6-15 and applying frequency response techniques:

   a. Sketch the asymptotes of the Bode plot and the phase angle plot.
   b. Obtain the ultimate gain, K_{cu}, and ultimate period, T_u.
   c. Tune the reset time of a PI controller by the controller synthesis method and determine the controller gain that would provide a gain margin of 2.

7-14. Consider the absorber presented in Problem 6-16. In part (a) of the problem, a feedback control loop was designed to control the exit concentration of ammonia. For this control loop:

   a. Sketch the asymptotes of the Bode plot and the phase angle plot.
   b. Obtain the ultimate gain, K_{cu}, and ultimate period, T_u.
   c. Tune a P controller for a phase margin of 45°.

7-15. Consider the block diagram shown in Fig. 7-41a. The input $N(s)$ represents noise that corrupts the output signal. If this process noise is significant, the control of the process may be difficult. To improve the control of noisy processes, filtering the feedback signal is usually done. A typical way to filter signals is by a filter

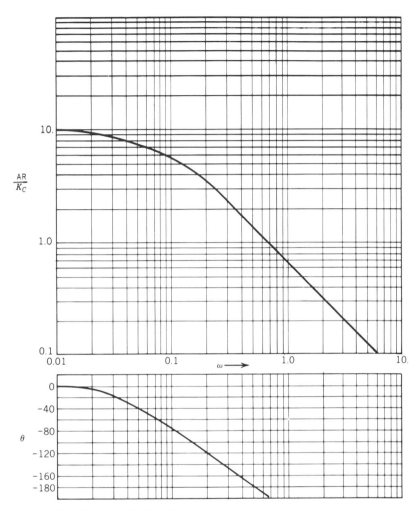

Figure 7-40.   Sketch for Problem 7-12.

device with a first-order transfer function. This device, either pneumatic, electronic, or digital, is installed between the transmitter and the controller as shown in Fig. 7-41*b*. The gain of the filter is one (1) and its time constant, called the filter time constant, is τ_F. Using frequency response techniques explain how τ_F affects the filtering of the noisy signal and the performance of the control loop. Specifically, plot the gain margin as a function of τ_F.

7-16.   Consider a thermal process with the following transfer fucntion for the process output versus controller output signal:

$$\frac{C(s)}{M(s)} = \frac{0.65\,e^{-0.35s}}{(5.1s + 1)(1.2s + 1)}$$

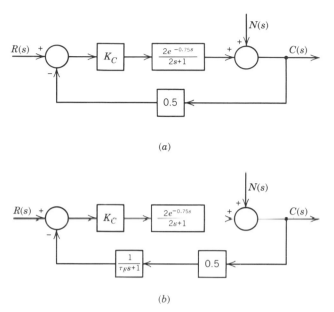

(a)

(b)

Figure 7-41. Sketch for Problem 7-15.

A sine wave of unity amplitude and a frequency of 0.80 rad/min is applied to the process (time constants and dead time are in minutes). Calculate the amplitude and phase lag of the sine wave out of the process (after the transient response dies out).

7-17. The symmetrical rectangular pulse shown in Fig. 7-42 has the advantage of averaging out the effect of nonlinearities on the result of the dynamic test.

   a. Derive the Fourier transform of the pulse, $Q(i\omega)$.

   b. Write the formulas for the magnitude, $|Q(i\omega)|$, and phase angle, $\angle Q(i\omega)$, as functions of frequency.

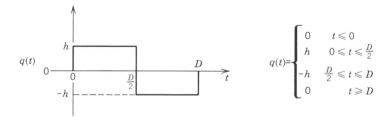

Figure 7-42. Sketch for Problem 7-17.

7-18. A ramp pulse of duration T_D and final amplitude x_0 is to be used in a pulse test of a process. Determine the Fourier transform integral of the pulse.

7-19.   The amplitude ratio versus frequency plot for a process results in the sketch given in Fig. 7-43. The phase angle plot does not reach a high-frequency asymptote, but becomes more negative as the frequency increases. At a frequency of 1.0 rad/min, the phase angle is -246 degrees. Postulate a transfer function for the process and estimate the gain, the time constants, and the dead time (if any).

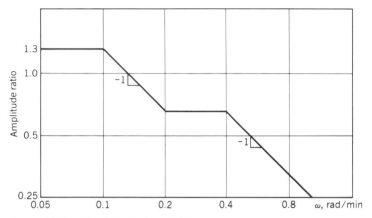

Figure 7-43.   Sketch for Problem 7-19.

7-20.   The Bode diagram shown in Fig. 7-44 is obtained for the transfer function of a tubular reactor temperature to the cooling water rate by the pulse testing method. Determine:

a.   The steady-state process gain.
b.   The time constant.
c.   An estimate of the dead time of the system.

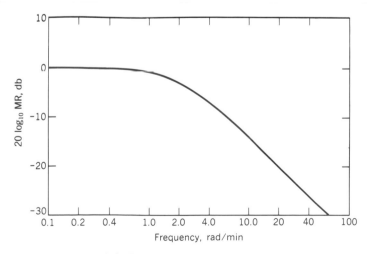

Figure 7-44.   Sketch for Problem 7-20.

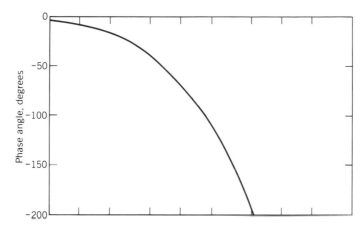

Figure 7-44. (*Continued*)

7-21. Derive the Fourier transforms, their magnitude ratio, and phase angle for the four pulses sketched in Fig. 7-35.

CHAPTER
8

Additional Control Techniques

The main emphasis of the previous chapters has been on the control of processes by the technique known as feedback control. Chapter 1 presented the principles of this technique along with some of its advantages and disadvantages. Chapter 5 presented the basic components and hardware necessary to implement a control system. Finally, in Chapters 6 and 7 different techniques to design and analyze single-loop feedback control systems were presented, discussed, and practiced. As mentioned in Chapter 1, feedback control is the most common technique used in the process industries.

In many processes it is possible and profitable to improve the performance attained with feedback control by applying other control techniques. The purpose of this chapter is to present some of the techniques that have been developed, and often used, to improve on the control performance provided by feedback control. These additional techniques require a greater amount of engineering and equipment than feedback control alone. Consequently, we will assume that an economic and technical feasibility study has been performed before these techniques are applied.

The techniques that will be presented in this chapter are ratio control, cascade control, feedforward control, override control, selective control, and multivariable control. Many actual industrial examples are shown throughout the chapter to aid the reader in the understanding of the principles and application of the techniques.

The implementation of the techniques presented in this chapter requires some amount of computing power. This required computing power has been obtained in the past by the use of computing relays, either pneumatic or electrical. In the last several years the advent of micro, mini, or large-scale computers has replaced many of these analog computing relays. We start the chapter with a discussion of analog computing relays and of microprocessor-based computing blocks.

8-1. COMPUTING RELAYS

As already mentioned, the implementation of the techniques presented in this chapter requires some amount of computing power. This required computing power is usually obtained by the use of computing relays, either pneumatic or electrical, or microprocessor-based computing blocks. The relays are black boxes that perform some mathematical manipulation of the signals. The blocks are the software equivalent of the computing relays. Some typical manipulations that can be performed by these analog computing relays or computing blocks are as follows:

1. Addition/subtraction. The output signal is obtained by adding and/or subtracting the input signals.

2. Multiplication/division. The output signal is obtained by multiplying and/or dividing the input signals.

3. Square root. The output signal is obtained by extracting the square root of the input signal.

4. High/low selector. The output signal is the highest/lowest of two or more input signals.

5. High/low limiter. The output signal is the input signal limited to some preset high/low limit value.

6. Function generator. The output signal is a function of the input signal. This function is usually approximated by a series of straight lines.

7. Integrator. The output signal is the time integral of the input signal. Another term used for integrator is "totalizer."

8. Linear lag. The output signal is the solution of a first-order differential equation for which the forcing function is the input signal. Mathematically, this calculation is described as follows:

$$\text{Output} = \left(\frac{1}{\tau s + 1} \right) \text{Input}$$

This calculation is often used as a filter for a noisy signal. The amount of filtering depends on the time constant, τ. The larger the time constant the more filtering is done.

9. Lead/lag. The output signal is the solution to the following transfer function:

$$\text{Output} = \left(\frac{\tau_{ld} s + 1}{\tau_{lg} s + 1} \right) \text{Input}$$

This equation is the same as Eq. (4-78); its behavior and the meaning of τ_{ld} and τ_{lg} were explained in Chapter 4. This calculation is often used in control schemes, such as feed-forward control, in which dynamic compensation is required.

With the advent of microprocessor-based control systems, the availability of more complex mathematical manipulations has increased tremendously. Table 8-1 shows some of the equations solved by the electronic computing relays of one manufacturer. Tables 8-2 and 8-3 show some of the equations solved by two different microprocessor-based control systems. It is interesting to notice that in all cases the constants are limited between some preset values. Since the constants are limited, they must be chosen such that the required mathematical calculation can be correctly performed. This section shows how to select, also called "scaling," the values of the constants. This scaling is also necessary to ensure compatibility between input and output signals.

Table 8-1. Computing Relays[a]

The signals in this table are given in fraction of span.

V_0 = output signal
V_1, V_2, V_3 = input signals

Add/subtract

$$V_0 = a_0(\pm a_1V_1 + V_2 \pm a_3V_3 \pm a_4V_4) + B_0$$

where

V_2 = reference input (0 to 1)
a_1, a_3, a_4 = 0.11 to 1.0
a_0 = 0.2 to 8.5
B_0 = −0.5 to 1.0

Multiply

$$V_0 = 4a_0(V_1V_2) + B_0$$

Divide

$$V_0 = \frac{a_0}{4}\left(\frac{V_1}{V_2}\right) + B_0$$

Multiply & Divide

$$V_0 = a_0\left(\frac{V_1V_2}{V_3}\right) + B_0$$

where (for the previous three relays)

a_0 = 0.1 to 8.0
B_0 = 0 to 0.5

Ratio Station

$$V_0 = \pm R(V_1 - B_i) + B_o$$

where

R = 0.3 to 3.0
B_i = input bias, 0 to 1
B_o = output bias, 0 to 1

Lead/lag

$$V_0 = a_0\left(\frac{1 + T_2s}{1 + T_1s}\right)(V_1 - B_i) + B_o$$

where

$$a_0 = 0.1 \text{ to } 2$$
$$B_i = 0 \text{ to } 1$$
$$B_o = 0 \text{ to } 1$$
$$T_1 = \text{lag time constant, seconds (0 to 3000)}$$
$$T_2 = \text{lead time constant, seconds (0 to 3000)}$$

Square Root

$$V_0 = \sqrt{V_1}$$

[a]These are examples of electronic computing relays marketed by Fisher Controls.

Table 8-2. Computing Blocks[a]

The signals in this table are given in percent of span.

OUT = output signal
X, Y, Z = input signals

Summer

$$\text{OUT} = K_x X + K_y Y + K_z Z + B_0$$

where

$$K_x, K_y, K_z = -9.999 \text{ to } +9.999$$
$$B_0 = -100\% \text{ to } +100\%$$

Multiplier/divider

$$\text{OUT} = \frac{K_A(K_x X + B_x)(K_y Y + B_y)}{(K_z Z + B_z)} + B_0$$

where

$$K_A = 0 \text{ to } 2$$
$$K_x, K_y, K_z = 0.1 \text{ to } 9.999$$
$$B_x, B_y, B_z, B_0 = -100\% \text{ to } 100\%$$

Sum of Square Roots

$$\text{OUT} = K_x \sqrt{X} + K_y \sqrt{Y} + K_z \sqrt{Z} + B_0$$

Square Root of Product

$$\text{OUT} = K_A \sqrt{X \cdot Y \cdot Z} + B_0$$

where (for both square roots)

$$K_A, K_x, K_y, K_z = -9.999 \text{ to } +9.999$$
$$B_0 = -100\% \text{ to } +100\%$$

Mass Flow

$$\text{OUT} = K_A \cdot \text{XSQRT} \cdot \sqrt{\frac{K_y Y + B_y}{K_z Z + B_z}}$$

where

$\quad$ XSQRT $=$ square root of the differential pressure.
$\qquad\qquad$ Must be supplied to the algorithm.
$\qquad K_A =$ 0 to 2
$\quad K_y, K_z =$ 0.1 to 1.00
$\quad B_y, B_z =$ 0.0% to 100%

Lead/lag Summer

$$\text{OUT} = \frac{K_A(sT_2 + 1)}{(sT_1 + 1)(sT_3 + 1)} (X - Y) + K_z Z + B_0$$

Lead/lag with Multiplier

$$\text{OUT} = \frac{K_A(sT_2 + 1)}{(sT_1 + 1)(sT_3 + 1)} (X - Y)(K_z Z) + B_0$$

where (for both lead/lags)

$\quad K_A =$ 0.1 to 99.99
$\quad K_z =$ -9.999 to $+9.999$
$\quad B_0 =$ -100% to $+100\%$
$\quad T_2 =$ lead time constant, min
$\qquad =$ 0.02 to 99.99; $\ 0 =$ off
$\quad T_1 =$ first lag time constant, min
$\qquad =$ 0 to 91.4; $\ 0 =$ off
$\quad T_3 =$ second lag time constant, min
$\qquad =$ 0 to 91.02; $\ 0 =$ off

External Ratio and Bias

$$\text{OUT} = Y(\text{effective ratio}) + (\text{effective bias})$$

$\quad$ effective ratio $= K_x X + B_x$ when there is a configured input X
$\qquad\qquad\qquad = $ RATIO when there is no configured input X

$\quad$ effective bias $= K_z Z + B_z$ when there is a configured input Z
$\qquad\qquad\qquad = $ BIAS when there is no configured input Z

where

$\quad K_x, K_z =$ -9.999 to $+9.999$
$\quad B_x, B_z =$ -100% to $+100\%$
$\quad$ RATIO $=$ -9.999 to $+9.999$
$\quad\ \ $ BIAS $=$ -100% to $+100\%$

Selector

$$OUT = \text{maximum of used inputs } X, Y, Z, M, A, C$$
$$OUT = \text{minimum of used inputs } X, Y, Z, M, A, C$$

where

$X, Y, Z, M, A, C = $ input signals

[a]These are examples of the computing blocks of the Honeywell TDC 2000 microprocessor-based control system (extended controller).

Table 8-3. Computing Blocks[a]

The signals in this table are given in percent of span.

1. Adder/subtractor
$$OUT = K_1X_1 + K_2X_2 + K_3X_3 + K_4X_4 + K_5$$

2. Multiplier
$$OUT - K_1[X_1X_2X_3X_4] + K_5$$

3. Divider
$$OUT - K_1\left[\frac{X_1 + K_2}{X_2 + K_3}\right] + K_5$$

4. Square root
$$OUT = K_1\sqrt{X_1 + K_2} + K_5$$

5. Mass flow computation
$$OUT - K_1\sqrt{\frac{K_2X_1X_2 + K_3}{X_3 + K_1}} + K_5$$

6. Ratio station
$$OUT = K_1X_1 + K_5$$

7. High selector
$$OUT = \text{Max of } (X_1, X_2, X_3, X_4) + K_5$$

8. Low selector
$$OUT = \text{Min of } (X_1, X_2, X_3, X_4) + K_5$$

where (for all blocks)

$X_1, X_2, X_3, X_4 = $ input
$-100 \le K_1, K_2, K_3, K_4 \le 100$
$-400\% < K_5 < +400\%$

[a]These are examples of the computing blocks of the Beckman MV8000 microprocessor-based systems.

The method used to scale is called the *unity scale method*. It is very simple to use and applies equally to analog instrumentation, pneumatic or electrical, and to microprocessor-based systems. The method consists of the following three steps:

1. Write the equation to be solved along with the range of each process variable. Assign each process variable a signal name.

2. Relate each process variable to its signal name by a normalized equation.

3. Substitute the set of normalized equations into the original equation and solve for the output signal.

Let us show the application of this method by a typical example.

Example 8-1.   Assume that it is necessary to calculate the mass flow rate of a certain gas as it flows through a process pipe, as shown in Fig. 8-1. This figure shows an orifice installed in the pipe. As presented in Appendix C, a simple equation for the calculation of mass flow through an orifice is the following:

$$\dot{m} = K[h\rho]^{1/2} \tag{8-1}$$

where

$$\dot{m} = \text{mass flow, lbm/hr}$$
$$h = \text{differential pressure across orifice, in. } H_2O$$
$$\rho = \text{density of gas, lbm/ft}^3$$
$$K = \text{orifice coefficient}$$

The density of this gas is given, around the operating conditions, by the following linearized equation (see Example 2-13):

$$\rho = 0.13 + 0.003(P - 30) - 0.00013(T - 500) \tag{8-2}$$

Thus, the equation that gives the mass flow is

$$\dot{m} = K[h(0.13 + 0.003(P - 30) - 0.00013(T - 500))]^{1/2} \tag{8-3}$$

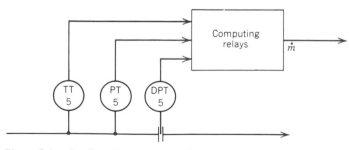

Figure 8-1.   Gas flow through process pipe.

For this process the ranges of the variables are the following:

Signal	Variable	Range	Steady State
$S1$	h	0–100 in. H_2O	50 in. H_2O
$S2$	T	300–700°F	500°F
$S3$	P	0– 50 psig	30 psig
$S4$	$\dot{m}$	0–700 lbm/hr	500 lbm/hr

The orifice coefficient is $K = 196.1\ \text{lbm/[hr(in.-lbm/ft}^3)^{1/2}]$

Equation (8-3) and the ranges given constitute Step 1 of the unit scale method. This information must be known by the process engineer. Note that a signal name was assigned to each process variable.

Step 2 calls for relating each process variable to its signal name by a normalized equation. This means that as the process variable varies between the low and high values of the range, the signal must vary between the values of 0 and 1. A simple equation to accomplish this is

$$\text{Signal} = \frac{\text{Process variable} - \text{Low value of range}}{\text{Span}} \tag{8-4}$$

Applying this equation we obtain

$$S1 = \frac{h}{100} \qquad \text{or} \qquad h = 100\ S1 \tag{8-5}$$

$$S2 = \frac{T - 300}{400} \qquad \text{or} \qquad T = 300 + 400\ S2 \tag{8-6}$$

$$S3 = \frac{P}{50} \qquad \text{or} \qquad P = 50\ S3 \tag{8-7}$$

and

$$S4 = \frac{\dot{m}}{700} \qquad \text{or} \qquad \dot{m} = 700\ S4 \tag{8-8}$$

Finally, to follow Step 3 substitute Eqs. (8-5) through (8-8) into Eq. (8-3) and solve for the output signal, $S4$.

$$700\ S4 = 196.1[100\ S1(0.13 + 0.003(50\ S3 - 30)$$
$$- 0.00013(300 + 400\ S2 - 500))]^{1/2}$$

Using algebra this equation can be simplified to

$$S4 = 1.08[S1(S3 - 0.35\ S2 + 0.44)]^{1/2} \tag{8-9}$$

This is now the normalized equation to be implemented with the computing relays or blocks.

The computing relays of Table 8-1 are now used to implement Eq. (8-9). Since there is no one single relay to implement this equation, it will be done by parts. The first part is to calculate the term in parentheses, with the use of an add/subtract unit, as follows:

$$V_0 = a_0(\pm a_1 V_1 + V_2 \pm a_3 V_3 \pm a_4 V_4) + B_0$$

or

$$V_0' = (S3 - 0.35\ S2) + 0.44$$

Let $V_1 = S3$, $V_3 = S2$, and $V_4 = 0$. Then, matching the last two equations, we get

$$a_0 = 1; \quad a_1 = 1; \quad V_2 = 0; \quad a_3 = 0.35; \quad B_0 = 0.44$$

The sign selection for a_1 must be positive and the sign selection for a_3 negative. The output of this first unit is now multiplied by signal $S1$ and by the factor $(1.08)^2$:

$$V_0 = 4a_0(V_1 V_2) + B_0$$

or

$$V_0 = (1.08)^2(S1)V_0'$$

Let $V_1 = S1$ and $V_2 = V_0'$. Then, matching the last two equations, we get

$$a_0 = \frac{(1.08)^2}{4} = 0.292; \quad B_0 = 0$$

Finally take the square root of the output signal from this last relay. Figure 8-2 shows the flow diagram of the required instrumentation.

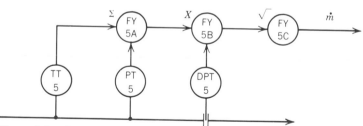

Figure 8-2. Gas mass flow calculation.

The output signal from FY5C, the square root extractor, is linearly related to the gas mass flow. As the mass flow varies between 0 and 700 lbm/hr, the signal will vary between 4 and 20 mA. This signal can now be used to perform any control or recording function.

Note that the signals in Fig. 8-1 are represented by solid lines. This is to emphasize the application of the unity scale method regardless of the type of signal. It is not necessary to specify an electrical, pneumatic, or digital signal when scaling. The unity scale method is applied in the same fashion for all three. This is the power of the method.

To complete this example let us show the implementation using the microprocessor-based system shown in Table 8-2. Equation (8-9) is still the one to be implemented. Before this implementation is done, notice that Table 8-2 specifies the output and input signals to be in percent of span. As it is, in Eq. (8-9) the signals $S1$, $S2$, $S3$, and $S4$ are in fraction of span, resulting from the unity scale method. Thus, we must first convert the signals to percent of span. This is easily done by multiplying both sides of Eq. (8-9) by 100%.

$$100(S4) = 1.08(100)[S1(S3 - 0.35\ S2 + 0.44)]^{1/2}$$
$$100(S4) = 1.08[100\ S1(100\ S3 - 0.35(100)S2 + 44)]^{1/2}$$

or

$$S4' = 1.08[S1'(S3' - 0.35\ S2' + 44)]^{1/2}$$

All the signals, $S1'$, $S2'$, $S3'$, and $S4'$, are now in percent of span.

As before, the term in parentheses is first implemented. The summer block is used:

$$OUT = K_x X + K_y Y + K_z Z + B_0$$

The term is

$$OUT' = S3' - 0.35\ S2' + 44$$

Let $X = S3'$ and $Y = S2'$. The input Z is not used. Then, matching the above two equations, we obtain

$$K_x = 1; \quad K_y = -0.35; \quad K_z = 0; \quad B_0 = 44\%$$

To complete the implementation, the following block is used:

$$OUT = K_A \sqrt{X \cdot Y \cdot Z} + B_0$$

The equation to be implemented is

$$S4' = 1.08 \sqrt{S1' OUT'}$$

Let $X = S1'$ and $Y = OUT'$. Since the input Z is not needed, it is set manually to a value of 100%. Then, matching the last two equations, yields

$$K_A = 1.08; \quad B_0 = 0\%$$

So by using this microprocessor-based system one manipulation has been saved. Only two blocks are required. As mentioned earlier, microprocessor-based systems usually have considerably more computation power than analog instrumentation. In addition, the setting of the constants is done faster and more accurately with these systems.

As shown, the unity scale method is rather simple and powerful. It is important to reemphasize that when using this method it is not necessary to consider whether we are working with analog, electrical, or pneumatic relays or computing blocks. The method is applied in the same manner for all of them; this is a result of the normalization. For example, in the case presented above, the steady-state flow is 0.714 fraction (500/700) of the span of the signal. If pneumatic instrumentation is used, the steady-state output from the last relay would be 11.57 psig, resulting from $3 + 0.714(12)$. If electronic instrumentation is used, the steady-state output from the last relay, FY5C in Fig. 8-2, would be 15.42 mA. If a microprocessor-based system is used, the steady-state output from the last block would be 0.714 fraction of the span plus the zero value.

One final comment is in order before finishing this section. It is always important to check the normalized equation, Eq. (8-9), before its implementation. This is easily done using the steady-state values. For example, the steady-state values for the signals, based on the normalized equations, are

$$\overline{S1} = 0.5; \quad \overline{S2} = 0.5; \quad \overline{S3} = 0.6; \quad \overline{S4} = 0.714$$

Substituting $\overline{S1}$, $\overline{S2}$, and $\overline{S3}$ into Eq. (8-9) yields

$$\overline{S4} = 1.08[0.5(0.6 - 0.35(0.5) + 0.44)]^{1/2}$$
$$\overline{S4} = 0.710$$

The difference between this calculated value of $\overline{S4}$ and the one obtained from the normalized equation is small (0.5%) and is mainly due to truncation errors. Certainly, in large-volume processes even this error can become significant. This may be an incentive to use computers, which have better accuracy than analog instrumentation. However, remember that the field instrumentation, sensors, and transmitters are still analog with accuracy limits.

The following sections present numerous control techniques that improve the control performance provided by feedback control. The diagrams showing their implementation have been drawn using analog instrumentation symbols. The reader must remember, however, that the implementation can also be accomplished using microprocessor-based systems. Furthermore, the use of these new systems *simplifies* tremendously the implementation. The principles of the techniques presented are the same no matter what type of system is used to implement them.

8-2. RATIO CONTROL

A very common control technique in the process industries is *ratio control*. This section presents two industrial cases of ratio control to show its meaning and implementation. The first case is a simple one, but it clearly explains the need for ratio control.

To fix ideas assume that two liquid streams, A and B, are to be blended in some proportion or ratio, R. That is,

$$R = \frac{B}{A}$$

The process is shown in Fig. 8-3. An easy way of accomplishing this task is shown in Fig. 8-4. Each flow is controlled by a flow loop in which the set points to the controllers are set such that the liquids are blended in the correct ratio. However, suppose now that one of the flows, stream A, cannot be controlled, just measured. This flow, referred to as "wild flow," is usually manipulated to control something else, like level or temperature, upstream. The controlling task is now more difficult. Somehow stream B must vary as stream A varies to maintain the blending in the correct ratio. Two possible ratio control schemes are shown in Fig. 8-5.

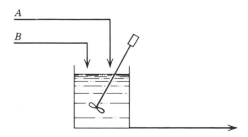

Figure 8-3. Blending of two liquid streams.

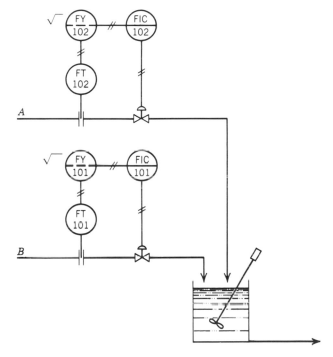

Figure 8-4. Control of blending of two liquid streams.

The first scheme, shown in Fig. 8-5a, consists of measuring the wild flow and multiplying it by the desired ratio, in FY102B, to obtain the required flow of stream B. Mathematically this is shown as follows:

$$B = RA$$

The output of the multiplier, or ratio station, FY102B is the required flow of stream B and, therefore, it is used as the set point to the flow controller of stream B, FIC101. So as the flow of stream A varies, the set point to the flow controller of stream B will vary accordingly to maintain both streams at the required ratio. Notice that if a new ratio between the two streams is required, the new R value must be set in the multiplier, or

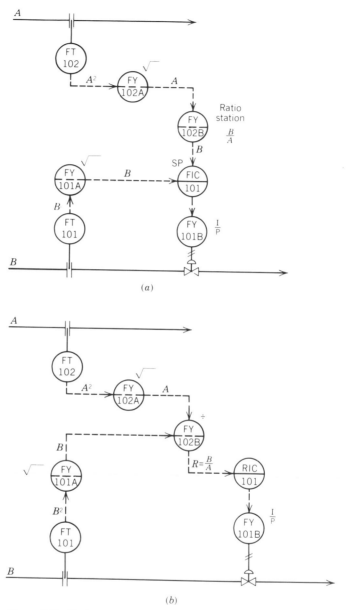

Figure 8-5. Ratio control of blending system.

ratio station. Also notice that the set point to the flow controller of stream B is set from another device, not from the front panel of the controller. Consequently, as explained in Section 5-3, the controller must have the remote/local switch set to remote.

The second ratio control scheme, shown in Fig. 8-5b, consists of measuring both streams and dividing them, in FY102B, to obtain the actual ratio flowing through the system. The calculated ratio is then sent to a controller, RIC101, which will manipulate the flow of stream B to maintain set point. The set point to this controller is the required ratio and is set from the front panel.

In this example, differential pressure sensors have been used to measure the flows. As shown in Appendix C, the output of this sensor-transmitter is related to the square of the flow. Therefore, square root extractors have been used to obtain the flow. However, as mentioned in Appendix C, most manufacturers today provide differential pressure transmitters with an internal square root extractor. Thus, the output signal from this transmitter is already linearly related to the flow and there is no need for the square root extractors. Both control schemes could have been implemented without the square root extractors; however, their use tends to make control loops behave more linearly, resulting in a more stable system.

Both control schemes are used in industry; however, the scheme shown in Fig. 8-5a is preferred because it is a more linear system than the one shown in Fig. 8-5b. This is demonstrated by analyzing the mathematical manipulations in both schemes. In the first scheme FY102B solves the following equation:

$$B = RA$$

The gain of this device, that is, how much its output changes per change in flow of stream A, is given by

$$\frac{\partial B}{\partial A} = R$$

which is a constant value. In the second scheme FY102B solves the following equation:

$$R = \frac{B}{A}$$

Its gain is given by

$$\frac{\partial R}{\partial A} = \frac{B}{A^2}$$

so as the flow of stream A changes this gain also changes, yielding a nonlinearity.

One final fact about this blending process is that, even if both flows can be controlled, the implementation of ratio control may still be more convenient than the control system shown in Fig. 8-4. Figure 8-6 shows a ratio control scheme for this case. If the total flow must be increased, the operator needs to change only one flow, the set point to FIC101. In the control system of Fig. 8-4 the operator needs to change two flows, the set points to both FIC101 and FIC102.

In Figs. 8-5 and 8-6 we have started to use arrowheads to indicate the direction of the transmission signals (information flow). Even though this is not standard, it makes these complex diagrams a lot easier to follow. We also use the abbreviation SP to indicate the set point signal to a controller (remember, the local/remote option in the controller must be set to remote).

The scheme shown in Fig. 8-5a is very common in the process industries. Manufacturers have developed, mainly for microprocessor-based systems, a controller that accepts a signal, multiplies the signal by a number (a ratio), and uses the result as the

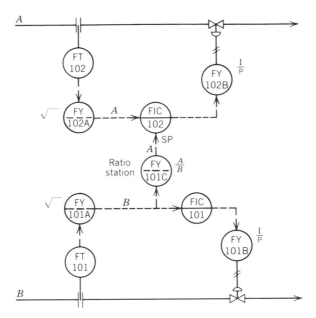

Figure 8-6.   Ratio control of blending system—manipulating both loops.

set point. This means that the ratio station FY102B in Fig. 8-5a would be included in the controller itself. Similarly, FY101C could be included in FIC102 in Fig. 8-6.

As noted from this blending system, we are starting to develop more complex control schemes than the simple feedback control. It is helpful in developing these schemes to remember that every signal must have a physical significance. In Figs. 8-5 and 8-6 we have labeled each signal with its significance. For example, in Fig. 8-5a the output signal from FT102 is related to the square of the flow stream A, A^2. The output from the square root extractor FY102A is then the flow of stream A. If this signal is then multiplied by the ratio B/A, the output signal from FY102B is the required flow of stream B. Even though it is not standard, we will continue to label signals with their significance throughout the chapter. We recommend that the reader do the same.

Example 8-2.   Another common example of ratio control used in the process industries is the control of the air/fuel ratio to a boiler or furnace. Air is introduced in a quantity in excess of that stoichiometrically required to ensure complete combustion of the fuel. The excess air introduced is dependent on the type of fuel and equipment used. However, the greater the amount of excess air introduced, the greater the energy losses through the stack gases. Therefore, its control is most important for proper economical operation.

The flow of combustibles is usually used as the manipulated variable to maintain the pressure of the steam produced in the boiler at some desired value. Figure 8-7 shows one way to control the steam pressure as well as the air/fuel ratio control scheme. This scheme is called parallel positioning control[1, 11, 12] with manually adjusted air/fuel ratio. The steam pressure is transmitted by PT101 to the pressure controller PIC101, this controller manipulates the fuel valve to maintain pressure. Simultaneously, the controller

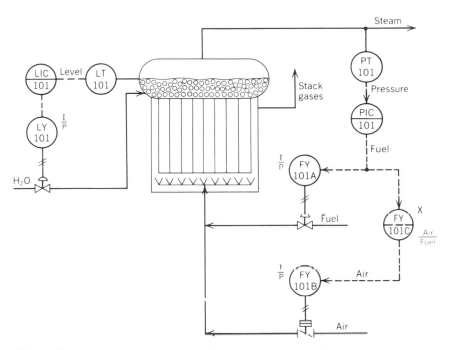

Figure 8-7. Parallel positioning control with manually adjusted air/fuel ratio.

also manipulates the air damper through the ratio station FY101C. This ratio station sets the air/fuel ratio required.

The control scheme shown in Fig. 8-7 does not actually maintain an air/fuel flow ratio but rather maintains a ratio of the signals to the final control elements. The flow through these elements depends on these signals and on the pressure drop across them. Consequently, any pressure fluctuation across the valve or air damper changes the flow, even though the opening has not changed, and this in turn will affect the combustion process and steam pressure. A better control scheme to avoid these types of disturbances is shown in Fig. 8-8 and is called full metering control[1]. The air/fuel ratio is still manually adjusted. In this scheme the pressure controller sets the flow of fuel and the air flow is ratioed from the fuel flow. The flow loops correct for any flow disturbances.

An interesting extension to these control schemes is the following one. Since the amount of excess air is so important to the economical operation of boilers, it has been proposed to analyze the stack gases, also called combustion or flue gases, for excess O_2. Based on this analysis the air/fuel ratio can then be adjusted, or trimmed. This new control scheme is shown in Fig. 8-9, which shows an analyzer transmitter, AT101, and a controller, AIC101. This controller maintains the required excess O_2 in the stack gases by biasing, in FY101D, the output signal from the ratio station that is the set point to the air flow controller. Another possible way of controlling the excess O_2 is by allowing AIC101 to set the required ratio. In this case the output from FY101A would be multiplied by the output signal from AIC101. Figure 8-9 also shows the use of high and low limiters, FY101E and FY101F. These two units are used mainly for safety reasons. They ensure that the air flow set point will always be between some preset high and low values.

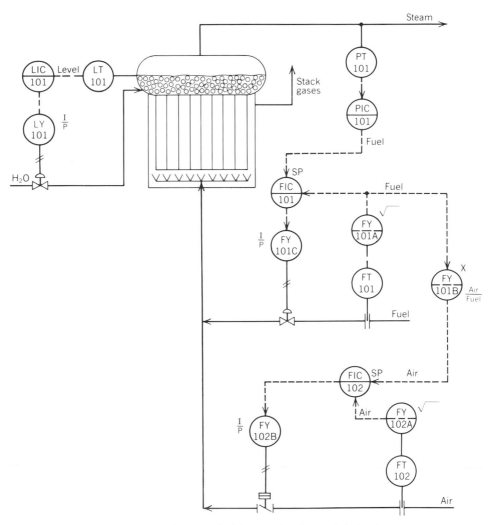

Figure 8-8. Full metering control with manually adjusted air/fuel ratio.

In the control scheme shown in Fig. 8-8, the air flow always follows the fuel flow. That is, the pressure controller will change the fuel flow first and then the air flow will follow the fuel. The most common way to implement air/fuel ratio control in the process industries is by the method known as "cross-limiting control." This implementation is such that when the steam pressure decreases, requiring more combustibles, the air flow increases first and then the fuel flow follows. When the steam pressure increases, requiring less combustibles, the fuel flow decreases first and then the air flow follows. This control strategy ensures that during transients the combustible mixture is always enriched in air. This provides complete combustion of the fuel, minimizing the probability of "smoking" the stack gases and of any other dangerous condition resulting from pockets of pure fuel in the combustion chamber. The implementation of this strategy is the subject of one of the exercise problems at the end of the chapter.

This section has shown two applications of ratio control. As mentioned at the beginning of the section, ratio control is a common technique used in the process industries. It is simple and easy to use. Along with the principles of ratio control, the applications have also shown the use of computing relays such as square root extractors and high and low limiters. In developing and explaining these control schemes, we have sometimes used more blocks than required in actual practice; an example of this is shown in Fig. 8-9. The mathematical calculations done by FY101B and FY101D could all be

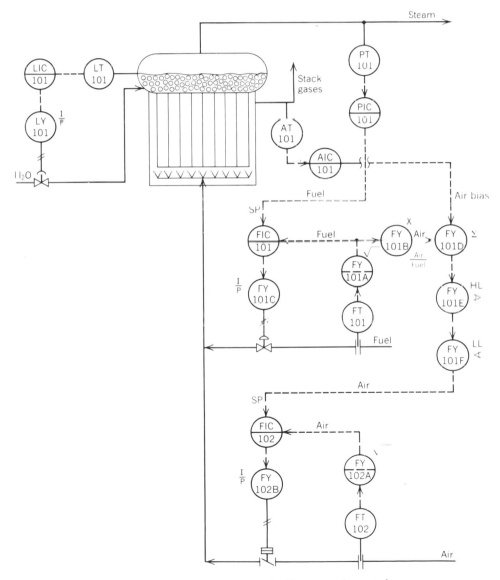

Figure 8-9. Full metering control with oxygen trim control.

done in FY101D. All the first relay, FY101B, is doing is multiplying a signal by a constant. The resulting signal is then added to another signal in the second relay, FY101D. All of this could easily be accomplished with a summer only.

8-3. CASCADE CONTROL

Cascade control is a very common, profitable, and useful control technique in the process industries. This section presents its principles and implementation using two practical cases. Examples of cascade control systems can be found in most processes.

Example 8-3.   Consider the catalyst regeneration process[2] shown in Fig. 8-10. The process, as its name implies, regenerates the catalyst of a chemical reactor. The catalyst is used in a reactor in which a hydrocarbon is dehydrogenated. Over a period of time, carbon (C) is deposited over the catalyst, poisoning it. When this happens, the catalyst looses its activity and must be regenerated. The regeneration step consists of burning the deposited carbon by blowing hot air over the catalyst bed. The oxygen in the air reacts with the carbon to form gaseous CO_2:

$$C + O_2 \rightarrow CO_2 \uparrow$$

When all of the carbon has been burned, the catalyst is again ready to be used. This is a batch process; however, the burning of carbon may last several hours.

   During the regeneration of the catalyst, an important variable to control is the temperature of the catalyst bed, T_C. A very high temperature may destroy the qualities of the catalyst, while a low temperature will result in a long burning time. The temperature

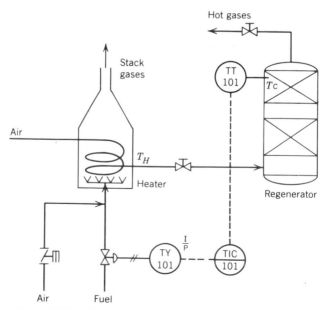

Figure 8-10.   Catalyst regeneration system—simple feedback control.

of the bed is controlled by the fuel flow to the air heater (a small furnace), as shown in Fig. 8-10. The air/fuel ratio controls have not been shown for the sake of simplicity. Similarly, only one temperature sensor has been shown, although in actual practice either an average temperature is obtained and used as the controlled variable or the highest temperature in the bed is used as the controlled variable. Section 8-5 shows an example in which a highest temperature is selected as the controlled variable.

Although the control scheme as shown in Fig. 8-10 works, it must be recognized that there are several system lags in series. The heater itself presents lags such as the combustion chamber and the tubes. The regenerator can have a significant lag depending on its volume and the properties of the catalyst. All of these system lags will yield a slow (long time constants and dead time) feedback control loop.

Suppose that a disturbance enters the heater, such as a change in inlet air temperature or a change in combustion efficiency. Any of these disturbances will affect the outlet air temperature from the heater, T_H. A change in T_H will eventually result in a change in catalyst bed temperature. With so many lags in this system, it will take a considerable amount of time for the control loop to feel a change in T_H. Because of these lags the simple temperature control loop shown will tend to overcompensate, resulting in sluggishness, cycling, and in general slow control.

A superior method or strategy of control is the application of a cascade control system as shown in Fig. 8-11. In this control scheme the temperature T_H is measured and used as an intermediate controlled variable. Therefore, the scheme consists of two sensors, two transmitters, two controllers, and one final control element. This instrumentation results in two control loops. One loop controls the outlet air temperature from the heater, T_H. The other loop controls the catalyst bed temperature, T_C. Of the two controlled variables, the catalyst bed temperature is still the important one. The heater outlet temperature is used only as an intermediate variable to satisfy the catalyst bed temperature requirements.

The way the scheme works is as follows. The controller TIC101 looks at the catalyst bed temperature, T_C, and decides how to manipulate the heater outlet temperature, T_H, to maintain T_C at its set point. This decision is passed on to controller TIC102 in the form of a set point. This controller then manipulates the fuel flow to maintain T_H at the value desired by TIC101. If one of the disturbances mentioned earlier enters the heater, T_H will deviate away from its set point and controller TIC102 will start to take corrective action right away, *before* T_C changes. What has been accomplished is to divide the total lag of the system in two in order to compensate for disturbances before they affect the primary controlled variable.

In general, the controller that controls the primary controlled variable, TIC101 in this case, is referred to as the master controller, outer controller, or primary controller. The controller that controls the secondary controlled variable is usually referred to as the slave controller, inner controller, or secondary controller. The terminology of primary/secondary is commonly preferred because for systems with more than two cascaded loops it extends naturally.

In designing a cascade control scheme, the *most* important consideration is that the inner or secondary loop *must* be faster than the outer or primary loop, a requirement that makes sense. This consideration can be extended to any number of cascaded loops. In a

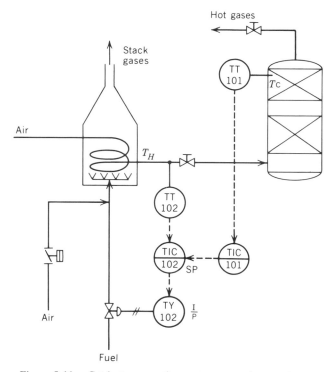

Figure 8-11. Catalyst regeneration system—cascade control.

system with three cascaded loops, the tertiary loop must be faster then the secondary loop and this must be faster than the primary loop.

Let us now look at the block diagram representation of a cascade control system. This will help to further understand this important strategy. Figure 8-12 shows the block diagram representation of the feedback control loop shown in Fig. 8-10. Simple transfer functions have been chosen to represent the system. Figure 8-13 shows the block diagram for the cascade system shown in Fig. 8-11. As shown in this last block diagram, the secondary loop starts to compensate for any disturbance, that is, the inlet air temperature, $T_{\text{air in}}$, that affects the secondary controlled variable, T_H, before its effect is felt by the primary controlled variable, T_C.

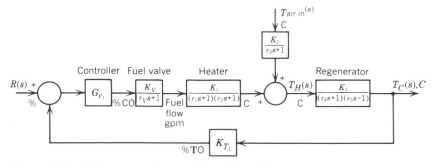

Figure 8-12. Block diagram of system shown in Fig. 8-10.

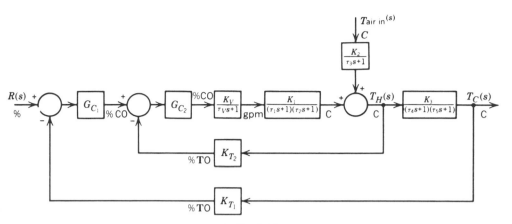

Figure 8-13. Block diagram of system shown in Fig. 8-11.

Note that the implementation of a cascade scheme changes the characteristic equation of the process control system and, consequently, changes its stability. Let us take this example to study the effect of implementing a cascade system on the overall loop stability. Let

$$\tau_V = 0.2 \text{ min} \qquad K_V - 3 \text{ gpm}/\%CO \qquad K_1 - 1 \text{ C/gpm}$$
$$\tau_1 = 3 \text{ min} \qquad \tau_2 = 1 \text{ min} \qquad K_3 = 0.8 \text{ C/C}$$
$$\tau_4 = 4 \text{ min} \qquad \tau_5 = 1 \text{ min} \qquad K_{T_1} = 0.5 \%TO/C$$
$$K_{T_2} = 0.5 \%TO/C$$

Assume that all controllers are proportional only.

Applying the direct substitution method shown in Chapter 6 or frequency response techniques shown in Chapter 7 to the feedback control loop of Fig. 8-12 yields

$$K_{cu_1} = 4.33 \%CO/\%TO \qquad \text{and} \qquad \omega_u = 0.507 \text{ rad/min}$$

where %TO stands for % transmitter output and %CO stands for controller output.

To determine the ultimate gain and frequency of the primary controller of the cascade scheme, the tuning of the secondary controller must first be obtained. This can be done by determining the ultimate gain of the inner loop of Fig. 8-13:

$$K_{cu_2} = 17.06 \%CO/\%TO$$

and using Ziegler–Nichols' suggestion

$$K_{C_2} = 0.5 K_{cu_2} = 8.53 \%CO/\%TO$$

Using block diagram algebra the secondary loop can be reduced to only one block, as shown in Fig. 8-14 (the reader should verify that this is the case). From this block diagram the following is determined:

$$K_{cu_1} = 7.2 \%/\% \qquad \text{and} \qquad \omega_u = 1.54 \text{ rad/min}$$

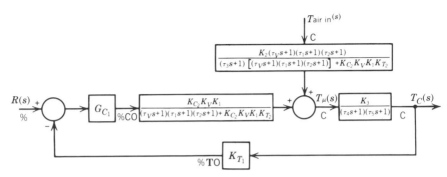

Figure 8-14.   Reduced block diagram of Fig. 8-13.

Comparing the results it is noticed that the cascade scheme has a greater ultimate gain, or limit of stability (7.2 %CO/%TO vs. 4.33 %CO/%TO), than the simple feedback control loop. The ultimate frequency of the cascade scheme is also greater (1.54 cycles/min vs. 0.507 cycles/min), indicating a faster process response.

In general, when correctly applied, the cascade scheme makes the overall loop more stable and faster responding. The methods of analysis are the same as for simple loops. The inner loop is first reduced to a single block using block diagram algebra. The procedure continues then just as before.

Two important questions still remain concerning how to put the cascade scheme into automatic operation and how to tune the controllers. The answers to both questions are the same: from inside out. That is, the innermost loop is first tuned and put into automatic while the other loops are in manual. Then we continue moving out in this manner. For the process shown in Fig. 8-11, TIC102 is first tuned and then set in automatic, while TIC101 is in manual. If the reverse is done, that is, if TIC101 is first set in automatic, nothing will happen because TIC102 will not be able to respond to the requests (set points) of TIC101. What will happen though, is that if TIC101 has reset action, it will probably wind up since its output will have no effect on the controlled variable—that is, the loop is open. Therefore, if any controller in a cascade scheme has reset action, it should have reset windup protection to avoid this problem.

The tuning of the controllers of cascade loops is, of course, more complex than for simple loops. The previous paragraphs have shown how to determine the ultimate gain and ultimate period for each loop. Once these terms are obtained, the Ziegler-Nichols method can be used to tune controllers. The reader must remember, however, that a minimum of three lags, or some amount of dead time, is required for a feedback loop to have an ultimate gain. The single loop tuning methods of Chapter 6 could also be applied to each loop after the loops inside of it have been tuned. The process reaction curve for each loop can be obtained, again after the loops inside of it have been tuned. However, because of interactions between the loops, the process curve may oscillate, making the determination of the loop dynamics, τ and t_0, difficult. Therefore, the tuning formulas of Chapter 6 may not give the satisfactory results in some cascade control systems. The best recommendation is to be careful and use common sense. A well tuned cascade system can be very rewarding and profitable.

Example 8-4. Consider the heat exchanger control system shown in Fig. 8-15. In this system, the outlet process fluid temperature is controlled by manipulating the steam valve position. Notice that the steam flow is not being manipulated. The steam flow depends on the steam valve position *and* the pressure drop across the valve. If a pressure surge in the steam pipe occurs, that is, the pressure upstream of the valve increases, the steam flow will change. The temperature control loop shown can compensate for this disturbance only after the process temperature has deviated away from set point.

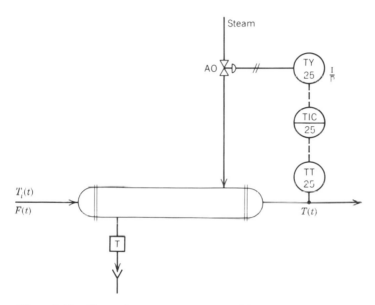

Figure 8-15. Heat exchanger temperature control loop.

Two cascade schemes that help this temperature control, when steam pressure surges are important disturbances, are shown in Fig. 8-16. Figure 8-16*a* shows a cascade scheme in which a flow loop has been added; the temperature controller resets the flow controller set point. Any flow changes are now compensated by the flow loop. The physical significance of the output signal from the temperature controller is the flow of steam required to maintain the temperature at set point. The cascade scheme shown in Fig. 8-16*b* accomplishes the same control but now the secondary variable is the steam pressure in the exchanger shell side. Any change in steam flow quite rapidly affects the shell side pressure. Any pressure change is then compensated for by the pressure loop. This pressure loop also compensates for disturbances in the heat content (latent heat) of the steam, since the pressure in the shell side is related to the condensing temperature and thus to the heat transfer rate in the exchanger. If a pressure sensor can be installed in the exchanger, then this last scheme is less expensive in implementation. It does not require an orifice with its associated flanges, which can be expensive. Both cascade schemes are very common in process industries. Can the reader say which of the two schemes gives better initial response to disturbances in inlet process temperature, $T_i(t)$?

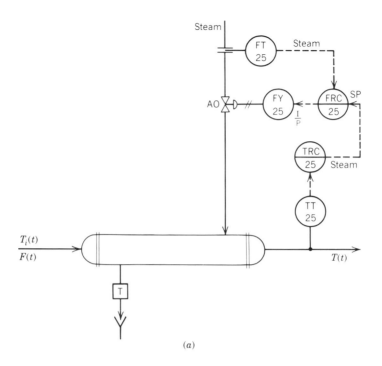

(a)

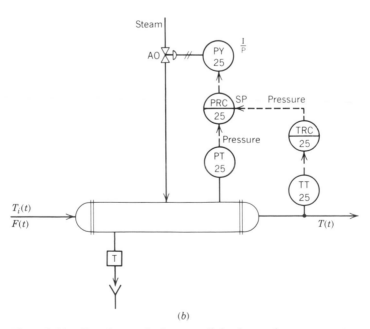

(b)

Figure 8-16. Cascade control schemes applied to heat exchanger temperature control.

Before leaving this section, it is important to say a word about the action of the controllers of a cascade system. For the scheme shown in Fig. 8-16a, the flow controller is reverse acting. This is decided in the same manner as shown in Chapter 5, that is, by the process requirements and action of the control valve. The temperature controller is also reverse acting. This is decided by the process requirements; that is, if the outlet temperature increases, the process requires that the steam flow decrease. Therefore, the temperature controller must decrease its output to the flow controller.

Finally, another very simple example of a cascade control system is that of a positioner on a control valve. The positioner acts as the inner controller of the cascade scheme. Positioners are discussed in Appendix C.

8-4. FEEDFORWARD CONTROL

This section presents the principles and application of one of the most profitable control schemes: feedforward control. Earlier chapters examined the advantages of feedback control, a very simple technique that compensates for any disturbance affecting the controlled variable. As the different disturbances $(D_1, \ldots, D_n)$ enter the process, the controlled variable deviates from set point and the feedback control system compensates by manipulating another input to the process, the manipulated variable, as shown in Fig. 8-17. Chapter 6 showed how to tune feedback control systems and how they behave under upset conditions.

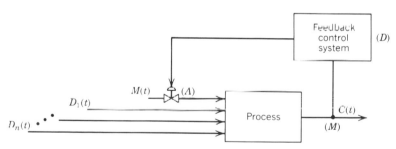

Figure 8-17. Feedback control.

The main disadvantage of feedback control systems, as already learned in previous chapters, is that in order for it to compensate for entering disturbances, the controlled variable must first deviate from set point. Feedback control acts upon an error between the set point and the controlled variable. This means that once a disturbance enters the process, it must propagate through the process and force the controlled variable to deviate from set point *before* corrective action can be taken to compensate for the disturbance. Thus perfect control, defined as no deviation of the controlled variable from set point in spite of disturbances, cannot be achieved with feedback control.

Many processes can stand some temporary deviation in the controlled variable. However, there are many other processes in which this deviation must be minimized to such an extent that feedback control alone may not provide the required control performance. For these cases feedforward control may prove most helpful.

Feedforward control compensates for disturbances *before* they affect the controlled variable. Specifically, feedforward control measures the disturbances before they enter the process and calculates the required value of the manipulated variable to maintain the controlled variable at its desired value or set point. If the calculation is done correctly, the controlled variable should remain undisturbed. Figure 8-18 depicts the concept of feedforward control[3].

Let us now use a process example to follow the different steps needed to design a feedforward control system.

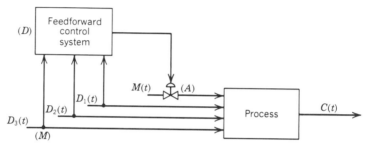

Figure 8-18.   Feedforward control.

A Process Example

Consider the simulated process shown in Fig. 8-19. In this process different flows are mixed in a three-tank system. In tank 1 streams $q_5(t)$ and $q_1(t)$ are mixed. The overflow from this tank goes into tank 2, where it is mixed with stream $q_2(t)$. The overflow from tank 2 flows into tank 3, where it is mixed with stream $q_7(t)$. In this process it is required to control the mass fraction (mf) of component A, $x_6(t)$, in the outlet stream from tank 3. The manipulated variable in this process is stream $q_1(t)$; all other stream flows and mass fractions are possible disturbances. Table 8-4 presents the process data and steady-state values of all variables. Figure 8-20a shows the response of the controlled variable to a change in stream $q_2(t)$ from 1000 gpm to 1500 gpm; an optimally tuned PI feedback controller is used. The major disturbance in this process is considered to be $q_2(t)$. Let us now see how to use feedforward techniques to improve on this control and minimize the deviation of $x_6(t)$.

The first step in designing a feedforward control system is to develop a steady-state mathematical model of the process. This model is an equation relating the manipulated variable to the controlled variable and disturbances. In this example, the manipulated variable is stream $q_1(t)$, the controlled variable is $x_6(t)$, and $q_2(t)$ is assumed to be the *major disturbance*. It is important to understand this last point. Feedforward control is used to compensate for the major disturbances, that is, those that occur most frequently and cause the greatest deviations in the controlled variable. The instrumentation and

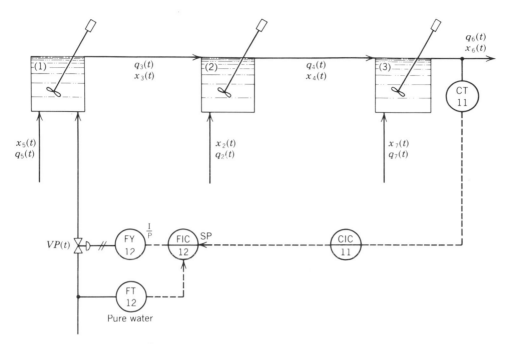

Figure 8-19. Process example.

engineering efforts required do not usually justify the feedforward compensation of minor disturbances. The compensation for these minor disturbances is shown later.

The required model of this process is obtained by the application of mass balances to the process. Writing a steady-state mass balance around the entire process gives

$$\bar{q}_5\rho + q_1(t)\rho + q_2(t)\rho + \bar{q}_7\rho - q_6(t)\rho = 0$$

In this equation the steady-state values of streams $q_5(t)$ and $q_7(t)$ have been used because they are considered minor disturbances. The term ρ is the density, mass/gal, of the streams, which is considered to be the same for all streams. Therefore

$$q_1(t) = q_6(t) - q_2(t) - \bar{q}_5 - \bar{q}_7$$

or

$$q_1(t) = q_6(t) - q_2(t) - 1000 \tag{8-10}$$

A steady-state mass balance on component A around the entire process gives us another needed relation:

$$\bar{q}_5\bar{x}_5 + q_2(t)\bar{x}_2 + \bar{q}_7\bar{x}_7 - q_6(t)x_6(t) = 0 \tag{8-11}$$

In this equation the steady-state values of the mass fractions, $\bar{x}_5$, $\bar{x}_7$, and $\bar{x}_2$, have been used because they are also considered minor disturbances. This last equation yields

$$q_6(t) = \frac{1}{x_6(t)}[\bar{q}_5\bar{x}_5 + \bar{q}_7\bar{x}_7 + \bar{x}_2 q_2(t)]$$

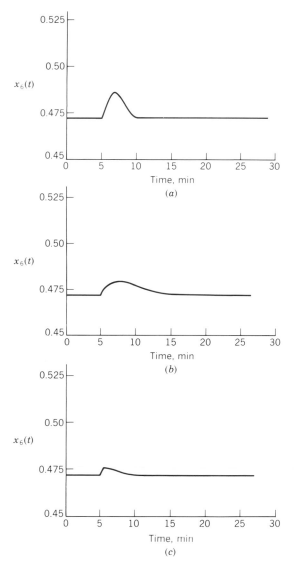

Figure 8-20. Response of $x_6(t)$ to a change in flow $q_2(t)$ from 1000 gpm to 1500 gpm.

or

$$q_6(t) = \frac{1}{x_6(t)} [850 + 0.99q_2(t)] \qquad (8\text{-}12)$$

Substituting Eq. (8-12) into (8-10) gives

$$q_1(t) = \frac{1}{x_6^{\text{set}}} [850 + 0.99q_2(t)] - q_2(t) - 1000 \qquad (8\text{-}13)$$

Table 8-4. Process Information and Steady-State Values

Information

Tank volumes: $V_1 = V_2 = V_3 = 7000$ gal
Concentration Transmitter Range: 0.3–0.7 mass fraction
The control valve has linear characteristics and the pressure drop across it can be considered constant.
The range of the $q_1(t)$ flow transmitter is 0–3800 gpm. The transmitter and valve dynamics can be considered negligible.
The density of all streams can also be considered similar.

	Steady-State Values	
Stream	Flow, gpm	Mass Fraction (mf)
1	1900	0.0000
2	1000	0.9900
3	2400	0.1667
4	3400	0.4088
5	500	0.8000
6	3900	0.4710
7	500	0.9000

In this equation we have replaced the term $r_6(t)$ by x_6^{set}. By this replacement we calculate the value of $q_1(t)$ required to force $x_6(t)$ to the set point x_6^{set}.

Equation (8-13) relates the manipulated variable to the major disturbance and to the controlled variable. It is the equation to be implemented for control and constitutes what is called the "feedforward controller." By solving this equation the controller determines the required flow $q_1(t)$ needed for a certain disturbance flow $q_2(t)$ and a certain set point, x_6^{set}.

Before implementing the feedforward controller, it is wise to check the equation. That is, substituting the steady-state value of the disturbance, $\bar{q}_2 = 1000$ gpm, and of the controlled variable, $\bar{x}_6 = 0.4718$ mf, yields a flow of

$$\bar{q}_1 = 1900 \text{ gpm}$$

which is the correct steady-state flow. Thus, we feel confident in our controller.

Figure 8-21 shows the implementation of the feedforward control system. The feedforward controller itself is implemented by FT11, HIC11, FY11A, and FY11B. The instrumentation used is the one shown in Table 8-3 even though electrical signal convention is used. Notice that the required set point, x_6^{set}, is generated in a "hand indicating controller," HIC11. This unit generates a signal set manually by the operator and related to the set point. That is, the low value of the range of the signal is equal to the low value of the range of the concentration transmitter, 0.3 mf. The high value of the range of the signal is equal to 0.7 mf. Thus, the operator is able to adjust the set point. This control scheme is called "steady-state feedforward control."

Figure 8-20b shows the response of the process under steady-state feedforward control for a change in stream $q_2(t)$ from 1000 gpm to 1500 gpm. This figure can be compared to Fig. 8-20a to appreciate the improvement obtained by implementation of feedforward control. The response curve still shows that the controlled variable does not stay constant

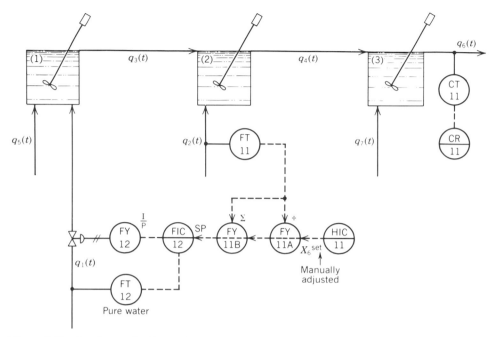

Figure 8-21. Steady-state feedforward control scheme.

(as expected!) but, rather, shows a transient error. This transient error is present because a dynamic imbalance exists between the effects of the disturbance, $q_2(t)$, and the manipulated variable, $q_1(t)$, on the controlled variable, $x_6(t)$. That is, the outlet composition responds faster to a change in stream $q_2(t)$ than to a change in stream $q_1(t)$. One of the objectives of feedforward control should be to balance the manipulated variable against the disturbance. It would be desirable to speed up the response of $x_6(t)$ to a change in $q_1(t)$ since it is slower than the response of $x_6(t)$ to a change in $q_2(t)$. This can be done by the use of a "lead/lag unit." Lead/lag units are presented in the next subsection. Figure 8-22 shows the control scheme with the lead/lag unit installed. This scheme is now referred to as "feedforward control with dynamic compensation." Figure 8-20c shows the response of the process to a change in $q_2(t)$ from 1000 gpm to 1500 gpm. Notice that the transient error has diminished significantly. Perfect balance has not been achieved; however, a marked improvement over the uncompensated response has been attained.

The improvement of the feedforward with dynamic compensation over the steady-state feedforward is due to the different ways the manipulated variable, $q_1(t)$, responds in each case. Figure 8-23a shows how $q_1(t)$ responds to a change in $q_2(t)$. This is the way determined by the feedforward controller, Eq. (8-13). Figure 8-23b shows how $q_1(t)$ responds to a change in $q_2(t)$ when the lead/lag unit is installed. Initially, $q_1(t)$ changes by a greater amount than necessary for steady-state compensation and then decays exponentially to the final steady-state value. By reacting in this way, $q_1(t)$ has more "mus-

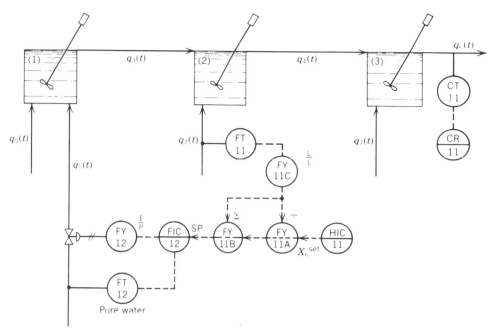

Figure 8-22. Feedforward control with dynamic compensation.

cle" to move the controlled variable faster. This type of response of $q_1(t)$ to change in $q_2(t)$ is due to the lead/lag unit.

It is important to mention at this time that the "settings" of a lead/lag unit are usually empirical. They could also be obtained analytically from a detailed mathematical analysis. Later on in this section we devote some space to an explanation of the operation of a lead/lag unit and how to obtain the information to set the unit.

Before proceeding with this example, it is important to mention that not all feedforward schemes require dynamic compensation. Therefore, it is recommended that when initially implementing feedforward control, the steady-state portion be first tested. If transient errors are present, the scheme may need dynamic compensation and the engineer is justified in implementing it. Certainly, engineering judgment will give an indication of whether transient errors should be expected. In the example presented, the disturbance enters the process closer to the controlled variable than to the manipulated variable. That is, there is one time constant (tank) less between the controlled variable and disturbance than between the controlled and manipulated variables. Therefore, we should not be surprised if a transient error exists. The magnitude of this error depends on the difference between the dynamic effects.

The feedforward control scheme implemented thus far compensates for changes in $q_2(t)$ only. This is because in the development of the feedforward controller, it was assumed that $q_2(t)$ was the only major disturbance. All other inputs to the process were assumed to be minor disturbances. That is, these inputs do not change much or not often

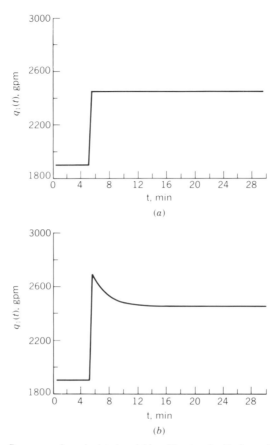

Figure 8-23.   Response of manipulated variable without and with dynamic compensation.

enough to warrant the effort and capital needed to compensate for them. Therefore, if any of these minor disturbances enters the process, the control scheme shown in Fig. 8-22 will not compensate for it and an offset in the controlled variable will result. As an example, Fig. 8-24a shows the response of the feedforward control scheme of Fig. 8-22 when $q_7(t)$ changes from 500 gpm to 600 gpm. The controlled variable reaches a new value and stays there. The control scheme does not compensate for this disturbance. A simple way to correct for this offset is by manually changing the output from HIC11. That is, once the operator notices that the outlet composition is above the set point, the decision is made to increase the flow of pure water. This is easily accomplished by decreasing the output from HIC11. By doing this, the feedforward controller increases the set point to the water flow controller, FIC12. This operation continues until the offset disappears. Thus, the operator can correct for any disturbance not compensated for by the feedforward controller.

   This procedure to compensate for any minor disturbance is simple and works weil. However, the disadvantage is that it requires the intervention of the operator. It would

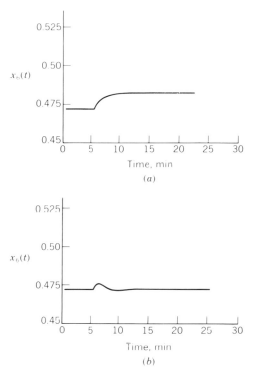

Figure 8-24. Response of control systems to disturbances in $q_7(t)$, (a) without feedback trim, (b) with feedback trim.

be much better if the compensation could be done automatically, without the need for operator intervention. Figure 8-25 shows such a scheme; it is called "feedforward control with dynamic compensation and feedback trim." In this scheme, the feedback controller, CIC11, replaces the operator. Note that the actual desired outlet composition is now the set point to CIC11. This controller then decides on the signal to the feedforward controller "x_6^{set}," to maintain set point. Figure 8-24b shows the response of this new control scheme when $q_7(t)$ changes from 500 to 600 gpm. As shown, the feedback compensation now brings the controlled variable back to set point.

To summarize, the feedforward controller compensates for the major disturbances. The feedback trim is needed for several reasons, among which are the facts that we don't always measure and compensate for all possible disturbances, that the feedforward controller equation is not exact, and that the instruments may encounter drifting, to name a few. For these reasons *feedforward control is implemented with feedback trim*.

The location of the feedback trim is important and must be considered in the overall control scheme. Figure 8-25 shows that the feedback trim was intoduced after the lead/lag unit. Had the lead/lag unit been installed after the introduction of the feedback trim, this compensation would then go through the unit and result in an unnecessary transient "bumping" of the process. Since there is no need to dynamically balance this feedback compensation against anything, there is no need to pass it through the lead/lag unit.

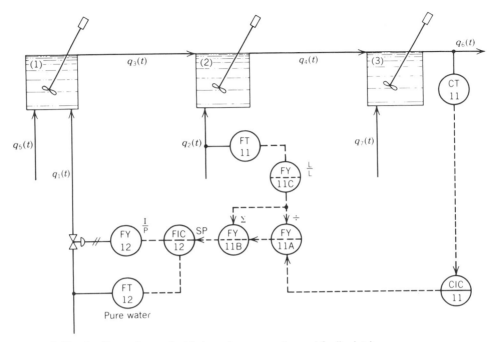

Figure 8-25. Feedforward control with dynamic compensation and feedback trim.

Lead/Lag Unit

Before continuing with more feedforward control, let us quickly review the lead/lag unit. This unit solves the following transfer function, as shown in Eq. (4-78):

$$\frac{Y(s)}{X(s)} = \frac{\tau_{ld}s + 1}{\tau_{lg}s + 1} \tag{4-78}$$

where

τ_{ld} = lead time constant, min
τ_{lg} = lag time constant, min

Sometimes this transfer function is multiplied by a gain factor, K, as shown in Table 8-3. In the time domain, the expression that describes the response $Y(t)$ to a unit step change in forcing function is given by Eq. (4-79):

$$Y(t) = \left[1 + \frac{\tau_{ld} - \tau_{lg}}{\tau_{lg}} e^{-t/\tau_{lg}} \right] \tag{4-79}$$

Fig. 8-26 shows the response graphically for the case of $\tau_{lg} = 1$ and different ratios of τ_{ld}/τ_{lg}. It is worthwhile to repeat what was said in Chapter 4 about this unit. First and most important, the initial amount of response is dependent on the ratio τ_{ld}/τ_{lg}. This initial response will be equal to the product of τ_{ld}/τ_{lg} and the magnitude of the step change. Fig. 8-23b shows the response of $q_1(t)$ to a change in $q_2(t)$ under dynamic feedforward

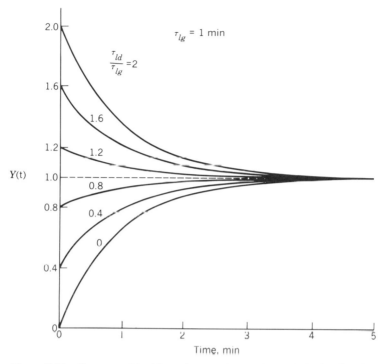

Figure 8-26. Response of lead/lag unit to a unit step change in forcing function.

control. The lead/lag unit used in the example shown has a τ_{ld}/τ_{lg} ratio greater than 1. Specifically, $\tau_{ld}/\tau_{lg} = 3.3/2.29 = 1.44$. Next, the final amount of change in output from the lead/lag unit is equal to the magnitude of the change in input. Finally, the exponential rate of decay, or increase, in output is only a function of the lag time constant, τ_{lg}.

As shown in the previous section, lead/lag units are used to compensate for dynamic imbalances in the process. To implement a unit the lead time constant, τ_{ld}, and the lag time constant, τ_{lg}, must be specified. These two time constants are part of the personality of each process. The following section shows how to obtain an initial estimate of τ_{ld} and τ_{lg}. Lead/lag units can be bought commercially as analog units or in microprocessor-based systems.

Block Diagram Design of Linear Feedforward Control

The first step in implementing feedforward control is to develop a model, or equation, of the process. This model must relate the manipulated variable to the controlled variable and all major disturbances. The process example presented at the beginning of this section shows how this is done for the process shown in Fig. 8-19. For this process the model was developed analytically, starting with basic mass balance equations. However, what if the required model cannot be developed analytically because the process is too difficult to describe or there are too many unknown parameters? This is the subject of this section. The process shown in Fig. 8-19 will be used to demonstrate the technique. A partial block diagram for this process is shown in Fig. 8-27a. Since the flow loop, once tuned, is fast and stable, Fig. 8-27a can be simplified as shown in Fig. 8-27b.

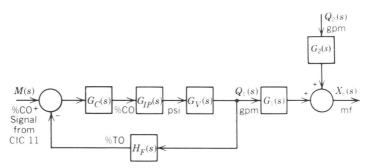

$G_1(s)$ = transfer function of process relating the outlet mass fraction to the flow of stream 1
$G_2(s)$ = transfer function of process relating the outlet mass fraction to the flow of stream 2
$G_v(s)$ = transfer function of valve relating the output flow to the pneumatic signal
$G_{IP}(s)$ = transfer function of I/P transducer, FY12
$G_c(s)$ = transfer function of flow controller, FIC12
$H_F(s)$ = transfer function of flow transmitter, FT12

Figure 8-27a. Partial block diagram of process shown in Fig. 8-19.

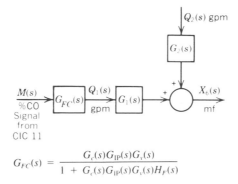

$$G_{FC}(s) = \frac{G_c(s)G_{IP}(s)G_v(s)}{1 + G_c(s)G_{IP}(s)G_v(s)H_F(s)}$$

Figure 8-27b. Simplified block diagram of Fig. 8-27a.

To implement feedforward control, the disturbance must be measured and used to calculate the manipulated variable in order to keep the controlled variable constant. Figure 8-28 shows the block diagram of this feedforward control scheme. Using block diagram algebra

$$X_6(s) = [G_2(s) + H_2(s)\, G_F(s)\, G_{FC}(s)\, G_1(s)]\, Q_2(s) \tag{8-14}$$

The objective now is to design the feedforward controller, $G_F(s)$, so that when $q_2(t)$ varies, $x_6(t)$ remains constant. If $x_6(t)$ is constant, then $X_6(s) = 0$ and from the last equation the feedforward controller can be obtained as

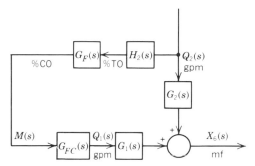

$H_2(s)$ = transfer function of the flow sensor-transmitter sensing the flow of stream 2.
$G_F(s)$ = transfer function of the feedforward controller.

Figure 8-28. Block diagram of the feedforward control scheme.

$$G_F(s) = -\frac{G_2(s)}{H_2(s)G_{FC}(s)G_1(s)} \qquad (8\text{-}15)$$

This is the feedforward controller equation to be implemented.

To gain more insight into this method, let us follow this example in more detail. Suppose that in the process shown in Fig. 8-19 the concentration controller, CIC11, is set in manual and a step change in set point to the flow controller, FIC12, is introduced. A record of the change in outlet mass fraction, $x_6(t)$, from the last tank is made. From these data, using the procedure presented in Chapter 6, the following model is developed:

$$G_{P1}(s) - G_{FC}(s)G_1(s) = \frac{K_{P1}e^{-t_{01}s}}{\tau_1 s + 1} \qquad (8\text{-}16)$$

By step testing stream 2 in the same fashion and recording $x_6(t)$, the following transfer function is developed:

$$G_2(s) - \frac{K_{P2}e^{-t_{02}s}}{\tau_2 s + 1}, \qquad \frac{mf}{gpm} \qquad (8\text{-}17)$$

Assuming that the flow sensor-transmitter has negligible dynamics, we obtain

$$H_2(s) = K_{T_2}, \qquad \frac{\%TO}{gpm} \qquad (8\text{-}18)$$

Substituting Eqs. (8-16), (8-17), and (8-18) into Eq. (8-15) yields

$$G_F(s) = \left(\frac{-K_{P2}}{K_{T_2}K_{P1}}\right)\left(\frac{\tau_1 s + 1}{\tau_2 s + 1}\right)e^{-(t_{02}-t_{01})s}$$

The first term in parentheses is a pure gain, with units (%CO/%TO); it represents how much the output from the feedforward controller, should change for a change in the flow transmitter output. This, of course, will change stream 1 when stream 2 changes, which is the idea behind feedforward control. The second term is the dynamic compensator lead/lag unit. The settings are given by $\tau_{ld} = \tau_1$ and $\tau_{lg} - \tau_2$. Finally, the third term, an exponential term, is another dynamic compensator. It can be considered a "dead time compensator." However, there are two possible problems with this term. The first problem is that the implementation of this compensation with analog instrumentation is impractical. With the use of computers this implementation is rather simple. Many of the microprocessor-based systems in use today have a dead time computing block. A second problem is encountered if the term $t_{0_2} - t_{0_1}$ is negative, yielding a positive exponential. Since the term is no longer dead time, it is interpreted as predicting the future, which is not possible. Therefore, the last term has been commonly dropped and the feedforward controller equation becomes

$$G_F(s) = \left(\frac{-K_{P_2}}{K_{T_2}K_{P_1}}\right)\left(\frac{\tau_1 s + 1}{\tau_2 s + 1}\right) \tag{8-19}$$

With today's computer instrumentation, if the term $t_{0_2} - t_{0_1}$ is positive, then the dead time compensator can be implemented.

Feedback compensation is usually added to the feedforward scheme. Figure 8-29 shows the block diagram of the complete scheme, and Fig. 8-30 shows the instrumentation diagram. In Fig. 8-30 the multiplier FY11A multiplies the signal from FT11 by $-K_{P_2}/K_{T_2}K_{P_1}$ and FY11B performs the lead/lag compensation. In most cases both calculations can be performed in a single computing relay, as shown in Table 8-2. The feedback compensation is added to the feedforward controller in FY11C.

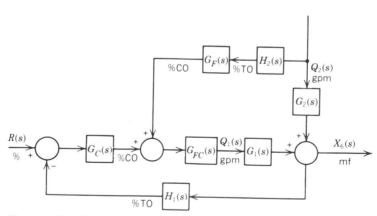

Figure 8-29. Block diagram of feedforward control with feedback trim.

The significance of the feedback trim in this feedforward control scheme is different from its significance when the feedforward controller was developed analytically. In the present case, the feedback trim is used to bias (up or down) the output from the feedforward controller to compensate for the minor disturbances. A comment about the scaling of the summer, FY11C, where the feedback is introduced is important. This summer, also sometimes called a "bias station," solves the following equation:

$$OUT = \text{feedback signal} + \text{feedforward signal} + \text{bias}$$

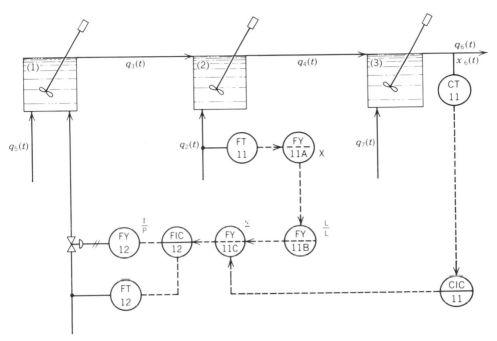

Figure 8-30. Feedforward control with dynamic compensation and feedback trim. Block diagram approach.

To be more specific and to show how to calculate the value of the bias, consider the use of the summer shown in Table 8-2:

$$OUT = K_x X + K_y Y + K_z Z + B_0$$

Let the feedback signal be the X input and the feedforward signal the Y input, the Z input is not used. At steady-state $q_2(t)$ is 1000 gpm. Assume that the range of the sensor-transmitter measuring this flow is 0–2500 gpm; therefore, the steady-state value of the flow is 40% of its span. Under this condition, the steady-state output from the feedforward controller, the input to the summer, is $(-K_{P_2}/K_{T_2}K_{P_1})$ 40%. What is done in the summer is to use the bias (B_0) to cancel the feedforward signal. Thus, the bias term is set at $-(-K_{P_2}/K_{T_2}K_{P_1})$ 40%. Since $q_1(t)$ at steady-state is 1900 gpm and the range of its transmitter is 0–3800 gpm, as indicated in Table 8-4, the output signal from the summer must be 50% of its span. This forces the output signal from the feedback controller to be 50% of its span at steady state.

In this example, it is easy to realize, using engineering principles, that the sign of K_{P_2} is positive while the sign of K_{P_1} is negative. This means that the term $-K_{P_2}/K_{T_2}K_{P_1}$ is a positive quantity, indicating that if $q_2(t)$ increases, $q_1(t)$ should also increase.

In the summer, the value of the scale factors K_x and K_y can both be set to $+1$ while the value for the bias (B_0) is then set to $-(-K_{P_2}/K_{T_2}K_{P_1})$ 40%. The sign of the scale factor of the feedforward signal (K_y in this case) can be considered the "action" of the feedforward controller. The engineer must be sure that all of the signs are correct before setting the complete control system in automatic.

With the new microprocessor-based systems there is sometimes no need to use a bias summer. Many microprocessor-based controllers have been implemented to allow the biasing to be done within the feedback controller. That is, these controllers accept the feedforward signal as another input. This signal is then used to bias the decision of the feedback controller. The result of this manipulation becomes the controller output. This output is the result of the feedforward/feedback scheme. This capability saves one block—another advantage of the new technology.

As shown, the block diagram design of feedforward controllers is rather simple and practical. Before closing this section, however, we would like to stress the following points:

1. The difference between the block diagram approach and the equation approach, the one presented at the beginning of this section in the process example, is that the former results in a linear feedforward controller and the latter usually results in nonlinear computations. Whenever possible, the equation approach is preferred over the block diagram approach. The reason for this is that performance of the linear controllers lessens outside the operating conditions where they were evaluated. When this happens, the feedback controller must then compensate more often since then the process nonlinearities act as disturbances. The controllers resulting from the equation approach can compensate for the nonlinearities. The dynamic compensation, lead/lag unit, is *linear* and *empirical* in both approaches.

2. The gains used in the block diagram approach were obtained by process testing. They could also have been obtained by linearization of the balance equations. For example, K_{P_2} could have been obtained by rearranging Eq. (8-11) as follows:

$$x_6(t) = \frac{1}{q_6(t)} [\bar{q}_5 \bar{x}_5 + q_2(t)\bar{x}_2 + \bar{q}_7 \bar{x}_7]$$

and then

$$K_{P_2} = \left. \frac{\partial x_6(t)}{\partial q_2(t)} \right|_{ss} = \frac{\bar{x}_2}{\bar{q}_6}$$

Since in the block diagram approach we are dealing with deviation variables, a bias term has to be added, which is not predicted by the design. The feedback trim is usually introduced in the unit that provides the bias (FY11C in Fig. 8-30).

Figure 8-29 demonstrates an interesting and important fact about feedforward control. Notice that the feedforward path, $H_2(s)G_F(s)$, is *not* part of the characteristic equation. This means that the addition of feedforward control does not affect the stability of the control loop.

Two Other Examples

Example 8-5. An interesting and challenging process for which the control techniques learned thus far are used is in the control of the liquid level in a boiler drum. Figure 8-31 shows a schematic of a boiler drum. The control of the level in the drum is very important. A high level may result in carrying over of water, and maybe impurities, into the steam system; a low level may result in tube failure due to overheating for lack of water in the boiling surfaces.

Figure 8-31 shows steam bubbles flowing upwards through the water; this presents an important effect. The specific volume (volume/mass) of the bubbles is very large, and therefore these bubbles displace the water. This results in a higher apparent level than the level due to water only. The presence of these bubbles also presents a problem under

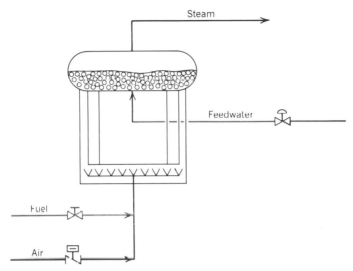

Figure 8-31. Boiler drum and accessories.

transient conditions. Consider the situation when the pressure in the steam header drops because of an increased demand for steam by the users. Because of this condition, a certain quantity of water will flash into steam bubbles. These new bubbles will tend to increase the apparent level in the drum. The drop in pressure also causes the volume of the existing bubbles to expand. This further increases the apparent level. This surge in level resulting from a decrease in pressure is called *swell*. An increase in steam header pressure, because of a decreased demand for steam by the users, has the opposite effect on the apparent level and is called *shrink*.

The swell/shrink phenomena combined with the importance of maintaining a good level, as explained earlier, makes the level control even more critical. The following paragraphs develop some of the level control schemes presently used in industry.

The drum level control is accomplished by manipulating the flow of feedwater. Figure 8-32 shows the simplest type of level control, referred to as "single-element control." A standard differential pressure sensor-transmitter is used. This control scheme relies only on the drum level measurement and, therefore, it must be reliable. Under prolonged transients the swell/shrink phenomena do not render a reliable measurement; consequently, a control scheme that compensates for these phenomena is required.

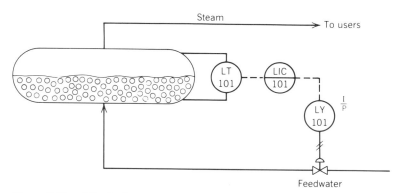

Figure 8-32. "Single-element control" scheme.

The new control scheme, called "two-element control," is shown in Fig. 8-33 and is essentially a feedforward/feedback control system. The idea behind this scheme is that the major reason for level changes is changes in steam flow and that for every pound of steam produced, a pound of feedwater should enter the drum, that is, there should be a mass balance. The signal output from FY101A provides the feedforward part of the scheme, while LIC101 provides the feedback compensation for any unmeasured flows such as blowdown.

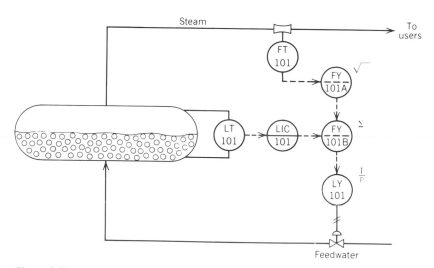

Figure 8-33. "Two-element control" scheme.

The "two-element control" scheme works quite well in many industrial boiler drums. However, there are some systems that exhibit variable pressure drop across the feedwater valve. The "two-element control" scheme does not directly compensate for this disturbance, and, consequently, it upsets drum level balance control by momentarily upsetting the mass. The "three-element control" scheme, shown in Fig. 8-34, provides the required

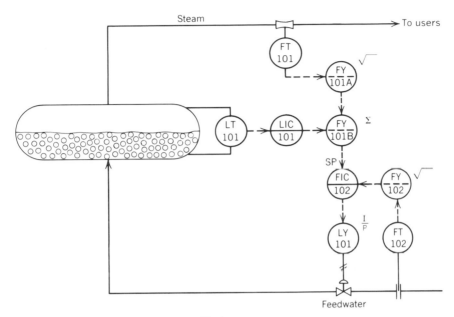

Figure 8-34. "Three-element control" scheme.

compensation. This scheme provides a tight mass balance control during transients. It is interesting to note that all that has been added to the "two-element control" scheme is a cascade control system.

The boiler drum level has provided a realistic example where the cascade and feedforward control schemes are used to improve the performance provided by feedback control. In this particular example, the use of these schemes is almost mandatory to avoid mechanical and process failures. Every step taken to improve the control was justified. Otherwise, there is no need to complicate matters. The reader is referred to References 11 and 12 for another complete discussion of this subject.

Example 8-6. We now present another realistic example that has proven to be a successful industrial application of feedforward control. The example is concerned with the temperature control in the rectifying section of a distillation column. Figure 8-35 shows the bottom of the column and the control scheme originally proposed and implemented. This column uses two reboilers. One of the reboilers, R-10B, uses a condensing process stream as a heating medium and the other reboiler, R-10A, uses condensing steam. For efficient energy operation, the operating procedure calls for using as much of the condensing process stream as possible. This stream must be condensed anyway, and thus serves as a "free" energy source. The steam flow is used to control the temperature in the column.

After startup of this column, it was noticed that the process stream serving as heating medium experienced changes in flow and in pressure. These changes acted as disturbances to the column and, consequently, the temperature controller needed to continually com-

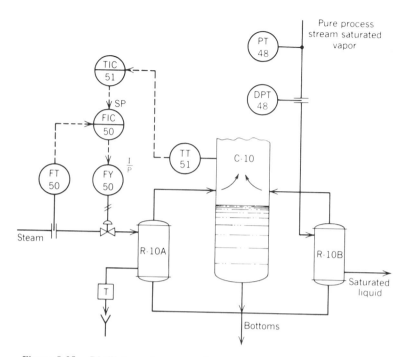

Figure 8-35. Distillation column control.

pensate for these distubances. The time constants and dead time in the column and reboilers complicated the temperature control. After the problem was studied, it was decided to use feedforward control. A pressure transmitter and a differential pressure transmitter had been installed in the process stream, and from them the amount of energy given off by the stream in condensing could be calculated. Using this information the amount of steam required to maintain the temperature at set point could also be calculated and, thus, corrective action could be taken before the temperature deviated from set point. This is a perfect application of feedforward control.

Specifically, the procedure was as follows. Since the process stream is saturated, the density, ρ, is a function of pressure only. Therefore, using a thermodynamic correlation, the density, ρ, of the stream can be obtained:

$$\rho = f_1(P) \tag{8-20}$$

Using this density and the differential pressure, h, obtained from the transmitter DPT48, the mass flow of the stream can be calculated from the orifice equation:

$$\dot{m} = K\sqrt{h\rho}, \qquad \text{lbm/hr} \tag{8-21}$$

Also, knowing the stream pressure and using another thermodynamic relation, the latent heat of condensation, λ, can be obtained:

$$\lambda = f_2(P), \qquad \text{Btu/lbm} \tag{8-22}$$

Finally, multiplying the mass flow rate times the latent heat, the energy, q_1, given off by the process stream in condensing is obtained:

$$\dot{q}_1 = \dot{m}\lambda, \qquad \text{Btu/hr} \tag{8-23}$$

Fig. 8-36 shows the implementation of Eqs. (8-20) through (8-23) and the rest of the feedforward scheme. Block PY48A performs Eq. (8-20), block PY48B performs Eq. (8-21), block PY48C performs Eq. (8-22) and block PY48D performs Eq. (8-23). Therefore, the output of relay PY48D is $\dot{q}_1$, the energy given off by the condensing process stream.

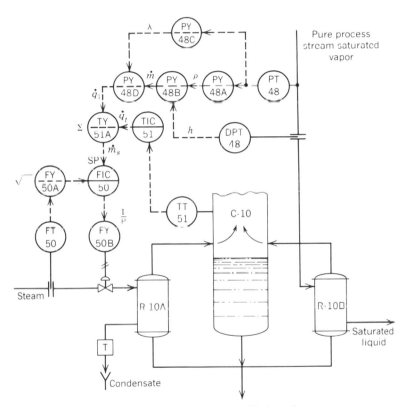

Figure 8-36. Feedforward control scheme for distillation column.

To complete the control scheme, the output of the temperature controller was considered to be the total energy required, $\dot{q}_t$, to maintain the temperature at its set point. Subtracting $\dot{q}_1$ from $\dot{q}_t$ the energy required from the steam, $\dot{q}_s$, was determined:

$$\dot{q}_s = \dot{q}_t - \dot{q}_1 \tag{8-24}$$

Finally, dividing $\dot{q}_s$ by the latent heat of condensation of the steam, h_{fg}, the required flow of steam, $\dot{m}_s$, was obtained:

$$\dot{m}_s = \frac{\dot{q}_s}{h_{fg}} \tag{8-25}$$

Block TY51A performs Eqs. (8-24) and (8-25); and its output is the set point to the flow controller FIC50. In Eq. (8-25) we are assuming that h_{fg} is constant.

Several things must be noted in this feedforward scheme. First, the model of the process is not one equation but, rather, several equations. This model was obtained using several process engineering principles. This makes process control fun, interesting, and challenging. Second, the feedback trim is an integral part of the control strategy. This compensation is q_t or total energy required to maintain set point in TIC51. Finally, the control scheme shown in Fig. 8-36 does not show a dynamic compensation or lead/lag unit. This unit may be installed later if needed.

Inverse Response

The boiler drum of Example 8-5 presents a phenomenon that can be seen in a number of different processes. As explained in the example, the level that can be recorded results from the mixture of water and bubbles and therefore is higher than the level due to water only. Now assume that it is desired to determine the gain, time constant, and dead time of the level loop and that it has been decided to do this by the step testing procedure learned in Chapter 6. Accordingly, the feedwater valve is opened to permit more water into the drum and to record the level dynamics. This is shown in Fig. 8-37. The figure shows that the drum level first drops and then increases to reach a new operating condition.

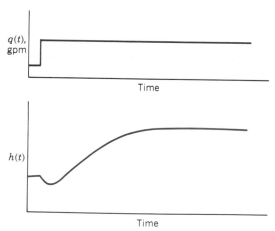

Figure 8-37. Response of drum level.

This type of response is referred to as "inverse response." Certainly this type of response is not magic but, rather, there is a physical explanation to it. What happens is that as soon as cooler water enters the drum, a certain amount of bubbles are condensed, leaving the space occupied by them void and thus reducing the apparent level. The level eventually increases because more water is entering and new bubbles are formed. Certainly, this process response is another reason why the two- or three-element control schemes, shown in Figs. 8-33 and 8-34, are so necessary.

There are several examples in the literature[9] of systems with inverse response. Chapter 9 presents the modeling of a chemical reactor that exhibits this response.

The transfer function that describes this type of response requires a positive zero:

$$\frac{H(s)}{Q(s)} = \frac{K_1(-\tau_1 s + 1)}{\tau_2 s + 1} \tag{8-26}$$

Since the evaluation of K_1, τ_1, and τ_2 is not easily done, many times the response is approximated by a first-order lag plus dead time:

$$\frac{H(s)}{Q(s)} = \frac{Ke^{-t_0 s}}{\tau s + 1}$$

The evaluation of the parameters is shown in Fig. 8-38. This type of system is usually quite nonlinear and, therefore, the parameters, K_1, τ, and t_0, are very sensitive to operating conditions.

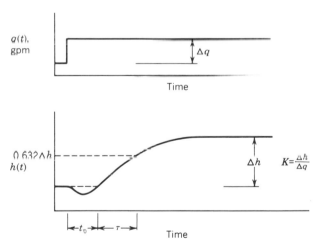

Figure 8-38. Evaluation of drum level parameters.

Feedforward Control Summary

As mentioned earlier and shown in this section, feedforward control is one of the most powerful control techniques available. However, feedforward control is not the solution to all control problems. Its implementation requires more engineering than simple feedback control and, consequently, it must be economically justified before it is implemented. When applied correctly, the economic rewards and self-satisfaction of a job well done are significant.

8-5. OVERRIDE AND SELECTIVE CONTROL

Override control is usually used as protective control to keep process variables within certain limits. Another protective control scheme is that of interlock control. Interlock controls are used primarily to protect against equipment malfunction. When a malfunction is detected, the interlock system usually shuts down the process. The action of override control is not so drastic. It maintains the process in operation but under a safer condition. In this section an example of override control is presented to show its principles and usage. Interlock control systems are not presented, but References (4 and 5) are given for their study.

Consider the process shown in Fig. 8-39. A hot saturated liquid enters the tank and from there is pumped under flow control back to the process. Under normal operation the level in the tank is at height h_1, as shown in the figure. If under any circumstances the liquid level drops to height h_2, the liquid will not have enough net positive suction

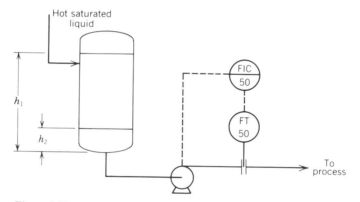

Figure 8-39. Tank and flow control loop.

head (NPSH) and cavitation at the pump will result. It is therefore necessary to design a control scheme that avoids this condition. This new control scheme is shown in Fig. 8-40.

The level in the tank is now measured and controlled. It is important to notice the action of the controllers and final control element. The variable-speed pump is such that as the energy (current in this case) input to it increases, it pumps more liquid. Therefore the FIC50 is a reverse-acting controller while the LIC50 is a direct-acting controller. The output of each controller is connected to a low selector relay, LS50, and from there its output goes to the pump.

Under normal operating conditions the level is at h_1, which is above the set point to the level controller; consequently, the controller will try to speed up the pump as much as possible, increasing its output to 20 mA. The output of the flow controller may be 16 mA and, consequently, the low selector switch selects this signal to manipulate the pump speed. This is the desired operating condition.

Let us now suppose that the flow of hot saturated liquid slows down and the level

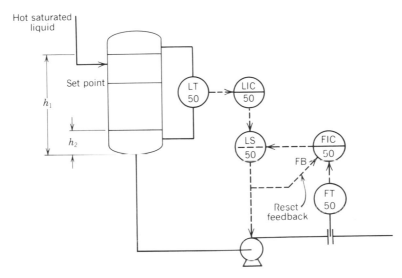

Figure 8-40. Override control scheme.

in the tank starts to drop. As soon as the level drops below the set point on the level controller, this controller will try to slow down the pump by reducing its output. As the level continues to drop, the output of the level controller also continues to drop. When it drops below the output of the flow controller, the low selector selects the output of the level controller to manipulate the pump. It can be said that the level controller "overrides" the flow controller.

An important consideration in designing an override control system is that if any of the controllers has integral mode of control, it *must* also have reset windup protection. The output of the controller must stop at 20 mA, not at a higher value. Figure 8-40 shows a reset feedback (FB) connection to the flow controller FIC50. This reset feedback is a method to implement reset windup protection, as explained in Section 6-5. The output from the controller in Fig. 6-29*b* is assumed to go to the pump. In Fig. 8-40 the output from the low selector is the one that goes to the pump and, thus, we use this signal as the reset feedback. The level controller in this example is proportional only. If it had been a PI controller, then it would have also required reset windup protection. Otherwise, it might never override the flow controller before cavitation starts.

As mentioned at the start of this section and as seen in this example, override control is used as a protective scheme. As soon as the process returns to normal operating conditions, the override scheme returns automatically to its normal operating status.

Selective control is another interesting control scheme used in the process industries. Two examples are presented to show its principles and implementation. Consider the exothermic catalytic plug flow reactor shown in Fig. 8-41. This figure also shows the typical temperature profile along the reactor. It is desired to control the temperature in the reactor at its hottest spot. This is shown in the figure. However, as the catalyst in the reactor ages or conditions change, the hot spot will move. In this case we would like to design a control scheme so that its measured variable "moves" as the hot spot moves.

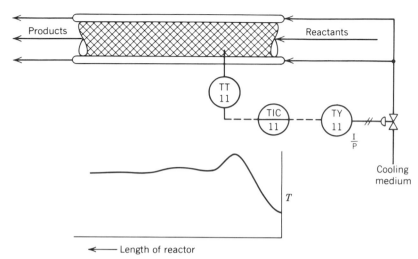

Figure 8-41. Temperature control of a plug-flow reactor.

This is shown in Fig. 8-42. With this scheme the high selector always selects the transmitter with the highest output, and in so doing the controller process variable is always
the highest temperature.

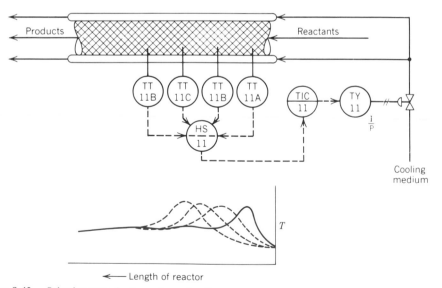

Figure 8-42. Selective control scheme for a plug-flow reactor.

In implementing this control scheme an important consideration is that all temperature
transmitters must have the same range so that their output signals can be compared on
the same basis. Another possible important consideration is to install some kind of

indication as to which transmitter is giving the highest signal. If the hot spot moves past the last transmitter, TT11D, this may be an indication that it is time to either regenerate or change the catalyst. The length of reactor left for the reaction is probably not enough to obtain the desired conversion.

Another interesting process where selective control can improve the operation is shown in Fig. 8-43. In this process a furnace heats an oil (Dowtherm) to provide an energy source to several process units. Each individual unit manipulates the flow of oil

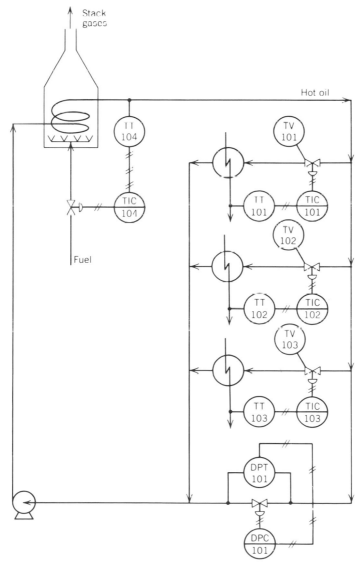

Figure 8-43.   Hot oil system.

required to maintain its controlled variable at set point. In addition, the outlet oil temperature from the furnace is also controlled by manipulating the fuel flow. To simplify the diagram, the air/fuel ratio control system is not shown; however, the reader must remember that it must be implemented. A bypass control loop, DPC101, is provided to ensure the flow of oil back to the furnace.

Suppose that it is noticed that the temperature control valve in each unit is not open very much, that is, TV101 is 20% open, TV102 is 15% open, and TV103 is 30% open. This indicates that the hot oil temperature provided by the furnace is quite high. Consequently, not much oil flow is necessary and much of it will bypass the users. This situation is very energy inefficient because in order to obtain a higher oil temperature, a higher quantity of fuel must be burned. Also, much of the energy provided by the fuel is lost to the surroundings in the piping system and through the stack gases.

The most efficient operation is the one that maintains the oil leaving the furnace at a temperature just hot enough to provide the necessary energy to the users with hardly any flow through the bypass valve. In this case most of the temperature control valves would be mostly open. Figure 8-44 shows a selective control scheme that provides this type of operation. The strategy consists in controlling the oil temperature leaving the furnace so that it is just hot enough (that is, as cool as possible) to keep the temperature control valves mostly open. It does this by first selecting the most open valve using a high selector, TY101, and then controlling it with the valve position controller, VPC101, at some desired opening, say 90% open. The controller does this control job by manipulating the set point of the furnace temperature controller.

The most open valve is being selected by comparing the pneumatic signals to each valve. Consequently, for this comparison to be correct, all of the valves must have the same characteristics.

This control scheme shows that with a bit of logic, a process operation can be significantly improved, in this case by minimizing the energy losses. The example also shows the ease of implementation of the control strategy.

8-6. MULTIVARIABLE PROCESS CONTROL

Up to this point in our study of automatic process control, only processes with a single controlled and manipulated variable have been considered. These processes are often referred to as single-input, single-output (SISO) processes. Often, however, processes in which more than one variable must be controlled are encountered; these are named multivariable processes or multiple-input, multiple-output (MIMO) processes. Some examples of these processes are shown in Fig. 8-45.

Figure 8-45a depicts a blending system for which it is necessary to control the outlet flow and outlet mass fraction of component A. To accomplish this objective the two control valves that regulate the flows of streams A and S are used. Figure 8-45b shows a chemical reactor for which it is necessary to control the outlet temperature and composition. The manipulated variables in this process are the cooling water flow and the outlet process flow. Figure 8-45c shows an evaporator with the level and outlet concen-

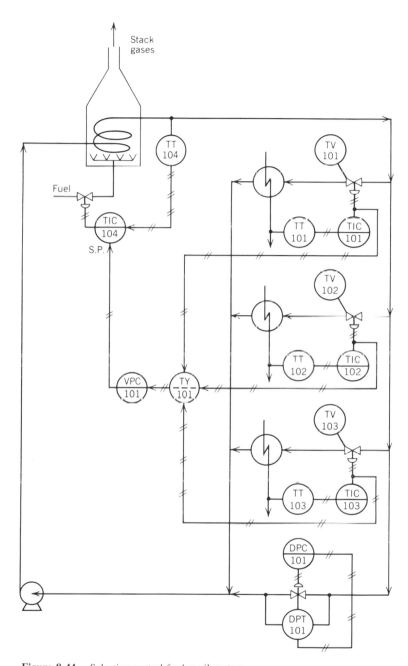

Figure 8-44. Selective control for hot oil system.

tration as controlled variables and with the outlet process flow and steam flow as manipulated variables. Figure 8-45*d* shows a paper-drying machine. The controlled variables
are the moisture and dry basis weight (fibers/area) of the final paper product. The two

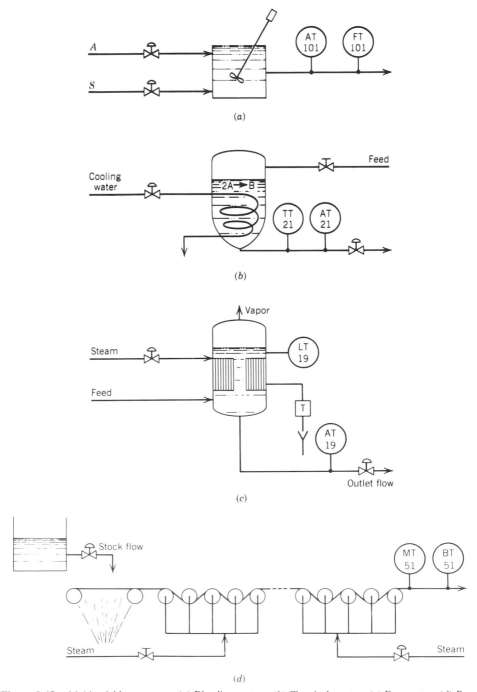

Figure 8-45. Multivariable processes. (*a*) Blending system. (*b*) Chemical reactor. (*c*) Evaporator. (*d*) Paper machine. (*e*) Distillation column.

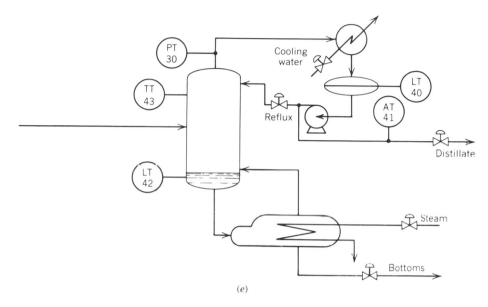

(e)

Figure 8-45. (*Continued*)

manipulated variables in this process are the stock flow to the machine and the steam flow to the last set of heated drums. Finally, Figure 8-45e depicts a typical distillation column with the necessary controlled variables: column pressure, distillate composition, accumulator level, base level, and tray temperature. To accomplish this control five manipulated variables are used: cooling water flow to the condenser, distillate flow, reflux flow, bottoms flow, and steam flow to the reboiler.

The above examples show that the control of these processes can be quite complex and challenging to the process engineer. There are usually three questions the engineer must ask when faced with a control problem of this type:

1. Which is the "best" pairing of controlled and manipulated variables?

2. How much interaction exists between the different control loops and how does this interaction affect the stability of the loops?

3. Can something be done to reduce the interaction between loops?

This section shows how to answer these questions using simple, proven techniques. To start, a short explanation of signal flow graphs (SFG) is presented. SFG is a convenient, easy method to reduce complicated block diagrams to transfer functions; it will prove helpful in answering question 2. The individual questions are then discussed.

Signal Flow Graphs (SFG)

Signal flow graphs (SFG) are graphical representations of the transfer functions that describe control systems. For complex systems it is much easier to develop the closed-loop transfer function and characteristic equation by using SFG techniques than by using the block diagram methods learned in Chapter 3. This section presents the SFG technique. For a more detailed presentation of SFG the reader is referred to Reference 6.

Figure 8-46a shows the graphical representation of a transfer function in block diagram and in signal flow graph. Figure 8-46b shows the representation of a feedback control system in both forms.

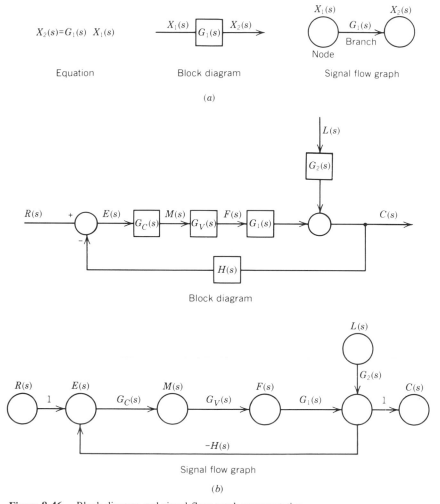

Figure 8-46.   Block diagram and signal flow graph representation.

It is important to define some terms before proceeding with the rules of SFG.

● A *node* represents the variable.
● A *branch*, shown as a line with an arrow, represents the transfer function relating the nodes it joins.

- A *path* is a continuous unidirectional succession of branches along which no node is passed more than once.
- The *path transfer function* is the product of the branch transfer functions encountered in traversing the path.
- A *loop* is a path that originates and terminates at the same node.

Referring to the SFG of Fig. 8-46b, we find these values:

Path:

$$R(s) \rightarrow E(s) \rightarrow M(s) \rightarrow F(s) \rightarrow C(s) \rightarrow C(s)$$

$$\text{Path transfer function} = G_C(s)G_V(s)G_1(s)$$

Loop:

$$E(s) \rightarrow M(s) \rightarrow F(s) \rightarrow C(s) \rightarrow E(s)$$

$$\text{Loop transfer function} = -G_C(s)G_V(s)G_1(s)H(s)$$

Figure 8-47 shows some rules necessary for signal flow graph algebra.

With these rules and the previous definitions, the required transfer function from the signal flow graph can be determined. The *Mason's Gain Formula* is used for this purpose.

$$T = \frac{\sum_i P_i \triangle_i}{\triangle} \tag{8-21}$$

where

T = transmittance (transfer function) between input and output nodes
P_i = product of the transfer function in the ith forward path between input and output nodes
$\triangle$ = signal flow graph determinant or characteristic equation
$$= 1 - (-1)^{k+1} \sum_k \sum_j L_{jk}$$
$$= 1 - \sum_j L_{j1} + \sum_j L_{j2} - \sum_j L_{j3} + \ldots$$
L_{j1} = product of all of the transfer functions in the jth single loop
L_{j2} = product of all of the transfer functions in the jth set of 2 nontouching loops
L_{j3} = product of all of the transfer functions in the jth set of 3 nontouching loops
$\vdots$

L_{jk} = product of all of the transfer functions in the jth set of k nontouching loops
$\triangle_i$ = $\triangle$ evaluated with loops touching P_i eliminated

Touching loops are those loops that have at least one node in common.

Let us look at several examples on the use of SFG to obtain the desired transfer functions.

Example 8-7. Consider the simple block diagram shown in Fig. 8-46b. Using signal flow graph determine the closed-loop transfer functions

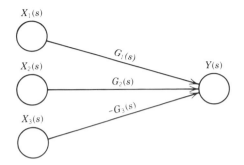

1.   Addition rule

$$Y(s) = G_1(s)X_1(s) + G_2(s)X_2(s)$$
$$- G_3(s)X_3(s)$$

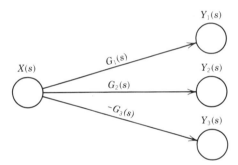

2.   Transmission rule

$$Y_1(s) = G_1(s)X(s)$$
$$Y_2(s) = G_2(s)X(s)$$
$$Y_3(s) = -G_3(s)X(s)$$

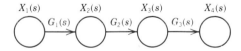

3.   Multiplication rule

$$X_4(s) = G_3(s)X_3(s) = G_3(s)G_2(s)X_2(s) = G_3(s)G_2(s)G_1(s)X_1(s)$$

Figure 8-47.   Rules for signal flow graph algebra.

$$\frac{C(s)}{R(s)} \quad \text{and} \quad \frac{C(s)}{L(s)}$$

The signal flow graph of this block diagram is also shown in Fig. 8-46*b*.

For the first required transfer function

$$T = \frac{C(s)}{R(s)}$$

there is only one forward path between $R(s)$, the input node, and $C(s)$, the output node. There is also only one loop; therefore

Path

$$P_1 = G_C(s)G_V(s)G_1(s)$$

Loop

$$L_{11} = -G_C(s)G_V(s)G_1(s)H(s)$$
$$\Delta = 1 + G_C(s)G_V(s)G_1(s)H(s)$$

Since the only loop is touching the forward path $\Delta_1 = 1$. Then we can determine that

$$\sum_i P_i \Delta_i = G_C(s)G_V(s)G_1(s)$$

Finally we find

$$\frac{C(s)}{R(s)} = \frac{G_C(s)G_V(s)G_1(s)}{1 + G_C(s)G_V(s)G_1(s)H(s)}$$

which is the expected result.

For the other required transfer function

$$T = \frac{C(s)}{L(s)}$$

again there is only one path between the input node, $L(s)$, and the output node, $C(s)$, and the only loop in this system is touching the path, so

Path

$$P_1 = G_2(s)$$

Loop

$$L_{11} = -G_C(s)G_V(s)G_1(s)H(s)$$
$$\Delta = 1 + G_C(s)G_V(s)G_1(s)H(s)$$
$$\Delta_1 = 1$$

Then we find that $\sum_i P_i \Delta_i = G_2(s)$

And finally we obtain

$$\frac{C(s)}{L(s)} = \frac{G_2(s)}{1 + G_C(s)G_V(s)G_1(s)H(s)}$$

which, again, is the expected result.

Example 8-8. Consider the block diagram of the cascade control system shown in Fig. 8-48a and obtain

$$\frac{C_2(s)}{R(s)}$$

To determine

$$T = \frac{C_2(s)}{R(s)}$$

the first step is to draw the signal flow graph as shown in Fig. 8-48b. There is only one path between $R(s)$ and $C_2(s)$ and this path is touching the two loops in this system. Furthermore, both loops are touching each other.

Path

$$P_1 = G_{C2}(s)G_{C1}(s)G_V(s)G_1(s)G_2(s)$$

Loops

$$L_{11} = -G_{C1}(s)G_V(s)G_1(s)H_1(s)$$

$$L_{21} = -G_{C2}(s)G_{C1}(s)G_V(s)G_1(s)G_2(s)H_2(s)$$

$$\triangle = 1 + G_{C1}(s)G_V(s)G_1(s)H_1(s)$$
$$+ G_{C2}(s)G_{C1}(s)G_V(s)G_1(s)G_2(s)H_2(s)$$

Since both loops are touching the only path, $\triangle_1 = 1$. Then we find

$$\sum_i P_i \triangle_i = G_{C2}(s)G_{C1}(s)G_V(s)G_1(s)G_2(s)$$

And finally we obtain

$$\frac{C_2(s)}{R(s)} = \frac{G_{C2}(s)G_{C1}(s)G_V(s)G_1(s)G_2(s)}{1 + G_{C1}(s)G_V(s)G_1(s)H_1(s) + G_{C2}(s)G_{C1}(s)G_V(s)G_1(s)G_2(s)H_2(s)}$$

Example 8-9. Consider the block diagram describing a 2×2 multivariable system shown in Fig. 8-49a. To simplify the block diagram we can define the following:

$$G_{P11}(s) = G_{V1}(s)G_{11}(s)H_1(s)$$

$$G_{P21}(s) = G_{V1}(s)G_{21}(s)H_2(s)$$

$$G_{P12}(s) = G_{V2}(s)G_{12}(s)H_1(s)$$

$$G_{P22}(s) = G_{V2}(s)G_{22}(s)H_2(s)$$

The block diagram is then drawn as shown in Fig. 8-49b. Determine

$$\frac{C_1(s)}{R_1(s)} \quad \text{and} \quad \frac{C_2(s)}{R_1(s)}$$

As usual the block diagram must first be redrawn into signal flow graph form; this is shown in Fig. 8-50.

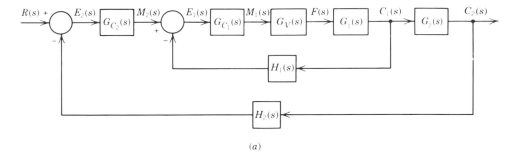

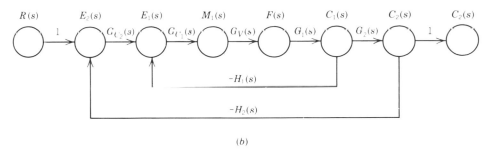

Figure 8-48. Block diagram and signal flow graph of a cascade control system.

First obtain

$$T = \frac{C_1(s)}{R_1(s)}$$

In this system there are several possible paths and loops.

Paths

$$P_1 = G_{C1}(s)G_{P11}(s)$$

$$P_2 = -G_{C1}(s)G_{P21}(s)G_{C2}(s)G_{P21}(s)$$

Loops

$$L_{11} = -G_{C1}(s)G_{P11}(s)$$

$$L_{21} = -G_{C2}(s)G_{P22}(s)$$

$$L_{31} = G_{C1}(s)G_{P21}(s)G_{C2}(s)G_{P12}(s)$$

$$L_{12} = G_{C1}(s)G_{P11}(s)G_{C2}(s)G_{P22}(s)$$

Loops L_{11} and L_{21} are the familiar feedback loops. Loop L_{31} is more complicated and goes through both feedback loops. Loop L_{12} develops from the multiplication of loops L_{11} and L_{21} since they are nontouching loops. Then we get

$$\triangle = 1 + G_{C1}(s)G_{P11}(s) + G_{C2}(s)G_{P22}(s) - G_{C1}(s)G_{P21}(s)G_{C2}(s)G_{P12}(s)$$
$$+ G_{C1}(s)G_{P11}(s)G_{C2}(s)G_{P22}(s)$$

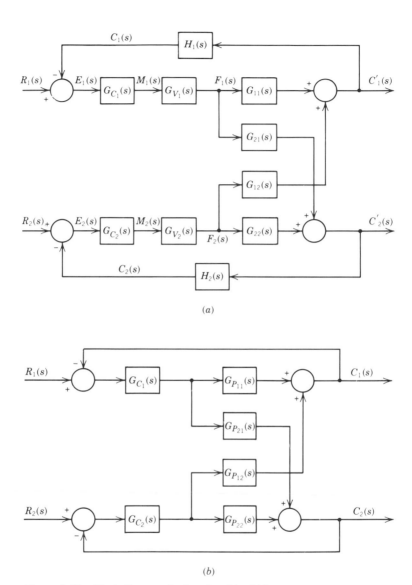

Figure 8-49. Block diagram of a 2 × 2 multivariable system.

Using algebra this △ term can also be written as follows:

$$\triangle = [1 + G_{C1}(s)G_{P11}(s)] \, [1 + G_{C2}(s)G_{P22}(s)] - G_{C1}(s)G_{P21}(s)G_{C2}(s)G_{P12}(s)$$

This final step was not necessary; however, it will prove helpful during the discussion of multivariable control stability later in this chapter. Continuing, we find

$$\triangle_1 = 1 + G_{C2}(s)G_{P22}(s)$$

$$\triangle_2 = 1$$

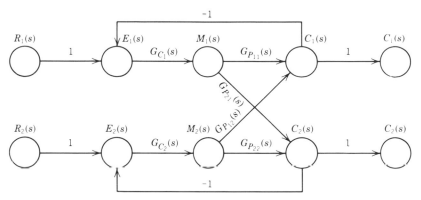

Figure 8-50. Signal flow graph of a 2 × 2 multivariable system.

Then we see

$$\sum_i P_i \Delta_i = G_{C1}(s)G_{P11}(s)\,[1 + G_{C2}(s)G_{P22}(s)] - G_{C1}(s)G_{P21}(s)G_{C2}(s)G_{P12}(s)$$

and finally we get

$$\frac{C_1(s)}{R_1(s)} = \frac{G_{C1}(s)G_{P11}(s)\,[1 + G_{C2}(s)G_{P22}(s)] - G_{C1}(s)G_{P21}(s)G_{C2}(s)G_{P12}(s)}{[1 + G_{C1}(s)G_{P11}(s)]\,[1 + G_{C2}(s)G_{P22}(s)] - G_{C1}(s)G_{P21}(s)G_{C2}(s)G_{P12}(s)}$$

For the other desired transfer function

$$T = \frac{C_1(s)}{R_2(s)}$$

the loops remain the same but the paths change.

Path

$$P_1 = G_{C2}(s)G_{P12}(s)$$

Then $\Delta_1 = 1$ and

$$\sum_i P_i \Delta_i = G_{C2}(s)G_{P12}(s)$$

Finally we see that

$$\frac{C_1(s)}{R_2(s)} = \frac{G_{C2}(s)G_{P12}(s)}{[1 + G_{C1}(s)G_{P11}(s)]\,[1 + G_{C2}G_{P22}(s)] - G_{C1}(s)G_{P21}(s)G_{C2}(s)G_{P12}(s)}$$

As shown by these examples, the signal flow graph technique is a simple, powerful tool. It provides a way to obtain closed-loop transfer functions that would be rather difficult to obtain using block diagram algebra.

Pairing Controlled and Manipulated Variables

The first question posed at the beginning of this section related to how to pair controlled and manipulated variables. Oftentimes it is simple to decide on the pairing, but sometimes, it is more difficult. Examples of a more difficult case are the blending system shown in

Fig. 8-45a and the chemical reactor shown in Fig. 8-45b. For these examples there is a technique that has proven to be successful in numerous industrial processes. To present the fundamentals of the technique a 2×2 process, shown in Fig. 8-51, is used. After this is done, the method is extended to an $n \times n$ process. (In this notation, the first n is the number of controlled variables, and the second n is the number of manipulated variables.)

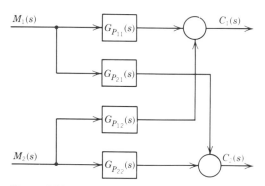

Figure 8-51.   Partial block diagram of a 2×2 multivariable process.

To begin with, it makes sense to control each controlled variable with the manipulated variable that has the greatest "influence" on it. In this context influence and gain have the same meaning; consequently, to make a decision we must find all gains (4 gains for a 2×2 system) of the process. Specifically, the following are the open-loop steady-state gains of interest:

$$K_{11} = \left.\frac{\Delta c_1}{\Delta m_1}\right|_{m_2} \qquad K_{12} = \left.\frac{\Delta c_1}{\Delta m_2}\right|_{m_1}$$

$$K_{21} = \left.\frac{\Delta c_2}{\Delta m_1}\right|_{m_2} \qquad K_{22} = \left.\frac{\Delta c_2}{\Delta m_2}\right|_{m_1}$$

where the notation K_{ij} is the gain relating the ith controlled variable to the jth manipulated variable.

The four gains can be arranged in the form of a matrix to give a more graphical description of their relationship to the controlled and manipulated variables. This matrix is called the steady-state gain matrix (SSGM) and is shown in Fig. 8-52.

	m_1	m_2
C_1	K_{11}	K_{12}
C_2	K_{21}	K_{22}

Figure 8-52.   Steady-state gain matrix (SSGM).

From this SSGM the combination of the controlled and manipulated variables that yields the largest absolute number in each row may appear to be the one that should be chosen. That is, if $|K_{12}|$ is larger than $|K_{11}|$, then m_2 is chosen to control c_1. However, this method of choosing the pairing of controlled and manipulated variables is not quite correct because the gains, K_{ij}, may all have different units; consequently, their comparison is not possible. The matrix, as it stands, is dependent on units and, therefore, is not useful for this purpose.

A technique developed by Bristol[7] has been proposed to normalize the SSGM, making it independent of the units. This technique, called the relative gain matrix (RGM), or interaction measure, has proven to yield good results.

The terms in the RGM as proposed by Bristol are defined as follows:

$$\mu_{11} = \frac{\left.\frac{\partial c_1}{\partial m_1}\right|_{m_2}}{\left.\frac{\partial c_1}{\partial m_1}\right|_{c_2}} \approx \frac{\left.\frac{\Delta c_1}{\Delta m_1}\right|_{m_2}}{\left.\frac{\Delta c_1}{\Delta m_1}\right|_{c_2}} = \frac{K_{11}}{K'_{11}}$$

Or in general

$$\mu_{ij} = \frac{\left.\frac{\partial c_i}{\partial m_j}\right|_{m}}{\left.\frac{\partial c_i}{\partial m_j}\right|_{c}} \tag{8 28}$$

Let us make sure we understand the meaning and significance of all terms in Eq. (8-28). The numerator

$$\left.\frac{\partial c_i}{\partial m_j}\right|_{m}$$

is the steady-state gain, K_{ij}, defined above. That is, this is the gain of m_j on c_i when all other manipulated variables are kept constant. The denominator

$$\left.\frac{\partial c_i}{\partial m_j}\right|_{c}$$

is another type of gain, K'_{ij}. This is the gain of m_j on c_i when all other loops are closed and assuming that all other controllers have integral action and, therefore, all other controlled variables are returned to their set points. Thus we can write

$$\mu_{ij} = \frac{\text{gain when the other loops are open}}{\text{gain when the other loops are closed}} = \frac{K_{ij}}{K'_{ij}}$$

As seen from this definition, the relative gain term is a dimensionless quantity that can be used to decide on the pairing of controlled and manipulated variables.

Once the relative gain terms have been evaluated, the relative gain matrix shown in Fig. 8-53 can be formed. From this matrix, the pairing of controlled and manipulated variables can be chosen. We will try to obtain a better understanding of the relative gain terms before the pairing rule is presented.

A property of the relative gain matrix is that the summation of all terms in each

	m_1	m_2
C_1	μ_{11}	μ_{12}
C_2	μ_{21}	μ_{22}

Figure 8-53. Relative gain matrix (RGM).

column must also equal one. This means that for a 2×2 system only one term has to be evaluated and the others can be obtained using this property. In a 3×3 system, which has nine terms in the matrix, only four independent terms need to be evaluated; the other five terms are then obtained using simple arithmetic. This property will be proven later in this section.

An understanding of the meaning of μ_{ij} should be obtained before proceeding into its evaluation. From the definition of μ_{ij}, notice that essentially it is a measure of the effect of closing all other loops on the process gain for a given controlled and manipulated variable pair. The numerical value of μ_{ij} is thus a measure of the interaction between the control loops.

If $\mu_{ij} = 1$ it means that the process gain is the same with all other loops open or closed. This indicates no interaction, or possibly offsetting interaction, between the particular loop and all other loops. The more μ_{ij} deviates from 1, the greater is the interaction between the loops.

If $\mu_{ij} = 0$ it may be due to one of two possibilities. First, the open loop gain, $\dfrac{\partial c_i}{\partial m_j}\bigg|_m$ may be zero. In this case c_i is not affected by m_j at least in terms of open loops. Second, the closed-loop gain is so large that $\mu_{ij} \approx 0$. This means that in order to keep the other controlled variables constant, the other loops are interacting significantly with the loop in question. In any case, either of the two possibilities indicates that c_i should not be controlled by m_j.

If $\mu_{ij} = \infty$ the closed-loop gain is zero or very small. This means that when the other loops are in automatic, the loop in question cannot be controlled because m_j affects c_i very little or not at all. The only way to control this loop is with the other loops in manual; this is certainly not a desirable situation!

In the last three paragraphs the extreme cases of the multivariable problem have been presented. In general, values of relative gains close to one represent controllable combinations of controlled and manipulated variables. Values of relative gains approaching zero or infinity represent uncontrollable combinations.

With this background the rule to pair controlled and manipulated variables can be presented. This *pairing rule*, first presented by Bristol[7] and then modified by Koppel[16], is as follows:

Always pair on positive RGM elements that are closest to 1.0. Check the resulting pairings for stability using Niederlinski's theorem. If the pairings yield an unstable system choose other positive pairings with values closest to 1.0. Avoid negative pairings if possible.

Niederlinski's theorem[17] provides a convenient way to check for the stability of the proposed pairings. To use this theorem assume that the proposed pairings are the diagonal elements, $m_i - c_i$, in the SSGM. The theorem then states:

The closed-loop system resulting from the pairing $m_1 - c_1, m_2 - c_2, \ldots, m_n - c_n$ is unstable if

$$\frac{|SSGM|}{\prod_{i=1}^{n} K_{ii}} < 0$$

where $|SSGM|$ is the determinant of the SSGM and K_{ii} are the diagonal elements in the matrix.

The proposed pairing rule is easy and convenient to use. Realize that only steady-state gain information is needed. This is certainly an advantage since this information can usually be obtained during the design stage of the process. Thus, it does not require the plant or process to be in operation. In the next section we go over different ways to obtain the steady-state gains. For an expanded presentation of the pairing rule and the complete subject of multivariable control, the reader is encouraged to read the monograph by McAvoy.[15]

To close this presentation let us look at two possible RGM's and understand what the terms μ_{ij} are telling us about the control system. Consider the following RGM:

	m_1	m_2
c_1	0.2	0.8
c_2	0.8	0.2

The terms $\mu_{11} = \mu_{22} = 0.2 = 1/5$ indicate that for this pairing the gain of each loop increases by a factor of 5 when the other loop is closed. The terms $\mu_{12} = \mu_{21} = 0.8 = 4/5$ indicate that for this pairing the gain increases only by a factor of 1.25. This expalins why the pairing of $c_1 - m_2$ and $c_2 - m_1$ is the correct one.

Consider another RGM:

	m_1	m_2
c_1	2	-1
c_2	-1	2

The terms $\mu_{11} = \mu_{22} = 2 = 1/0.5$ indicate that the gain of each loop is cut in half when the other loop is closed. The terms $\mu_{12} = \mu_{21} = -1$ indicate that the gain of each loop changes sign when the other loop is closed. Certainly, this last case is undesirable because it means that the action of the controller depends on whether the other loop is closed or open. This explains why the correct pairing is $c_1 - m_1$ and $c_2 - m_2$.

Obtaining the Process Gains and Relative Gains.   The blending process shown in Fig. 8-45a is used to review the methods to obtain the open-loop gains and to present the evaluation of the relative gains.

To obtain the open-loop steady-state gains analytically, the equations that describe the process are written first. From these equations the gains, K_{ij}, are then evaluated. In the blending system,[10] the outlet flow, F, and the outlet mass fraction of component A, x, are to be controlled. A steady-state total mass balance provides the expression for the outlet flow:

$$F = A + S \tag{8-29}$$

A steady-state mass balance on component A provides the other required expression:

$$Fx = A$$

or

$$x = \frac{A}{F}$$

Substituting Eq. (8-29) into the equation for x yields:

$$x = \frac{A}{A + S} \tag{8-30}$$

In this 2×2 system there are four open-loop gains: K_{FA}, K_{FS}, K_{xA}, and K_{xS}. From Eq. (8-29) the first two can be evaluated as

$$K_{FA} = \left.\frac{\partial F}{\partial A}\right|_S = 1$$

and

$$K_{FS} = \left.\frac{\partial F}{\partial S}\right|_A = 1$$

From Eq. (8-30) the last two gains can be evaluated as

$$K_{xA} = \left.\frac{\partial x}{\partial A}\right|_S = \frac{S}{(A + S)^2} = \frac{1 - x}{F}$$

and

$$K_{xS} = \left.\frac{\partial x}{\partial S}\right|_A = \frac{-A}{(A + S)^2} = \frac{-x}{F}$$

Consequently, the steady-state gain matrix can be shown to be

	A	S
F	1	1
x	$\dfrac{1 - x}{F}$	$\dfrac{-x}{F}$

For this blending process the development of the descriptive set of equations and the evaluation of the gains were fairly simple. For some processes these are not easily done. Examples are the distillation column, the paper machine, and the chemical reactor shown in Fig. 8-45. For these processes the describing set of equations is complex and, consequently, the evaluation of the gains becomes a difficult task. Fortunately, however, the design of many processes is usually done by steady-state computer simulations.[13] From these simulations it is usually simple to evaluate the required gains. For a 2 × 2 system three computer runs suffice to obtain the four gains. In these cases, the differential of the variables to obtain the gains cannot be used but, rather, the following approximation is used:

$$K_{ij} \approx \frac{\Delta c_i}{\Delta m_j}$$

If for any reason it is impossible to use the analytical or computer simulation methods, then there is another method to obtain K_{ij}'s when the process exists. This method consists of obtaining the necessary data to evaluate the gains. The technique used to obtain the open-loop gains can be any of the identification techniques learned so far, for instance, step testing (Chapter 6).

Once the open-loop gains have been obtained, the evaluation of the other gains, K'_{ij}, and the relative gain terms, is fairly straightforward. A simple method for any 2 × 2 method is presented first and then extended to any higher-order system.

The effect of a change in the manipulated variables on c_1 can be expressed as follows:

$$\Delta c_1 = K_{11}\Delta m_1 + K_{12}\Delta m_2 \tag{8-31}$$

Similarly, on c_2 we have

$$\Delta c_2 = K_{21}\Delta m_1 + K_{22}\Delta m_2 \tag{8-32}$$

To obtain the gain

$$\left.\frac{\partial c_1}{\partial m_1}\right|_{c_2}$$

or

$$\left.\frac{\Delta c_1}{\Delta m_1}\right|_{c_2}$$

Δc_2 in Eq. (8-32) is set equal to zero:

$$0 = K_{21}\Delta m_1 + K_{22}\Delta m_2$$

or

$$\Delta m_2 = -\frac{K_{21}}{K_{22}}\Delta m_1$$

Substituting this expression for Δm_2 into Eq. (8-31) yields

$$\Delta c_1 = K_{11}\Delta m_1 - \frac{K_{12}K_{21}}{K_{21}}\Delta m_1$$

and finally we obtain

$$K'_{11} = \left.\frac{\Delta c_1}{\Delta m_1}\right|_{c_2} = \frac{K_{11}K_{22} - K_{12}K_{21}}{K_{22}} \tag{8-33}$$

Similarly, for the gain

$$\left.\frac{\partial c_2}{\partial m_2}\right|_{c_1}$$

or

$$\left.\frac{\Delta c_2}{\Delta m_2}\right|_{c_1}$$

Δc_1 in Eq. (8-31) is set equal to zero:

$$0 = K_{11}\Delta m_1 + K_{12}\Delta m_2$$

or

$$\Delta m_1 = -\frac{K_{12}}{K_{11}}\Delta m_2$$

Substituting this expression for Δm_1 into Eq. (8-32) yields

$$\Delta c_2 = -\frac{K_{21}K_{12}}{K_{11}}\Delta m_2 + K_{22}\Delta m_2$$

and finally we obtain

$$K'_{22} = \left.\frac{\Delta c_2}{\Delta m_2}\right|_{c_1} = \frac{K_{22}K_{11} - K_{21}K_{12}}{K_{11}} \tag{8-34}$$

Following the same procedure the other two K'_{ij} gains can be evaluated as

$$K'_{12} = \left.\frac{\Delta c_1}{\Delta m_2}\right|_{c_2} = \frac{K_{21}K_{12} - K_{11}K_{22}}{K_{21}} \tag{8-35}$$

and

$$K'_{12} = \left.\frac{\Delta c_2}{\Delta m_1}\right|_{c_1} = \frac{K_{21}K_{12} - K_{11}K_{22}}{K_{12}} \tag{8-36}$$

It is interesting, and important, to note that the K'_{ij} gains can be evaluated simply by a combination of open-loop gains. Once these gains are obtained, the evaluation of the relative gains is as follows:

$$\mu_{11} = \frac{\left.\dfrac{\partial c_1}{\partial m_1}\right|_{m2}}{\left.\dfrac{\partial c_1}{\partial m_1}\right|_{c2}} \approx \frac{\left.\dfrac{\Delta c_1}{\Delta m_1}\right|_{m2}}{\left.\dfrac{\Delta c_1}{\Delta m_1}\right|_{c2}} = \frac{K_{11}}{\dfrac{K_{11}K_{22} - K_{12}K_{21}}{K_{22}}}$$

$$\mu_{11} = \frac{K_{11}K_{22}}{K_{11}K_{22} - K_{12}K_{21}} \tag{8-37}$$

Similarly, the other relative gains are obtained as

$$\mu_{12} = \frac{K_{12}K_{21}}{K_{21}K_{12} - K_{11}K_{22}} \tag{8-38}$$

$$\mu_{21} = \frac{K_{21}K_{12}}{K_{21}K_{12} - K_{11}K_{22}} \tag{8-39}$$

and

$$\mu_{22} = \frac{K_{11}K_{22}}{K_{11}K_{22} - K_{12}K_{21}} \tag{8-40}$$

Finally, the relative gain matrix becomes

	m_1	m_2
c_1	$\dfrac{K_{11}K_{22}}{K_{11}K_{22} - K_{12}K_{21}}$	$\dfrac{K_{12}K_{21}}{K_{21}K_{12} - K_{11}K_{22}}$
c_2	$\dfrac{K_{21}K_{12}}{K_{21}K_{12} - K_{11}K_{22}}$	$\dfrac{K_{11}K_{22}}{K_{11}K_{22} - K_{12}K_{21}}$

From this matrix, using the pairing rule presented earlier, the correct combination of controlled and manipulated variables is chosen.

It is easily shown from the matrix that the terms in each row and each column add up to one. The dimensionality consistency of each term is also easily shown.

Applying the relative gain matrix to the blending process example yields

	A	S
F	x	$1 - x$
x	$1 - x$	x

The pairing will depend on the value of x. If $x > 0.5$, the correct pairing is $F - A$ and $x - S$. If $x < 0.5$, the correct pairing becomes $F - S$ and $x - A$. A value of $x = 0.5$ yields all μ's equal to 0.5. This value for a 2×2 system indicates the highest degree of interaction.

The steps taken above to develop the relative gain matrix for a 2×2 system required only simple algebra. For a higher-order system the same procedure could be followed; however, more algebraic steps must be taken to reach a final solution. For these higher-order systems matrix algebra can be used to simplify the development of the relative gain matrix. The procedure, as proposed by Bristol[7], is as follows:

Calculate the transpose of the inverse of the steady-state matrix and multiply each term of the new matrix by the corresponding term in the original matrix. The terms thus obtained are the terms of the "interaction measure (matrix)" or relative gain matrix.

This procedure might look out of reach for those unfamiliar with matrix algebra. But given the tremendous utility of this method, it is worthwhile to surpass this difficulty. There are digital computer programs that can easily crank out the necessary numbers. This procedure yields the same relative gain matrix as the one previously shown when applied to any 2×2 system.

The Interaction Index. Nisenfeld and Schultz[8] define the interaction index for a multivariable system in which manipulated variable m_j is used to control variable c_i as follows:

$$I_{ij} \equiv \left| \frac{1 - \mu_{ij}}{\mu_{ij}} \right| \tag{8-41}$$

where the bars stand for the absolute value of the quantity between them. *These authors state that an interaction index less than one avoids instability due to loop interaction.*

To obtain a better understanding of this index, consider the blending system shown in Fig. 8-54. The flow controller adjusts stream A and the composition controller adjusts stream S.

A particular interaction index for this process is

$$I_{FA} = \left| \frac{1 - \mu_{FA}}{\mu_{FA}} \right| = \left| \frac{K_{FS} K_{xA}}{K_{FA} K_{xS}} \right|$$

If a set point change in flow rate takes place, a deviation ΔF is detected by the flow controller. The corrective action by this controller would be $\Delta A = \dfrac{\Delta F}{K_{FA}}$ (in magnitude) in order to move the flow rate to the new set point. Because of interaction, this change in stream A causes a change in composition given by $\Delta x = K_{xA} \Delta A = K_{xA} \Delta F / K_{FA}$. This change in composition is detected by the composition controller, which takes corrective action to restore the composition back to its set point. This action on stream S has a magnitude $\Delta S = \Delta x / K_{xS} = K_{xA} \Delta F / K_{FA} K_{xS}$. Finally, and again because of interaction, this change in stream S causes a deviation in flow rate of magnitude $\Delta F' = K_{FS} \Delta S = K_{FS} K_{xA} \Delta F / K_{FA} K_{xS} = I_{FA} \Delta F$. As we continue, we realize that a full cycle has now been completed and will be repeated until all deviations die out. Note that for this to be the case, the interaction index, I_{FA}, must be less than unity in order for the deviations in flow, $\Delta F'$, after each complete cycle to be less than the deviation, ΔF, at the start of the cycle. If the interaction index is greater than unity, the deviations will increase after each cycle until the controller output reaches saturation, a highly undesirable situation.

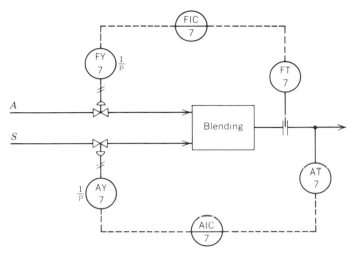

Figure 8-54. Blending system.

For higher-order systems, the interaction indices may be computed for each possible pairing by the equation

$$I_{ij} = \left| \frac{1 - \mu_{ij}}{\mu_{ij}} \right|$$

Those combinations yielding values greater than unity can be rejected.

Positive and Negative Interactions. Positive interaction is the interaction experienced when all relative gain terms are positive. It is interesting to see under what conditions this type of interaction results. A lot can be learned from the expression for μ_{11} for a 2×2 system:

$$\mu_{11} = \frac{K_{11}K_{22}}{K_{11}K_{22} - K_{12}K_{21}} = \frac{1}{1 - \dfrac{K_{12}K_{21}}{K_{11}K_{22}}} \tag{8-42}$$

If there is an odd number of positive K's, then the value of μ_{11} will be positive and, furthermore, its numerical value will be between 0 and 1. It is fairly simple, and left to the reader, to prove that this is a general rule.

Positive interaction is the most common type of interaction in multivariable control systems. In these systems the control loops "help" each other. To understand what we mean by this, consider the blending system shown in Fig. 8-54 and its block diagram shown in Fig. 8-55. For this system the gains of the control valves are positive since both valves are air-to-open. The gains K_{11}, K_{21}, and K_{12} are also positive, while the gain K_{22} is negative. The flow controller, $G_{C1}(s)$, is reverse acting while the analyzer controller, $G_{C2}(s)$, is direct acting. Assume now that the set point to the flow controller decreases ($\downarrow$) and the flow controller, $G_{C1}(s)$, will in turn lower ($\downarrow$) its output. This will cause the output from $G_{V1}(s)$ and $G_{11}(s)$ to decrease ($\downarrow$). Since K_{21} is positive the output from

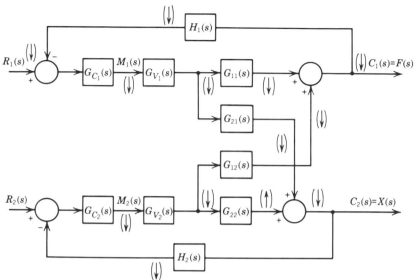

The following notation is used:

$C_i(s)$ = Laplace transform of controlled variable i

$M_i(s)$ = Laplace transform of manipulated variable i, output of controller i

$H_i(s)$ = transfer function of sensor-transmitter i

$G_{Vi}(s)$ = transfer function of valve i

$G_{ij}(s)$ = transfer function describing how flow of stream j affects controlled variable i

Figure 8-55. Block diagram and notation of blending system.

$G_{21}(s)$ also decreases (↓) when the output from $G_{V1}(s)$ decreases, resulting in lowering the analysis, $X(s)$ (↓). When this happens the analysis controller, $G_{C2}(s)$, decreases (↓) its output. This will cause the output from $G_{V2}(s)$ and $G_{12}(s)$ to decrease (↓) and the output from $G_{22}(s)$ to increase (↑). Fig. 8-55 shows the arrows that indicate the direction in which the output from each block moves. The figure clearly shows that the outputs from $G_{11}(s)$ and $G_{12}(s)$ both decrease. This is what we mean by "both loops help each other."

When there are an even number of positive values of K or an equal number of positive and negative values of K, the value of μ_{11} will either be less than 0 or greater than 1. In either case there will be some μ_{ij} with negative values in the same row and column. The interaction in this case is said to be a negative interaction. It is important to realize that for a relative gain term to be negative the signs of the open-loop and closed-loop gains must be different. This means that the gain of the feedback controller *must reverse* when the other loops are closed. For this type of interaction the control loops "fight" each other. This is left for you to prove.

Controller Tuning for Interacting Systems. The tuning of the controllers of interacting systems is a difficult task. Shinskey[9] recommends that each controller be tuned with the other controllers in manual and then their tuning be readjusted when the other controllers are transferred to automatic. This readjustment could be based on the relative gain term for that loop. Specifically, multiply the gain of the controller by the relative gain term (multiply the proportional band by the reciprocal of the relative gain term). This derives from the fact that the relative gain terms are equal to the ratio of the open-loop gain to the closed-loop gain.

For the blending system previously discussed, if the desired concentration is $x = 0.6$, the relative gain terms are $\mu_{FA} = \mu_{xS} = 0.6$, and the proper pairing is F-A and x-S. With the composition controller in manual, if the tuning of the flow controller results in a controller gain of 0.5 (200% PB), it should be readjusted to $(0.6)(0.5) = 0.3$ (333% PB). A similar readjustment should be done to the gain of the composition controller obtained by tuning it with the flow controller in manual.

It is interesting to note that the terms in the relative gain matrix are useful, not only to help decide upon the pairing of controlled and manipulated variables, but also to correct controller tuning parameters to account for interaction.

Interaction and Stability

The question of how interaction between control loops affects their stability is now answered, by using the blending system shown in Fig. 8-54. The interaction existing in this process is shown graphically in Fig. 8-55. Notice that any change in the output from controller 1 affects not only $C_1(s)$ but also $C_2(s)$. The same is true for any change in output from controller 2.

As explained in Chapters 6 and 7, the roots of the characteristic equation define the stability of control loops. For the system shown in Fig. 8-55, the characteristic equations for each individual loop are

$$1 + G_{C1}(s)G_{V1}(s)G_{11}(s)H_1(s) = 0 \tag{8-43}$$

and

$$1 + G_{C2}(s)G_{V2}(s)G_{22}(s)H_2(s) = 0 \tag{8-44}$$

As already learned, each loop is stable if the roots of the particular characteristic equation have negative real parts. To analyze the stability of the *complete* control system shown in Fig. 8-55 the characteristic equation for the complete system must be determined. As shown in Example 8-9, this characteristic equation is

$$[1 + G_{C1}(s)G_{V1}(s)G_{11}(s)H_1(s)][1 + G_{C2}(s)G_{V2}(s)G_{22}(s)H_2(s)]$$
$$- G_{C1}(s)G_{V1}(s)G_{C2}(s)G_{V2}(s)G_{12}(s)G_{21}(s)H_1(s)H_2(s) = 0 \tag{8-45}$$

The terms in brackets are the characteristic equations of the individual loops. By analyzing this equation the following conclusions for a 2×2 system can be reached:

1. The roots of the characteristic equation for each individual loop are *not* the roots of the characteristic equation for the complete system. Therefore, it is possible for the complete system to be unstable even though each individual loop is stable. "Complete system" means that both loops are in automatic at the same time.

2. For interaction to affect the stability of the complete system, it must work both ways. That is, each manipulated variable *must* affect both controlled variables. If either $G_{12}(s) = 0$ or $G_{21}(s) = 0$, the last term of the characteristic equation disappears. Therefore, if each individual loop is stable, the complete system is also stable. When the interaction works both ways, the system is said to be fully coupled. When the interaction works only one way, the system is said to be partially, or half, coupled. Interaction is not a problem in a half-coupled system.

3. The interaction effect on one loop can be eliminated by interrupting the other loop; this is easily done by ''switching'' a controller to manual. Suppose that controller 2 is switched to manual. This has the effect of setting $G_{C2}(s) = 0$, leaving the following characteristic equation:

$$1 + G_{C1}(s)G_{V1}(s)G_{11}(s)H_1(s) = 0$$

which is the same as if only one loop existed. This is one of the reasons why many controllers in industry are in manual. Manual changes in the output of controller 2 become simple disturbances to loop 1.

Usually, however, it is not necessary to be this drastic to yield a stable system. By simply lowering the gain and increasing the reset time, thus slowing the reset action, the same effect, that is, stabilizing the complete system, can be accomplished while both controllers remain in automatic. The effect of doing this is to move all of the roots of the characteristic equation to the negative side of the real axis.

The preceding paragraphs have described how to analyze the effect of interaction on the stability of control loops. This can be done fairly simply once the characteristic equation for the system is known. The reader must remember that these comments have been made with respect to a 2×2 system, which is probably the most common multivariable control system. For higher-order systems the same procedure must be followed; however, the conclusions obtained from the characteristic equation may not be as simply determined.

Decoupling

There is still one more question to answer: Can something be done to reduce or eliminate the interaction between interacting loops? That is, can a control system be built to decouple the interacting, or coupled, loops? Decoupling can be a profitable, realistic possibility when applied carefully. The relative gain matrix provides an indication of when decoupling could be beneficial. The closer in value the terms of the matrix are to each other, the greater the interaction between the loops. For existing systems, operating experience is usually enough to decide.

There are several types of decouplers and ways to design them. This section presents one of the easiest and most practical methods.

Consider the general 2×2 interacting control system shown in Fig. 8-56. This block diagram shows graphically the interaction between the two loops. To circumvent

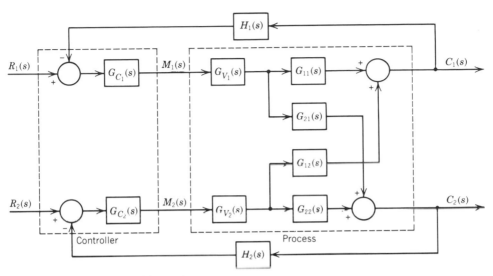

Figure 8-56. General 2 × 2 interacting system.

this interaction a decoupler may be designed and installed as shown in Fig. 8-57. The decoupler should be designed such that the process-decoupler combination yields two control loops that appear to be independent. To put it in mathematical terms

$$\frac{\partial C_1}{\partial M_2}\bigg|_{M_1} = 0 \qquad \text{and} \qquad \frac{\partial C_2}{\partial M_1}\bigg|_{M_2} = 0$$

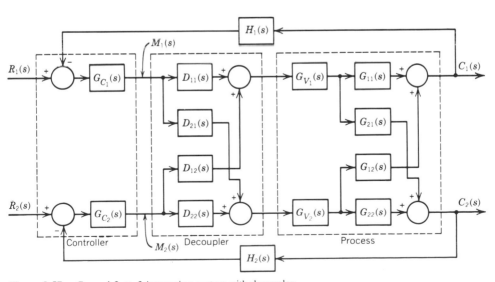

Figure 8-57. General 2 × 2 interacting system with decoupler.

That is, the decoupler should be designed such that a change in the output of controller 1, $M_1(s)$, produces a change in $C_1(s)$ but not in $C_2(s)$. Similarly, a change in the output of controller 2, $M_2(s)$, should produce a change in $C_2(s)$ but not in $C_1(s)$.

Another way of interpreting the decoupler is by considering it as part of the controllers, making the controller–decoupler combination an interacting controller. Thus, the idea is to design an interacting controller to produce a noninteracting system.

The block diagram of Fig. 8-57 can be simplified by combining the transfer functions of the process as follows:

$$G_{Pij}(s) = G_{Vj}(s)G_{ij}(s)$$

This new block diagram is shown in Fig. 8-58. An open-loop block diagram of the decoupler and process is shown in Fig. 8-59. From this last block diagram the decouplers for a 2 × 2 system can be designed.

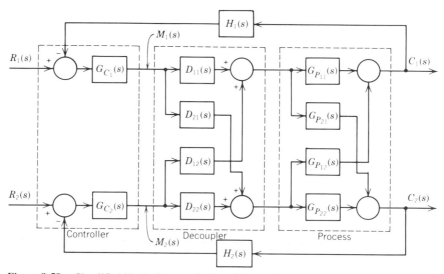

Figure 8-58. Simplified block diagram of a 2 × 2 interacting system with decoupler.

The output from the second controller, $M_2(s)$, affects the first controlled variable, $C_1(s)$, as given by the following equation:

$$C_1(s) = [D_{12}(s)G_{P11}(s) + D_{22}(s)G_{P12}(s)]M_2(s) \qquad (8\text{-}46)$$

Similarly, $M_1(s)$ affects $C_2(s)$ as given by the following equation:

$$C_2(s) = [D_{21}(s)G_{P22}(s) + D_{11}(s)G_{P21}(s)]M_1(s) \qquad (8\text{-}47)$$

As it is, we now have two equations, (8-46) and (8-47), and four unknowns, $D_{11}(s)$, $D_{12}(s)$, $D_{22}(s)$, and $D_{21}(s)$. Therefore, we have two degrees of freedom, which means that two of the unknowns must be set before calculating the rest. A common procedure is to set them to unity. Usually, the elements chosen are $D_{11}(s)$ and $D_{22}(s)$. Following this procedure the two equations become

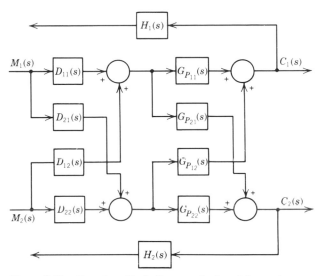

Figure 8-59. Open-loop block diagram of a 2 × 2 interacting system with decoupler.

$$C_1(s) = [D_{12}(s)G_{P11}(s) + G_{P12}(s)]M_2(s) \qquad (8\text{-}48)$$

and

$$C_2(s) = [D_{21}(s)G_{P22}(s) + G_{P21}(s)]M_1(s) \qquad (8\text{-}49)$$

The objective now is to design $D_{12}(s)$ so that when the output from the second controller changes, the first controlled variable remains constant. If this controlled variable is to remain constant, its deviation variable is zero, $C_1(s) = 0$. Then from Eq. (8-48) we get

$$0 = [D_{12}(s)G_{P11}(s) + G_{P12}(s)]M_2(s)$$

and finally we obtain

$$D_{12}(s) = -\frac{G_{P12}(s)}{G_{P11}(s)} \qquad (8\text{-}50)$$

Similarly, $D_{21}(s)$ can be designed so that when $M_1(s)$ changes $C_2(s)$ remains at zero. From Eq. (8-49) we obtain

$$D_{21}(s) = -\frac{G_{P21}(s)}{G_{P22}(s)} \qquad (8\text{-}51)$$

Equations (8-50) and (8-51) are the decoupler design equations for a 2 × 2 system. Remember that $D_{11}(s)$ and $D_{22}(s)$ are set to unity. Let us now use an example to show in more detail the calculations and implementations.

Example 8-10. Consider the evaporator shown in Fig. 8-60. In this evaporator an aqueous solution of NaOH is concentrated from 0.2 to 0.5 NaOH mass fraction. There are two controlled variables in this evaporator, the liquid level and the outlet NaOH mass fraction, and two manipulated variables, the feed flow and the product flow. The data

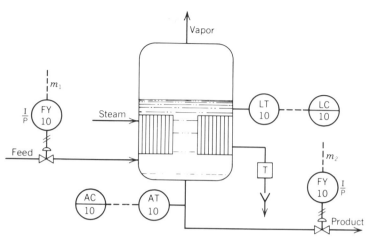

Figure 8-60. Evaporator.

shown in Table 8-5 were obtained from dynamic testing. The level transmitter has a range of 1–5 m and the concentration transmitter has a range of 0.2–0.8 NaOH mass fraction. Decide on the correct pairing of variables and design a decoupler system. Show the instrumentation diagram required to implement the decoupler. All the instrumentation, except the valves, is electronic.

The first thing to do is to draw a general open-loop block diagram of this system, as shown in Fig. 8-61a. From the available data notice that the block diagram can be simplified as shown in Fig. 8-61b. From the information given in Table 8-6 and using Fit 3 as shown in Chapter 6, the following transfer functions are determined:

$$G_{PL1}(s) = \frac{0.072\,e^{-0.3s}}{2.7s + 1}, \quad \frac{\text{m}}{\%\ \text{controller 1 output}}$$

$$G_{Px1}(s) = \frac{-0.00336\,e^{-0.15s}}{1.05s + 1}, \quad \frac{\text{mass fraction}}{\%\ \text{controller 1 output}}$$

$$G_{PL2}(s) = \frac{0.036\,e^{-0.03s}}{2.97s + 1}, \quad \frac{\text{m}}{\%\ \text{controller 2 output}}$$

$$G_{Px2}(s) = \frac{-0.00144\,e^{-0.4s}}{1.65s + 1}, \quad \frac{\text{mass fraction}}{\%\ \text{controller 2 output}}$$

With this information we can now decide on the correct pairing and design the decoupler systems. The steady-state gain matrix is as follows:

	m_1	m_2
L	0.072	0.036
x	−0.00336	−0.00144

By applying Eqs. (8-37) through (8-40) we obtain the relative gain matrix:

Table 8-5. Evaporator Dynamic Data

Time, min	m_1, %CO[a]	x, mass fraction	L, m
0	50	0.500	3.00
0	75	0.500	3.00
0.25	75	0.493	3.10
0.50	75	0.477	3.22
0.75	75	0.464	
1.00	75	0.453	3.48
1.25	75	0.447	
1.50	75	0.439	
2.00	75	0.427	3.72
2.50	75	0.419	
3.00	75	0.416	3.14
4.00	75	0.416	4.48
5.00	75	0.416	4.50
6.00	75	0.416	4.62
7.00	75	0.416	4.70
8.00	75	0.416	4.76
9.00	75	0.416	4.79
10.00	75	0.416	4.80
11.00	75	0.416	4.80

Time, min	m_2, %CO[a]	x, mass fraction	L, m
0	50	0.500	3.00
0	75	0.500	3.00
0.5	75	0.497	3.12
1.0	75	0.488	3.25
1.5	75	0.483	3.35
2.0	75	0.479	3.42
3.0	75	0.470	3.57
4.0	75	0.466	3.70
5.0	75	0.465	3.80
6.0	75	0.464	3.85
7.0	75	0.464	3.89
8.0	75	0.464	3.90
9.0	75	0.464	3.90

[a]%CO means % controller output

$$
\begin{array}{c|cc}
 & m_1 & m_2 \\
\hline
L & -6 & 7 \\
x & 7 & -6
\end{array}
$$

which yields the pairing: $L\text{-}m_2$ and $x\text{-}m_1$.

To design the decoupler system we follow Eqs. (8-50) and (8-51). In order to avoid any problems in following these equations, the correct closed-loop block diagram is developed, as shown in Fig. 8-62. From Eq. (8-51) we get

$$
D_{x2}(s) = -\frac{G_{Px2}(s)}{G_{Px1}(s)} = -\left(\frac{0.00144}{0.00336}\right)\left(\frac{1.05s + 1}{1.65s + 1}\right)e^{-0.25s}
$$

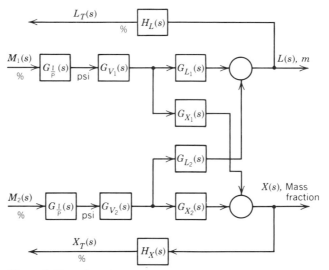

Figure 8-61a. Open-loop block diagram of evaporator.

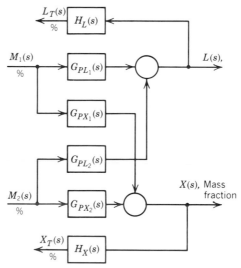

Figure 8-61b. Simplified open-loop block diagram of evaporator.

From Eq. (8-50) we get

$$D_{L1}(s) = -\frac{G_{PL1}(s)}{G_{PL2}(s)} = -\left(\frac{0.072}{0.036}\right)\left(\frac{2.97s + 1}{2.7s + 1}\right)e^{-0.27s}$$

We now have the two decoupler equations. Notice that both of them consists of a gain term, a lead-lag term, and a dead time compensation. As mentioned when we were discussing feedforward control, the implementation of the dead time compensation with

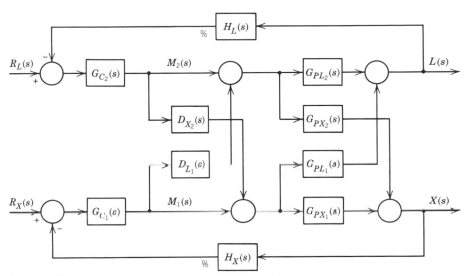

Figure 8-62. Closed-loop block diagram of evaporator with decoupler.

analog instrumentation is rather difficult. On the other hand, it is rather easy with today's microprocessor-based controllers. The decoupler equations usually implemented are

$$D_{x2}(s) = -0.428 \left(\frac{1.05s + 1}{1.65s + 1} \right) \tag{8-52}$$

and

$$D_{L1}(s) = -2 \left(\frac{2.97s + 1}{2.7s + 1} \right) \tag{8-53}$$

Fig. 8-63 shows the implementation of this decoupler system.

Example 8-10 has shown the implementation of a decoupler system for a 2 × 2 process. If the process is completely decoupled, the level loop should not affect the composition loop and vice versa. A way to check this is by deriving the transfer functions of the decoupled system using signal flow graphs. This will be the subject of one of the problems at the end of the chapter.

So far we have shown how to design decoupler systems for a 2 × 2 processes, but what about for a 3 × 3 or more complex processes? We now proceed to show how to design these systems.

Consider a process with n controlled and n manipulated variables. The desired decoupled relationship between the controlled variables, $C(s)$, and controller outputs, $M(s)$, can be written in matrix form as follows:

$$
\begin{bmatrix} C_1(s) \\ C_2(s) \\ \vdots \\ C_n(s) \end{bmatrix}
=
\begin{bmatrix}
N_1(s) & 0 & \cdots & 0 \\
0 & N_2(s) & \cdots & 0 \\
\vdots & \vdots & & \vdots \\
0 & 0 & \cdots & N_n(s)
\end{bmatrix}
\begin{bmatrix} M_1(s) \\ M_2(s) \\ \vdots \\ M_n(s) \end{bmatrix}
\tag{8-54}
$$

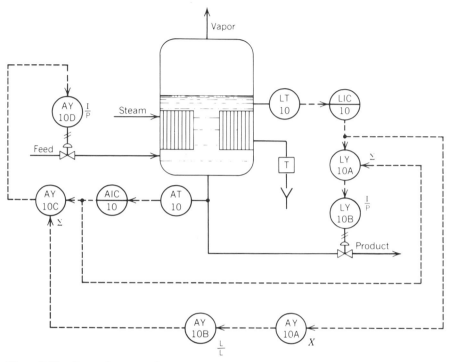

Figure 8-63. Decoupler system for evaporator.

where $N_1(s)$, $N_2(s)$, . . . , $N_n(s)$ are any desired functions.

The first matrix to the right of the equal sign represents the process-decoupler combination. More specifically this matrix is

$$
\begin{bmatrix}
N_1(s) & 0 & \dots & 0 \\
0 & N_2(s) & \dots & 0 \\
\vdots & \vdots & & \vdots \\
0 & 0 & \dots & N_n(s)
\end{bmatrix}
=
\begin{bmatrix}
G_{P11}(s) & \dots & G_{p1n}(s) \\
G_{P21}(s) & \dots & G_{P2n}(s) \\
\vdots & & \vdots \\
G_{Pn1}(s) & \dots & G_{Pnn}(s)
\end{bmatrix}
\begin{bmatrix}
D_{11}(s) & \dots & D_{1n}(s) \\
D_{21}(s) & \dots & D_{2n}(s) \\
\vdots & & \vdots \\
D_{n1}(s) & \dots & D_{nn}(s)
\end{bmatrix}
\tag{8-55}
$$

or

$$ \overline{N} = \overline{G_P} \cdot \overline{D} $$

Premultiplying both sides by $\overline{G}_P^{-1}$ yields the decoupler matrix:

$$ \overline{D} = \overline{G}_P^{-1} \cdot \overline{N} \tag{8-56} $$

In this equation there are $n^2 + n$ unknown parameters but only n^2 equations. Therefore, there are n degrees of freedom that must be set before calculating the other parameters. As mentioned earlier, the degrees of freedom usually chosen are the diagonal elements of the decoupler matrix, $D_{ii}(s)$, which are set to unity.

Before finishing with this subject of decoupling, it is important to point out that the decoupler design is equivalent to designing a feedforward controller in which each manipulated variable is the major disturbance to the other loops. In the example just shown, the term $D_{21}(s)$ serves as the feedforward controller for the composition loop. Notice that the terms in $D_{21}(s)$ are exactly like the terms in many feedforward controllers: a gain and a lead/lag, that is, steady-state and dynamic compensation. The term $D_{12}(s)$ serves as the feedforward controller for the level loop.

Finally, it is also important to realize that, unlike feedforward control, the decouplers are part of the characteristic equation of the control loops. Thus, they affect the stability of the control loop.

8-7. SUMMARY

This chapter has presented several techniques to aid in the control of processes. All of these techniques, with the exception of multivariable control, are quite common in industry today. Multivariable control requires one more degree of sophistication and understanding than the other techniques and, consequently, it has lagged in application. The reader is referred to Reference 15 for a detailed presentation of this important subject.

All of the techniques have been presented and explained without regard to the instrumentation used to implement them. Both analog and digital instrumentation are used; however, with the increasing use of microprocessor-based systems, the implementation is becoming easier and, consequently, application of these techniques will become even more popular. It is also important to realize that none of these techniques completely replaces feedback control. Some kind of feedback control is always needed to complete the control. As was mentioned at the beginning of the chapter, the application of these techniques requires a greater amount of engineering and equipment than just feedback control and, consequently, a justification must exist before they are applied. In addition, the application of these techniques requires a greater amount of training of the operating personnel. If the operator does not understand the techniques, it will be rather difficult to apply them. The best recommendation is to develop the simplest control technique that provides the required control performance.

REFERENCES

1. O'Meara, J. E., "Oxygen Trim for Combustion Control," *Instrumentation Technology*, March 1979.

2. Johnson, M. L., "Dehydrogenation Reactor Control," *Instrumentation Technology*, December 1976.

3. *Instrument Engineers Handbook*—Vol. II, Bela G. Liptak, Editor, Chilton, 1970.

4. Becker, J. V., and R. Hill, "Fundamentals of Interlock Systems," *Chemical Engineering Magazine*, October 15, 1979.

5. Becker, J. V., "Designing Safe Interlock Systems," *Chemical Engineering Magazine*, October 15, 1979.

6. DiStefano, J. J., A. R. Stubberud, and T. J. Williams, *Feedback and Control Systems*, Schaum, New York.

7. Bristol, E. H., "On a New Measure of Interaction for Multivariable Process Control," *Trans. IEEE*, January 1966.

8. Nisenfeld, A. E., and H. M. Schultz, "Interaction Analysis Applied to Control System Design," *Instrumentation Technology*, April 1971.

9. Shinskey, F. G., *Distillation Control for Productivity and Energy Conservation*, McGraw-Hill, New York, 1977.

10. Shinskey, F. G., *Process Control Systems*, McGraw-Hill, New York, 1979.

11. Scheib, T. J., and T. D. Russell, "Instrumentation Cuts Boiler Fuel Costs," *Instrumentation and Control Systems*, November 1981.

12. Congdon, P., "Control Alternatives for Industrial Boilers," *InTech*, December 1981.

13. Wang, J. C., "Relative Gain Matrix for Distillation Control—A Rigorous Computational Procedure," Instrument Society of America, 1979.

14. Hulbert, D. G., and E. T. Woodburn, "Multivariable Control of a Wet Grinding Circuit," *AIChE J*, March 1983.

15. McAvoy, T., "Interaction Analysis: Principles and Applications," Instrument Society of America, Research Triangle Park, N.C., 1983.

16. Koppel, L. B., "Input-Output Pairing in Multivariable Control," submitted to *AIChE J*, January 1982.

17. Niederlinski, A., "A Heuristic Approach to the Design of Linear Multivariable Control Systems," *Automatica*, Vol. 7, No. 691, 1971.

PROBLEMS

8-1. Consider the piping system shown in Fig. 8-64, in which natural gas flows into a process. It is necessary to meter the flow of gas and total it so that every 24 hours the total amount is known. It is proposed to add the output of both transmitters and then total (integrate) the result. The flow rate through each meter, as differential pressure, is given by the following:

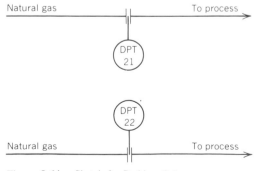

Figure 8-64. Sketch for Problem 8-1.

$$DPT21: Q(MSCFH) = 44.73 \sqrt{h}$$

$$DPT22: Q(MSCFH) = 48.10 \sqrt{h}$$

where h is the differential pressure in inches of H_2O. Both transmitters have a range of 0–100 in. H_2O. Specify the instrumentation required to calculate the total flow rate, using Table 8-2. Determine the scale factors.

8-2. Consider the heat exchanger shown in Fig. 8-15, in which a process fluid is heated by condensing steam. A control scheme calls for calculating the energy transferred to the process fluid using the following equation:

$$Q = F\rho C_p(T - T_i)$$

The following information is known:

Variable	Range	Steady-State
F	0–50 gpm	30 gpm
T	50–120°F	80°F
T_i	25–60°F	50°F
Q	$0\text{–}300{,}000 \dfrac{\text{Btu}}{\text{hr}}$	$182{,}070 \dfrac{\text{Btu}}{\text{hr}}$

The density and heat capacity are assumed to be constant with a product of 202.3 Btu/gpm-°F-hr. Using Table 8-1, specify the instrumentation required to calculate the energy transferred. Specify the scale factors.

8-3. Figure 8-65 shows the reflux to the top of a distillation column. The "internal reflux computer" computes the set point, L_E^{set}, of the external reflux flow controller so as to maintain the internal reflux, L_I, at a set value, L_I^{set}. The internal reflux

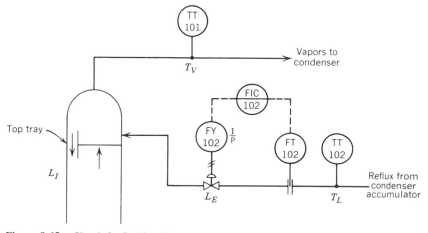

Figure 8-65. Sketch for Problem 8-3.

is greater than the external reflux because of the condensation of vapors on the top tray, which is required to bring the subcooled reflux at T_L up to its bubble point, T_V. An energy balance on the top tray yields the working equation

$$(L_I - L_E)\lambda = L_E C_{P_L}(T_V - T_L)$$

Show all of the instrumentation required for the internal reflux computer and compute the scaled coefficients, using Table 8-3. Design specifications are as follows:

$$C_{P_L} = 0.76 \text{ Btu/lbm-°F}; \qquad \lambda = 285 \text{ Btu/lbm}$$

Transmitter	Range	Normal Value
FT102(L_E)	0–5000 $\dfrac{\text{lbm}}{\text{hr}}$	3000 $\dfrac{\text{lbm}}{\text{hr}}$
TT102(T_L)	100–300°F	195°F
TT101(T_V)	150–250°F	205°F

8-4.   Consider the system, shown in Fig. 8-66, to dilute a 50% by weight NaOH solution to a 30% by weight solution. The NaOH valve is manipulated by a controller not shown in the diagram. Since the flow of the 50% NaOH solution can vary frequently, it is desired to design a ratio control scheme to manipulate the flow of H_2O to maintain the required dilution. The nominal flow of the 50% NaOH solution is 200 lbm/hr. The flow element used for both streams is such that the output signal from the transmitters is linearly related to the volumetric flow. Furthermore, you may assume that the density of each stream is fairly constant and, therefore, that the output signal from each transmitter is also related to the mass flow. Specify the range of each transmitter, in mass flow units, so that the nominal flow is about midscale in the range. Specify the computing relays, as well as their constants, required to implement the ratio control scheme. Use the computing blocks shown in Table 8-2.

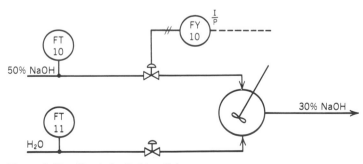

Figure 8-66.   Sketch for Problem 8-4.

8-5. In the production of paper it is necessary to mix some components in a given proportion so as to form a stock that will be supplied to a paper machine to produce the final sheet with the desired characteristics. Consider the process shown in Fig. 8-67. For a particular formulation the final mixture must contain 47 mass % of hardwood slurry, 50 mass % of pine slurry, 2 mass % of additive, and 1 mass % of dye. The nominal system must be designed for a possible maximum production of 2000 lbm/hr.

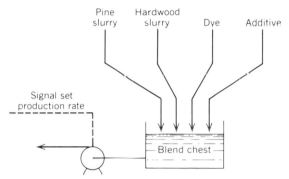

Figure 8-67. Sketch for Problem 8-5.

a. Specify the range of each flowmeter. The flowmeters used in this application are magnetic flowmeters. Their output signal is linearly related to the mass flow rate.

b. Design a control system to control the level in the blend chest and at the same time maintain the correct formulation. It is also required to know at the end of the day the total amount of mass of each stream that has been added to the blend chest. Show the necessary instrumentation to implement this control scheme. Using the computing blocks given in Table 8-3, scale the necessary instrumentation.

8-6. Example 8-2 shows a control scheme (Fig. 8-8) to control the air/fuel ratio to a boiler or furnace. In this scheme, as explained in the example, the air flow always lags the fuel flow.

a. Design a control scheme in which the fuel flow always moves first and then the air flow follows.

b. The most common way to implement this control is by the scheme known as "cross-limiting control." This implementation is such that when the steam pressure drops, requiring more combustible, the air flow increases first and then the fuel follows. When the steam pressure increases, requiring less combustibles, the fuel flow decreases first and then the air flow follows. Design the control system to accomplish this objective.

Note: This is a challenging problem. As a hint, you may consider the use of high and low selectors to select which signal to send to the fuel and air flow controllers.

8-7.   Consider the block diagram, shown in Fig. 8-68, of a cascade control system. As explained in this chapter, the tuning of the primary controller is not straightforward; however, frequency response techniques provide a way to tune the controller.

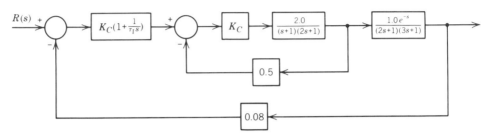

Figure 8-68.   Sketch for Problem 8-7.

    a.   Tune the secondary controller, P, to obtain a damping factor of 0.707.

    b.   Tune the primary controller, PI, by the Ziegler-Nichols method.

  Hint: Using frequency response techniques, determine the ultimate gain and ultimate period of the primary control loop.

8-8.   Consider the exothermic reactor shown in Fig. 8-69. The diagram shows the control of the temperature of the reactor by manipulating the cooling water valve.

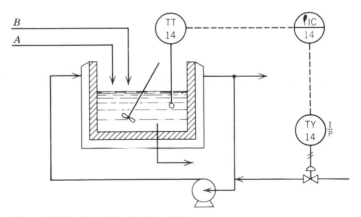

Figure 8-69.   Sketch for Problem 8-8.

    a.   Design the control scheme required to control the reactants to the reactor. The flows of A and B can both be measured and controlled. The required ratio between these flows is 2.5 gpm of B/gpm of A. The flowmeter of A has been calibrated between 0 and 40 gpm and that of B between 0 and 200 gpm. Show and scale the necessary instrumentation, using Table 8-1.

b. Operating experience has shown that the inlet temperature of the cooling water varies somewhat. Because of the lags in the system, that is, cooling jacket, metal wall, and reactor volume, this disturbance usually results in cycling of the reactor temperature. The engineer in charge of this unit has been wondering whether some other control scheme can aid him in improving the temperature control. Design a control scheme to help him.

c. Operating experience has also shown that under some infrequent conditions the cooling system does not provide enough cooling. In this case the only way to control the temperature is by slowing down the flow of reactants. Design a control scheme to do this automatically. The scheme must be such that when the cooling capacity returns to normal the scheme of part (b) is reestablished.

8-9. Consider the drying process shown in Fig. 8-70. In this process wet paper stock is being dried to produce the final paper product. The drying is done using hot air; this air is heated in a heater in which fuel is burned to provide the energy. The controlled variable is the moisture of the paper leaving the drier. Fig. 8-70 shows the original control scheme proposed and installed.

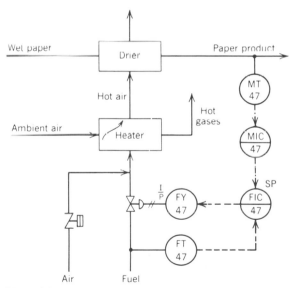

Figure 8-70. Sketch for Problem 8-9.

a. A few weeks after start-up, the process engineer noticed that even though the moisture controller was keeping the moisture within certain limits from set point, the oscillations were more than he really liked. After searching for possible causes, and making sure that the moisture controller (MIC47) was well-tuned, he found that the hot air temperature leaving the heater varied more than originally thought during the design stage. He attributed these variations to daily changes in ambient temperature and possible disturbances in the combustion chamber of the heater. Design a control scheme to maintain the hot air temperature at the desired value to help maintain moisture set point.

b. The control scheme just described significantly helped the moisture control. A few weeks later, however, the operators complained that every once in a while the moisture would go out of set point considerably but the control scheme would eventually bring it back to set point. This disturbance would require that the paper produced during this period be reworked and, therefore, it represented a production loss. After searching through the production logs, the process engineer discovered that changes in inlet moisture were the cause of this disturbance. Design a feedforward control scheme that compensates for these disturbances. There is a moisture transmitter, with a range of 5–20 mass %, available to measure the inlet moisture.

8-10. Consider the control scheme for the solid drying system shown in Fig. 8-71. The major disturbance to this process is the moisture content of the incoming solids. For this disturbance the control system responds quite slowly. It is desired to

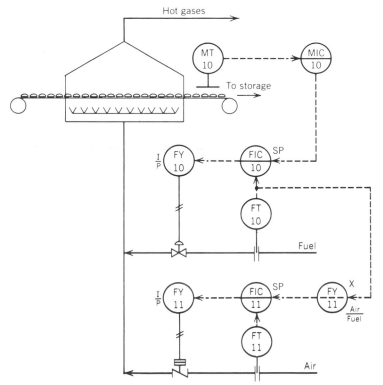

Figure 8-71.   Sketch for Problem 8-10.

implement a feedforward system to improve this control. After some initial work, the following data have been obtained:

Step change in inlet moisture $= +2\%$

Time, min	Exit Moisture, %	Time, min	Exit Moisture, %
0	5.0		
0.5	5.0	5.0	6.6
1.0	5.1	5.5	6.7
1.5	5.2	6.0	6.8
2.0	5.4	6.5	6.9
2.5	5.7	7.0	7.0
3.0	5.9	7.5	7.0
3.5	6.1	8.0	6.9
4.0	6.3	8.5	7.0

Step change in output signal from moisture controller, MIC-10, $= +4$ mA.

Time, min	Exit Moisture, %	Time, min	Exit Moisture, %
0	5.0	5.0	3.81
0.5	5.0	5.5	3.70
1.0	4.95	6.0	3.55
1.5	4.93	6.5	3.45
2.0	4.85	7.0	3.35
2.5	4.70	7.5	3.25
3.0	4.60	8.5	3.10
3.5	4.40	9.5	3.03
4.0	4.20	10.5	2.99
4.5	4.00	11.5	3.00

The feedback moisture analyzer has a range of 1%–7% moisture. There is another analyzer with a range of 10%–15% moisture. The latter analyzer can be used to measure the inlet moisture. These analyzers have been used before in this particular process and have proven to be reliable. Both sensor-transmitters are electronic with time constants of about 15 seconds.

a. Based on process engineering principles (mass balances, energy balances, etc.), develop a feedforward control scheme. The statement of the problem has not provided all of the necessary information. Assume that you can obtain this information from the plant's files. Show the implementation of this scheme.
b. Draw a complete block diagram for this process. Include all known transfer functions.
c. Develop a feedforward control scheme using the block diagram approach. Show the implementation of this scheme.

8-11. Consider the process shown in Fig. 8-10. After startup, operating experience showed that it is difficult to maintain outlet temperature from the heater constant because of frequent changes in the regeneration air mass flow rate. These changes are caused by the variations in inlet pressure and temperature.

    a. Design a feedforward control scheme to compensate for changes in regeneration air mass flow. Show all the required instrumentation. Use the computing blocks given in Table 8-2.

    b. The feedforward control scheme designed in (a) is a steady-state compensator. Explain what kind of data are required to apply dynamic compensation and how you would go about obtaining the data.

Operating Data

Air flow $= 10000 \dfrac{\text{lbm}}{\text{hr}}$ Inlet temperature $= 70°F$

Outlet air temperature $= 600°F$ Inlet pressure $= 30$ psig

Cp of air $= 7$ Btu/lb mole-°F

Inlet air orifice coefficient: $K = 2085 \dfrac{\text{lbm/hr}}{\left(\text{in.} \dfrac{\text{lbm}}{\text{ft}^3}\right)^{1/2}}$

Fuel heating value $= 18,000$ Btu/lbm

The inlet air conditions will be measured by the following sensor-transmitters:

Temperature: 25°F–125°F
Pressure: 0 psig–45 psig
Differential pressure transmitters: 0 in.–150 in.

The outlet temperature from the heater is measured by a sensor-transmitter with a range of 400°F–800°F. The fuel flow rate is measured by a sensor-transmitter with a range of 0–200 lbm/hr.

8-12. Design a feedforward controller for controlling the outlet temperature of the furnace sketched in Fig. 8-72. The controller must compensate for variations in feed rate and feed temperature by manipulating the set point on the fuel flow controller. Show all of the instrumentation required using Table 8-3. The following information is known:

Gas specific heat: $C_p = 0.24$ Btu/lbm-°F
Fuel heat of combustion: $\Delta H_C = -22,000$ Btu/lbm
Inlet gas temperature: $\overline{T}_i = 85°F$
Outlet gas temperature: $\overline{T}_o = 930°F$
Furnace efficiency $= 0.80$
Range of FT42: 0–15,000 lbm/hr
Range of TT42: 0–120°F

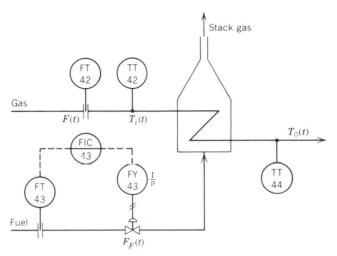

Figure 8-72. Sketch for Problem 8-12.

Range of FT43: 0–200 lbm/hr
Range of TT44: 700–1100°F

For the sake of simplicity, the combustion air is not shown.

8-13. Let us propose that some of the process data, such as the fuel heat of combustion or the gas specific heat, of the furnace of Problem 8-12 are not known. Thus, it is necessary to design the feedforward controller using the block diagram approach. The following data are obtained from the step tests on the furnace:

Variable	Step Change	Change in T_o	Time Constant	Dead Time
F	750 lbm/hr	14°F	35 s	5 s
T_i	10°F	13°F	10 s	15 s
F_F^{set}	5% of range	20°F	68 s	8 s

a. Draw the complete block diagram including all transfer functions.
b. Design the feedforward controller to compensate for variations in feed rate and feed temperature by manipulating the set point on the fuel flow controller. Include the dynamic compensator.
c. If it is decided to compensate for only one of the disturbances, which one would it be and why? How would your design be modified?

8-14. For the sketched stripping section of a distillation column shown in Fig. 8-73, the objective is to maintain the bottom's purity at a desired value. This objective is commonly attained by controlling the temperature in one of the trays (the column pressure is assumed constant) using the steam flow to the reboiler as the manipulated variable. A usual "major" disturbance is the feed flow to the column.

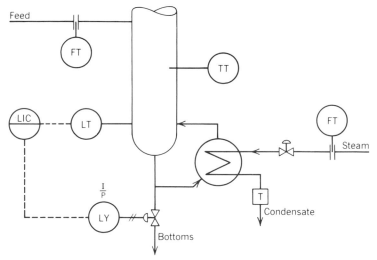

Figure 8-73.   Sketch for Problem 8-14.

   a.   Sketch a feedforward/feedback control scheme to compensate for this major disturbance; describe it briefly.

   b.   Briefly describe the dynamic tests that you would perform on the column in order to tune the feedback controller and the feedforward controller.

8-15.   Consider the vacuum filter shown in Problem 6-15. Using the information given in that problem, design a feedforward/feedback control scheme to compensate for changes in inlet moisture. You may assume that there is a moisture transmitter with a range of 60% to 95%. This transmitter has negligible dynamics. Show the complete implementation using the computing blocks of Table 8-2. Specify the scale factor.

8-16.   Consider the evaporator system of Problem 6-20. Design a feedforward/feedback control scheme to compensate for density changes in the solution entering the first effect. The density meter to measure the inlet density has a range of 55–75 lbm/ft^3 and negligible dynamics. Use the computing relays of Table 8-1 to implement the scheme.

8-17.   Consider the process shown in Problem 6-22 to dry phosphate pebbles. As mentioned in the problem, an important disturbance to the process is the moisture of the inlet pebbles. Using the information provided, design a feedforward/feedback control scheme to compensate for this disturbance. There is a moisture transmitter available to measure the inlet moisture. This transmitter has a range of 12% to 16% and negligible dynamics. Use the computing blocks of Table 8-3 and specify the scale factors.

8-18.   Consider the furnace shown in Fig. 8-74. The outlet temperature of the hydrocarbon is to be controlled by the fuel to the heater. There are two types of fuel

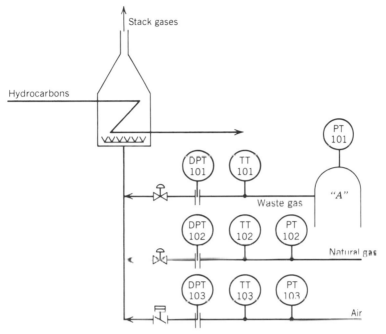

Figure 8-74.   Sketch for Problem 8-18.

available. One is a waste gas, which is produced in another process unit. The other fuel is natural gas. The best operating procedure is to use as much waste gas as available; this conserves natural gas. The waste gas flow is used to control the pressure in the process unit where it is produced.

It can be assumed that the net heating value of the waste gas is about one-fifth of that of the natural gas. The required air/fuel ratios are, on a mass basis

$$\frac{air}{waste\ gas} = R_1 \quad \text{and} \quad \frac{air}{natural\ gas} = R_2$$

a.   Design a simple control system to control the pressure in process unit A.

b.   Design a control scheme to control the outlet temperature of the hydrocarbons. Since all fuels are gases, it is recommended to compensate for temperature and pressure variations. It can be assumed that there is never enough waste gas to heat the hydrocarbon flow, that is, some natural gas is always needed.

c.   For safety reasons it is necessary to design a control scheme such that in case of loss of burner flame, the natural gas and air flows cease; the waste gas flow may continue. Available for this job is a burner switch whose output is 20 mA as long as the flame is present, and whose output drops to 4 mA as soon as the flame stops. Design this control scheme into the previous one.

8-19.   Consider the reactor shown in Fig. 8-75, in which the irreversible and complete liquid reaction $A + B \rightarrow C$ occurs. Product C is the raw material for several

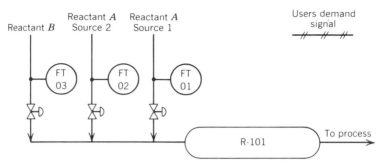

Figure 8-75. Sketch for Problem 8-19.

process units downstream from the reactor. Depending on the number of units operating and on their production rate, the production required from the reactor can vary between 4000 and 20,000 kmole/hr of product C. Reactant A is available from two sources. Because of a long-term contract, source 1 is the least expensive. However, the contract is written with two limitations: a maximum instantaneous rate of 16,800 kmole/hr and a maximum monthly consumption of 3.456×10^6 kmole. If either of these limitations is exceeded, a very high penalty must be paid. In this case, it is less expensive to use the excess from source 2. You may assume that the densities of each reactant, A and B, and of product C do not vary much and, therefore, they can be assumed constant.

Design a control system that will preferentially use reactant A from source 1 and will not allow to exceed any contractual limitations. The feed ratio of A to B is 2:1.

8-20. Figure 8-76 shows a system to filter an oil before processing. The oil enters a header in which the pressure is controlled by manipulating the inlet valve. From the header, the oil is distributed to four filters. The filters consist of a shell with tubes inside similar to heat exchangers. The tube's wall is the filter medium through which the oil must flow to be filtered. The oil enters the shell and flows through the filter medium into the tubes. As time passes, the filter starts to build up a cake and, consequently, the oil pressure required for flow increases. If the pressure increases too much, the walls may collapse. Thus, at some point the filter is taken out of service and cleaned. Under normal conditions, the total oil flow can be handled by three filters.

a. Design a flow control system to set the oil flow through this system.
b. Design a control system so that as the oil pressure in each filter increases above some predetermined value, the oil flow to that filter starts to decrease. The total oil flow through the system must still be maintained.

8-21. Consider the block diagrams shown in Figs. 8-77a and 8-77b. Using signal flow graph technique, obtain $C(s)/R(s)$ for each of them.

8-22. Given the block diagram shown in Fig. 8-78 of a cascade control system with interacting lags, determine the transfer functions, $C_2(s)/R(s)$ and $C_2(s)/L(s)$.

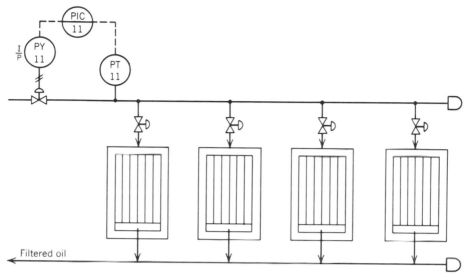

Figure 8-76. Sketch for Problem 8-20.

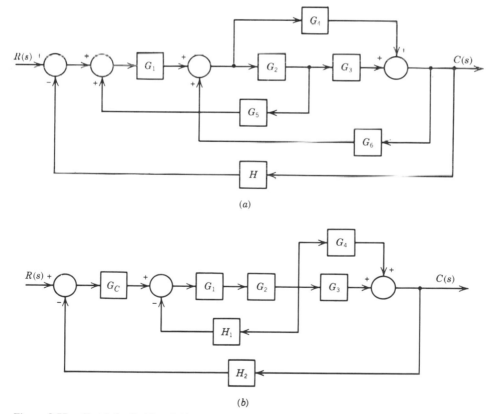

(a)

(b)

Figure 8-77. Sketch for Problem 8-21.

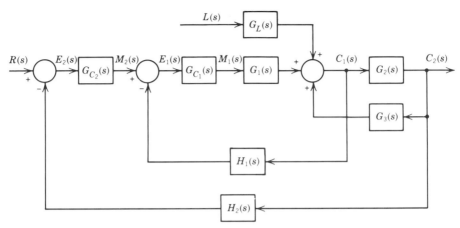

Figure 8-78.   Sketch for Problem 8-22.

8-23.   As shown in Section 8-6, distillation columns are typical examples of multivariable systems. In an interesting article by Wang,[13] which is recommended for reading, the steady-state gains relating the controlled variables (distillate composition, y, and bottoms composition, x) and the manipulated variables (distillate rate, D, and boil-up rate, or steam rate to the reboiler, V) for a particular column were obtained from a flow sheet simulation program. These steady-state gains are as follows:

	D	V
y	-0.205991×10^{-2}	0.912422×10^{-5}
x	10.272573×10^{-4}	-0.644203×10^{-5}

Based on these data decide on the correct pairing.

8-24.   An article by D. G. Hulbert and E. T. Woodburn[14] presents multivariable control of a wet-grinding circuit. Figure 8-79 shows the schematic of the circuit. For this particular circuit the following variables were decided to be controlled: the torque required to turn the mill (TOR), the density of cyclone feed (DCF), and the flow rate from the mill (FML). As explained in the article, the selection of these variables was based on considerations of observability, controllability, and importance from a metallurgical point of view. TOR and FML were regarded as being descriptive of the conditions within the mill and DCF as relating to conditions within the cyclone. TOR and DCF were measured directly but FML was calculated from measurements giving a mass balance around the sump. However, to simplify the diagram of Fig. 8-79, a transmitter giving FML is assumed.

   The manipulated variables for this system are the feed rate of solids (SF), the feed rate of water to the mill (MW), and the feed rate of water to the sump (SW). The feed rate of solids is manipulated by the speed of the belt conveying the solids to the mill (an electrical signal is used).

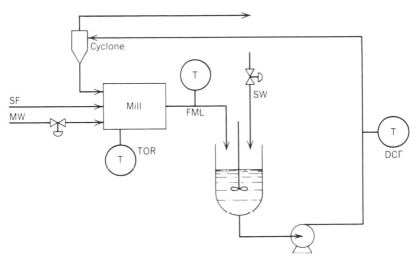

Figure 8-79. Sketch for Problem 8-24.

By means of open-loop testing, the following transfer functions relating the controlled and manipulated variables were obtained:

	SF (kg-s^{-1})	MW (kg-s^{-1})	SW (kg-s^{-1})
TOR (Nm)	$\dfrac{119}{217s + 1}$	$\dfrac{153}{337s + 1}$	$\dfrac{-21}{10s + 1}$
FML (m^3-s^{-1})	$\dfrac{0.00037}{500s + 1}$	$\dfrac{0.000767}{33s + 1}$	$\dfrac{-0.00005}{10s + 1}$
DCF (kg-m^{-3})	$\dfrac{930}{500s + 1}$	$\dfrac{-667e^{-320s}}{166s + 1}$	$\dfrac{-1033}{47s + 1}$

All time constants and dead times are in seconds. Based on this information choose the correct pairing and design a decoupler for this system. Show all the instrumentation required.

8-25. Show that the system of Fig. 8-60 is indeed decoupled. This can be done by deriving the closed loop transfer function, $C_1(s)/R_2(s)$, and substituting the design transfer functions for the decouplers assuming that perfect decoupling is possible.

8-26. Consider the distillation column shown in Fig. 8-80. The following transfer functions have been determined by pulse testing a computer model of the column:

$$X_D(s) = G_{P11}(s)R(s) + G_{P12}V(s)$$

$$= \text{overhead composition, heavy key}$$

$$X_B(s) = G_{P21}(s)R(s) + G_{P22}V(s)$$

$$= \text{bottom's composition, heavy key}$$

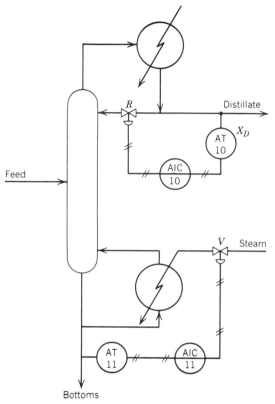

Figure 8-80. Sketch for Problem 8-26.

$$G_{P11}(s) = -6.0\left(\frac{1 - 0.78s}{1 + 0.25s}\right)\frac{\%}{\%}$$

$$G_{P12}(s) = 1.5\left(\frac{1 - 0.22s}{1 + 0.25s}\right)\frac{\%}{\%}$$

$$G_{P21}(s) = -5.0\left(\frac{1 + 0.86s}{1 + 0.25s}\right)\frac{\%}{\%}$$

$$G_{P22}(s) = 7.0\left(\frac{1 + 0.09s}{1 + 0.25s}\right)\frac{\%}{\%}$$

$R(s)$ is the reflux rate; $V(s)$ is the steam rate. Both rates are given in % of range.

a. Calculate the steady-state measure of interaction for this system. Do the composition control loops reinforce or fight each other? Are the controlled and manipulated variables properly paired?
b. Design the decouplers for this system. Briefly discuss any problem with the implementation of these decouplers. What do you suggest?

8-27. Consider the 2 × 2 process shown in Fig. 8-81. The following transfer functions have been determined by pulse tests:

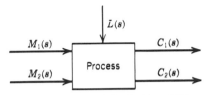

Figure 8-81. Sketch for Problem 8-27.

$$C_1(s) = G_{P11}(s)M_1(s) + G_{P12}(s)M_2(s) + G_{P1L}(s)L(s)$$
$$C_2(s) = G_{P21}(s)M_1(s) + G_{P22}(s)M_2(s) + G_{P2L}(s)L(s)$$

where

$$G_{P11}(s) = \frac{0.81}{(1.2s + 1)(0.63s + 1)} \frac{\%}{\%}$$

$$G_{P12}(s) = \frac{1.2}{(2.3s + 1)(1.1s + 1)} \frac{\%}{\%}$$

$$G_{P21}(s) = \frac{1.1e^{-0.3s}}{1.2s + 1} \frac{\%}{\%}$$

$$G_{P22}(s) = \frac{0.6e^{-s}}{2s + 1} \frac{\%}{\%}$$

$$G_{P1L}(s) = \frac{0.5}{2.0s + 1} \frac{\%}{\%}$$

$$G_{P2L}(s) = \frac{1.5}{1.8s + 1} \frac{\%}{\%}$$

a. Obtain the correct pairing for this system.
b. Draw the complete block diagram and the signal flow graph for this system.
c. Determine the closed-loop transfer functions,

$$\frac{C_1(s)}{L(s)} \quad \text{and} \quad \frac{C_2(s)}{L(s)}$$

d. Design a decoupler for this system and show its implementation.

CHAPTER
9

Modeling and Simulation of Process Control Systems

Mathematical modeling and computer simulation are indispensable tools for analyzing and designing control systems for complex nonlinear processes. They complement nicely the tools for linear systems analysis that we have studied in the preceding chapters of this book.

A question that comes up at this point is: When should we use computer simulation in designing a control system? There are several considerations in arriving at such a decision. First we must consider how critical the performance of the control system is for the safe and profitable operation of the process. For example, the control system for a large centrifugal compressor is usually critical enough to merit simulation, while that of a simple level controller or pressure regulator would not be. Our second consideration is how confident we are regarding the performance of the control system. This usually depends on our experience and familiarity with the particular control application. For example, an experienced control engineer may not bother with simulating the temperature control of a continuous stirred tank reactor, while the same simulation project can be quite exciting and informative to a senior student in his first control course. The third consideration is the time and effort required to carry out the simulation. This may range from a few hours for a relatively simple process to several man-months for a complex process that is being simulated for the first time. Other considerations include the availability of computing facilities, experienced personnel, and sufficient process data to carry out the simulation.

There arc three major steps in performing the dynamic simulation of a process:

1. Development of a mathematical model of the process and its control system.
2. Solution of the model equations.
3. Analysis of the results.

The first three sections of this chapter are devoted to the first of these steps. They include the development of the mathematical models of two complex processes: a multicomponent distillation column and a process furnace. The balance of the chapter is devoted to the numerical solution of the model equations. Although we will not formally discuss the third step, the importance of properly analyzing the results of the simulation cannot be overstated. Without it, the entire simulation effort is wasted. This analysis should include a verification of the simulation results whenever possible.

9-1. DEVELOPMENT OF COMPLEX PROCESS MODELS

The basic principles of writing the equations that describe the response of process variables with time were presented in Chapter 3. To summarize, the fundamental conservation equation is written in general form as

$$\left\{ \begin{array}{c} \text{Rate of} \\ \text{accumulation} \\ \text{of conserved} \\ \text{quantity} \end{array} \right\} = \left\{ \begin{array}{c} \text{Rate of} \\ \text{conserved} \\ \text{quantity} \\ \text{IN} \end{array} \right\} - \left\{ \begin{array}{c} \text{Rate of} \\ \text{conserved} \\ \text{quantity} \\ \text{OUT} \end{array} \right\} \tag{9-1}$$

The conserved quantity can be total mass, mass of a component, energy, and momentum. The rate in and out terms must account for all of the mechanisms by which the quantity conserved enters or exits the *control volume* or portion of the universe over which the "balance" is performed. For example, all of the conserved quantities listed can flow in and out of the control volume (convection), energy can enter and exit by heat conduction and radiation, components can be transferred by diffusion, and momentum can be generated or destroyed by mechanical forces. In the case of chemical reactions, the rate of reaction must be accounted for as an input term for reaction products and an output term for reactants.

The rate of accumulation in Eq. (9-1) is always of the form

$$\left\{ \begin{array}{c} \text{Rate of} \\ \text{accumulation} \\ \text{of quantity} \\ \text{conserved} \end{array} \right\} = \frac{d}{dt} \left\{ \begin{array}{c} \text{Total quantity} \\ \text{conserved in} \\ \text{the control} \\ \text{volume} \end{array} \right\} \tag{9-2}$$

where t is time. This means that the mathematical models consist of a set of simultaneous first-order differential equations or, in its simplest form, a single first-order differential equation, with the time as the independent variable. In addition, the model may contain algebraic equations that arise from physical property and rate expressions and from balance equations in which the accumulation term is neglected.

In order to express the total quantity conserved and rates in terms of process variables (e.g., temperatures, pressures, compositions), these variables must be relatively uniform throughout the control volume. When this requirement is satisfied in a model in which the process is divided into a number of control volumes of finite size, or "lumps," the model is said to be a "lumped-parameter model." In the other hand, "distributed-parameter models" result when the process variables vary continuously with position. In this case the balance equations must be applied to each *point* in the process and the mathematical model consists of partial differential equations with time and position as independent variables. Even then, each equation is always first-order in the time variable. The only way for the equations to be higher than first-order in the time variable is when equations are combined to eliminate variables. The fact that the equations are first-order in time is stressed here because it guides the design of computer programs used to simulate processes. This will become apparent in Section 9-5.

In developing the mathematical model it is important to keep in mind the maximum number of independent balance equations that apply to each control volume (or point) of the process. These are, for a system with N components, given by

> N mass balances
> 1   energy balance
> 1   momentum balance in each direction of interest, up
>   to three

The N independent mass balances can be N component balances, or one total mass balance and $N - 1$ component balances. The momentum balance is usually of no use in process simulation because it introduces as unknowns the reaction forces on the equipment and pipe walls, which are seldom of interest. A more useful balance is Bernoulli's equation extended to include friction, shaft work, and accumulation of kinetic energy.

In addition to the balance equations, other equations are written separately to express physical properties (e.g., density, enthalpy, equilibrium coefficients) and rates (e.g., reaction, heat transfer, mass transfer) in terms of process variables (e.g., temperature, pressure, composition).

We will next illustrate the concepts summarized above in the development of two models, one of a multicomponent distillation column and one of a process furnace. The first one will serve as an example of a complex lumped-parameter model, while the second will serve as an example of a distributed-parameter model. In order to keep the notation simple, we will not indicate explicitly that the variables are functions of time, for example, we will write simply T for $T(t)$.

Our method of developing mathematical models will be the one introduced in Chapter 3, that is:

1. Write the balance equations.

2. Count the new variables (unknowns) that appear in each equation, so as to keep track of the number of variables and equations.

3. Keep introducing relationships until there is an equal number of equations and variables, and all the variables of interest have been considered.

The order in which we will write the balance equations will usually be as follows:

> Total mass balance
> Component (or element) balances
> Energy balance
> Mechanical energy balance (if relevant)

9-2. DYNAMIC MODEL OF A DISTILLATION COLUMN

Dynamic models of distillation columns are among the most complex encountered in process control systems for a single unit operation. The complexity of the model derives from the large number of highly nonlinear differential equations that must be solved in order to study the dynamic response of the temperature and composition on each tray of the column and the composition of the products. For example, a 100-tray column with a five-component feed will require the solution of about 600 differential equations—five component balances and one enthalpy balance for each of the 100 trays—not counting the equations required to simulate the condenser, the reboiler, and the control system. In addition, one phase-equilibrium relationship must be established for each component on each tray, and the relationships for tray hydraulics, enthalpy, density, and other physical properties must be established for each tray. In most cases these relationships are highly nonlinear functions of temperature, pressure, and composition.

Let us consider the distillation column sketched in Fig. 9-1 with an N-component feed, a total condenser, and a thermosyphon reboiler. We are interested in the response of the product compositions x_D and x_B with time. The two manipulated variables are the steam flow to the reboiler and the distillate product rate. The reflux rate is manipulated by the accumulator level controller and the bottoms product rate by the bottom level controller. This is a common control arrangement. We will assume that the pressure in the column is not controlled, and that it is essentially constant from the bottom to the top, that is, that the pressure drop from tray to tray is negligible.

Although we are interested only in the composition and rates of the product streams, these depend on the conditions on the trays, the reboiler, and the condenser. This forces

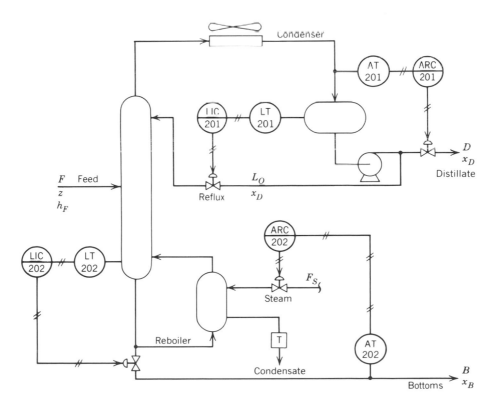

Figure 9-1. Sketch of distillation column.

us to divide the column into a number of control volumes, one for each tray, one for the reboiler, and one for the condenser. We then must write the N mass balances and the enthalpy balance equations for each of these control volumes and solve all of these equations simultaneously, together with the additional equations that describe the control system.

Tray Equations

A typical tray, the jth tray from the top, is sketched in Fig. 9-2. The first equation we write is the total mass balance, which, in the absence of chemical reactions, can be written in molar units. Assuming that because of its low density, accumulation of mass in the vapor phase is negligible compared to that in the liquid phase, the total mass balance is:

$$\frac{dM_j}{dt} = L_{j-1} + V_{j+1} - L_j - V_j \tag{9-3}$$

where

M_j is the hold-up of liquid on tray j, kgmoles
L_j is the liquid rate from tray j in kgmoles/s
V_j is the vapor rate from tray j in kgmoles/s

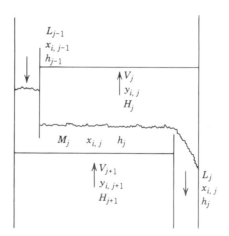

Figure 9-2. Typical tray of distillation column.

In writing the component mass balance for each tray we assume that the liquid on the tray is perfectly mixed so that the properties of the liquid leaving the tray are the same as those of the liquid on the tray. Without this lumped-parameter approximation we would have to write the balances at each point on each tray and the resulting equations would be partial differential equations. A relevant question is what to do with the liquid in the downcomer. We can assume that it is perfectly mixed with the liquid on the tray, treat it as a perfectly mixed tank (a lag on the properties of the liquid), as a nonmixed pipe (dead time), or neglect it altogether. Notice that the only difference between the first assumption and the last is in whether or not the moles of liquid in the downcomer are part of the liquid hold-up on the tray (M_j).

The ith component balance on the jth tray is then

$$\frac{d}{dt}(M_j x_{i,j}) = L_{j-1} x_{i,j-1} + V_{j+1} y_{i,j+1} - L_j x_{i,j} - V_j y_{i,j} \tag{9-4}$$

where

$x_{i,j}$ is the mole fraction of component i in the liquid on tray j
$y_{i,j}$ is the mole fraction of component in the vapor leaving tray j

Equation (9-4) applies to $N - 1$ components, as the sum of the mole fractions must add up to unity:

$$\sum_{i=1}^{N} x_{i,j} = 1 \tag{9-5}$$

At this point we have, for each tray, $N + 1$ equations, counting Eq. (9-5), which is not a balance equation, and $2N + 3$ variables, that is, $2N$ liquid and vapor compositions, the liquid and vapor rates, and the moles of liquid on the tray. Neglecting heat losses, the energy balance around the jth tray is given by

$$\frac{d}{dt}(M_j h_j) = L_{j-1} h_{j-1} + V_{j+1} H_{j+1} - L_j h_j - V_j H_j \tag{9-6}$$

where

h_j is the molar enthalpy of the liquid on tray j, J/kgmole
H_j is the molar enthalpy of the vapor leaving tray j, J/kgmole

In Eq. (9-6), as well as in other energy balances to follow, we assume that the enthalpy of the liquid, h_j, is essentially equal to its internal energy. Rigorously, we should use the internal energy, rather than the enthalpy, in the accumulation term.

We now have $(N + 2)$ equations per tray and $(2N + 5)$ variables, with the liquid (h_j) and vapor (H_j) enthalpies being the two new variables. As we haved used up all of the relevant conservation equations on the tray, we must turn to thermodynamics and other relationships in order to calculate the remaining variables. We can obtain the vapor composition from the Murphree tray efficiency relationship:

$$\eta_M = \frac{y_{i,j} - y_{i,j+1}}{y_{i,j}^* - y_{i,j+1}} \tag{9-7}$$

where

η_M is the Murphree tray efficiency (assumed constant)
$y_{i,j}^*$ is the mole fraction of component i in the vapor in equilibrium with the liquid leaving tray j

Equation (9-7) applies to each component on each tray, which gives us N additional equations per tray, while at the same time introducing N new variables, the equilibrium mole fractions $y_{i,j}^*$. These we obtain from the N vapor-liquid equilibrium relationships:

$$y_{i,j}^* = K_i(T_j, P, x_{1,j}, x_{2,j}, \ldots, x_{N,j})x_{i,j} \qquad (9\text{-}8)$$

where

K_i is the equilibrium coefficient for component i
T_j is the temperature on tray j, K
P is the column pressure, N/m^2

We can write Eq. (9-8) for each component on each tray, giving us N additional equations per tray and only one new variable, the temperature T_j, as the pressure is common to all the trays. An additional equation is obtained from the fact that the vapor mole fractions must add up to unity:

$$\sum_{i=1}^{N} y_{i,j} = 1 \qquad (9\text{-}9)$$

At this point we have, for each tray, $3N + 6$ variables and $(3N + 3)$ equations. A relationship between the moles of liquid on the tray and the liquid rate from the tray can be obtained from the tray hydraulics. A popular equation is Francis's weir formula:

$$L_j = k\rho_j \left(\frac{M_j - M_0}{A\rho_j} \right)^{1.5} \qquad (9\text{-}10)$$

where

M_0 is the liquid hold-up at zero flow, kgmoles
ρ_j is the molar liquid density, kgmole/m^3
A is the cross-sectional area of the tray, m^2
k is a dimensional coefficient, m$^{1.5}$/s

The final relationships are obtained from physical property correlations:

$$h_j = h(T_j, P, x_{1,j}, x_{2,j}, \ldots, x_{N,j}) \qquad (9\text{-}11)$$

$$H_j = H(T_j, P, y_{1,j}, y_{2,j}, \ldots, y_{N,j}) \qquad (9\text{-}12)$$

$$\rho_j = \rho(T_j, P, x_{1,j}, x_{2,j}, \ldots, x_{N,j}) \qquad (9\text{-}13)$$

This gives us a total of $3N + 7$ equations with $3N + 7$ variables per tray; thus we have one equation for calculating each variable. Of these, $N + 1$ equations are first-order ordinary differential equations and the rest are algebraic equations.

Feed and Top Trays

Although the tray equations apply to every tray, the feed and top tray equations are slightly different. At the feed tray we have one additional input term, the feed. This means that the following rate terms must be added to the right-hand side of the balance equations:

Total mass $\qquad$ F kgmoles/s to Eq. (9-3)
Component mass $\qquad$ $F\, z_i$ kgmoles/s to Eq. (9-4)
Energy $\qquad$ $F\, h_F$ J/s to Eq. (9-6)

where

z_i is the mole fraction of component i in the feed
h_F is the molar enthalpy of the feed, J/kgmole

The top tray equations are the same as the other tray equations except that the liquid flowing into the tray is the reflux. In our notation this means that the term L_0 is the reflux rate, $x_{i,0}$ is the mole fraction of component i in the reflux, and h_0 is the molar enthalpy of the reflux.

Reboiler

A sketch of the reboiler is shown in Fig. 9-3. We assume that the rate of recirculation through the thermosyphon reboiler is high compared with the bottoms rate so that the liquid in the bottom of the column is well mixed and at the same composition as the liquid in the reboiler tubes. The total mass balance is then

$$\frac{d}{dt} M_B = L_{NT} - B - V_{NT+1} \tag{9-14}$$

where

M_B is the hold-up of liquid in the column bottom, including the reboiler tubes, kgmoles
L_{NT} is the liquid rate from tray NT, the last tray, kgmoles/s
B is the bottoms product rate, kgmoles/s
V_{NT+1} is the vapor rate into the bottom tray, kgmoles/s

for $N - 1$ of the components we can write the component mass balances:

$$\frac{d}{dt} (M_B x_{i,B}) = L_{NT} x_{i,NT} - B x_{i,B} - V_{NT+1}\, y_{i,NT+1} \tag{9-15}$$

where

$x_{i,B}$ is the mole fraction of component i in the column bottom
$y_{i,NT+1}$ is the mole fraction of component i in the vapor stream entering the bottom tray

For the Nth component we use the fact that the sum of the mole fractions must be unity:

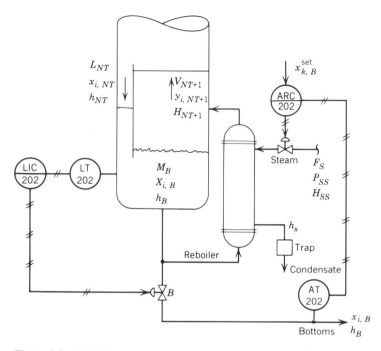

Figure 9-3. Distillation column bottom with thermosyphon reboiler.

$$\sum_{i=1}^{N} x_{i,B} = 1 \qquad (9\text{-}16)$$

In the energy balance we must take into consideration the heat transfer capacity of the reboiler. We do this by expressing the heat rate as a function of the temperature difference:

$$\frac{d}{dt}(M_B h_B) = L_{NT} h_{NT} - B h_B - V_{NT+1} H_{NT+1} + U_R A_R (T_S - T_B) \qquad (9\text{-}17)$$

where

h_B is the molar enthalpy of the liquid in the column bottom, J/kgmole
H_{NT+1} is the molar enthalpy of the vapor entering the bottom tray, J/kgmole
U_R is the overall heat transfer coefficient of the reboiler, J/s-m²-K
A_R is the reboiler heat transfer area, m²
T_S is the temperature of the steam outside the reboiler tubes, K
T_B is the temperature in the reboiler and column bottom, K

The vapor leaving the reboiler is assumed to be in equilibrium with the liquid in the column bottom. Thus the vapor mole fractions are given by the N equilibrium relationships:

$$y_{i,NT+1} = K_i(T_B, P, x_{1,B}, x_{2,B}, \ldots, x_{N,B}) x_{i,B} \qquad (9\text{-}18)$$

We can also make use of the fact that the sum of the vapor mole fractions is unity:

$$\sum_{i=1}^{N} y_{i,NT+1} = 1 \tag{9-19}$$

At this point we have, for the reboiler, $2N + 3$ equations and $2N + 7$ variables: M_B, B, V_{NT+1}, $X_{i,B}$, $y_{i,NT+1}$, h_B, H_{NT+1}, T_B, and T_S. An additional equation results from the relationship between the bottoms product rate and bottoms hold-up. This is established by the proportional level controller (LIC202), assuming a linear control valve:

$$B = K_{BC} \left(\frac{M_B - M_{B0}}{A_B \rho_B} \right) \frac{R_{max}}{R_L} \tag{9-20}$$

where

K_{BC} is the dimensionless controller gain
M_{B0} is the hold-up at zero flow, kgmoles
ρ_B is the molar liquid density, kgmoles/m^3
A_B is the cross-sectional area of the column, m^2
B_{max} is the bottoms rate when the level controller output is at its maximum, kgmoles/s
R_L is the range of the level transmitter, m

Thermodynamic relationships give us these physical properties:

$$h_B = h(T_B, P, x_{1,B}, x_{2,B}, \ldots, x_{N,B}) \tag{9-21}$$

$$H_{NT+1} = H(T_B, P, y_{1,NT+1}, y_{2,NT+1}, \ldots, y_{N,N+1}) \tag{9-22}$$

$$\rho_B = \rho(T_B, P, x_{1,B}, x_{2,B}, \ldots, x_{N,B}) \tag{9-23}$$

This leaves us with one more variable to calculate, the steam temperature T_S. For this we must define a new control volume, the steam chest outside the reboiler tubes. For the balances on the steam chest we assume that there is no accumulation of condensate, that is, the steam trap removes the condensate at the same rate at which it is produced. We also assume that the steam is saturated in the chest and that the reboiler tubes are at about the same temperature as the condensing steam; that is, we neglect the resistance to heat transfer on the condensing steam side of the reboiler tubes. The accumulation of energy is then concentrated in the reboiler tubes, since the mass of the steam vapors is small compared to the mass of metal in the tubes. Subject to this assumption, the energy balance on the steam chest is given by

$$C_{MR} \frac{d}{dt} T_S = F_S [H_{SS} - h_S(T_S)] - U_R A_R (T_S - T_B) \tag{9-24}$$

where

C_{MR} is the heat capacitance of the reboiler tubes, J/K
F_S is the flow rate of steam, kg/s

H_{SS} is the inlet enthalpy of the steam, J/kg

$H_S(T_S)$ is the enthalpy of the condensate as it leaves through the steam trap, J/kg

Notice that in Eq. (9-24) we have neglected the accumulation of steam in the chest by assuming that the flow rate of condensate out is the same as that of steam into the chest.

The flow rate of steam is calculated from a model of the control valve:

$$F_S = C_{VS}(VP_S) \sqrt{P_{SS} - P_S} \qquad (9\text{-}25)$$

where

C_{VS} is the capacity factor of the steam valve, kg/s (N/m²)½

VP_S is the steam valve position in fraction of lift

P_{SS} is the steam supply pressure, N/m²

P_S is the pressure in the steam chest, N/m²

The valve position is either an input variable to the column or the manipulated variable used to control the mole fraction of one of the components in the bottoms product. In the latter case it is calculated from a model of the composition controller (ARC202):

$$VP_S = f_{AC}(x_{k,B} - x_{k,B}^{set}) \qquad (9\text{-}26)$$

where

f_{AC} is the analyzer controller function

$x_{k,B}$ is the mole fraction of the key component in the bottoms product

$x_{k,B}^{set}$ is the set point on the mole fraction of the key component

In Eq. (9-26) we have not shown the detailed model of the composition sensor and transmitter, which should be a part of it. These details are given later in the model of the overhead composition controller, which forms part of the condenser accumulator model.

The pressure in the steam chest is a function of the temperature, assuming that the steam is saturated as it condenses:

$$P_S = P_S(T_S) \qquad (9\text{-}27)$$

This completes the model of the steam chest in which three additional variables F_S, VP_S, and P_S, one enthalpy balance equation, and three algebraic equations were introduced to calculate the state variable T_S.

Condenser Model

A sketch of the total condenser is given in Fig. 9-4. The vapor stream into the condenser is the vapor leaving the top tray of the column (tray number 1), and the liquid out of the accumulator drum is split into the distillate product (D) and the reflux to the column (L_0), which is the liquid input to the top tray. The distillate product rate is manipulated by the

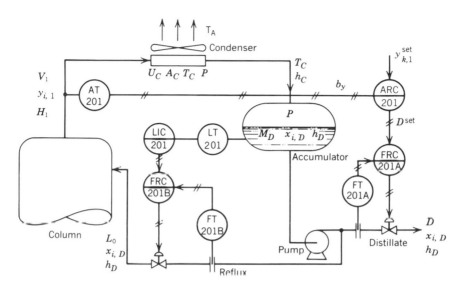

Figure 9-4. Air-cooled total condenser for distillation column.

overhead product composition controller (ARC201), and the reflux rate is manipulated by a proportional level controller on the accumulator drum (LIC201).

In a column with a total condenser, the pressure is determined solely by the heat balance. That is, if heat is supplied in the reboiler at a higher rate than it is removed in the condenser, the pressure in the column rises with time. This has the effect of increasing the temperatures in all of the trays and in the condenser, causing an increase in the rate of heat removed in the condenser until the heat balance is again satisfied at a new (higher) steady-state pressure. This self-regulating mechanism is present even when the pressure is controlled; thus the column pressure can be controlled by manipulating the heat transfer rate in the condenser or in the reboiler. For simplicity, we shall assume that the pressure in our column is not controlled and that the condenser is at maximum capacity. This has the effect of keeping the column pressure at the lowest point that the condenser capacity permits, which usually results in a higher separation of the components.

In modeling the condenser we neglect the accumulation of mass in the vapor phase. This means that the liquid entering the accumulator drum is of the same composition and flow rate as the vapors entering the condenser. Thus we do not need material balances around the condenser control volume. However, as the enthalpy and temperature change on condensation, we need an energy balance:

$$C_{MC} \frac{d}{dt} T_C = V_1(H_1 - h_C) - U_C A_C(T_C - T_A)$$

(9-28)

where

T_C is the temperature in the condenser, K
T_A is the temperature of the cooling air outside the condenser tubes, K

h_C is the molar enthalpy of the liquid leaving the condenser, J/kgmole
C_{MC} is the heat capacitance of the condenser tubes, J/K
U_C is the condenser heat-transfer coefficient, J/s-m²-K
A_C is the condenser heat-transfer area, m²

In writing Eq. (9-28) we have made several simplifying assumptions. First we have assumed that the accumulation of energy takes place only in the tube walls. Our second assumption is the lumped-parameter assumption that both the condenser tubes and the cooling medium outside it are at uniform temepratures T_C and T_A, respectively. A third assumption is that the heat-transfer resistance in the condensing side of the condenser tubes is negligible compared with the heat transfer resistance on the air side. Using the assumption that the liquid leaving the condenser is at the same composition as the vapor from the column, we can calculate the enthalpy of the liquid leaving the condenser by

$$h_C = h(T_C, P, y_{1,1}, y_{2,1}, \ldots, y_{N,1}) \tag{9-29}$$

and the pressure in the column as the bubble-point pressure of this liquid at the condenser temperature by

$$P = P_{BP}(T_C, y_{1,1}, y_{2,1}, \ldots, y_{N,1}) \tag{9-30}$$

where

P_{BP} is the bubble-point pressure as a function of temperature and composition, N/m²

Notice how the combination of Eqs. (9-28) and (9-30) models the pressure regulation effect of the heat balance discussed earlier. In the model, an increase in the vapor rate from the column, V_1, causes a rise of temperature T_C, which in turn causes an increase in the rate of heat removal from the column, Eq. (9-28), and in the column pressure, Eq. (9-30). If the pressure were to be controlled by manipulating the cooling air flow through the condenser, U_C and T_A would become variables in Eq. (9-28) and balances on the air side of the condenser would be necessary to calculate T_A. This refinement is left as an exercise for the student.

Condenser Accumulator Drum

We now turn our attention to the condenser accumulator drum of Fig. 9-4. At steady state, the rate, composition, and enthalpy of the liquid streams from the accumulator—the reflux and the distillate product—are the same as those of the liquid from the condenser. However, the liquid in the accumulator constitutes a time lag to changes in rate, composition, and enthalpy. This is why in our dynamic model we must include the balances around the accumulator. The total mass balance gives us the lag to changes in vapor rate:

$$\frac{d}{dt} M_D = V_1 - L_0 - D \tag{9-31}$$

where

M_D is the hold-up of liquid in the accumulator, kgmoles
L_0 is the reflux rate, kgmoles/s
D is the distillate product rate, kgmoles/s

In Eq. (9-31) the rate of liquid into the accumulator is the same as the vapor rate into the condenser, because we neglected the accumulation of mass in the condenser.

The $N - 1$ component mass balances on the accumulator give us the lag to changes in vapor composition:

$$\frac{d}{dt} (M_D x_{i,D}) = V_1 y_{i,1} - L_0 x_{i,D} - D x_{i,D} \tag{9-32}$$

where

$x_{i,D}$ is the mole fraction of component i in the accumulator liquid

The mole fraction of the Nth component is obtained from the fact that the mole fractions must add up to unity:

$$\sum_{i=1}^{N} x_{i,D} = 1 \tag{9-33}$$

The lag to enthalpy changes is modeled by the energy balance:

$$\frac{d}{dt} (M_D h_D) = V_1 h_C - L_0 h_D - D h_D \tag{9-34}$$

where

h_D is the molar enthalpy of the liquid in the accumulator, J/kgmole

At this point we have $N + 2$ equations and $N + 4$ variables: M_D, $x_{i,D}$, h_D, L_0, and D. The equation for the reflux rate is given by the proportional accumulator level controller (LIC201). In its simplest form, assuming perfect control of reflux flow (FRC201B)

$$L_0 = K_{LC0} L_{0\max} \left(\frac{M_D - M_{D0}}{M_{D\max} - M_{D0}} \right) \tag{9-35}$$

where

$L_{0\max}$ is the reflux rate when the controller output is at its maximum, kgmoles/s
K_{LC0} is the gain of the level controller (dimensionless)
M_{D0} is the hold-up of liquid in the accumulator when the reflux rate is zero, kgmoles
$M_{D\max}$ is the hold-up of liquid in the accumulator when the level is maximum, kgmoles

In Eq. (9-35) we have neglected variations in liquid density. By contrast, Eq. (9-20) offers an alternative model for the level controller that considers variations in density. Either model can be used for either level controller.

The final equation needed to complete the model of the accumulator drum is the equation for the distillate rate D. This equation requires a model of the overhead composition control loop. For simplicity, we simulate the analyzer (AT201) as a first-order lag:

$$\frac{d}{dt} b_y = \frac{1}{\tau_{AT}} \left[\frac{y_{k,1} - y_0}{y_{max} - y_0} - b_y \right] \tag{9-36}$$

where

b_y is the normalized signal from the analyzer
$y_{k,1}$ is the mole fraction of the key component in the vapor from the top tray
y_0 is the lower limit of the analyzer calibrated range (mole fraction)
y_{max} is the upper limit of the analyzer calibrated range
τ_{AT} is the time constant of the first-order lag, s

Notice that the analyzer signal b_y varies from zero to unity as the mole fraction varies from y_0 to y_{max}.

The analyzer controller (ARC201) is modeled as a PID (proportional/integral-derivative) controller (see Example 9-4):

$$D^{set} = f_{AC} (b_y, y_{k,1}^{set}, K_{AC}, \tau_{IAC}, \tau_{DAC}) \tag{9-37}$$

where

D^{set} is the controller output and the set point of the distillate flow controller (FRC201A), kgmoles/s
$y_{k,1}^{set}$ is the set point to the analyzer controller (ARC201)
K_{AC} is the analyzer controlelr gain
τ_{IAC} is the analyzer controller integral time, s
τ_{DAC} is the analyzer controller derivative time, s

Finally, the flow controller can be assumed to be fast enough to keep the distillate equal to the set point at all times:

$$D = D^{set} \tag{9-38}$$

However, limits must be imposed on the distillate to ensure that it is always positive and less than the maximum flow that can be controlled by the flow controller:

$$0 \leqslant D \leqslant D_{max} \tag{9-39}$$

where

D_{max} is the maximum distillate flow that can be measured by the flow transmitter (FT201A), kgmoles/s

This completes the model of the column. In the model we have divided the column into NT control volumes, one for each tray. Additional control volumes were used for the reboiler, its steam chest, the condenser, and its accumulator drum. The many differential and algebraic equations that we have written from basic physical and chemical principles are necessary to compute the process variables: flow rates, compositions, and temperatures.

Initial Conditions

In order to simulate the column we need the initial conditions for all of our *state variables*. The state variables are all those variables that appear in the derivatives of the differential equations. The name derives from the fact that these are the variables that uniquely define the state of the model at any instant of time. Since all of our differential equations are first order, we need only one initial condition per differential equation. Our model state variables are as follows:

The moles of liquid on each tray:	M_j,	$j = 1, 2, \ldots, NT$
the column bottom:	M_B	
and the accumulator drum:	M_D	
The liquid mole fractions on each tray:	$x_{i,j}$,	$i = 1, 2, \ldots, N - 1$
		$j = 1, 2, \ldots, NT$
the column bottom:	$x_{i,B}$,	$i = 1, 2, \ldots, N - 1$
and the accumulator drum:	$x_{i,D}$,	$i = 1, 2, \ldots, N - 1$
The liquid molar enthalpies on each tray:	H_j,	$j = 1, 2, \ldots, NT$
the column bottom:	h_B	
and the accumulator drum:	h_D	
The temperature in the steam chest:	T_S	
and in the condenser:	T_C	
The output of the analyzer transmitter (AT201):	b_y	
The outputs of the composition controllers:	VP_S and D^{set}	

The initial conditions for these variables are determined by the type of run that we want to simulate. The most common type of run in control studies of continuous systems is the study of the response of the system to changes in the input variables (e.g., disturbances and set points) from some design steady-state conditions. For this type of run, the initial values of the state variables must satisfy the model equations at steady state, that is, with all the derivative terms set to zero. For a model as complex as the one we have presented here, we need a computer program to solve systems of nonlinear algebraic equations in order to compute the initial steady-state values of the state variables. This is because when the derivative terms are set to zero, the differential equations become algebraic equations. Popular programs to solve sets of nonlinear algebraic equations include Newton-Raphson[1] and quasi-Newton methods.[2]

Another type of simulation run is the startup of the column. In this case we may assume that the initial conditions are those corresponding to the trays, column bottom, and accumulator being full of liquid at the feed composition and enthalpy. For the tray hold-ups we assume zero initial liquid flow rates, that is, minimum hold-up. The temperatures throughout the column, reboiler, and condenser may be assumed to be that of the feed, and the pressure equal to the bubble-point pressure of the feed at this temperature. Many variations of these conditions may also be assumed. The major difficulty with simulating startup is deciding on and implementing the sequence with which the various column inputs are brought up to their design values. This can be time consuming if the simulation is not carried out in an interactive fashion, that is, with complete interaction of the engineer with the computer solution of the equations.

Input Variables

The input variables to our model are the following:

Feed flow rate:	F,
mole fraction:	z_i, $\quad i = 1, 2, \ldots, N - 1$,
and enthalpy:	h_F
Steam flow rate:	F_S
supply pressure:	P_{SS}
and enthalpy:	H_{SS}
Condenser cooling air temperature:	T_A
Composition controller set points:	$x_{k,B}^{set}$ and $y_{k,1}^{set}$

In the disturbance type of simulation run each of these variables is changed from its design value, usually as a step or ramp function, and the time response of the output and internal variables is analyzed. In some runs more than one variable may be changed at a time. In a startup simulation the variables are brought up to their design values in a sequence that is designed with the help of the simulation to achieve minimum startup time, minimum energy consumption, or minimum loss of off-specification product.

Summary

In this section we have developed a mathematical model of a multicomponent distillation column. Features of the model include variable column pressure, a simple tray hydraulics model, and simple dynamic models of the reboiler and the condenser. In the step-by-step development of the model, we have seen the use of material and energy balances, phase equilibrium, hydraulic models, thermodynamic relationships, and models of the control system for arriving at the complete model of the column.

9-3. DYNAMIC MODEL OF A FURNACE

In the preceding example we divided the distillation column into a number of control volumes and assumed that the compositions, temperatures, and other variables were uniform throughout each of the control volumes. This means that the properties were "lumped" in each control volume. In this example we will look into the model of a furnace in which the properties are distributed and the system cannot conveniently be divided into control volumes of uniform properties. This is, thus, an example of a distributed parameter model.

The system to be modeled is a furnace, shown in Fig. 9-5, which is used to heat a process gas by burning fuel in the firebox outside of the tube. We shall assume that the temperature in the firebox is uniform, but the temperature of the gas in the tube and that of the tube walls are functions of position, measured as the distance Z from the entrance to the tube. The outlet process temperature is controlled by a feedback controller (TRC42) that manipulates the flow of fuel to the burners, F_H. Possible disturbances are the flow, F, and inlet temperature, T_0, of the process fluid, and the heat content of the fuel, H_F.

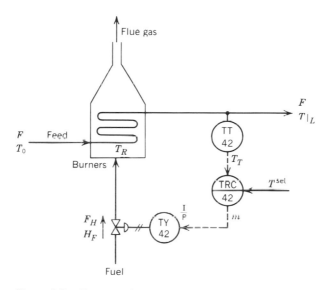

Figure 9-5. Furnace to heat process fluid.

As the temperature of the gas in the tube varies continuously with the distance Z from the entrance to the tube, we must select as our control volume a section of the tube of a short enough length ΔZ that the properties may be assumed to be uniform within it. Such a control volume is shown in Fig. 9-6. Mass balances on this control volume would tell us that the flow, F, of the process fluid is constant with position if we neglect the accumulation of mass in the control volume; we can safely do this as our main interest is to model the heat-transfer effects. We shall also neglect heat conduction along the tube wall in the direction of flow. Under these restrictions, the energy balance on the gas in the control volume of Fig. 9-6 is given by

$$\frac{\pi D_i^2}{4} \rho \Delta Z \, C_v \frac{dT}{dt} = F \, C_p \, T|_Z - F \, C_p \, T|_{Z+\Delta Z} + h_i \, \pi D_i \Delta Z \, (T_W - T) \qquad (9\text{-}40)$$

where

T is the temperature of the process fluid, K
T_W is the temperature of the tube wall, K
F is the flow of process fluid, kgmole/s
D_i is the inside diameter of the tube, m

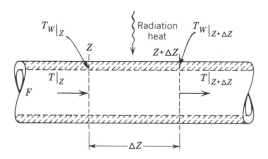

Figure 9-6. Section of furnace tube.

h_i is the inside film coefficient of heat transfer, J/s-m²-K

ρ is the molar density of the fluid, assumed constant, kgmole/m³

C_v is the molar heat capacity of the fluid at constant volume, assumed constant,
      J/kgmole-K

C_p is the molar heat capacity of the fluid at constant pressure, assumed constant,
      J/kgmole-K

ΔZ is the length of the control volume, m

In Eq. (9-40) we had to differentiate between the temperature entering the control volume, $T|_Z$, and the temperature leaving the control volume, $T|_{Z+\Delta Z}$, in the enthalpy terms because the difference between these two terms is taken. However, we did not need this differentiation in the accumulation and heat transfer rate terms because the element is assumed to be small enough so that an average value can accurately represent the temperature in the element. The same applies to the temperature of the wall T_W. The heat capacity at constant pressure, C_p, is used in the enthalpy flow terms because they include the flow work, while the heat capacity at constant volume, C_v, is used in the accumulation term because there is no flow work associated with it. The assumption that the physical properties are constant is made for simplicity. They could just as well be assumed to be functions of temperature and evaluated at T.

Equation (9-40) represents the energy balance around a control volume of length ΔZ positioned at any distance from the entrance to the tube. It can be reduced to an energy balance around any *point* on the tube by dividing by the volume of the control volume and taking limits as $\Delta Z \to 0$. We do this next:

$$\rho C_v \frac{\partial T}{\partial t} = \frac{4}{\pi D_i^2} F C_p \frac{T|_Z - T|_{Z+\Delta Z}}{\Delta Z} + \frac{4h_i}{D_i}(T_W - T) \tag{9-41}$$

where the time derivative has been changed to a partial because it is taken at a fixed point on the tube and temperature is a function of both position and time. Taking limits on Eq. (9-41) as $\Delta Z \to 0$, we get:

$$\rho C_v \frac{\partial T}{\partial t} = -\frac{4}{\pi D_i^2} F C_p \frac{\partial T}{\partial Z} + \frac{4h_i}{D}(T_W - T) \tag{9-42}$$

where we have substituted the identity:

$$\frac{\partial T}{\partial Z} = \lim_{\Delta Z \to 0} \frac{T|_{Z+\Delta Z} - T|_Z}{\Delta Z} \tag{9-43}$$

Equation (9-42) is a partial differential equation that represents the energy balance per unit volume at any point on the tube at any instant of time. Since it is a single equation with two variables, T and T_W, we need another equation. We shall get it from an energy balance on the tube metal wall of the control volume in Fig. 9-6. Neglecting heat conduction along the tube, we obtain

$$\rho_M \frac{\pi}{4} (D_o^2 - D_i^2) \Delta Z\, C_M \frac{dT_W}{dt} = \epsilon\sigma\, \pi D_o \Delta Z\, (T_R^4 - T_W^4) - h_i\, \pi D_i\, \Delta Z (T_W - T) \tag{9-44}$$

where

ρ_M is the density of the tube metal, assumed constant, kg/m^3
C_M is the specific heat of the metal, assumed constant, J/kg-K
D_o is the outside diameter of the tube, m
T_R is the radiating temperature of the firebox, assumed uniform, K
ϵ is the emissivity of the tube surface, assumed constant
σ is the Stefan-Boltzman constant, J/s-m^2-K^4

In modeling the heat transfer by radiation in Eq. (9-44) we have assumed that the area of the firebox is much larger than that of the tube and that the firebox completely encloses the tube. In order to convert Eq. (9-44) into a partial differential equation that applies to a point on the tube, all we need to do is divide it by the volume of the element:

$$\rho_M C_M \frac{\partial T_W}{\partial t} = \frac{4\epsilon\sigma D_o}{D_o^2 - D_i^2} (T_R^4 - T_W^4) \quad \frac{4h_i D_i}{D_o^2 - D_i^2} (T_W - T) \tag{9-45}$$

In obtaining Eq. (9-45) we have introduced a new variable, T_R. An equation for it can be obtained from an energy balance on the firebox:

$$M_B C_B \frac{dT_R}{dt} = F_H H_F \eta_F - \int_0^L \epsilon\sigma\pi D_o\, (T_R^4 - T_W^4)\, dZ \tag{9-46}$$

where

M_B is the effective mass of the firebox, kg
C_B is the specific heat of the firebox, assumed constant, J/kg-K
F_H is the fuel flow rate, kg/s
H_F is the heating value of the fuel, J/kg
η_F is the efficiency of the furnace, assumed constant
L is the constant length of the furnace tube, m

Notice that in Eq. (9-46) we needed to integrate the heat-transfer rate to the tube walls over the length of the tube. The efficiency of the furnace, η_F, accounts for the heat losses through the walls and in the stack gases. This is a simplification that saves us from having to develop a more detailed model of the firebox.

The fuel rate, F_H, is calculated from a model of the temperature control system, as follows:

Control Valve, Neglecting Actuator Lag

$$F_H = C_{VH} f_V(m) \sqrt{\Delta P} \tag{9-47}$$

Temperature Controller (TRC42), as in Example 9-4

$$m = F_{TC} (T^{set}, T_T, K_c, \tau_I, \tau_D) \tag{9-48}$$

Temperature Transmitter (TT42), Assumed a First-Order Lag

$$\frac{dT_T}{dt} = \frac{1}{\tau_T} (T|_L - T_T) \tag{9-49}$$

where

C_{VH} is the valve capacity factor, kg/s-Pa$^{1/2}$
$f_V(m)$ is the valve characteristic function
ΔP is the pressure drop across the valve, Pa
T^{set} is the temperature set point, K
T_T is the transmitter signal, K
K_c is the proportional gain
τ_I is the integral time, s
τ_D is the derivative time, s
τ_T is the sensor time constant, s

This completes the development of the model of the furnace. It consists of two partial differential equations, Eqs. (9-42) and (9-45), one integro-differential equation, Eq. (9-46), and the model of the temperature control loop, Eqs. (9-47) through (9-49). In terms of the initial and boundary conditions, we need:

$T_0(t)$	the entrance temperature as a function of time
$T(0,Z)$	the initial temperature profile in the furnace
$T_W(0,Z)$	the initial tube wall temperature profile
$T_R(0)$	the initial temperature of the firebox
$T_T(0)$	the initial transmitter signal

The input variables to the model are, in addition to $T_0(t)$

$F(t)$	the process fluid flow
$H_F(t)$	the heat content of the fuel
$T^{set}(t)$	the set point of the temperature controller

Let us look next at methods for solving the model equations, given that they are partial differential equations.

9-4. SOLUTION OF PARTIAL DIFFERENTIAL EQUATIONS

Although there are packaged programs that can handle partial differential equations directly (see Section 9-6), one common way to handle them is to discretize the position variables so that each partial differential equation is converted into several ordinary differential equations. The advantage of this procedure is that it results in a model that can be solved by standard programs for ordinary differential equations. We shall demonstrate the procedure by discretizing the model equations for the furnace.

Our first step is to divide the tube length L into N increments of length ΔZ, where

$$\Delta Z = \frac{L}{N} \tag{9-50}$$

In order to simplify this presentation we will assume that the increments are of uniform length, although they do not have to be.

A section of the tube showing two increments is given in Fig. 9-7. The only derivative with respect to Z that must be discretized in our model of the furnace appears in Eq. (9-42). We have the option of three finite difference approximations of the derivative at point j:

Forward difference $\quad \left.\dfrac{\partial T}{\partial Z}\right|_j = \dfrac{T_{j+1} - T_j}{\Delta Z} \tag{9-51}$

Backward difference $\quad \left.\dfrac{\partial T}{\partial Z}\right|_j = \dfrac{T_j - T_{j-1}}{\Delta Z} \tag{9-52}$

Central difference $\quad \left.\dfrac{\partial T}{\partial Z}\right|_j = \dfrac{T_{j+1} - T_{j-1}}{2\Delta Z} \tag{9-53}$

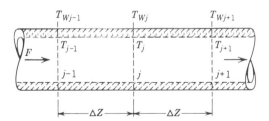

Figure 9-7. Discretization of tube into sections of length ΔZ.

Of these, the central difference is the most accurate in terms of the truncation error of the Taylor series expansion of the $T(Z)$ function. Let us, then, try it first and substitute it into Eq. (9-42):

$$\rho C_v \frac{dT_j}{dt} = \frac{4}{\pi D_i^2} F C_p \left(\frac{T_{j-1} - T_{j+1}}{2\Delta Z} \right) + \frac{4h_i}{D_i} (T_{Wj} - T_j) \tag{9-54}$$

where

T_j is the fluid temperature at the interface between the jth and the $(j + 1)$th increment, K.

There is a basic difficulty with Eq. (9-54) in its ability to model the physical phenomenon of heat transfer in the furnace. To understand this difficulty we must realize that the first term on the right-hand side of Eq. (9-54) represents the heat transfer by convection into our increment of furnace tube. However, in the actual furnace, the propagation of temperature variation by convection can occur only from left to right, that is, in the direction of flow. In Eq. (9-54) this is not the case, as the downstream temperature T_{j+1} has an effect on the upstream temperature T_j through the convection term. It is easy to show that the forward difference approximation, Eq. (9-51), also results in an unreal model. This leaves us with the backward difference approximation as the only one that results in a physically accurate model. Substituting Eq. (9-52) into Eq. (9-42), we obtain

$$\rho C_v \frac{dT_j}{dt} = \frac{4}{\pi D_i^2 \Delta Z} F C_p (T_{j-1} - T_j) + \frac{4h_i}{D_i} (T_{Wj} - T_j) \tag{9-55}$$

Equation (9-55) is not only a more accurate representation of the propagation of temperature variations by convection but is also numerically more stable. This is because the convection term adds to the self-regulation of temperature T_j in the heat transfer term. In other words, an increase in T_j causes a decrease in the rate of change, dT_j/dt; this decrease in rate is greater when calculated by Eq. (9-55) than when calculated by Eq. (9-54). By comparing Eq. (9-55) with Eq. (9-41) we can see that the former could have been derived by directly performing the energy balance on the jth increment of the tube length while assuming that the temperature is uniform throughout the increment.

The discretization of the furnace model is completed by writing Eqs. (9-45) and (9-46) as follows:

$$\rho_M C_M \frac{dT_{Wj}}{dt} = \frac{4\epsilon\sigma D_o}{D_o^2 - D_i^2} (T_R^4 - T_{Wj}^4) - \frac{4h_i D_i}{D_o^2 - D_i^2} (T_{Wj} - T_j) \tag{9-56}$$

$$M_B C_B \frac{dT_R}{dt} = F_H H_F \eta_F - \sum_{j=1}^{N} \epsilon\sigma\pi D_o \Delta Z (T_R^4 - T_{Wj}^4) \tag{9-57}$$

where T_{Wj} is the tube wall temperature at the interface between the jth and $(j + 1)$th increments. In Eq. (9-57) the sum of the heat radiation terms approximates the integral of Eq. (9-46).

Had we considered heat transfer by conduction along the tube, we would have gotten terms with the second partial derivatives with respect to Z in Eqs. (9-42) and (9-45). In this case central difference approximations would have been physically accurate for these derivatives because temperature variations are propagated in both directions by the conduction terms.

In summary, a distributed parameter model can be converted into a set of first-order ordinary differential equations by proper discretization of the spatial derivatives. Each of the original partial differential equations is converted by this procedure into a number of ordinary differential equations in which time is the only independent variable.

9-5. COMPUTER SIMULATION OF DYNAMIC PROCESS MODELS

Once the model equations have been derived, the next step in the dynamic simulation of a physical system is the solution of the equations. When a digital computer is used to solve the equations, three general approaches can be taken for programming the model equations:

1. Use of some simple numerical integration method for solving the equations.
2. Use of a general-purpose subroutine package for the solution of differential equations.
3. Use of a simulation language for the simulation of continuous systems.

The first of these approaches is presented here to provide the student with a tool for solving simple models without having to learn a new programming package or language. The other two approaches are recommended for the solution of more complex models by students in advanced courses and by practitioners in industry. The use of analog computers for process simulation will not be discussed because it requires more space than we can devote in this book for its proper treatment. However, instructors are encouraged to use preprogrammed analog simulations in class demonstrations and assignments.

As we saw in the preceding sections, the dynamic process model, even for distributed systems, can be transformed into a set of first-order ordinary differential equations and auxiliary algebraic equations. In general the differential equations can be written in the form

$$\frac{dx_i}{dt} - f_i(x_1, x_2, \ldots x_n, t) \qquad \text{for } i - 1, 2, \ldots, n \qquad (9\text{-}58)$$

where

x_i are the model state variables, e.g., temperatures, compositions
f_i are the derivative functions that result when the model equations are solved for the derivatives
n is the number of differential equations

All general approaches to solving dynamic models presume that the model equations are in the form of Eq. (9-58). In order to solve these equations we must know the initial values of all the state variables, that is, $x_i(t_0)$, where t_0 is the initial time. Although not explicitly indicated in Eq. (9-58), we also need the input or *forcing functions* that drive the model equations. When the derivative functions, f_i, are very complex, it is often convenient to express them as several simpler algebraic equations, in which case an auxiliary variable is generated for each algebraic equation.

Continuous Stirred Tank Reactor Simulation—An Example

In order to fix ideas, let us look at a relatively simple model of a continuous stirred tank reactor (CSTR). A sketch of the jacketed reactor is given in Fig. 9-8. Assuming that both the reactor and the jacket are perfectly mixed, that the volumes and physical properties are constant, and neglecting heat losses, the model equations are

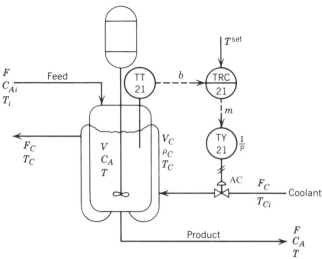

Figure 9-8.   Sketch of continuous stirred tank reactor.

Balance on Mass of Reactant A

$$\frac{dC_A}{dt} = \frac{F}{V}(C_{Ai} - C_A) - kC_A^2 \tag{9-59}$$

Energy Balance on Reactor Contents

$$\frac{dT}{dt} = \frac{F}{V}(T_i - T) - \frac{\Delta H_R}{\rho C_p}kC_A^2 - \frac{UA}{V\rho C_p}(T - T_C) \tag{9-60}$$

Energy Balance on Jacket

$$\frac{dT_C}{dt} = \frac{UA}{V_C\rho_C C_{pC}}(T - T_C) - \frac{F_C}{V_C}(T_C - T_{Ci}) \tag{9-61}$$

Reaction Rate Coefficient

$$k = k_0 e^{-\frac{E}{R(T + 273.16)}} \tag{9-62}$$

Lag in Temperature Sensor (TT21)

$$\frac{db}{dt} = \frac{1}{\tau_T}\left(\frac{T - T_M}{\Delta T_T} - b\right) \tag{9-63}$$

Proportional-Integral Feedback Controller (TRC21)

$$\frac{dy}{dt} = \frac{1}{\tau_I}(m - y) \tag{9-64}$$

$$m = y + K_c\left(\frac{T^{set} - T_M}{\Delta T_T} - b\right) \tag{9-65}$$

Limits on Controller Output Signal

$$0 \le m \le 1 \tag{9-66}$$

Equal Percentage Control Valve (Air to Close)

$$F_C = F_{C\max} \alpha^{-m} \tag{9-67}$$

where

C_A is the concentration of the reactant in the reactor, kgmole/m^3

C_{Ai} is the concentration of the reactant in the feed, kgmole/m^3

T is the temperature in the reactor, C

T_i is the temperature of the feed, C

T_C is the jacket temperature, C

T_{Ci} is the coolant inlet temperature

b is the transmitter signal on a scale of 0 to 1

F is the feed rate, m^3/s

V is the reactor volume, m^3

k is the reaction rate coefficient, m^3/kgmole-s

ΔH_R is the heat of reaction, assumed constant, J/kgmole

ρ is the density of the reactor contents, kgmole/m^3

C_p is the heat capacity of the reactants, J/kgmole-C

U is the overall heat-transfer coefficient, J/s-m^2-C

A is the heat transfer area, m^2

V_C is the jacket volume, m^3

ρ_C is the density of the coolant, kg/m^3

C_{pC} is the specific heat of the coolant, J/kg-C

ΔT_T is the transmitter calibrated range, C

F_c is the coolant rate, m^3/s

T_M is the lower limit of the transmitter range, C

τ_T is the time constant of the temperature sensor, s

τ_I is the controller integral time, s

y is the controller reset feedback variable

m is the controller output signal on a scale of 0 to 1

K_c is the controller gain, dimensionless

$F_{C\max}$ is the maximum flow through the control valve, m^3/s

α is the valve rangeability parameter

k_0 is the Arrhenius frequency parameter, m^3/s-kgmole

E is the activation energy of the reaction, J/kgmole

R is the ideal gas law constant, 8314.39 J/kgmole-K

In this model of the reactor and its temperature controller the state variables are C_A, T, T_C, b, and y. The auxiliary variables r_A, m, and F_C can be calculated, along with the derivative functions, from the values of the state variables at any point in time. The input variables to the model are F, C_{Ai}, T_i, T_{Ci}, and T^{set}. A point worth noting is that some of the auxiliary variables are more relevant in analyzing the controller performance than some of the state variables. For example, the controller output, m, or coolant rate, F_C, are of greater interest than the jacket temperature, T_C, and the reset feedback variable, y.

The model of the proportional-integral (PI) controller is the "reset-feedback" implementation of the integral action. For a more detailed discussion see Section 6-5 and

Example 9-4. The transmitter signal, b, and the controller output signal, m, are normalized, that is, expressed as fraction of range. This makes the model valid for both electronic, digital, and pneumatic instrumentation. Notice that in Eq. (9-65) the controller set point must be normalized by the same formula used to normalize the temperature in the transmitter equation, Eq. (9-63).

In order to simulate the reactor we must determine the model parameters and initial conditions. In practice the model parameters are obtained from equipment specifications and from piping and instrumentation diagrams. Let us work with the following parameters for our reactor:

$V = 7.08$ m^3	$\Delta H_R = -9.86 \times 10^7$ J/kgmole
$\rho = 19.2$ kgmole/m^3	$U = 3550$ J/s-m^2-C
$C_p = 1.815 \times 10^5$ J/kgmole-C	$V_C = 1.82$ m^3
$A = 5.40$ m^2	$C_{pC} = 4184$ J/kg-C
$\rho_C = 1000$ kg/m^3	$E = 1.182 \times 10^7$ J/kgmole
$k_0 = 0.0744$ m^3/s-kgmole	$F_{Cmax} = 0.020$ m^3/s
$\tau_T = 20$ s	$T_M = 80$C
$\alpha = 50$	$\Delta T_T = 20$C

If the purpose of the simulation is to tune the controller at the design operating conditions, the initial conditions are taken at the design operating point. An important requirement is that *the initial conditions satisfy the model equations at steady state*, that is, all the derivatives calculated from the model equations must be exactly zero at the initial values of the state variables. Since we have one model equation for each state and auxiliary variable, the number of design specifications must not exceed the number of input variables. In this example the input variables at design conditions are as follows:

$$F = 7.5 \times 10^{-3} \text{ m}^3\text{/s} \qquad\qquad T^{\text{set}} = 88.0\text{C}$$
$$C_{Ai} = 2.88 \text{ kgmoles/m}^3 \qquad\qquad T_i = 66.0\text{C}$$
$$T_{Ci} = 27.0\text{C}$$

We can now use the model equations to calculate all the other initial values of the state and auxiliary variables. The order of calculations is as follows:

From Equation Number	Calculation
(9-65)	$b = \dfrac{T^{\text{set}} - T_M}{\Delta T_T} = 0.40$
(9-63)	$T = b\Delta T_T + T_M = 88.0$C
(9-62)	$k = 1.451 \times 10^{-3}$ m^3/kgmole-s
(9-59)	$C_A = 1.133$ kgmole/m^3
(9-60)	$T_C = 50.5$C
(9-61)	$F_C = 7.392 \times 10^{-3}$ m^3/s
(9-67)	$m = 0.2544$ (dimensionless)
(9-64)	$y = 0.2544$ (dimensionless)

Notice that the only way to satisfy Eqs. (9-63), (9-64), and (9-65) at steady state is for the reactor temperature to be at the set point. This is because the controller has integral action.

Having the model equations, the parameter values, and the initial conditions, we are ready to program the equations on the computer.

Numerical Integration by Euler's Method

The simplest numerical method for the solution of ordinary differential equations is Euler's method. It consists of assuming that the derivative functions are constant over a small *integration interval* Δt. The outline of a program to solve the set of equations of the form of Eq. (9-58) by Euler's method is as follows:

1. Initialization: set $t = t_0$ and $x_i = x_i(t_0)$ for $i = 1, 2, \ldots, n$.
2. Calculate all derivative functions, f_i, using the model equations:

$$f_i = f_i(x_1, x_2, \ldots, x_n, t) \qquad \text{for } i = 1, 2, \ldots, n \qquad (9\text{-}68)$$

3. Calculate the values of the state variables after a time increment Δt (Euler's formula). For $i = 1, 2, \ldots, n$, calculate

$$x_i|_{t+\Delta t} = x_i|_t + f_i \Delta t \qquad (9\text{-}69)$$

$$\text{Let } t = t + \Delta t \qquad (9\text{-}70)$$

4. If t is less than t_{max}, repeat from Step 2. Otherwise exit.

An essential feature of this program is that all of the derivative functions are calculated in Step 2 before any of the state variables is incremented in Step 3. This ensures that all of the derivative functions correspond to the state of the system at time t, as they must.

Before we can run the program outlined above, we must select an initial time t_0, final time t_{max}, and integration interval Δt. We must also decide the frequency with which the relevant simulation variables are going to be printed.

Duration of Simulation Runs

The duration in "problem time" of each simulation run is $t_{max} - t_0$. The units of this quantity are determined by the units of the rate and time constants in the model equations. For example, in the reactor model equations all rates are per second and all time constants are in seconds, thus, the units of time are seconds.

In most simulations the initial time t_0 can be set to zero. The only exception to this rule is in the very rare cases in which the model parameters are functions of time, for instance, because of fouling of heat-transfer surfaces and the like.

Once the value of t_0 is set, the value of t_{max} determines the duration of each simulation run. This duration should be long enough for the response of the system to be completed, but not so long that the response will be compressed in a very small fraction of the total duration of the run. Thus, the correct value of t_{max} depends on the speed of response of the simulated process. For fast processes t_{max} need be only a few seconds, while for slow processes t_{max} may be on the order of hours. Time responses for runs that are too long (compressed), too short (incomplete), and just right are illustrated in Fig. 9-9.

What determines the speed of response of the process? Strictly speaking, the dominant eigenvalue, this is, the reciprocal of the longest process time constant, controls the time

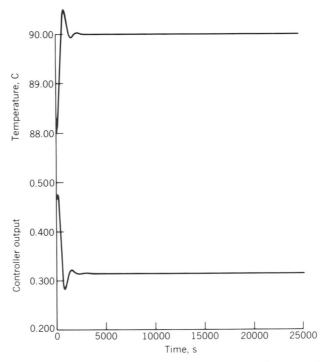

Figure 9-9a.   Reactor response to a 2C rise in set point. Example of a run that is too long.

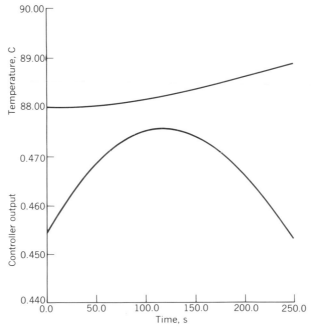

Figure 9-9b.   Reactor response to a 2C rise in set point. Example of a run that is too short.

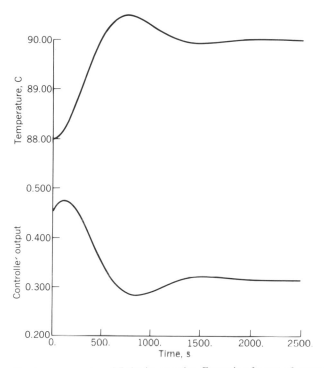

Figure 9-9c.   Reactor response to a 2C rise in set point. Example of a run of correct duration.

required to complete the response. Unfortunately, eigenvalues are difficult to determine for complex nonlinear process models such as those developed earlier in this chapter. On the other hand, it is sometimes possible to estimate the longest time constant from either familiarity with similar processes or engineering intuition. For example, for the reactor we are considering here the longest time constant is probably of the order of magnitude of the reactor residence time, V/F, or about 1000 s. Once an estimate of the longest time constant is available, the run duration can be set to about five times the time constant. This rule of thumb is based on the fact that for a first-order process the response is complete in five time constants; we might expect higher-order processes to take longer, and closed-loop responses to be completed in a shorter time. In many cases we have no convenient way of estimating the run duration and must select it by trial and error. The one thing to remember on the selection of t_{max} is this:

> Regardless of the method used to estimate the duration of simulation runs, it should always be adjusted based on observation of the responses obtained in the first few runs.

It is seldom a good idea to keep the run duration much longer than it needs to be. This results in a waste of computer time.

Selection of Integration Interval

The integration interval, Δt, affects the accuracy of the numerical integration of the differential equations and the computer time required to perform the computations. The effect on computer time is simply that the number of computations is inversely proportional to the integration interval, that is, proportional to the number of integration "steps," N:

$$N = \frac{t_{max} - t_0}{\Delta t} \tag{9-71}$$

With regard to the accuracy of the numerical integration, it is a common misconception that the solution is more accurate the shorter the integration interval Δt. Although, as we shall see shortly, it is theoretically true that the *truncation error* is larger the longer the integration interval, in practice, because of the limited precision of computer calculations, there is a limit on how small the integration interval can be. Below this limit the *round-off error* increases with decreasing integration interval. The round-off error is the one incurred because a finite number of significant digits is carried by the computer in performing calculations. Although the minimum size of the integration interval can be reduced by specifying that the computations be done in "double precision," that is, doubling the number of significant digits carried by the computer, it is seldom good practice to waste computer time by running with an integration interval that is much smaller than that required by the truncation error. In other words, our attitude should be one of selecting an integration interval near the maximum allowed by the required accuracy of the numerical integration calculations.

The truncation error of a numerical integration procedure is the one incurred when the continuous derivatives are approximated at discrete values of the independent variable. For example, in Euler's method, we assume that the values of the derivatives at time t are good approximations of the continuous derivatives from t to $(t + \Delta t)$. This results in Eq. (9-69), which can be thought of as a Taylor series expansion of the function $x_i(t)$ "truncated" after the first derivative term. Euler's formula, being the simplest, results in the greatest truncation error for a given integration interval Δt. As we mentioned earlier, the truncation error increases with the integration interval Δt. If the integration interval is large enough, the numerical solution will become unstable, that is, will produce an unstable response even for a stable process. A related danger is that in computing an unstable process response the cumulative effect of truncation error can render the results useless after a few integration steps.

As in selecting the duration of the simulation runs, the integration interval should be adjusted based on the observed accuracy of the first few runs. A simple procedure is to run the same case at different integration intervals and check that the results agree within a tolerable error, that is, within an acceptable number of significant digits such as four or five. The longest integration interval that gives acceptable results should be selected.

For Euler's method and a well-behaved process model, a good estimate of the integration interval is one that requires between 1000 and 5000 steps to complete a simulation run. A well-behaved model is one in which all of the eigenvalues (time constants) are of about the same order of magnitude. In contrast, a *stiff* model is one in which the ratio of the largest to the smallest eigenvalue (or time constants) is large. We will discuss stiffness in a separate section at the end of this chapter.

Display of Simulation Results

The results of a dynamic simulation are usually the time responses of the model variables. In a numerical solution these responses are calculated as the values of the state and auxiliary variables at each step of the numerical integration procedure. These results can be displayed in either tabular or graphical form. Of these the graphical form is more informative in studying the response of control systems. It can be in the form of low-, medium-, or high-resolution graphs. Low-resolution graphs require only a regular character printer and have a resolution of about 6 to 10 points per inch. Medium-resolution graphs require a dot matrix printer or screen and have a resolution of about 80 points per inch. High-resolution graphs require a special digital plotter or graphics terminal and are of quality comparable to that of engineering drawings. The computer generation of graphs usually requires packaged plotting subprograms. Figure 9-9 presents examples of high-resolution graphs while Fig. 9-10 shows an example of a low-resolution graph.

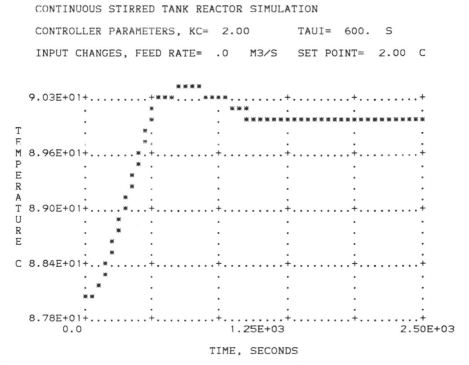

Figure 9-10.   Reactor response to a 2C rise in set point. Example of a low-resolution graph using the printer.

When printing the simulation results in tabular form, it is poor practice to print the variables at each integration step. This not only wastes paper and the trees used to produce it, but also makes the results difficult to read and interpret. The recommended approach is to print the model variables at intervals of time Δt_P, the printing interval. This interval should be a multiple of the integration interval Δt. The printing interval should be selected so that a total of about 50 entries are printed for a complete simulation run. In other words

$$\Delta t_P \approx \frac{t_{max} - t_0}{50} \tag{9-72}$$

The number 50 is selected because a typical page of computer output holds about 50 lines. This selection is arbitrary and could be changed to 25 or 100 depending on the degree of resolution required on the time response.

When searching for errors in the computer program, it usually becomes necessary to print the variables at every integration step. In this case Δt_P can be temporarily set equal to Δt, the integration interval. Paper can be conserved by setting t_{max} equal to about 50 Δt for the error-detection runs.

Example Results for Euler's Method

A listing of the FORTRAN program to simulate the continuous stirred tank reactor by Euler's method is given in Fig. 9-11. The integration interval is 0.25 s, the duration of

```
C
C        PROGRAM FOR SIMULATION OF CONTINUOUS STIRRED TANK REACTOR
C
C        METHOD:  EULER INTEGRATION
C
C        VARIABLES IN INPUT DATA
C
C        FIRST LINE
C
C            TO      INITIAL TIME OF THE RUN, S
C            TMAX    FINAL TIME OF THE RUN, S
C            DTIME   INTEGRATION INTERVAL, S
C            DPRNT   PRINTING INTERVAL, S
C
C        SECOND LINE
C
C            KC      CONTROLLER PROPORTIONAL GAIN, DIMENSIONLESS
C            TAUI    CONTROLLER INTEGRAL TIME, S
C            DF      CHANGE IN FEED RATE, M3/S
C            DTSET   CHANGE IN SET POINT, C
C
C        NOTE: A ZERO OR NEGATIVE VALUE OF TMAX STOPS EXECUTION.
C
         IMPLICIT REAL*8 (A-H,O-Z)
         REAL*8 M, K, KO, KC
C
C        DATA STATEMENTS FOR PARAMETER VALUES
C
         DATA V, U, A / 7.08, 3550., 5.40 /
         DATA KO, E, DHR, RHO / 0.0744, 1.182D7, -9.86D7, 19.2/
         DATA CP, VC, RHOC, CPC / 1.815D5, 1.82, 1000., 4184. /
         DATA R / 8314.39 /
         DATA TAUT, ALPHA, FCMAX, TM, DTT / 20., 50., 0.020, 80., 20. /
         DATA TO, TMAX, DTIME, DPRNT / 0., 2500., 0.5, 50. /
C
C        DATA STATEMENTS FOR DESIGN CONDITIONS
C
         DATA FO, CAI, TI, TCI, TSETO / 7.55D-3, 2.88, 66., 27., 88. /
```

Figure 9-11. Listing of FORTRAN program for simulation of continuous stirred tank reactor. Euler integration method.

```
C
C          READ AND PRINT INPUT DATA
C
  1    WRITE(6,200)
 200   FORMAT(' ENTER TO, TMAX, DTIME, DPRNT')
       READ  ( 5, * ) TO, TMAX, DTIME, DPRNT
       IF ( TMAX .LT. 0. ) STOP
C
       WRITE ( 4, 300 ) TO, TMAX, DTIME
 300   FORMAT(//10X,'RUN PARAMETERS, TO=',1PG10.3,5X,'TMAX=',G10.3,5X
      F           'DTIME=',G10.3)
C
       WRITE(6,210)
 210   FORMAT(' ENTER KC, TAUI, DF, DTSET')
       READ  ( 5, * ) KC, TAUI, DF, DTSET
C
       WRITE ( 6, 100 ) KC, TAUI, DF, DTSET
       WRITE ( 4, 100 ) KC, TAUI, DF, DTSET
 100   FORMAT(//10X,'CONTINUOUS STIRRED TANK REACTOR SIMULATION'/
      1        /' 10X,'CONTROLLER PARAMETERS, KC=',G10.3,
      2           5X,'TAUI=',G10.3,' S'/
      3         / 10X,'INPUT CHANGES, FEED RATE=',G10.3,' M3/S',
      4           5X,'SET POINT=',G10.3,' C')
C
C          CALCULATE STEADY-STATE INITIAL CONDITIONS AND CONSTANTS
C
       TIME = TO
       B  = ( TSETO - TM ) / DTT
       T  = TSETO
       K  = KO * DEXP ( - E / ( R * (T+273.16) ) )
       CA = ( -FO+ DSQRT( FO**2 + 4.* K * V * FO * CAI ) ) / ( 2.*K*V )
       TC = T - ( FO * RHO * CP *(TI - T) - V * DHR * K * CA**2 ) /
      1         ( U * A )
       FC = U * A * ( T - TC ) / ( RHOC * CPC * ( TC - TCI ) )
       M  = - DLOG ( FC / FCMAX ) / DLOG ( ALPHA )
       Y  = M
C
       EOR = E / R
       DHRC = DHR / ( RHO * CP )
       UAIC = U * A / ( V * RHO * CP )
       UAOC = U * A / ( VC * RHOC * CPC )
C
C          PRINT HEADER FOR OUTPUT TABLE AND INITIAL CONDITIONS
C
       WRITE ( 6, 110 )
       WRITE ( 4, 110 )
 110   FORMAT(//10X,'TIME, S',8X,'T, C',11X,'M',9X,'FC, M3/S',
      1           /X, (A, KMOL/M3'/)
       WRITE ( 6, 120 ) TIME, T, M, FC, CA
       WRITE ( 4, 120 ) TIME, T, M, FC, CA
 120   FORMAT(10X,5G12.4)
C
C          INITIALIZATION FOR EULER INTEGRATION
C
       F = FO + DF
       TSET = TSETO + DTSET
       NSTEPS = ( TMAX - TO ) / DTIME + 0.99
       NPSTES = DPRNT / DTIME + 0.5
       KP     = NPSTPS
C
       DO 90 I=1,NSTEPS
C
C          CALCULATE AUXILIARY VARIABLES
C
       K  = KO * DEXP( - EOR / ( T + 273.16 ) )
       M  = DMAX1( 0.D0, DMIN1( 1.D0,
      F                         Y + KC * ( ( TSET-TM )/DTT - B ) ) )
       FC = FCMAX * ALPHA ** ( - M )
C
C          EVALUATE DERIVATIVE FUNCTIONS
C
       DCA = F * ( CAI - CA ) / V - K * CA**2
       DT  = F * ( TI - T ) / V - DHRC * K * CA**2 - UAIC * (T-TC)
       DTC = UAOC * ( T - TC ) - FC * ( TC - TCI ) / VC
       DB  = ( (T-TM)/DTT - B ) / TAUT
       DY  = ( M - Y ) / TAUI
C
C          INCREMENT STATE VARIABLES BY EULER INTEGRATIION
C
       CA = CA + DCA * DTIME
       T  = T  + DT  * DTIME
       TC = TC + DTC * DTIME
       B  = B  + DB  * DTIME
       Y  = Y  + DY  * DTIME
       TIME = TO + I * DTIME
```

Figure 9-11. (*Continued*)

```
C
C                PRINT INTERMEDIATE RESULTS
C
       KP = KP - 1
       IF ( KP .GT. 0 ) GOTO 90
          KP = NPSTPS
          WRITE ( 6, 120 ) TIME, T, M, FC, CA
          WRITE ( 4, 120 ) TIME, T, M, FC, CA
90     CONTINUE
       GOTO 1
       END
```

Figure 9-11.   *(Continued)*

the run is 2500 s, and the printing interval is 50 s. A table of printed results for the reactor response to a 2C rise in set point is given in Fig. 9-12. The graphs in Fig. 9-9 are for the same run. An obvious advantage of the table over the graphs is that we can display more variables. As shown in Fig. 9-9, the controlled and manipulated variables are the important variables to plot when analyzing control system performance.

The program listed in Fig. 9-11 is designed for interactive use through a time-sharing terminal. This is why it prints messages on the terminal (unit 6) to prompt the user. Data are input through the terminal (unit 5) and results are displayed on the terminal and printed on the printer (unit 4). All other programs listed in this chapter are similarly designed.

Modified Euler Method

Euler's method is the simplest to program on a computer, but also the least efficient. The modified Euler method, based on the trapezoidal rule of numerical integration, offers a more accurate representation of the derivatives, that is, a smaller truncation error for a given integration interval. The outline of a program for the modified Euler method is as follows:

1. Initialization: set $t = t_0$ and $x_i = x_i(t_0)$ for $i = 1, 2, \ldots, n$.

2. Evaluation for the derivative functions:
    (a) For $i = 1, 2, \ldots, n$, calculate

$$f_i^{(1)} = f_i(x_1, x_2, \ldots, x_n, t) \tag{9-73}$$

    (b) For $i = 1, 2, \ldots, n$, calculate

$$f_i^{(2)} = f_i(x_1 + f_1^{(1)}\Delta t, \ldots, x_n + f_n^{(1)}\Delta t, t + \Delta t) \tag{9-74}$$

3. Calculate the state variables at $t + \Delta t$:
    For $i = 1, 2, \ldots, n$, calculate

$$x_i\big|_{t+\Delta t} = x_i\big|_t + \frac{1}{2}(f_i^{(1)} + f_i^{(2)})\Delta t \tag{9-75}$$

RUN PARAMETERS, TO= .0 TMAX= 2.50D+03   DTIME= .250

CONTINUOUS STIRRED TANK REACTOR SIMULATION

CONTROLLER PARAMETERS, KC= 2.00   TAUI= 600.   S

INPUT CHANGES, FEED RATE= .0   M3/S   SET POINT= 2.00   C

TIME, S	T, C	M	FC, M3/S	CA, KMOL/M3
.0	88.00	.2544	.7392D-02	1.133
50.00	88.05	.4687	.3197D-02	1.133
100.0	88.17	.4753	.3116D-02	1.133
150.0	88.36	.4741	.3131D-02	1.132
200.0	88.59	.4662	.3228D-02	1.132
250.0	88.85	.4531	.3398D-02	1.131
300.0	89.12	.4359	.3634D-02	1.131
350.0	89.38	.4161	.3928D-02	1.130
400.0	89.64	.3948	.4269D-02	1.129
450.0	89.88	.3733	.4643D-02	1.128
500.0	90.08	.3527	.5033D-02	1.127
550.0	90.24	.3339	.5417D-02	1.126
600.0	90.37	.3177	.5771D-02	1.126
650.0	90.45	.3045	.6076D-02	1.125
700.0	90.50	.2946	.6317D-02	1.124
750.0	90.51	.2879	.6485D-02	1.123
800.0	90.49	.2841	.6581D-02	1.123
850.0	90.46	.2830	.6611D-02	1.123
900.0	90.40	.2839	.6587D-02	1.122
950.0	90.34	.2864	.6522D-02	1.122
1000.	90.28	.2901	.6430D-02	1.122
1050.	90.21	.2944	.6322D-02	1.122
1100.	90.15	.2990	.6210D-02	1.122
1150.	90.10	.3035	.6101D-02	1.122
1200.	90.05	.3077	.6001D-02	1.123
1250.	90.01	.3115	.5912D-02	1.123
1300.	89.98	.3147	.5839D-02	1.123
1350.	89.96	.3173	.5780D-02	1.123
1400.	89.95	.3192	.5737D-02	1.123
1450.	89.94	.3205	.5707D-02	1.123
1500.	89.94	.3213	.5691D-02	1.123
1550.	89.95	.3216	.5685D-02	1.123
1600.	89.95	.3214	.5688D-02	1.123
1650.	89.96	.3209	.5698D-02	1.123
1700.	89.97	.3203	.5714D-02	1.124
1750.	89.98	.3194	.5733D-02	1.124
1800.	89.99	.3185	.5753D-02	1.124
1850.	90.00	.3176	.5773D-02	1.124
1900.	90.01	.3167	.5793D-02	1.123
1950.	90.02	.3160	.5811D-02	1.123
2000.	90.02	.3153	.5826D-02	1.123
2050.	90.02	.3147	.5838D-02	1.123
2100.	90.03	.3143	.5848D-02	1.123
2150.	90.03	.3140	.5854D-02	1.123
2200.	90.02	.3139	.5858D-02	1.123
2250.	90.02	.3138	.5859D-02	1.123
2300.	90.02	.3138	.5859D-02	1.123
2350.	90.02	.3139	.5857D-02	1.123
2400.	90.01	.3141	.5854D-02	1.123
2450.	90.01	.3142	.5850D-02	1.123
2500.	90.01	.3144	.5845D-02	1.123

Figure 9-12.   Reactor response to a 2C rise in set point. Table of printed results.

Let $t = t + \Delta t$ (9-76)

4. If t is less than or equal to t_{max}, repeat from Step 2. Otherwise exit.

In this procedure the derivative functions must be evaluated twice per integration step, once at time t and once at time $(t + \Delta t)$. In the second evaluation the state variables are incremented using the values of the derivative functions from the first evaluation. Finally, the increment in the state variable is calculated from the arithmetic average of the two evaluations of the derivative functions. Given that the evaluation of the derivative functions is the time-consuming step of the calculations, the modified Euler method takes about twice as many calculations per integration step as Euler's method. Its advantage is that the reduction in truncation error allows us to increase the integration interval by more than twice the value required by Euler's method while maintaining the same accuracy. This results in fewer total evaluations of the derivative functions since the resulting number of steps is less than half that required by Euler's method.

Since the same derivative functions must be evaluated twice in the program, it is convenient to put the calculations of the derivative functions in a subroutine, which is then called by the numerical integration method. Once this is done, model equations are separated from the numerical integration routine so that the latter can be changed without affecting the subprogram with the model equations. Similarly, the same numerical integration subprogram can be used with different model equation subprograms.

The subroutine for the modified Euler integration is listed in Fig. 9-13. It calls the subroutine DERIV twice per integration step, once for each evaluation of the derivative functions. The main program is listed in Fig. 9-14 and subroutine MODEL is listed in Fig. 9-15. Subroutine MODEL consists of three different sections, each with its own entry point and return. The first section is called by the main program under the subroutine name "MODEL." In this section we read the model parameters, calculate and set up the initial conditions and the number of differential equations to be integrated, and calculate constant terms in the model equations. The initial conditions and number of equations are passed back to the main program as arguments. The main program sets up the rest of the run parameters and calls subroutine INT for the numerical integration of the differential equations.

The second section of subroutine MODEL is called by subroutine INT under the name "DERIV." Its purpose is to evaluate the derivative functions from the model equations every time the integration subroutine calls it with a new set of state variable values. This is the second step of the modified Euler procedure outlined above. Note that although subroutine MODEL is called only once per run at the first entry point, the second entry point, DERIV, is called twice per integration step during a run. In other words, when DERIV is called, execution begins at the statement "ENTRY DERIV. . . ," not at the beginning of subroutine MODEL.

The third section of subroutine MODEL is called by subroutine INT under the name "OUTPUT." The purpose of this section is to print entries into the time response table. The integration subroutine INT calls OUTPUT at intervals of time Δt_P (DPRNT) after the first call to DERIV during the appropriate integration step. It is only at this point of each integration step that the state and auxiliary variables have values that correspond to

```
          SUBROUTINE INT( TO, X, TF, DT, DPRNT, N, XT, RK, F )
C
C         PURPOSE - TO INTEGRATE A SET OF ORDINARY DIFFERENTIAL EQUATIONS
C
C         METHOD  - MODIFIED EULER (TRAPEZOIDAL RULE, OR RUNGE-KUTTA
C                   ORDER 2)
C
C         VARIABLES IN ARGUMENT LIST
C
C         VARIABLE  TYPE  I/O  DIMENSION  DESCRIPTION
C
C            TO     R*8   I/O     -       INDEPENDENT VARIABLE
C                                          INPUT  - INITIAL VALUE
C                                          OUTPUT - FINAL VALUE
C            X      R*8   I/O     N       VECTOR OF STATE VARIABLES
C                                          INPUT  - INITIAL VALUES
C                                          OUTPUT - FINAL VALUES
C            TF     R*8   I       -       FINAL VALUE OF INDT. VARIABLE
C            DT     R*8   I       -       INTEGRATION INTERVAL
C            DPRNT  R*8   I       -       PRINTING INTERVAL
C            N      I     I       -       NUMBER OF DIFFTL. EQUATIONS
C            XT     R*8   W       N       WORKING STATE VECTOR
C            RK     R*8   W       N       WORKING STEP VECTOR
C            F      R*8   W       N       WORKING DERIVATIVE VECTOR
C
C         SUBROUTINES CALLED - DERIV - TO CALCULATE DERIVATIVE FUNCTIONS
C                              OUTPUT- TO PRINT INTERMEDIATE RESULTS
C
          IMPLICIT REAL*8 (A-H,O-Z)
          DIMENSION X(1), XT(1), RK(1), F(1)
C
C         INITIALIZATION FOR STEP INTEGRATION
C
          NSTEPS = ( TF - TO ) / DT + 0.99
          NPSTPS = DPRNT / DT + 0.5
          KP     = 0
C
          DO 90 I=1,NSTEPS
C
C              FIRST DERIVATIVE EVALUATION
C
               CALL DERIV( N, TO, X, F, 1 )
C
C              PRINT INTERMEDIATE RESULTS EVERY NPSTPS
C
               KP = KP - 1
               IF ( KP .GT. 0 ) GOTO 10
                  KP = NPSTPS
                  CALL OUTPUT
   10          CONTINUE
C
C              SECOND DERIVATIVE EVALUATION
C
               T = TO + DT
               DO 20 J=1,N
                  RK(J) = F(J) * DT
                  XT(J) = X(J) + RK(J)
   20          CONTINUE
               CALL DERIV( N, T, XT, F, 2 )
C
C              MODIFIED EULER STEP ON STATE VARIABLES
C
               TO = TO + DT
               DO 30 J=1,N
                  X(J) = X(J) + 0.5 * ( RK(J) + F(J) * DT )
   30          CONTINUE
   90     CONTINUE
C
C         PRINT FINAL VALUES
C
          CALL DERIV( N, TO, X, F, 0 )
          CALL OUTPUT
          RETURN
          END
```

Figure 9-13. Listing of FORTRAN subroutine for modified Euler integration method (Runge-Kutta order 2).

```
C
C          PROGRAM FOR GENERAL PURPOSE SIMULATION OF DYNAMIC SYSTEMS
C
C          BY ARMANDO B. CORRIPIO              ON MAY 11, 1983
C
C          VARIABLES IN INPUT DATA
C
C          VARIABLE   TYPE   DESCRIPTION
C
C             TO       R*8   INITIAL TIME OF THE RUN
C             TMAX     R*8   FINAL TIME OF THE RUN
C             DTIME    R*8   INTEGRATION INTERVAL
C             DPRNT    R*8   PRINTING INTERVAL FOR INTERMEDIATE RESULTS
C
C          TO STOP, ENTER A ZERO OR NEGATIVE VALUE FOR TMAX
C
C          SUBROUTINES CALLED
C
C             MODEL  - TO ACCEPT INPUT DATA AND SET UP INITIAL CONDITIONS
C             INT    - FOR NUMERICAL INTEGRATION OF THE DIFFERNTIAL EQNS.
C
           IMPLICIT REAL*8 (A-H,O-Z)
           DIMENSION X(100), XT(100), RK(100), F(100)
C
           DATA TO, TMAX, DTIME, DPRNT / 0., 10., 7.8125E-3, .25 /
C
C          READ AND PRINT RUN PARAMETERS
C
  1        WRITE(6,200)
  200      FORMAT(' ENTER TO, TMAX, DTIME, DPRNT. (TMAX = 0 TO STOP).' )
           READ ( 5, * ) TO, TMAX, DTIME, DPRNT
           IF( TMAX .LE. 0D0 ) STOP
C
           WRITE ( 6, 300 ) TO, TMAX, DTIME
           WRITE ( 4, 300 ) TO, TMAX, DTIME
  300      FORMAT(/7X,'RUN PARAMETERS, TO=',1PG10.3,2X,'TMAX=',G10.3,2X,
          F                'DTIME=',G11.5)
C
C          ACCEPT MODEL INPUT DATA AND INITIALIZE
C
           TIME = TO
           CALL MODEL( N, TIME, X, F, DTIME )
C
C          CALL INTEGRATION SUBROUTINE
C
           CALL INT( TIME, X, TMAX, DTIME, DPRNT, N, XT, RK, F )
C
           GOTO 1
           END
```

Figure 9-14. Listing of FORTRAN program to be used with a general-purpose integration routine (INT) for sets of ordinary differential equations.

the true state of the system, since the state variables are set to approximate values for the second evaluation. This is also the reason why DERIV must be called ahead of OUTPUT at the completion of the run (in subroutine INT) for printing the final conditions.

Notice that in this programming structure the main program and integration subroutines can be used with any model subroutine. All of the statements that are specific to the model are isolated in subroutine MODEL. The use of multiple entry points on the subroutine allows us to transfer the values of parameters and variables from one section

```
      SUBROUTINE MODEL( NODE, TIME, XS, DXS, DTIME )
C
C        CONTINUOUS STIRRED TANK REACTOR SIMULATION
C
C        THIS SUBROUTINE HAS THREE ENTRY POINTS. THE PURPOSE OF EACH IS
C
C           MODEL  - TO ACCEPT INPUT DATA AND COMPUTE CONSTANT COEFFNTS.
C                    AND INITIAL CONDITIONS.
C           DERIV  - TO COMPUTE THE DERIVATIVE FUNCTIONS FROM THE STATE
C                    VARIABLES.
C           OUTPUT - TO PRINT INTERMEDIATE VALUES OF SELECTED VARIABLES.
C
C        VARIABLES IN ARGUMENT LIST
C
C        VARIABLE  TYPE  I/O  DIMENSION  DESCRIPTION
C
C           NODE    I     O      -       NUMBER OF DIFFRNTL. EQUATIONS
C           TIME    R*8   I      -       INDEPENDENT VARIABLE
C           XS      R*8   O/I   NODE     STATE VARIABLE VECTOR
C           DXS     R*8   O     NODE     VECTOR OF DERIVATIVE FUNCTIONS
C           DTIME   R*8   I      -       INTEGRATION INTERVAL
C
C        VARIABLES IN INPUT DATA
C
C           KC        CONTROLLER PROPORTIONAL GAIN, DIMENSIONLESS
C           TAUI      CONTROLLER INTEGRAL TIME, S
C           DF        CHANGE IN FEED RATE, M3/S
C           DTSET     CHANGE IN SET POINT, C
C
      IMPLICIT REAL*8 (A-H,O-Z)
      REAL*8 M, K, KO, KC
      DIMENSION XS(1), DXS(1)
C
C        DATA STATEMENTS FOR PARAMETER VALUES
C
      DATA V, U, A / 7.08, 3550 , 5.40 /
      DATA KO, E, DHR, RHO / 0.0744, 1.182D7, -9.86D7, 19.2/
      DATA CP, VC, RHOC, CPC / 1.815D5, 1.82, 1000., 4184. /
      DATA R / 8314.39 /
      DATA TAUT, ALPHA, FCMAX, TM, DTT / 20., 50., 0.020, 80., 20. /
C
C        DATA STATEMENTS FOR DESIGN CONDITIONS
C
      DATA FO, CAI, TI, TCI, TSETO / 7.55D-3, 2.88, 66., 27., 88. /
C
C        READ IN CONTROLLER PARAMETERS AND INPUT CHANGES
C
      WRITE(6,210)
  210 FORMAT(' ENTER KC, TAUI, DF, DTSET')
      READ  ( 5, * ) KC, TAUI, DF, DTSET
      WRITE ( 6, 100 ) KC, TAUI, DF, DTSET
      WRITE ( 4, 100 ) KC, TAUI, DF, DTSET
  100 FORMAT(//10X,'CONTINUOUS STIRRED TANK REACTOR SIMULATION'/
     1       / 10X,'CONTROLLER PARAMETERS, KC=',G10.3,
     2         5X,'TAUI=',G10.3,' S'/
     3       / 10X,'INPUT CHANGES, FEED RATE=',G10.3,' M3/S',
     4         5X,'SET POINT=',G10.3,' C')
C
C        CALCULATE STEADY-STATE INITIAL CONDITIONS AND CONSTANTS
C
      B  = ( TSETO - TM ) / DTT
      T  = TSETO
      K  = KO * DEXP ( - E / ( R * (T+273.16) ) )
      CA = ( -FO+ DSQRT( FO**2 + 4.* K * V * FO * CAI ) ) / (2.*K*V )
      TC = T - ( FO * RHO * CP * (TI - T) - V * DHR * K * CA**2 ) /
     1         ( U * A )
      FC = U * A * ( T - TC ) / ( RHOC * CPC * ( TC - TCI ) )
      M  = - DLOG ( FC / FCMAX ) / DLOG ( ALPHA )
      Y  = M
C
      XS(1) = CA
      XS(2) = T
      XS(3) = TC
      XS(4) = B
      XS(5) = Y
      NODE  = 5
C
```

Figure 9-15. Listing of FORTRAN subroutine for dynamic simulation of a continuous stirred tank reactor. This routine is independent of the numerical integration method.

```
          EOR = E / R
          DHRC = DHR / ( RHO * CP )
          UAIC = U * A / ( V * RHO * CP )
          UAOC  = U * A / ( VC * RHOC * CPC )

    C
    C          PRINT HEADER FOR OUTPUT TABLE AND INITIAL CONDITIONS
    C
          WRITE ( 6, 110 )
          WRITE ( 4, 110 )
      110 FORMAT(//10X,'TIME, S',8X,'T, C',11X,'M',9X,'FC, M3/S',
         1          2X,'CA, KMOL/M3'/)
          WRITE ( 6, 120 ) TIME, T, M, FC, CA
          WRITE ( 4, 120 ) TIME, T, M, FC, CA
      120 FORMAT(10X,5G12.4)
    C
    C          INITIALIZATION OF INPUT VARIABLES
    C
          F = FO + DF
          TSET = TSETO + DTSET
    C
          RETURN
    C
    C          ENTRY TO CALCULATE DERIVATIVE FUNCTIONS
    C
          ENTRY DERIV( NODE, TIME, XS, DXS, NEVAL )
    C
    C          STORE STATE VARIABLE INTO LOCAL VARIABLES
    C
          CA = XS(1)
          T  = XS(2)
          TC = XS(3)
          B  = XS(4)
          Y  = XS(5)
    C
    C          CALCULATE AUXILIARY VARIABLES
    C
          K  = KO * DEXP( - EOR / ( T + 273.16 ) )
          M  = DMAX1( 0.DO, DMIN1( 1.DO,
         F                         Y + KC * ( (TSET-TM)/DTT - B ) ) )
          FC = FCMAX * ALPHA ** ( - M )
    C
    C          EVALUATE DERIVATIVE FUNCTIONS
    C
          DXS(1) = F * ( CAI - CA ) / V - K * CA**2
          DXS(2) = F*( TI  - T  ) / V - DHRC * K * CA**2 - UAIC*(T-TC)
          DXS(3) = UAOC * ( T - TC ) - FC * ( TC - TCI ) / VC
          DXS(4) = ( (T-TM)/DTT - B ) / TAUT
          DXS(5) = ( M - Y ) / TAUI
    C
          RETURN
    C
    C          ENTRY TO OUTPUT SELECTED VARIABLES
    C
          ENTRY OUTPUT
    C
          WRITE ( 6, 120 ) TIME, T, M, FC, CA
          WRITE ( 4, 120 ) TIME, T, M, FC, CA
    C
          RETURN
          END
```

Figure 9-15.   (*Continued*)

of the subroutine to the other without having to use COMMON statements or additional arguments.

For the continuous stirred tank reactor example, an integration interval of 2.5 s was required to match the four-significant-digit accuracy of the results of the Euler method. This value, compared with 0.25 s for Euler's method, means that the total number of function evaluations required by the modified Euler is one-fifth the number required by the simple Euler method.

Runge-Kutta-Simpson Method

The fourth-order Runge-Kutta requires four evaluations of the derivative functions per integration step. The outline of a program that uses the Runge-Kutta-Simpson method is as follows:

1. Initialization: set $t = t_0$ and $x_i = x_i(t_0)$ for $i = 1, 2, \ldots, n$.

2. Evaluation of derivative functions:
   (a) For $i = 1, 2, \ldots, n$, calculate

$$k_{1,i} = \Delta t\, f_i(x_1, \ldots, x_n, t) \tag{9-77}$$

   (b) For $i = 1, 2, \ldots, n$, calculate

$$k_{2,i} = \Delta t\, f_i(x_1 + \tfrac{1}{2}k_{1,1}, \ldots, x_n + \tfrac{1}{2}k_{1,n}, t + \tfrac{1}{2}\Delta t) \tag{9-78}$$

   (c) For $i = 1, 2, \ldots, n$, calculate

$$k_{3,i} = \Delta t\, f_i(x_1 + \tfrac{1}{2}k_{2,1}, \ldots, x_n + \tfrac{1}{2}k_{2,n}, t + \tfrac{1}{2}\Delta t) \tag{9-79}$$

   (d) For $i = 1, 2, \ldots, n$, calculate

$$k_{4,i} = \Delta t\, f_i(x_1 + k_{3,1}, \ldots, x_n + k_{3,n}, t + \Delta t) \tag{9-80}$$

3. Increment the state variables:
   For $i = 1, 2, \ldots, n$, let

$$x_i\big|_{t+\Delta t} = x_i\big|_t + \frac{1}{6}(k_{1,i} + 2k_{2,i} + 2k_{3,i} + k_{4,i}) \tag{9-81}$$

$$\text{Let } t = t + \Delta t \tag{9-82}$$

4. If t is less than or equal to t_{max}, repeat from Step 2. Otherwise exit.

Notice that the final increment in the state variables is an average of the four evaluations of the derivative functions, one at t, two at $(t + \tfrac{1}{2}\Delta t)$, and one at $(t + \Delta t)$. A subroutine for the Runge-Kutta-Simpson method is listed in Fig. 9-16. It can be substituted for the modified Euler subroutine of Fig. 9-13 without affecting the main program or subroutine MODEL of Figs. 9-14 and 9-15. This modularity is quite useful because it saves considerable reprogramming time and the associated effort of finding and correcting programming errors.

The fourth-order Runge-Kutta requires twice as many derivative evaluations per integration step as the modified Euler. Consequently, in order to save computing time we need to be able to more than double the integration interval Δt over that of modified Euler for the same accuracy of the results. In the case of the continuous stirred tank reactor, the Runge-Kutta-Simpson method required an integration interval of 25 s, which resulted in an 80% savings in total derivative evaluations over the modified Euler solution.

```
      SUBROUTINE INT( TO, X, TF, DT, DPRNT, N, XT, RK, F )
C
C         PURPOSE - TO INTEGRATE A SET OF ORDINARY DIFFERENTIAL EQUATIONS
C
C         METHOD  - RUNGE-KUTTA ORDER 4, SIMPSON'S RULE
C
C         VARIABLES IN ARGUMENT LIST
C
C         VARIABLE   TYPE   I/O   DIMENSION   DESCRIPTION
C
C           TO       R*8    I/O      -        INDEPENDENT VARIABLE
C                                             INPUT  - INITIAL VALUE
C                                             OUTPUT - FINAL VALUE
C           X        R*8    I/O      N        VECTOR OF STATE VARIABLES
C                                             INPUT  - INITIAL VALUES
C                                             OUTPUT - FINAL VALUES
C           TF       R*8    I        -        FINAL VALUE OF INDT. VARIABLE
C           DT       R*8    I        -        INTEGRATION INTERVAL
C           DPRNT    R*8    I        -        PRINTING INTERVAL
C           N        I      I        -        NUMBER OF DIFFTL. EQUATIONS
C           XT       R*8    W        N        WORKING STATE VECTOR
C           RK       R*8    W        N        WORKING STEP VECTOR
C           F        R*8    W        N        WORKING DERIVATIVE VECTOR
C
C         SUBROUTINES CALLED - DERIV - TO CALCULATE DERIVATIVE FUNCTIONS
C                              OUTPUT- TO PRINT INTERMEDIATE RESULTS
C
      IMPLICIT REAL*8 (A-H,O-Z)
      DIMENSION X(1), XT(1), RK(1), F(1)
C
C         INITIALIZATION FOR STEP INTEGRATION
C
      NSTEPS = ( TF - TO ) / DT + 0.99
      NPSTPS = DPRNT / DT + 0.5
      KP     = 0
C
      DO 90 I=1,NSTEPS
C
C         FIRST DERIVATIVE EVALUATION
C
      CALL DERIV( N, TO, X, F, 1 )
C
C         PRINT INTERMEDIATE RESULTS EVERY NPSTPS
C
      KP = KP - 1
      IF ( KP .GT. 0 ) GOTO 10
         KP = NPSTPS
         CALL OUTPUT
 10   CONTINUE
C
C         SECOND DERIVATIVE EVALUATION
C
      T = TO + 0.5 * DT
      DO 20 J = 1,N
         RK(J) = F(J) * DT
         XT(J) = X(J) + 0.5 * F(J) * DT
 20   CONTINUE
      CALL DERIV( N, T, XT, F, 2 )
```

Figure 9-16. Listing of FORTRAN subroutine for Runge-Kutta-Simpson integration of ordinary differential equations.

```
C
C              THIRD DERIVATIVE EVALUATION
C
        DO 30 J = 1,N
            RK(J) = RK(J) + 2. * F(J) * DT
            XT(J) = X(J) + 0.5 * F(J) * DT
 30     CONTINUE
        CALL DERIV( N, T, XT, F, 3 )
C
C              FOURTH DERIVATIVE EVALUATION
C
        T = TO + DT
        DO 40 J = 1,N
            RK(J) = RK(J) + 2. * F(J) * DT
            XT(J) = X(J) + F(J) * DT
 40     CONTINUE
        CALL DERIV( N, T, XT, F, 4 )
C
C              RUNGE-KUTTA-SIMPSON INTEGRATION STEP
C
        TO = TO + DT
        DO 50 J = 1,N
            RK(J) = RK(J) + F(J) * DT
            X(J)  = X(J) + ( RK(J) / 6. )
 50     CONTINUE
 90     CONTINUE
C
C              PRINT FINAL VALUES
C
        CALL DERIV( N, TO, X, F, 0 )
        CALL OUTPUT
        RETURN
        END
```

Figure 9-16. (*Continued*)

Summary

The numerical methods presented in this section can be used by students in the study of control systems for simple processes. When simulating more complex processes, we will probably want to use more sophisticated subroutines, which are usually available in the software libraries of most major computer installations. Some of these are described in the next section.

9-6.   SPECIAL SIMULATION LANGUAGES AND SUBROUTINES

A number of general-purpose numerical integration subroutines and simulation languages are available that can be used in simulating the dynamic behavior of process control systems. The numerical integration subroutines take the place of the modified Euler and Runge-Kutta-Simpson subroutines presented in the preceding section. The major advantages they offer are as follows:

1.   Automatic adjustment of the integration interval for meeting a specified tolerance on the truncation error.

2.   Numerical methods that are more efficient than the fourth-order Runge-Kutta-Simpson presented in the preceding section.

3. In some cases, methods for efficiently handling stiff systems of differential equations. (See Section 9-8 for a discussion of stiffness.)

The design of these general-purpose subroutines is similar to that of the modified Euler and Runge-Kutta-Simpson subroutines presented in the preceding section: A main program sets up the run parameters and initial conditions and calls the integration subroutine, which in turn calls a model subroutine that calculates the derivative functions. Intermediate printing of results must usually be done by the user in the main program—by repeated returns from the integration routine—or in the model subroutine.

A list of the most commonly available numerical integration subroutines is given in Table 9-1. In order to use them we must consult the users' manual of the particular package of which they form part.

A number of general-purpose simulation languages have also been developed for systems that are modeled by differential equations. Some of these languages have been specifically designed to simulate the dynamic response of chemical processes and their control systems. In addition to the advantages listed for integration subroutines, simulation languages offer the following advantages:

1. A set of modular subprograms for modeling specific instruments and dynamic response elements, e.g., switches, selectors, dead time, first-order lags.

2. Subprograms for iteratively solving algebraic equations in the model.

3. Features for facilitating run control and for printing and plotting of the simulation results.

A list of commonly available simulations languages is given in Table 9-2. As can be seen from the features, DSS/2 is designed to handle sets of partial differential equations, and a number of languages are process oriented. The manuals for these programs give the specific instructions on how to use them.

Table 9-1. Subroutines for the Solution of Ordinary Differential Equations

Name	Features
DVERK[a]	Automatic integration interval adjustment Runge-Kutta-Verner method
LSODE[b]	Automatic integration interval adjustment Implicit algorithm for stiff systems Uses Gear's method
LSODI[b]	Same as LSODE with the added capability of handling implicit algebraic equations along with the differential equations
PDECOL[b]	Same as LSODE, but for partial differential equations with two independent variables

[a] IMSL, Inc., 6th Floor NBC Building, 7500 Bellair Boulevard, Houston, Tex.
[b] Hindmarsh, Alan C., Mathematics and Statistics Section, L-300 Lawrence Livermore Laboratories, Livermore, Calif.

Table 9-2. Simulation Languages

Language	Features
CSMP[a]	FORTRAN precompiler Modules for dynamic and logic blocks Built-in printer-plots capability
ACSL[b]	Same as CSMP
DSS/2[c]	Can solve ordinary and partial differential equations
DYNSYL[d]	Process-oriented with process and control modules Real-time interactive operation Graphic output Implicit integration algorithm for stiff systems
DYFLO2[e]	Process-oriented with process and control modules
EPRI-MMS[f]	Process-oriented with power plant and control modules

[a] "Continuous System Modeling Program," IBM Corporation, 1133 Westchester Avenue, White Plains, N.Y.

[b] "Advanced Continuous Simulation Language," Mitchell and Gauthier, Associates, 1337 Old Marlboro Road, Concord, Mass.

[c] "Distributed System Simulator—Version 2," W.E. Schiesser, Lehigh University, Bethlehem, Pa.

[d] "DYNSYL: A General Purpose Dynamic Simulator for Chemical Processes," Chemical Engineering Division, Lawrence Livermore Laboratory, Livermore, Calif.

[e] "DYFLO2," R.G.E. Franks, E. I. duPont de Nemours and Company, Wilmington, Del.

[f] "EPRI-Modular Modeling System," Electric Power Research Institute, P.O. Box. 10412, Palo Alto, Calif.

9-7. CONTROL SIMULATION EXAMPLES

In simulating process control systems we normally encounter a number of dynamic elements or modules that are common to many systems. Examples of these modules are first- and second-order lags, dead time, proportional-integral-derivative controllers, lead/lag dynamic compensators, and others. In this section we will present a number of examples of such models and then put them together in the simulation of a feedback control loop.

Example 9-1(a). Simulation of First- and Second-Order Lags.
First-Order Lag:

$$\frac{Y(s)}{X(s)} = \frac{K}{\tau s + 1}$$

where

$Y(s)$ is the output
$X(s)$ is the input
K is the gain
τ is the time constant

Solution. In order to simulate a transfer function, we first clear fractions:

$$\tau s\, Y(s) + Y(s) = K\, X(s)$$

Assuming that the initial conditions are zero and making use of the real differentiation theorem of Laplace transforms, Eq. (2-3), we can write the equation in terms of the time variables:

$$\tau \frac{dy(t)}{dt} + y(t) = K\, x(t)$$

Finally, we solve for the highest derivative:

$$\frac{dy(t)}{dt} = \frac{1}{\tau}[K\, x(t) - y(t)]$$

This equation is in a form that is ready to be incorporated into a simulation program. The variable $x(t)$ is the input to the first-order lag.

Example 9-1(b).   Second-Order Lag.

$$\frac{Y(s)}{X(s)} = \frac{K}{\tau^2 s^2 + 2\xi\tau s + 1}$$

where

> τ is the characteristic time constant
> ξ is the damping ratio

Solution.   Clear fractions:

$$\tau^2 s^2 Y(s) + 2\xi\tau s Y(s) + Y(s) = KX(s)$$

Invert to time domain assuming zero initial conditions:

$$\tau^2 \frac{d^2 y(t)}{dt^2} + 2\xi\tau \frac{dy(t)}{dt} + y(t) = Kx(t)$$

Solve for highest derivative:

$$\frac{d^2 y(t)}{dt^2} = \frac{1}{\tau^2}\left[Kx(t) - 2\xi\tau \frac{dy(t)}{dt} - y(t) \right]$$

This equation must now be transformed into two first-order differential equations by the following procedure:

Let

$$y_2(t) = \frac{dy(t)}{dt}$$

Then

$$\frac{dy_2(t)}{dt} = \frac{d}{dt}\left(\frac{dy(t)}{dt} \right) = \frac{d^2 y(t)}{dt^2}$$

Combination of these last three equations results in the following first-order differential equations, which can be readily incorporated into a simulation program:

$$\frac{dy(t)}{dt} = y_2(t)$$

$$\frac{dy_2(t)}{dt} = \frac{1}{\tau^2}[Kx(t) - 2\xi\tau y_2(t) - y(t)]$$

The variable $x(t)$ is the input variable, $y_2(t)$ is an intermediate variable, and $y(t)$ is the output of the second-order lag.

Example 9-1(c).   Two First-Order Lags in Series.

$$\frac{Y(s)}{X(s)} = \frac{K}{(\tau_1 s + 1)(\tau_2 s + 1)}$$

where τ_1 and τ_2 are the time constants.

This transfer function is equivalent to that of part (b) when the damping ratio ξ is equal to or greater than one. The relationships between the two sets of parameters are

$$\tau_1 = \frac{\tau}{\xi - \sqrt{\xi^2 - 1}} \qquad \tau_2 = \frac{\tau}{\xi + \sqrt{\xi^2 - 1}}$$

Solution.   A convenient way to simulate this transfer function is to break up the block diagram into two blocks in series as shown in Fig. 9-17. Each of the two blocks is now a first-order lag that can be simulated as in part (a) of Example 9-1(a).

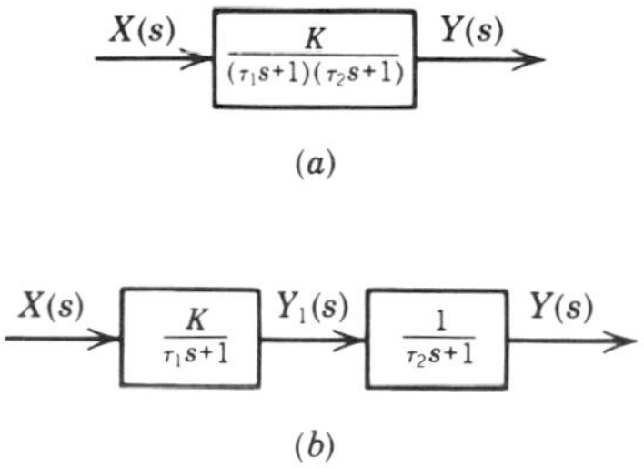

(a)

(b)

Figure 9-17.   Separating a second-order lag into two first-order lags in series. (a) Block diagrams of original transfer function. (b) Equivalent block diagram.

$$\frac{dy_1(t)}{dt} = \frac{1}{\tau_1}[Kx(t) - y_1(t)]$$

$$\frac{dy(t)}{dt} = \frac{1}{\tau_2}[y_1(t) - y(t)]$$

In this set $x(t)$ is the input variable, $y_1(t)$ is an intermediate variable, and $y(t)$ is the output variable. The gain K can be placed in either of the two blocks without affecting the response of the output variable $y(t)$.

The programming of the time-domain equations for the first-order lags is illustrated in Example 9-5.

Example 9-2. Simulation of Dead Time, Transportation Lag, or Time Delay.

$$\frac{Y(s)}{X(s)} = e^{-t_0 s}$$

or, in the time domain

$$y(t) = x(t - t_0)$$

where t_0 is the dead time.

Solution. One way to simulate dead time is to store the input response $x(t)$ and "play it back" t_0 units of time later to generate the output $y(t)$. Since the stored response is not needed after it is played back, values of the response need only be stored for t_0 units of time, provided that the storage locations are refreshed constantly with input values. The way the dead-time simulation program works is as follows.

Assume that the value of the input is to be stored at each integration interval Δt. Then the number of values that must be stored is

$$k = \frac{t_0}{\Delta t}$$

A one-dimensional array Z of dimension k or greater is used to store the values. Then the program steps are as follows:

Step 1. Initialize all the k locations of Z with the initial value of x, x_0 (usually zero):

$$\text{For } i = 1, 2, \ldots, k, \quad \text{let } Z_i = x_0$$

Step 2. Initialize the index i of the table of values:

$$\text{Let } i = 1$$

Step 3. At each integration step

(a) Pick up the value Z_i from the table as the output and store the value of the input in that location:

$$\text{Let } y = Z_i$$
$$Z_i = x$$

(b) Increment the index:

$$\text{Let } i = i + 1$$

(c) When the index exceeds k, reset it to one to reuse the same storage locations:

$$\text{If } i \text{ is greater than } k, \text{ let } i = 1$$

Note that the second and successive times the program goes through the storage locations it picks the value that was stored there k steps before. Since each step represents Δt units of time, the value replayed as output is what the input was $k\Delta t$ or t_0 units of time before. The storage and playback process is illustrated graphically in Fig. 9-18.

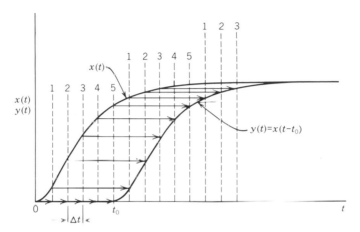

Figure 9-18. Dead time simulation by storing sampled values of the function and playing them back t_0 units of time later. In this case five storage locations are used ($k = 5$).

The logic of the dead-time simulation program gets somewhat complicated when it is to be used with a modified Euler or fourth-order Runge-Kutta numerical integration subroutine. This is because the dead-time simulation statements must be executed along with the evaluation of the derivative functions; this execution is done two or four times per integration step, while the *storage operation* of the dead-time program must be done only once per integration step. To circumvent this difficulty it is necessary for the integration subroutine to set a flag that tells the dead-time program to store the input value. The store flag should be set for the first derivative evaluation only, since it is then that the state variables have their most accurate values after the previous integration step.

The dead-time simulation procedure described in the preceding example is used in the computer simulation of a feedback control loop in Example 9-5.

Example 9-3. Simulation of a Lead/Lag Dynamic Compensation Unit.

$$\frac{Y(s)}{X(s)} = \frac{\tau_1 s + 1}{\tau_2 s + 1}$$

where τ_1 and τ_2 are the time constants of the lead and the lag, respectively.

Solution. Clear fractions:

$$\tau_2 s\, Y(s) + Y(s) = \tau_1 s\, X(s) + X(s)$$

Invert the Laplace transform, assuming zero initial conditions:

$$\tau_2 \frac{dy(t)}{dt} + y(t) = \tau_1 \frac{dx(t)}{dt} + x(t)$$

As is evident, this equation requires the derivative of the input variable $x(t)$. Since numerical differentiation is usually a very inaccurate operation, it should be avoided whenever possible.

Returning to the equation in the Laplace domain, let us collect terms with s after dividing by τ_2:

$$s \left[Y(s) - \frac{\tau_1}{\tau_2} X(s) \right] = \frac{1}{\tau_2} [X(s) - Y(s)]$$

Define

$$Y_1(s) = Y(s) - \frac{\tau_1}{\tau_2} X(s)$$

and substitute:

$$s\, Y_1(s) = \frac{1}{\tau_2} [X(s) - Y(s)]$$

Invert assuming zero initial conditions:

$$\frac{dy_1(t)}{dt} = \frac{1}{\tau_2} [x(t) - y(t)]$$

And, from the definition of $Y_1(s)$, the output variable is given by

$$y(t) = y_1(t) + \frac{\tau_1}{\tau_2} x(t)$$

These equations of a lead/lag unit are programmed in the simulation of a PID controller in Example 9-5.

Example 9-4.   Simulation of an Industrial Proportional-Integral-Derivative (PID) Controller.

$$E(s) = R(s) - C(s)$$

$$\frac{M(s)}{E(s)} = K_c \left(1 + \frac{1}{\tau_I s} \right) \left(\frac{\tau_D s + 1}{\alpha \tau_D s + 1} \right)$$

where

K_c is the controller gain
τ_I is the integral time
τ_D is the derivative time
$E(s)$ is the error signal
$R(s)$ is the set-point signal
α is the noise filter parameter (usually 0.05 to 0.10)

$M(s)$ is the controller output

$C(s)$ is the controlled variable

This transfer function fits the majority of the off-the-shelf analog controllers used in industry.

Solution. A block diagram of the controller is given in Fig. 9-19a. As in the case of the two first-order lags in series [Example 9-1(c)], it is convenient to separate the transfer functions into two blocks as illustrated in Fig. 9-19b. It is evident from this figure that the derivative section is a lead/lag unit in which the lead time constant is the derivative time and the lag is a fixed fraction α of the lead.

Usually it is considered undesirable to have the derivative action act on the set-point signal because it causes large derivative "kicks" on changes in set point by the operator. These kicks are avoided by placing the derivative unit on the measured variable, before the calculation of the error, as shown in Fig. 9-19c. This block diagram also shows the use of the *reset feedback* scheme to implement the integral action. As discussed in Section 6-5, it is easier to impose limits on the controller output when the reset feedback implementation is used. The equations that represent the block diagram of Fig. 9-19c are

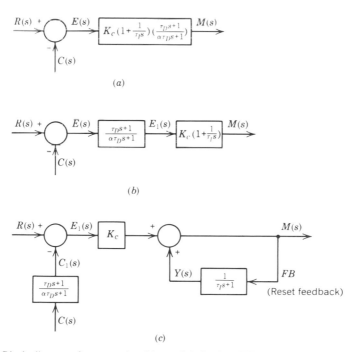

Figure 9-19. Block diagrams for proportional-integral-derivative (PID) controller. (*a*) Block diagram for industrial PID controller. (*b*) Block diagram after separation of the derivative block and the proportional-plus-integral block. (*c*) Block diagram in which the derivative block is on the process variable and the integral action is implemented through reset feedback.

Derivative (Lead/Lag) Unit

$$\frac{C_1(s)}{C(s)} = \frac{\tau_D s + 1}{\alpha \tau_D s + 1} \tag{A}$$

Error Calculation

$$E_1(s) = R(s) - C_1(s) \tag{B}$$

Proportional Action

$$M(s) = K_c E_1(s) + Y(s) \tag{C}$$

Reset Feedback

$$Y(s) = \frac{1}{\tau_I s + 1} M(s) \tag{D}$$

Equation (A) is that of a lead/lag unit and can be simulated as discussed in Example 9-3:

$$\frac{dc_2(t)}{dt} = \frac{1}{\alpha \tau_D} [c(t) - c_1(t)]$$

$$c_1(t) = c_2(t) + \frac{1}{\alpha} c(t)$$

Equations (B) and (C) can be inverted in a straightforward manner:

$$e_1(t) = r(t) - c_1(t)$$

$$m(t) = K_c e_1(t) + y(t)$$

Equation (D) is a first-order lag with unity gain and can be simulated as discussed in Example 9-1(a):

$$\frac{dy(t)}{dt} = \frac{1}{\tau_I} [m(t) - y(t)]$$

In this model of the controller $r(t)$ is the set-point (input) signal, $c(t)$ is the controlled variable (input) signal, $m(t)$ is the manipulated (output) variable, $e_1(t)$ is the error, and $c_1(t)$, $c_2(t)$, and $y(t)$ are intermediate variables. Note that $c_2(t)$ and $y(t)$ are the two state variables of the model.

The programming of the equations for the PID controller is illustrated in the following example.

Example 9-5.   Simulation of the Feedback Control Loop of Fig. 9-20a with the Following Transfer Functions.

Process

$$G(s) = \frac{Ke^{-t_0 s}}{(\tau_1 s + 1)(\tau_2 s + 1)}$$

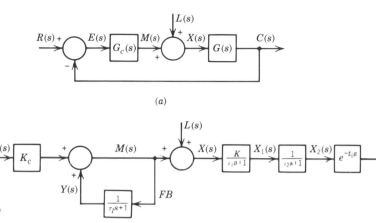

Figure 9-20. Block diagram for feedback control loop. (*a*) Block diagram of feedback control loop. (*b*) Modified block diagram for proportional-integral-derivative (PID) and second-order process.

PID Controller

$$G_c(s) = K_c \left(1 + \frac{1}{\tau_I s} \right) \left(\frac{\tau_D s + 1}{\alpha \tau_D s + 1} \right)$$

where the symbols have the same meaning as in Examples 9-1(c), 9-2, and 9-4, and $L(s)$ is a disturbance input.

Solution. In Fig. 9-20*b* we have separated the blocks of the feedback control loop into first-order components. We also moved the derivative element of the controller so that it does not act on the set-point signal and used the reset-feedback implementation of the integral action (see Example 9-4).

The PID controller can be simulated as in Example 9-4. Since the variable names are the same as in that example, we do not need to repeat the time-domain equations here. The equations for the rest of the loop are as follows:

$$x(t) = L(t) + m(t)$$

$$\frac{dx_1(t)}{dt} = \frac{1}{\tau_1}[K x(t) - x_1(t)]$$

$$\frac{dx_2(t)}{dt} = \frac{1}{\tau_2}[x_1(t) - x_2(t)]$$

The dead time is simulated by storing values of $x_2(t)$ and playing them back t_0 units of time later, as explained in Example 9-2, so that

$$c(t) = x_2(t - t_0)$$

These equations, together with the equations for the controller given in Example 9-4, constitute the model of the feedback control loop. A FORTRAN subroutine to model the loop is listed in Fig. 9-21. This subroutine, named MODEL, is to be used with the main program of Fig. 9-14 and the numerical integration subroutines of Figs. 9-13 or 9-16. Notice the use of argument IEVAL in the simulation of the dead time. The integration subroutine sets this argument to the sequence number of each derivative evaluation during each integration step. The value of IEVAL is used in the dead-time simulation statements so that values are stored only during the first derivative evaluation (IEVAL = 1) of each integration step. In order for the dead-time model to work properly, the integration interval, DTIME (Δt), must be an integer submultiple of the dead time, TO (t_0). Furthermore, in the subroutine, as listed in Fig. 9-21, the dead time is limited to a maximum of 100 times the integration interval.

```
      SUBROUTINE MODEL( NODE, TIME, XS, DX, DTIME )
C
C         PURPOSE - TO ACCEPT INPUT DATA, CALCULATE THE DERIVATIVES
C                   (ENTRY DERIV), AND PRINT THE RESULTS (ENTRY OUTPUT)
C                   FOR A PID FEEDBACK LOOP WITH A SECOND-ORDER PLUS
C                   DEAD TIME PLANT.
C
C         NOTES:  1. SECOND TIME CONSTANT CAN BE SET TO ZERO.
C
C                 2. DERIVATIVE TIME CAN BE SET TO ZERO.
C
C                 3. IF INTEGRAL TIME IS SET TO ZERO, A PURE INTEGRAL
C                    CONTROLLER WITH GAIN KC RESULTS.
C
      IMPLICIT REAL*8 (A-H,O-Z)
      REAL*8    M, K, KC, L
      DIMENSION XS(1), DX(1), Z(104)
C
      DATA IDIST / 'D' /
C
C         ACCEPT PROCESS PARAMETERS
C
      WRITE( 6, 200 )
  200 FORMAT(' ENTER K, TAU1, TAU2, AND TO.  TAU1 DEFAULTS TO 1.0')
      READ( 5, * )   K, TAU1, TAU2, TO
         IF ( TAU1 .LE. 0. ) TAU1 = 1.
      WRITE( 6, 210 ) K, TAU1, TAU2, TO
      WRITE( 4, 210 ) K, TAU1, TAU2, TO
  210 FORMAT(/10X,'PROCESS GAIN =',1PG9.1,5X,'TIME CONSTANTS',2G12.5/
     F                  /38X,'DEAD TIME',5X,G12.5 )
C
C         ACCEPT CONTROLLER PARAMETERS
C
      WRITE( 6, 220 )
  220 FORMAT(' ENTER KC, TAUI, AND TAUD.  INTEGRAL CONTR. IF TAUI=0')
      READ( 5, * ) KC, TAUI, TAUD
      ALPHA = 0.10
      WRITE( 6, 230 ) KC, TAUI, TAUD, ALPHA
      WRITE( 4, 230 ) KC, TAUI, TAUD, ALPHA
  230 FORMAT(/10X,'CONTROLLER GAIN =',1PG12.5/
     F           /10X,'INTEGRAL TIME   =',G12.5/
     O           /10X,'DERIVATIVE TIME =',G12.5,2X,'(ALPHA =',G12.5,')' )
C
C         ACCEPT TYPE OF INPUT, SET POINT OR DISTURBANCE.
C
      WRITE( 6, 240 )
  240 FORMAT(' SET POINT (SP) OR DISTURBANCE (D) CHANGE?' )
      READ ( 5, 250 ) INP
  250 FORMAT(A1)
      IF ( INP .NE. IDIST ) GOTO 10
         R = 0.
         L = 1.
         WRITE( 6, 260 )
         WRITE( 4, 260 )
```

Figure 9-21. Listing of FORTRAN subroutine for dynamic simulation of the feedback control loop of Fig. 9-20*b*.

```
 260        FORMAT(/12X,'RESPONSE TO DISTURBANCE STEP')
            GOTO 20
  10   L = 0.
       R = 1.
       WRITE( 6, 270 )
 270   FORMAT(/12X,'RESPONSE TO SET POINT STEP' )
  20   CONTINUE
C
C          INITIALIZE DEAD-TIME VECTOR Z AND INDEX ITO.
C
       KTO  = TO / DTIME + 0.5
       IF ( KTO .GT. 0 ) KTO  = KTO / ( (KTO + 99) / 100 )
       DO 30 I=1,KTO
  30      Z(I) = 0.
       ITO = 1
C
C          SET INITIAL CONDITIONS TO ZERO (DEVIATION VARIABLES).
C
       NODE = 8
       DO 40 I=1,NODE
  40      XS(I) = 0.
C
       DX2  = 0.
       DC2  = 0.
C
C          PRINT HEADER FOR RESPONSE TABLE.
C
       WRITE( 6, 280 )
       WRITE( 4, 280 )
 280   FORMAT(/10X,4X,'TIME',11X,'C',11X,'M',7X,'INPUT',10X,'X2',
      F            7X,'DERIV'/)
       RETURN
C
C          EVALUATION OF DERIVATIVE FUNCTIONS.
C
       ENTRY DERIV( NODE, TIME, XS, DX, IEVAL )
C
       X1    = XS(1)
       X2    = XS(2)
       C2    = XS(3)
       Y     = XS(4)
       IF ( TAU2 .LE. 0. ) X2 = X1
C
C          DEAD TIME SIMULATION. STORE ONLY ON FIRST EVALUATION.
C
       C   = X2
       IF ( TO .LE. 0. ) GOTO 50
          C   = Z(ITO)
          IF ( IEVAL .NE. 1 ) GOTO 50
             Z(ITO) = X2
             ITO    = ITO + 1
             IF ( ITO .GT. KTO ) ITO = 1
  50   CONTINUE
C
C          DERIVATIVE ACTION ON MEASUREMENT ONLY.
C
       C1  = C
       IF ( TAUD .LE. 0. ) GOTO 60
          C1  = C2 + C / ALPHA
          DC2 = ( C - C1 ) / ( ALPHA * TAUD )
  60   CONTINUE
C
C          PROPORTIONAL-INTEGRAL BY RESET FEEDBACK OR PURE INTEGRAL.
C
       E  = R - C
       E1 = R - C1
       IF ( TAUI .LE. 0. ) GOTO 70
          M   = KC * E1 + Y
          DY  = ( M - Y ) / TAUI
          GOTO 80
  70   M  = Y
       DY = KC * E
  80   CONTINUE
C
C          FIRST-ORDER LAGS.
C
       X   = M + L
       DX1 = ( K * X - X1 ) / TAU1
       IF ( TAU2 .GT. 0. ) DX2 = ( X1 - X2 ) / TAU2
```

Figure 9-21. (*Continued*)

```
C
C               SET VECTOR OF DERIVATIVES, INCLUDING IAE, ISE, ITAE & ITSE.
C
          DX(1) = DX1
          DX(2) = DX2
          DX(3) = DC2
          DX(4) = DY
C
          DX(5) = DABS( E )
          DX(6) = E**2
          DX(7) = TIME * DABS( E )
          DX(8) = TIME * E**2
C
          RETURN
C
C               PRINT ENTRIES ON RESPONE TABLE.
C
          ENTRY OUTPUT
          WRITE( 6, 350 ) TIME, C, M, X, X2, C1
          WRITE( 4, 350 ) TIME, C, M, X, X2, C1
  350     FORMAT(10X,1P6G12.4)
C
C               PRINT ERROR INTEGRALS, IAE, ISE, ITAE AND ITSE.
C
          IF ( IEVAL .NE. 0 ) GOTO 99
          WRITE( 6, 120 ) ( XS(I),I=5,8 )
          WRITE( 4, 120 ) ( XS(I),I=5,8 )
  120 FORMAT(/10X,'ERROR INTEGRALS - IAE  =',1PG12.5,4X,'ISE  =',G12.5/
     F          28X,'ITAE =',G12.5,4X,'ITSE =',G12.5)
C
  99      RETURN
          END
```

Figure 9-21. *(Continued)*

The subroutine of Fig. 9-21 allows for the second time constant, TAU2 (τ_2), and the derivative time, *TD* (τ_D), to be set to zero. The derivative noise filter parameter, ALPHA (α), is set to 0.10. The lag of the derivative unit, $\alpha\tau_D$, is usually the smallest time constant of the model. Because of this it is important that the integration interval, DTIME (Δt), be smaller than $\alpha\tau_D$. Otherwise the numerical integration will be unstable.

The subroutine of Fig. 9-21 is very useful for verifying the results of root locus and frequency response analysis (Chapter 7) and for testing the response of the loop for various combinations of loop parameters and tuning formulas (Chapter 6). Responses to both set point and disturbance inputs can be simulated. Only unit step changes are considered in this particular subroutine.

The response to a disturbance input is shown in Fig. 9-22. The tuning is for minimum IAE-disturbance (Table 6-3).

In order to facilitate controller tuning the subroutine of Fig. 9-21 calculates and prints the four error integrals defined in Chapter 6: IAE, ISE, ITAE and ITSE. This feature requires that four state variables, one for each error integral, be added to the four state variables needed to simulate the loop.

In this example we have illustrated the simulation of the dynamic models that were described in the preceding examples. These models are useful in the simulation of the sensors, controllers, and control valves that are usually associated with the dynamic model of the process itself.

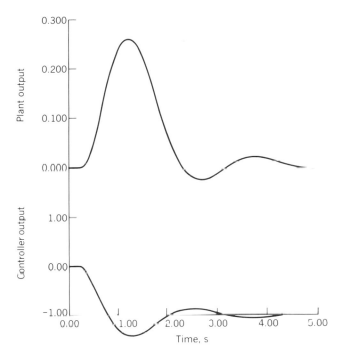

Figure 9-22. Response of feedback control loop of Fig. 9-20*b* to a unit step change in disturbance.
Process: $K = 1$, $\tau_1 = 1$ s, $\tau_2 = 0.50$ s, $t_0 = 0.25$ s.
Controller: $K_c = 2.82$, $\tau_I = 0.822$ s, $\tau_D = 0.261$ s.

9-8. STIFFNESS

As discussed in Section 9-5, a set of differential equations is said to be *stiff* if the ratio of its largest to its smallest eigenvalue is large. In Chapter 2 we defined the eigenvalues of a dynamic system as the roots of the characteristic equation of the system transfer function. The larger the stiffness of a model, the higher the number of integration steps and thus the computer time required to simulate the system on a digital computer. In this section we will look at the sources of stiffness that can be avoided when modeling a system. We will also discuss briefly some numerical integration methods that have been designed for efficiently handling stiff systems of differential equations.

Sources of Stiffness in a Model

A model consisting of n first-order ordinary differential equations is said to be of nth order. This is because it can be easily shown that, in the absence of dead-time elements, the *linearized* equations can be reduced to a single overall transfer function the denominator of which is an nth-degree polynomial in the Laplace transform variable s. It follows that the characteristic equation of such a system has n roots, which are the n eigenvalues of the system. Therefore, with each first-order ordinary differential equation we write when modeling the system, we add one eigenvalue to the response of the model.

Because of the interaction between the differential equations that constitute the model of a system, the response of each of the state variables may in principle be affected by every eigenvalue of the model equations. For the linearized system, the response can be represented by the following equation:

$$x_i(t) = b_1 e^{r_1 t} + b_2 e^{r_2 t} + \ldots + b_n e^{r_n t} + (\text{input terms}) \tag{9-83}$$

where

$x_i(t)$ is the response of the ith state variable

$b_1, b_2, \ldots, b_n$ are coefficients that depend on the system and its input forcing function

$r_1, r_2, \ldots, r_n$ are the eigenvalues of the system, 1/time

and the "input terms" depend on the specific input forcing function applied to the system.

Note: Equation (9-83) can be obtained by Laplace transforming the linearized system equations, algebraically eliminating all the state variables except $X_i(s)$, and inverting by partial fractions expansion as learned in Chapter 2.

If the system is stable, the real parts of all of the eigenvalues are negative (see Section 6-2). This means that each term of Eq. (9-83) exponentially decays to zero with time. The larger the magnitude of the negative real part of the eigenvalue, the faster the corresponding term of Eq. (9-83) drops to zero. Conversely, the term with the eigenvalue that has the smallest real part (in magnitude), is the one that takes the longest to decay to zero. Thus, this is the "dominant" eigenvalue that controls the time it takes for the model to reach steady state. As we shall see shortly, for explicit integration formulas, the largest eigenvalue imposes an upper stability limit on the integration interval. It follows that the number of integration steps (computations) required to solve the differential equations increases with the ratio of the largest eigenvalue to the smallest.

In a stiff set of differential equations, some eigenvalues are much larger in magnitude than the dominant eigenvalue. However, we note that the terms of Eq. (9-83) corresponding to those large eigenvalues must necessarily decay to zero much more quickly than the dominant eigenvalue, thus their dynamic effect on the overall response is short-lived. In fact, they last for a very small fraction of the duration of the response and can usually be neglected. This suggests that the stiffness of a model is reduced if the eigenvalues that are much larger in magnitude than the dominant eigenvalue are removed from the model equations.

A modeling assumption that allows us to remove eigenvalues from the set of model equations is the *quasi-steady-state assumption*. It consists of neglecting the accumulation term of a balance equation so that it becomes one of the algebraic equations of the model instead of one of the first-order differential equations. This reduces the order of the system by one and thus eliminates one eigenvalue. Let x_j be the state variable associated with the differential equation to be eliminated. The quasi-steady-state assumption consists of the following approximation in the model equation:

$$\frac{dx_j}{dt} = f_j(x_1, x_2, \ldots, x_n, t) \approx 0 \tag{9-84}$$

where

f_j is the derivative function
$x_1, x_2, \ldots, x_n$ are the state variables

Notice that the approximation indicated by Eq. (9-84) does not affect the steady-state solution of the model equations; only the transient effect is neglected. Notice also that variable x_j ceases to be a state variable and becomes an auxiliary variable of the model. The ideal situation is the one in which variable x_j can be calculated explicitly from Eq. (9-84). However, even if Eq. (9-84) has to be solved iteratively for x_j, a computational advantage results if the reduction in the stiffness of the system allows a significant increase in the integration interval. The reason for this is that, although the iterative solution method may require a few evaluations of Eq. (9-84) at each integration step, the stiff system requires the evaluation of *all* of the model equations at the increased number of steps required by the stiff model.

Now that we recognize that it may be possible to reduce stiffness by applying the quasi-steady-state assumption to some of the differential equations of the model, we face the problem of recognizing which equations introduce the large eigenvalues that cause stiffness. The solution of this problem is part of the art of simulation. Before we share some of our experience, we must stress the fact that it is not always possible to isolate the equations that cause the stiffness.

In our experience many of the first order differential equations that constitute the model act as first-order lags, that is, the state variable that appears in the derivative term lags the other variables of the model. If this is the case, there will be an eigenvalue associated with the differential equation under consideration, which is of the order of magnitude of the reciprocal of the time constant of the lag. An easy way to estimate such an eigenvalue is to take the partial derivative of the derivative function with respect to the state variable in the derivative term. In reference to Eq. (9-84), an estimate of the eigenvalue is given by

$$r_j \approx \frac{\partial f_j}{\partial x_j} \tag{9-85}$$

where r_j is the eigenvalue associated with the equation in question. The problem with this estimate is that it may be far off when the model equations are highly coupled, as are, for example, the balance equations in countercurrent flow equipment such as distillation columns (see Section 9-2).

The following are some examples of equations that can introduce large eigenvalues into the solution of the model equations. Note that in each case the dynamic effect under consideration is suspected to be much faster than the dominant eigenvalues of the system.

1. The valve or transmitter lag with a time constant of a few seconds when the process time constants are of the order of minutes or hours.
2. Accumulation in the vapor phase when accumulation in the liquid phase is also considered. The mass of the vapor phase is usually much smaller than that of the liquid phase.
3. The acceleration of a fluid in a short conduit relative to the accumulation of the fluid in tanks. This is analogous to the acceleration effect on a subcompact car when it collides with a fully loaded 18-wheeler: the subcompact may be assumed to instantaneously reach the velocity of the 18-wheeler on impact.

The following example illustrates the effect of the quasi-steady-state assumption on the dynamic response of a system.

Example 9-6.   Simulation of Tank Drainage by Gravity Through a Vertical Discharge Pipe.
A cylindrical tank (see Fig. 9-23) of diameter D_t and height H is initially full of water. It discharges through a vertical pipe of length L and diameter D_p. Both the tank

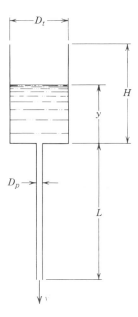

Figure 9-23.   Sketch of gravity drainage tank.

and the pipe discharge are opened to the atmosphere, and the frictional drop in the pipe, including contraction and expansion losses, is given by

$$\Delta p_f = k_f \frac{\rho v^2}{2} \tag{A}$$

where

k_f is a friction coefficient $= \dfrac{4fL}{D_p} + k_c + k_e$

Δp_f is the frictional pressure drop, N/m^2
ρ is the density of the water, kg/m^3
v is the average velocity in the pipe, m/s
f is Fanning's friction factor, assumed constant
k_c is the contraction loss coefficient
k_e is the expansion loss coefficient $(= 1.0)$

Determine the amount of time it takes for the tank to drain.

Solution. A mass balance around the tank results in the following equation:

$$0 - \rho \frac{\pi D_p^2}{4} v = \frac{d}{dt}\left(\rho \frac{\pi D_t^2}{4} y\right) \qquad \text{1 eqn., 2 unks. } (v, y) \quad \text{(B)}$$

where

D_p is the diameter of the pipe, m
D_t is the diameter of the tank, m
y is the level in the tank, m

As the density is constant, Eq. (B) simplifies to

$$\frac{dy}{dt} = -\left(\frac{D_p}{D_t}\right)^2 v \qquad\qquad \text{(C)}$$

A force balance on the pipe results in the following equation:

$$\text{sum of the forces} = \frac{d}{dt}\text{(momentum)}$$

$$\frac{\pi D_p^2}{4}(p - p_a) + \rho g \frac{\pi D_p^2}{4} L - \frac{\pi D_p^2}{4}\Delta p_f = \frac{d}{dt}\left(\rho \frac{\pi D_p^2}{4} L v\right) \qquad \text{(D)}$$

$$\text{2 eqns., 3 unks. } (p)$$

where

p is the pressure at the bottom of the tank, N/m^2
p_a is atmospheric pressure, N/m^2
g is the acceleration of gravity, 9.8 m/s^2
L is the length of the pipe, m

The pressure at the bottom of the tank is related to the level in the tank at any instant of time by the formula

$$p = p_a + \rho g y \qquad\qquad \text{3 eqns., 3 unks. } \quad \text{(E)}$$

We can now substitute Eqs. (A) and (E) into Eq. (D) to obtain

$$\frac{dv}{dt} = \frac{g}{L}\left(y + L - \frac{k_f}{2g}v^2\right) \tag{F}$$

Equations (C) and (F) are sufficient to determine the dynamic response of the level of water in the tank and the velocity in the pipe. The initial conditions are that the tank is full and the velocity in the pipe is zero:

$$y(0) = y_0 = H$$

$$v(0) = 0$$

We can solve Eqs. (C) and (F) to determine the time t_f required to empty the tank:

$$y(t_f) = 0$$

A possible problem is that the set of Eqs. (C) and (F) may be stiff; this will be the case if the time required for the fluid in the pipe to accelerate to an equilibrium value at a given level in the tank is significantly less than the time it takes for the level to drain.

If the acceleration term on the right-hand side of Eq. (D) is negligible, then

$$y + L - \frac{k_f}{2g}v^2 = \frac{L}{g}\frac{dv}{dt} \approx 0 \tag{G}$$

$$v \doteq \sqrt{2g(y + L)/k_f} \tag{H}$$

Thus, if the set of Eqs. (C) and (F) turns out to be severely stiff, we could replace Eq. (F) by its approximation, Eq. (H). Physically this is equivalent to neglecting the acceleration of the fluid in the pipe in the force balance equation, Eq. (D). This is the quasi-steady-state assumption.

Let the parameters be as follows:

Height of tank:	$H = 0.762$ m (30 in.)
Diameter of tank:	$D_t = 0.1397$ m (5.5 in.)
Diameter of pipe:	$D_p = 6.83 \times 10^{-3}$ m (0.269 in.)
Length of pipe:	$L = 0.4572$ m (18 in.)
Acceleration of gravity:	$g = 9.8$ m/s² (32.2 ft/s²)
Friction coefficient:	$k_f = 4.2$ (dimensionless)
Density of water:	$\rho = 1000$ kg/m³ (62.3 lb/ft³)

A listing of the FORTRAN subroutine MODEL to solve this problem is shown in Fig. 9-24. In this subroutine the two models of the tank are simulated simultaneously in order to compare the dynamic responses. The first two state variables correspond to y and v from Eqs. (C) and (F), while the third state variable corresponds to the level y (YP) from Eq. (C) when Eq. (H) is used to calculate the velocity v (VP).

The subroutine of Fig. 9-24 was combined with the Runge-Kutta-Simpson subroutine of Fig. 9-16 and the main program of Fig. 9-14 to produce the responses tabulated in Fig. 9-25. The results of two runs are tabulated. The first run presents a detailed response of the first 0.5 s of operation, showing how the velocity v calculated by the exact model starts at zero and accelerates up to and then lags the quasi-steady-state velocity VP. Herein

```
       SUBROUTINE MODEL( NODE, TIME, XS, DXS, DTIME )
C
C          GRAVITY DRAINAGE TANK SIMULATION
C
C          THIS SUBROUTINE SIMULATES AN EXACT MODEL THAT INCLUDES THE
C             INERTIA OF THE LIQUID IN THE DISCHARGE PIPE, AND A
C             QUASI-STEADY-STATE MODEL THAT NEGLECTS THE INERTIA.
C
C          VARIABLES IN INPUT DATA
C
C             DP       DIAMETER OF PIPE, M
C             DT       DIAMETER OF TANK, M
C             L        LENGTH OF PIPE, M
C             KF       DISCHARGE FRICTION COEFFICIENT IN VELOCITY HEADS
C             YO       INITIAL HEIGHT OF LIQUID IN TANK, M
C
       IMPLICIT REAL*8 (A-H,O-Z)
       REAL*8 L, KF
       REAL*4 XY
       COMMON/ PLOTD / MP, XY(205,3)
       DIMENSION XS(1), DXS(1)
C
       DATA G / 9.8 /, VO / 0. /
C
C          ACCEPT INPUT DATA
C
       MP = 0
       WRITE(6,210)
  210  FORMAT(' ENTER DP, DT, L, KF, YO')
       READ ( 5, * ) DP, DT, L, KF, YO
       WRITE ( 6, 100 ) DP, DT, L, KF
       WRITE ( 4, 100 ) DP, DT, L, KF
  100  FORMAT(//10X,'GRAVITY DRAINAGE TANK SIMULATION'/
      1          / 10X,'PIPE DIAMETER =',1PG12.5,' M',
      2              5X,'TANK DIAMETER =',G12.5,' M'/
      3          / 10X,'PIPE LENGTH   =',G12.5,' M',
      4              5X,'PIPE FRICTION =',G12.5 )
C
C          CALCULATE CONSTANTS AND SET INITIAL CONDITIONS
C
       C1 = ( DP / DT ) ** 2
       C2 = KF / ( 2. * G )
C
       XS(1) = YO
       XS(2) = VO
       XS(3) = YO
       NODE  = 3
C
C          PRINT HEADER FOR OUTPUT TABLE
C
       WRITE ( 6, 110 )
       WRITE ( 4, 110 )
  110  FORMAT(//10X,'TIME, S',8X,'Y, M',7X,'V, M/S',7X,'YP, M',
      1              6X,'VP, M/S'/)
  120  FORMAT(10X,5G12.5)
C
       RETURN
C
C                ENTRY TO CALCULATE DERIVATIVE FUNCTIONS
C
       ENTRY DERIV( NODE, TIME, XS, DXS, NEVAL )
C
C                STORE STATE VARIABLE INTO LOCAL VARIABLES
C
```

Figure 9-24. Listing of FORTRAN subroutine for simulating the gravity drainage tank.

```
             Y  = XS(1)
             V  = XS(2)
             YP = XS(3)
C
C
C            CALCULATE DERIVATIVES FOR EXACT (STIFF) MODEL
C
             DXS(1) = -C1 * V
             DXS(2) = ( G / L ) * ( Y + L - C2 * V ** 2 )
C
C            CALCULATE DERIVATIVE FOR QUASI-STEADY-STATE MODEL
C
             VP = DSQRT( ( YP + L ) / C2 )
             DXS(3) = -C1 * VP
C
             RETURN
C
C                ENTRY TO OUTPUT SELECTED VARIABLES
C
                 ENTRY OUTPUT
C
             WRITE ( 6, 120 ) TIME, Y, V, YP, VP
             WRITE ( 4, 120 ) TIME, Y, V, YP, VP
C
             MP = MP + 1
             XY(MP,1) = TIME
             XY(MP,2) = Y
             XY(MP,3) = V
C
             RETURN
          END
```

Figure 9-24. (*Continued*)

lies the difference between the two models. The second run shows how the two models result in essentially the same drainage times and levels in the long run. The difference here is that the exact model, being somewhat stiff, requires a much smaller integration interval (DTIME) than the quasi-steady-state model. The eigenvalues for the exact model at the initial conditions can be calculated by linearization of Eqs. (C) and (F). They are, respectively -0.0023 and -21.9 s^{-1}. These values represent a degree of stiffness of near 10,000.

The preceding example illustrates that the quasi-steady-state assumption can result in essentially the same dynamic results as the original model. The beauty of this happy fact is that the greater the degree of stiffness of the system, the smaller the error caused by the quasi-steady-state assumption.

Numerical Integration of Stiff Systems

The three numerical integration methods we considered in Section 9-5, Euler, modified Euler, and Runge-Kutta-Simpson, are said to be explicit methods. This is because the state variables are calculated explicitly at each integration step from the values resulting from the previous step. In that same section we mentioned that, for a large enough integration step, the numerical solution may become unstable even if the response of the model equations is dynamically stable. In this section we will briefly introduce a class of implicit numerical integration methods that do not have stability limitations.

RUN PARAMETERS, TO= .0   TMAX= .500   DTIME= 1.5625D-02

GRAVITY DRAINAGE TANK SIMULATION

PIPE DIAMETER = 6.8300D-03 M   TANK DIAMETER = .13790   M

PIPE LENGTH   = .45720   M   PIPE FRICTION = 4.2000

TIME, S	Y, M	V, M/S	YP, M	VP, M/S
.0	.76200	.0	.76200	2.3853
.62500D-01	.76188	1.4182	.76163	2.3849
.12500	.76161	2.0954	.76127	2.3846
.18750	.76126	2.3076	.76090	2.3842
.25000	.76091	2.3646	.76054	2.3839
.31250	.76054	2.3790	.76017	2.3835
.37500	.76018	2.3825	.75981	2.3831
.43750	.75981	2.3831	.75944	2.3828
.50000	.75945	2.3830	.75908	2.3824

RUN PARAMETERS, TO= .0   TMAX= 165.   DTIME= 1.5625D-02

GRAVITY DRAINAGE TANK SIMULATION

PIPE DIAMETER = 6.8300D-03 M   TANK DIAMETER = .13790   M

PIPE LENGTH   = .45720   M   PIPE FRICTION = 4.2000

TIME, S	Y, M	V, M/S	YP, M	VP, M/S
.0	.76200	.0	.76200	2.3853
5.0000	.73328	2.3573	.73292	2.3567
10.000	.70454	2.3287	.70419	2.3280
15.000	.67616	2.3001	.67581	2.2994
20.000	.64812	2.2714	.64778	2.2708
25.000	.62044	2.2428	.62010	2.2422
30.000	.59310	2.2142	.59278	2.2136
35.000	.56612	2.1856	.56580	2.1850
40.000	.53949	2.1570	.53918	2.1563
45.000	.51321	2.1283	.51291	2.1277
50.000	.48728	2.0997	.48699	2.0991
55.000	.46170	2.0711	.46141	2.0705
60.000	.43647	2.0425	.43619	2.0419
65.000	.41160	2.0139	.41133	2.0132
70.000	.38707	1.9852	.38681	1.9846
75.000	.36290	1.9566	.36264	1.9560
80.000	.33907	1.9280	.33883	1.9274
85.000	.31560	1.8994	.31536	1.8988
90.000	.29248	1.8708	.29225	1.8701
95.000	.26971	1.8421	.26949	1.8415
100.00	.24729	1.8135	.24707	1.8129
105.00	.22522	1.7849	.22501	1.7843
110.00	.20351	1.7563	.20330	1.7557
115.00	.18214	1.7277	.18195	1.7270
120.00	.16112	1.6990	.16094	1.6984
125.00	.14046	1.6704	.14028	1.6698
130.00	.12015	1.6418	.11998	1.6412
135.00	.10019	1.6132	.10002	1.6126
140.00	.80575D-01	1.5846	.80419D-01	1.5839
145.00	.61315D-01	1.5560	.61166D-01	1.5553
150.00	.42406D-01	1.5273	.42265D-01	1.5267
155.00	.23848D-01	1.4987	.23715D-01	1.4981
160.00	.56413D-02	1.4701	.55157D-02	1.4695
165.00	-.12214D-01	1.4415	-.12332D-01	1.4408

Figure 9-25. Results of two runs of the gravity drainage tank simulation program.

To fix ideas regarding numerical stability, let us consider the following first-order linear differential equation:

$$\frac{dx}{dt} = rx \tag{9-86}$$

where

x is the state variable
r is the eigenvalue, 1/time

The analytical solution of this equation subject to some initial condition x_0 is

$$x(t) = x_0 e^{rt} \tag{9-87}$$

Evidently, this response is stable if r has a negative real part (or is just a negative number). Now let us consider the numerical solution of Eq. (9-86) using Euler's method. From Eq. (9-69) we obtain

$$x|_{t+\Delta t} = x|_t + r\, x|_t\, \Delta t = (1 + r\Delta t)x|_t \tag{9-88}$$

where Δt is the integration interval, always real and positive. It is evident from Eq. (9-88) that if each value of x is going to be smaller in magnitude than the preceding one, the following condition must hold:

$$|1 + r\Delta t| < 1 \tag{9-89}$$

or if r is a real number

$$-2 < r\Delta t < 0 \tag{9-90}$$

Equation (9-89) is the condition of numerical stability of Euler's method, and Eq. (9-90) represents the limits of stability on $r\Delta t$ when it is a real number. For the case of r being a negative real number, the values of Δt for which the Euler numerical solution is stable are

$$\Delta t < \frac{2}{|r|} \tag{9-91}$$

Notice that the maximum value of Δt is inversely proportional to the magnitude of the eigenvalue r. Remember, however, that the value of Δt for which the numerical solution is *accurate*, that is, matches the real solution, Eq. (9-87), within a certain tolerance is usually much smaller than the stability limit given by Eq. (9-91).

In Section 9-5 we saw how the modified Euler and Runge-Kutta-Simpson formulas gave us the same accuracy as the Euler method, but with integration intervals that were an order of magnitude larger. However, in terms of the stability limits on Δt, those of the modified Euler and Runge-Kutta explicit methods are about the same as the limit for Euler's method given by Eq. (9-91). This is significant because, for a stiff set of differential equations, the stability limit on the integration interval is smaller than the overall accuracy limit! This is because the large eigenvalues that cause the solution to go unstable may have, as we saw earlier, little effect on the overall accuracy of the calculated response. In other words, in terms of the large eigenvalues that cause stiffness, we only need to be numerically stable, not accurate.

Implicit numerical integration methods have been developed for stiff systems of differential equations. A property of these methods is that they are not limited by stability on the size of the integration interval. This means that the only consideration in selecting the integration interval is the accuracy of the overall response. To illustrate one of the simplest methods, let us consider the numerical solution of Eq. (9-86) by the implicit modified Euler method. The formula for the numerical integration is

$$x|_{t+\Delta t} = x|_t + \frac{1}{2}[rx|_t \, \Delta t + rx|_{t+\Delta t} \, \Delta t] \tag{9-92}$$

The difference between this formula and the explicit formula, Eq. (9-75), is that it is implicit in the state variable $x|_{t+\Delta t}$, that is, the term $x|_{t+\Delta t}$ on the right-hand side, is not approximated by a different formula. In this simple case of a single first-order linear equation, the implicit solution is straightforward. Solving for $x|_{t+\Delta t}$ from Eq. (9-92) yields

$$x|_{t+\Delta t} = \left[\frac{1 + \frac{1}{2}r\Delta t}{1 - \frac{1}{2}r\Delta t} \right] x|_t \tag{9-93}$$

The condition of stability for this formula is

$$\left| \frac{1 + \frac{1}{2}r\Delta t}{1 - \frac{1}{2}r\Delta t} \right| < 1 \tag{9-94}$$

It can easily be shown that this condition is satisfied for all positive values of Δt, as long as the real part of the eigenvalue, r, is negative. In other words, the numerical solution is stable as long as the original system is stable!

The general form of the implicit modified Euler equation, the real counterpart to Eq. (9-75), is as follows:

For $i = 1, 2, \ldots, n$, calculate

$$x_i|_{t+\Delta t} = x_i|_t + \frac{1}{2}[f_i(x_1, x_2, \ldots, x_n, t)|_t \tag{9-95}$$
$$+ f_i(x_1, x_2, \ldots, x_n, t)|_{t+\Delta t}]$$

where f_i are the derivative functions. Notice that the implicit solution of this set of n nonlinear algebraic equations for the values of the state variables, $x_i|_{t+\Delta t}, i = 1, 2, \ldots, n$, requires an iterative solution, that is, the repeated evaluation of all of the model equations, at each integration step. The advantage of this method relies on reducing the total number of integration steps by a factor that is larger than the number of function evaluations per step. This advantage is obviously greater the stiffer the system, and nonexistent for nonstiff systems. Subroutine LSODE, an entry in Table 9-1, uses Gear's implicit method[3] for stiff systems. This is a variable-order variable-step-size method.

9-9. SUMMARY

In this chapter we have presented the development of dynamic models of process control systems and methods to simulate these systems on digital computers. We have also presented models of a multicomponent distillation column, a process furnace, a continuous stirred tank chemical reactor, and several dynamic elements that commonly appear in process control systems. The chapter includes a discussion of stiffness, its source, and methods to handle it when solving sets of differential equations.

REFERENCES

1. Conte, S. D. and C. deBoor, *Elementary Numerical Analysis*, 3rd ed., McGraw-Hill, New York, 1980, Chapters 5 and 8.
2. Broyden, C. G., "A Class of Methods for Solving Nonlinear Simultaneous Equations," *Math. of Comps.*, Vol. 19, No. 92, Oct. 1965, pp. 577–593.
3. Gear, C. W., *Numerical Initial Value Problems in Ordinary Differential Equations*, Prentice Hall, Englewood Cliffs, N.J., 1971.
4. Luyben, W. L., *Process Modeling, Simulation, and Control in Chemical Engineering*, McGraw-Hill, New York, 1973, Chapters 3 and 5.

PROBLEMS

9-1. You are assigned the job of checking whether or not the concentration control loop of Problem 6-24 can be started up in automatic. You decide to do this by programming the nonlinear model equations on the computer and simulating the startup procedure. Assume that both reactors are initially full of the feed solution of concentration 0.80 lbmole/gal, and that the reaction starts as soon as the solution is brought up to the operating temperature. As your assignment also includes the tuning of the concentration controller, you decide first to simulate the reactors operating at their design conditions. This will give you an opportunity to tune the controller and to check that your simulation matches the steady-state operating conditions.

(a) Write the model equations and program them for solution on a computer. Set the initial conditions to the design operating conditions and check that they constitute a steady state.
(b) Tune the composition controller for minimum IAE on a step change of 10% in reactants flow.
(c) Set the initial conditions to the startup conditions and simulate the startup.

Note: It is necessary, in order to prevent reset windup, to use the reset feedback implementation of the feedback controller (see Example 9-4), and to limit the controller output between 0 and 100% of range.

(d) Observe the effect of reset windup by setting the limits on the controller output at -25% and 125%.

Caution: Put limits on the control valve so that the flow still varies from zero to maximum, that is, do not let the flow be negative. The controller limits correspond to 0 and 24 mA, respectively, on the 4 to 20 mA range.

Discuss your results briefly.

9-2. The oil heater of Problem 6-26 needs to be run at half the design production rate and you are asked to check whether it will be necessary to change the tuning parameters of the level and temperature controllers. You decide to do this by simulating the nonlinear model equations on the computer and tuning the controllers at both the design flow and at half the design flow. The controllers are to be tuned for minimum IAE on disturbance inputs.

(a) Write the model equations and program them for solution on a computer. Verify that the design conditions constitute a steady state.
(b) Tune the controllers at full design flow for step changes of 5% in the outlet valve position.
(c) Reduce the flow by half by closing the outlet valve and observe the response.
(d) Set the initial conditions to the final steady-state conditions of part (c), the half-flow conditions, and tune the controllers at these conditions. Is there any difference? Briefly discuss your results.

9-3. A tank containing 20,000 liters of solution is used to neutralize an acid stream containing hydrochloric acid (HCl) with a reagent consisting of a 0.1 N solution of sodium hydroxide. The design flow and concentration of the HCl solution are 500 liters/min and 0.01 N.

A sketch of the tank and its pH controller is given in Fig. 9-26. The pH transmitter signal is linear with the pH of the solution in the tank, which is defined by

$$pH = -\log_{10} [H^+]$$

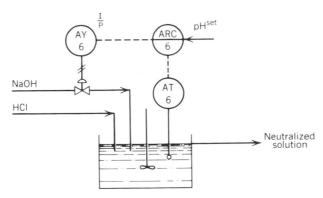

Figure 9-26. Sketch of waste water neutralization for Problem 9-3.

where $[H^+]$ is the hydrogen ion concentration, gmoles/liter. The pH electrode can be simulated by a linear first-order lag of 15 s. The controller is proportional-integral (PI) and the design pH set point is 7. The transmitter is calibrated for a pH range of 1 to 13. The control valve has a constant pressure drop of 10 psi and is sized for 100% overcapacity. The valve has equal percentage characteristics with a rangeability parameter of 50. A first-order lag with a time constant of 5 s can be used to simulate the valve actuator.

The neutralization reaction can be assumed to instantaneously reach equilibrium. Thus, the concentrations of hydrogen and hydroxyl ions are related by the water dissociation constant:

$$[H^+][OH^-] = 10^{-14} \text{ (gmoles/liter)}^2$$

The tank is initially at the steady-state design conditions.

(a) Write the equations necessary to model the tank and its control system assuming perfect mixing and constant volume of the solution in the tank. Calculate the initial conditions.
(b) Program the equations for solution on a computer, and tune the controller for minimum IAE (integral of the absolute value of the error) on a 0.005 gmoles/liter step change in HCl concentration.
(c) Obtain responses of the tuned controller for changes in set point of ± 0.5 pH units and for $\pm 10\%$ change in the flow of the HCl solution.

9-4. The evaporator shown in Fig. 9-27 is used to concentrate a solution of sugar from a feed mass fraction X_F to a product mass fraction X_P. Saturated steam at 20 psia is available as the heating medium and cooling water to the barometric condenser enters at T_W. The pressure in the evaporator may be assumed to be the vapor pressure of water at the condenser outlet temperature, T_C. The evaporator contents can be assumed to be perfectly mixed, and the calandria tubes may be assumed

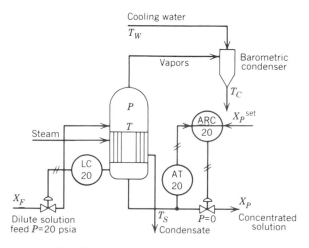

Figure 9-27. Single-effect evaporator for Problem 9-4.

to be at the same temperature as the condensing steam, T_S. Heat losses may be neglected.

The evaporator has a heat transfer area of 309.5 ft², with a capacity of 4.2 ft³ of solution per foot of level. The heat capacitance of the calandria tubes is 154 Btu/°F. The heat transfer coefficient is 150 Btu/hr-ft²-°F. The volume of solution below the heat transfer tubes is 4.68 ft³.

The control valves have linear characteristics and time constants of 5 s and are sized for 100% overcapacity. The pressure drop across the feed control valve is constant at 16 psi, and that across the product valve is equal to the pressure in the evaporator. This is because a barometric leg makes the pressure on the downstream side of the product valve zero absolute. The level sensor is a float with a range of 0 to 2 ft above the bottom of the tubes. The concentration transmitter (AT20) has a range of 0.3 to 0.8 mass fraction units and a time constant of 25 s. The level controller (LC20) is proportional only with a set point of 25% of range and is tuned for tight level control (proportional band between 1 and 10%). The concentration controller (ARC20) is proportional-integral (PI).

The design conditions are:

> Feed mass fraction: 0.10
> Feed temperature: 90°F
> Product mass fraction (set point): 0.70
> Condenser exit temperature: 105°F
> Cooling water inlet temperature: 90°F

The following correlations can be used to calculate the physical properties of the solution and vapor:

> Solution enthalpy: $H = (1 - 0.7x)(T - T_R)$ Btu/lb
> Solution density: $\rho = 62.384 + 23.14x + 11.27x^2$ lb/ft³
> Boiling point rise: BPR $= 1.13 - 6.59x + 26.9x^2$ °F

where x is the mass fraction of sugar
           T is the temperature, °F
           T_R is the reference temperature, °F

The heat capacity of liquid water is 1.0 Btu/lb-°F and that of water vapor is 0.46 Btu/lb-°F. These heat capacities may be assumed to be constant and used to correct the latent heat vaporization of water for temperature.

The Antoine equation may be used to calculate the vapor pressure of water as a function of temperature and vice versa:

$$\text{For water:} \quad \ln P° = 18.3036 - \frac{3816.44}{T + 227}$$

where $P°$ is in mm of mercury and T in C.

The evaporator is initially at the steady-state design conditions. Possible disturbances are the steam valve position, feed mass fraction, cooling water rate to the condenser, cooling water inlet temperature, and product composition set point.

(a) Write the model equations and calculate the design steady-state conditions. Program the equations for solution on a computer and verify that the initial conditions constitute a steady state.

(b) Tune the level controller for quarter-decay-ratio response by trial-and-error. Tune the composition controller using the Ziegler-Nichols ultimate gain method.

(c) Obtain responses to step inputs in composition set point and in each of the disturbances listed in the statement of the problem. Briefly discuss your results.

9-5.   A sketch of a forward-feed triple-effect evaporator is shown in Fig. 9-28. The inlet and outlet conditions, and each of the effects, are the same as in Problem 9-4. The heat transfer coefficients for each of the effects are 400, 300, and 130 Btu/hr-ft²-°F, respectively. At design conditions, there is a pressure drop of 1 psi in the vapor lines between effects 1 and 2, and between 2 and 3; this pressure drop may be assumed to vary proportional to the square of the vapor flow in the line.

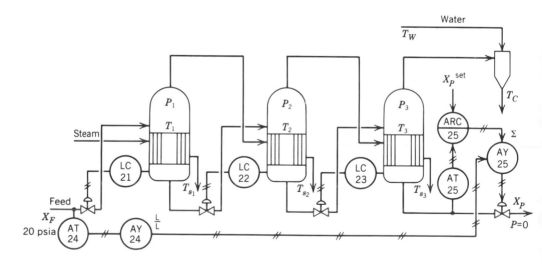

Figure 9-28.   Triple-effect evaporator for Problem 9-5.

In addition to the three level controllers and the product mass fraction controller, a feedforward controller with a lead/lag dynamic compensator is installed on this system to compensate for changes in feed composition.

(a) Write the model equations and program them for solution on a digital computer.

(b) Simulate the startup assuming that the initial concentrations and temperatures in the effects are those of the feed. Use the same controller tuning parameters determined in Problem 9-4.

Note: In order to prevent reset windup it is necessary to use the reset feedback implementation of the PI controller (see Example 9-4), and to limit the controller output at 0 and 100% of range.

(c) Using the final steady-state conditions from part (b) as initial conditions, do part (c) of Problem 9-4 for the multiple-effect evaporator. Also tune the feedforward controller. Briefly discuss your results.

9-6. A mixture of benzene and toluene is to be flashed in a steam-heated flash drum such as the one sketched in Fig. 9-29. The drum is to process 20,000 lb/hr of saturated liquid containing 40 mole% benzene. The drum is 5 ft in diameter and 5 ft high and is normally half full of liquid. Steam is available at 30 psia, saturated. A heating coil with an overall heat transfer coefficient of 200 Btu/hr-ft²-°F is to be designed to be capable of vaporizing the entire feed at design conditions. The coil is to be made of 2-in. schedule 40 steel pipe.

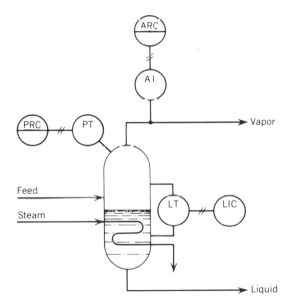

Figure 9-29. Steam-heated flash drum for Problem 9-6.

(a) Design a control system to control the pressure and liquid level in the drum and the composition of the vapor stream. The interaction between these control objectives should be minimized.
(b) Write the model equations, calculate the steady-state initial conditions, simulate the flash drum, tune the controllers, and study the performance of your

control system to variations in feed rate, composition, and temperature. The solution may be assumed to follow Raoult's Law.

9-7.   The distillation column sketched in Fig. 9-1 is fed with a mixture of benzene and toluene. At design conditions the feed rate is 3500 lbmoles/hr; its composition is 44 mole% benzene and is liquid at its bubble point. The column has six sieve trays, each holding 450 lbmoles of liquid and having a Murphree tray efficiency of 70%. The feed enters at the fourth tray.

The partial reboiler may be assumed to be an equilibrium separation stage. It has an overall heat transfer coefficient of 350 Btu/hr-ft²-°F and a heat transfer area of 2000 ft². Steam is available at 75 psia and saturated. The bottom level controller (LIC202) is a proportional controller with 100% proportional band. The level transmitter (LT202) is calibrated so that the hold-up of liquid in the reboiler and column bottom is 600 lbmoles at the low limit and 1000 lbmoles at the high limit. The set point is 50% of this range and the control valve on the bottom product is linear, sized for 100% overcapacity and with negligible lag.

The condenser may be assumed to maintain the column pressure constant at 1 atmosphere and the reflux at its bubble point. The condenser accumulator drum has a capacity that varies from 500 and 1500 lbmoles of liquid betwen the level transmitter (LT201) low and high limits. The level controller (LIC201) is proportional only with 100% proportional band and a set point of 50% of the range. The control valve on the reflux stream is linear, sized for 100% overcapacity, and with negligible lag.

At design conditions the distillate product is to have a composition of 82 mole% benzene and the bottoms product is to contain no more than 15% benzene. The bottoms composition is to be controlled by manipulating the steam flow to the reboiler and the distillate composition by manipulating the distillate flow. For sizing the control valves, the reflux ratio of 2.0 may be used.

(a) Design the two composition control loops, including transmitter ranges, control valve sizes, and controller modes. You may assume that both compositions can be measured continuously with a (first-order) lag of one minute in the composition transmitters, AT201 and AT202.

(b) Write the model equations assuming equimolar overflow, that is, liquid and vapor rates do not change from tray to tray. You may also assume perfect mixing in each tray, the reboiler, and accumulator drum, and constant physical properties. For benzene-toluene at atmospheric pressure the relative volatility may be assumed constant at 2.43, that is,

$$\frac{y^*}{x} \frac{(1 - x)}{(1 - y^*)} = 2.43$$

where y^* is the mole fraction of benzene in the vapor, which is in equilibrium with the liquid of mole fraction x.

(c) Simulate the startup of the column in which all of the trays, the bottom, and the accumulator drum are initially filled with a liquid of the same composition and temperature as the feed.

(d) Using the final steady-state conditions from part (c) as the initial conditions, study the performance of the control system to changes in feed rate and mole fraction and in composition set points. Tune the composition controllers by the controller synthesis method.

Note: It may be necessary to iterate between parts (c) and (d) of this problem, as the startup simulation is a convenient way to have the program calculate the steady-state conditions, while the controllers should be tuned at the steady-state operating conditions.

9-8. Simulate the furnace modeled in Section 9-3. The furnace tube is a 4-in. schedule 40 steel pipe with a length of 120 ft. The fluid being heated is air, with a heat capacity of 6.96 Btu/lbmole-°F at constant pressure, and of 4.98 Btu/lbmole-°F at constant volume. The fuel is natural gas with a heating value of 990 Btu/ft³ (at 60°F, 30 in. Hg), and the furnace efficiency is 78%.

At design conditions the entrance temperature of the air is 80°F and its flow rate is 85 lbmoles/hr. The set point on the outlet temperature is 1000°F. The inside coefficient of heat transfer may be assumed to be constant at 100 Btu/hr-ft²-°F, and the emissivity of the tube surface as constant at 0.75. The density of steel is 480 lb/ft³ and its specific heat 0.12 Btu/lb-°F. The effective mass of the firebox is 1750 lb and its specific heat is 0.32 Btu/lb-°F.

The temperature transmitter (TT42 in Fig. 9-5) has a calibrated range of 500 to 1500°F, and a time constant of 50 s. The control valve has a design pressure drop of 15 psi, has equal percentage characteristics with a rangeability parameter of 50, and is sized for 100% overcapacity based on the design conditions. The valve actuator has a negligible time lag.

Divide the tube into 5, 10, and 20 sections of equal length and compare the results of an uncontrolled ($K_C = 0$) response to a 10% increase in air flow from the initial steady-state conditions. Then, using the number of sections that appears to give acceptable accuracy, tune the temperature controller (TRC42 in Fig. 9-5) and study the response of the controller to step changes in feed flow, inlet temperature, and temperature set point. A startup simulation may be used to obtain the steady-state temperature profiles of the gas and the tube, i.e., the initial conditions.

9-9. For Problem 9-8 study the performance of a well-tuned feedforward controller consisting fo a gain and lead/lag dynamic compensator. The feedforward controller measures the flow of air to the furnace and its output is added to the output of the feedback controller.

9-10. The steam heater of Fig. 9-30 is used to heat a process stream having a heat capacity of 4200 J/kg-C and a density of 850 kg/m³. It consists of 86 tubes arranged in two passes (43 tubes per pass). The tubes are 1-in. (0.0254 m) 18 gage copper tubes (density = 8920 kg/m³, specific heat = 394 J/kg-C) 4.88 m long. The steam condensing coefficient may be assumed to be high enough that the tubes are at the same temperature as the condensing steam. Accumulation of condensate may be neglected. The steam is available at 350 kN/m² and saturated. The inside coefficient of heat transfer is 1400 J/s-m²-C.

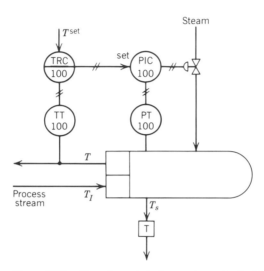

Figure 9-30. Cascade control of steam heater for Problem 9-10.

At the design conditions 25 kg/s of the process fluid are to be heated from 60 to 80C.

The control system is a cascade loop with the temperature controller manipulating the set point of the shell pressure controller. The pressure in the shell is the vapor pressure of water at the steam condensing temperature. It can be calculated using the Antoine equation for water (see Problem 9-4).

The temperature transmitter (TT100) calibrated range is 50 to 150C, and that of the pressure transmitter (PT100) is 0 to 500 kN/m² gage. The temperature controller (TRC100) is a proportional-integral-derivative (PID) controller and the pressure controller (PIC100) is proportional-integral (PI). The control valve is equal-percentage with a rangeability parameter of 50 and is sized for 100% overcapacity based on design conditions; its actuator has a time constant of 5 s. The temperature sensor has a time constant of 35 s and the pressure sensor lag is negligible.

(a) Write the model equations, dividing the heat exchanger tube into 10 sections of equal length. Program the equations for solution on a computer and simulate the startup in order to obtain the design steady-state temperature profiles

(b) Tune the pressure controller for 5% overshoot by the controller synthesis method. Then tune the temperature controller for minimum IAE on disturbance inputs. Observe the response to $\pm 40\%$ changes in process fluid flow.

(c) Remove the pressure controller, having the output of the temperature controller directly operate the valve. Tune the temperature controller as in part (b) and compare results.

9-11. In the recovery of DMF (dimethyl-formamide) from aqueous solutions by distillation, a vaporizer is used to separate heavy components that would foul up the distillation column. The vaporizer is sketched in Fig. 9-31. The feed consists of a mixture of 5–16 weight% DMF, 1% heavies, and the balance is water, which is the most volatile component. The liquid purge contains all of the heavies in the feed, since they are nonvolatile; 4% by weight of the feed is purged at design conditions. Saturated steam at 60 psia is available to heat the thermosyphon reboiler, and the rate of recirculation is high enough for the contents of the vaporizer to be assumed to be perfectly mixed. At design conditions the flow rate of solution to the vaporizer is 2000 lb/hr at 25C and 10% by weight DMF. Equipment characteristics and physical properties are given below. The vapor may be assumed to be in equilibrium with the liquid.

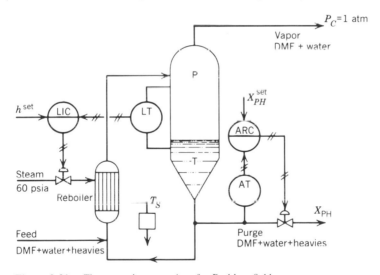

Figure 9-31. Thermosyphon vaporizer for Problem 9-11.

Volume of vaporizer: designed for 1 min retention time based on feed rate.

Cross-sectional area: designed for 2.0 ft/s vapor velocity.

Reboiler: boiling coefficient of heat transfer (limiting) is 240 Btu/hr-ft^2-°F. Area is 150 ft^2. Tubes are 1-in. 16 BWG steel tubes, and can be assumed to be at the condensing steam temperature.

Pressure drop from vaporizer to column: 0.5 psi at design conditions, proportional to the square of the vapor flow. The column operates at atmospheric pressure.

Level controller: proportional with 100% proportional band.

Level transmitter: range of 10 in. with the design level at 50% of range. Negligible lag.

Purge composition transmitter: 0 to 100 weight% range; time constant of 45 s.

Purge composition controller: proportional-integral (PI) with a set point of 25 weight% of heavy component.

Control valves: sized for 100% overcapacity; linear characteristics; negligible actuator lag.

Physical properties:

	Water	DMF	Heavies
Molecular weight	18	73	200 (avg)
Latent heat of vaporization			
at 25C, Btu/lb	1050	261	—
Relative volatility	54.0	1.0	0
Specific heat, Btu/lb-°F			
Liquid	1.0	0.53	0.45
Vapor	0.46	0.36	—
Specific gravity of liquid	1.0	0.94	0.90

The vapor-liquid equilibrium may be modeled by assuming that relative volatiltiy is constant and that the water partial pressure follows Raoult's Law. The vapor pressure of water can be determined from Antoine's equation, given in Problem 9-4.

(a) Write the model equations, calculate the design steady-state conditions, and size the vaporizer. Program the equations for computer solution and verify that the initial conditions constitute a steady state.

(b) Tune each of the controllers for quarter-decay-ratio response while keeping the other one in manual. Determine the measure of interaction (see Chapter 8) between the two loops. Correct the tuning parameters if necessary.

(c) Obtain responses to step changes in feed rate and composition (of heavies).

APPENDIX
A
Instrumentation Symbols and Labels

This appendix presents the symbols and labels used in this book for the instrumentation diagrams. Most companies have their own methods, and even though most of them are similar, they are not completely alike. The symbols and labels used in this book follow closely the standard published by the Instrument Society of America (ISA).[1] The appendix presents just the information needed for this book. For more complete information the reader is referred to the ISA standard.

In general, the instrument identification, also referred to as the tag number, is of this form:

I	R C	− 1 0 1	A
First letter	Succeeding letters	Loop number	Suffix (not required)
Functional identification		Loop identification	

The meanings of some identification letters are given in Table A-1.

The symbols used to designate the function of relays will usually be written next to the instrument label. Some of the most common symbols are presented in Table A-2. A summary of other special abbreviations is given in Table A-3. Finally, Table A-4 presents the instrument symbols.

Figure A-1 shows an example of a flow loop. The flow element, FE-10, is an orifice plate with flange taps. The element is connected to an electronic flow transmitter, FT-10. The output of the transmitter goes to a square root extractor, FY-10A, and from here the signal goes to a flow indicator controller, FIC-10. The signal from FY-10A also goes to a low flow alarm, FLA-10. The output from the controller goes to an I/P transducer, FY-10B, to convert the electronic signal to a pneumatic signal. The output from the transducer then goes to the flow control valve, FCV-10. The controller and the alarm are both board-mounted instruments while the square root extractor is mounted behind the panel. All of the other instruments are field mounted.

Table A-1. Meanings of Identification Letters

	First letter	Succeeding letters
A	Analysis	Alarm
B	Burner flame	
C	Conductivity	Control
D	Density or specific gravity	
E	Voltage	Primary element
F	Flow rate	
H	Hand (manually initiated)	High
I	Current	Indicate
J	Power	
K	Time or time schedule	Control station
L	Level	Light or low
M	Moisture or humidity	Middle or intermediate
O		Orifice
P	Pressure or vacuum	Point
Q	Quantity or event	
R	Radioactivity or ratio	Record or print
S	Speed or frequency	Switch
T	Temperature	Transmit
V	Viscosity	Valve, damper, or louver
W	Weight or force	Well
Y		Relay or compute
Z	Position	Drive

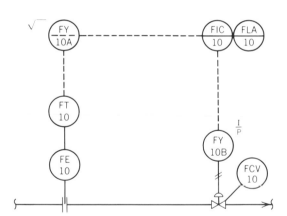

Figure A-1. Flow loop.

Note: In most of the instrument diagrams in the book, the control valve label has been omitted for the sake of simplicity. The label for the element is also omitted for the same reason. Only the label for the transmitter is shown. The reader can assume that the label for the transmitter also represents the element.

Table A-2. Function Designations for Relays

Symbol	Function	
1. 1–0 or ON–OFF	Two-position device	
2. Σ or ADD	Addition and/or subtraction	
3. Δ or DIFF	Subtraction of signals only	
4. $\pm$	Bias	
5. AVG	Average	
6. % or 1:3 or 2:1 (typical)	Gain or attenuate (input:output)	
7. $\times$	Multiply	
8. $\div$	Divide	
9. $\sqrt{\ }$ or SQ. RT.	Square root	
10. X^n or $X^{1/n}$	Raise to power	
11. $f(x)$	Characterize or function generator	
12. 1:1	Boost	
13. $>$	High select	
14. $<$	Low select	
15. $\not>$	High limit	
16. $\not<$	Low limit	
17. REV	Reverse	
18. $\int$	Integrate (time integral)	
19. D or d/dt	Derivative or rate	
20. L/L	Lead/lag unit	
21. E/P or P/I (typical)	For input/output sequences:	
	Designation	Signal
	E	Voltage
	H	Hydraulic
	I	Current
	P	Pneumatic
	R	Resistance
	O	Electromagnetic or sonic
22. A/D or D/A	A	Analog
	D	Digital

Table A-3. Summary of Special Abbreviations

Abbreviation	Meaning
A	Analog
AS	Air supply
D	Derivative control
	Digital signal
DIR	Direct acting
DEC	Decrease
ES	Electric supply
FC	Fail closed
FI	Fail indeterminate
FL	Fail locked
FO	Fail open
GS	Gas supply
I	Current signal
	Interlock
INC	Increase
M	Motor actuator
NS	Nitrogen supply
P	Pneumatic signal
	Proportional control
	Purge or flushing device
R	Reset control
	Resistance
REV	Reverse acting
RTD	Resistance temperature detector
S	Solenoid actuator
S.P.	Set point
SS	Steam supply
T	Trap
WS	Water supply

Table A-4. General Instrument Symbols—Balloons

(1)

Locally, or field,
mounted

(2)

Board, or control
room, mounted

(3)

Mounted behind
the board

(4)

Pneumatic-operated
globe valve

(5)

Pneumatic-operated
butterfly valve,
damper or louver

(6)

Hand-actuated
control valve

(7)

Control valve
with positioner

(8)

Motor

(9)

Solenoid

(10)

Single-acting
cylinder

(11)

Double-acting
cylinder

(12)

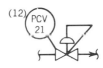

Pressure-reducing
regulator, self-
contained

(13)

Back pressure
reducing regulator,
self-contained

(14)

Pressure relief or safety
valve, angle pattern

(15)

Pressure relief or
safety valve,
straight through
pattern

(16)

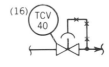

Temperature
regulator filled
system type

(17)

Three-way valve
FO to path A-C

(18)

Orifice plate with
flange or corner taps

(19)

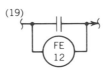

Orifice plate with vena
contracta, radius, or
pipe taps

(20)

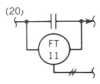

Orifice plate with
vena contracta,
radius, or pipe
taps connected to
differential pressure
transmitter

Table A-4. (continued).

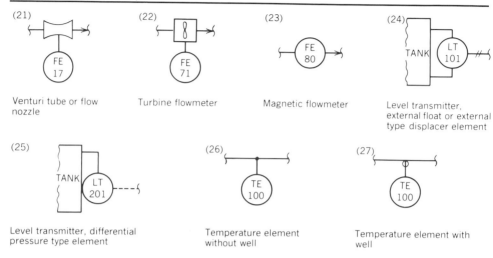

| (21) Venturi tube or flow nozzle — FE 17 | (22) Turbine flowmeter — FE 71 | (23) Magnetic flowmeter — FE 80 | (24) Level transmitter, external float or external type displacer element — TANK LT 101 |

| (25) Level transmitter, differential pressure type element — TANK LT 201 | (26) Temperature element without well — TE 100 | (27) Temperature element with well — TE 100 |

REFERENCES

1. "Instrumentation Symbols and Identification," Instrument Society of America, Standard ISA-S5-1, January 31, 1975, Research Triangle Park, N.C.

APPENDIX
B
Case Studies

This appendix presents a series of design case studies that provide the reader with an opportunity to design process control systems from scratch. It must be recognized that the first step in designing control systems for process plants is deciding which process variables must be controlled. This decision should be made by the process engineer who designed the process and the instrument or control engineer who will design the control system and specify the instrumentation. This is certainly very challenging and requires team effort. The second step is the actual design of the control system. In the case studies that follow, the first step has been done. It is the second step that is the subject of these case studies. The reader must remember that there might be more than one way to design the control systems.

Case I.   Ammonium Nitrate Prilling Plant Control System[1]

Ammonium nitrate is a major fertilizer. The flow sheet shown in Fig. B-1 shows the process for its manufacture. A weak solution of ammonium nitrate (NH_4NO_3) is pumped from a feed tank to an evaporator. At the top of the evaporator there is an ejector vacuum system. The air fed to the system controls the vacuum drawn. The concentrated solution is pumped to a surge tank and then fed into the top of a prilling tower. The development of this tower is one of the major postwar developments in the fertilizer industry. In this tower the concentrated solution of NH_4NO_3 is dropped from the top against a strong updraft of air. The air is supplied by a blower at the bottom of the tower. The air chills the droplets in spherical form and removes part of the moisture, leaving damp pellets or prills. The pellets are then conveyed to a rotary dryer where they are dried. They are then cooled, conveyed to a mixer for the addition of an antisticking agent (clay or diatomaceous earth), and bagged for shipping.

A.   Draw the necessary instrumentation to implement the following:
   1.   Record the flow of weak solution of NH_4NO_3 to the evaporator.
   2.   Control the level in the evaporator.
   3.   Control the pressure in the evaporator. This can be accomplished by manipulating the flow of air to the exit pipe of the evaporator.

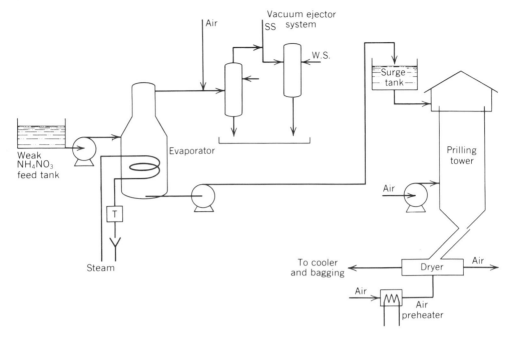

Figure B-1. Ammonium nitrate prilling plant.

4. Control the level in the surge tank.
5. Control the temperature of the dried pellets leaving the dryer.
6. Control the density of the strong solution leaving the evaporator.
   Be sure to show all necessary alarms and specify the action of valves and controllers.

B. How would you control the production rate out of this unit?

C. If the production from this unit varies often, it may also be desired to vary the air flow through the prilling tower. How would you implement this?

D. A difficult loop to tune is the temperature loop of the dried pellets. Therefore, the following data were obtained by changing the temperature controller output by + 10%.

Time (min)	Temperature (°F)
0	200
1	200
3	202
5	208
6.5	214
8.5	218
11.0	220
13.0	222
15.0	221.8
17.0	223.0

The temperature transmitter for this loop has a range of 100°F to 300°F. Tune a PI controller by the controller synthesis method and a PID controller by the minimum IAE method.

REFERENCES

1. The Foxboro Co. Application Engineering Data AED 288-3, January 1972.

Case II. Natural Gas Dehydration Control System

Consider the process shown in Fig. B-2. The purpose of this process is to dehydrate the natural gas entering the absorber; this is accomplished by the use of a liquid dehydrant (glycol). The glycol enters the top of the absorber and flows down the tower countercurrent to the gas, picking up the moisture in the gas. From the absorber, the glycol flows through a heat exchanger into the stripper. In the reboiler, at the base of the stripper, the glycol is stripped of its moisture, which is boiled off as steam. This steam leaves the top of the stripper and is condensed and used for the water reflux. This water reflux is used to condense the glycol vapors that might otherwise be exhausted along with the steam.

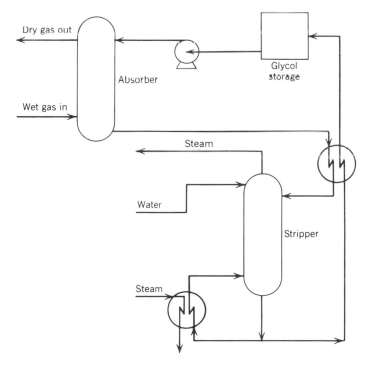

Figure B-2. Natural gas dehydration system.

The process engineer who designed the process has decided that the following must be controlled.

1. The liquid level at the bottom of the absorber.
2. The water reflux into the stripper.
3. The pressure in the stripper.
4. The temperature in the top third of the stripper.
5. The liquid level at the bottom of the stripper.
6. Efficient absorber operation at various throughputs.

Draw a complete instrumentation diagram showing the needed instrumentation to accomplish the desired control. The instrumentation to be used (except the valves) is electronic (4–20 mA).

Case III. Sodium Hypochlorite Bleach Preparation Control System[1]

Sodium hypochlorite (NaOCL) is formed by

$$2NaOH + Cl_2 \rightarrow NaOCL + H_2O + NaCl$$

The flow sheet in Fig. B-3 shows the process for its manufacture. The process is as follows.

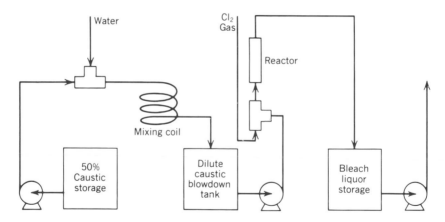

Figure B-3. Sodium hypochlorite bleach process.

Dilute caustic (NaOH) is continuously prepared, to a set concentration, by water dilution of 50% caustic solution. This diluted caustic solution is stored in an intermediate tank. From this tank, the solution is then pumped to the hypochlorite reactor. Chlorine gas is introduced into the reactor for the reaction.

A. Prepare a detailed instrument diagram to accomplish the following:
1. Control the level in the dilution tank.
2. Control the dilution of the caustic solution. The concentration of this stream is to be measured by a conductivity cell. When the dilution of this stream decreases, the output from this cell increases.

3. Control the level in the bleach liquor storage tank.
4. Control the ratio of excess NaOH/available Cl_2 in the outlet stream from the hypochlorite reactor. This ratio is measured by an ORP (oxidation-reduction potential) technique. As the ratio increases, the ORP signal also increases.

Be sure to show all the needed alarms, recorders, and indicators. All instrumentation (except valves) is electronic (4–20 mA). Specify the action of valves and controllers. Briefly discuss your design.

B. How would you set the production rate from this unit?

C. For safety reasons, when there is no flow of caustic solution from the dilute caustic tank to the reactor, the flow of chlorine must also be stopped. Design this scheme and explain it.

D. Explain in detail, with numerical example, how you would tune the ORP controller using the step-testing procedure and the minimum IAE method for a PI controller.

REFERENCE

1. The Foxboro Co. Application Engineering Data. January 1972.

Case IV.   Control Systems in the Sugar-Refining Process

The process units shown in Fig. B-4 form part of the process to refine sugar. Raw sugar is fed to the process through a screw conveyor. Water is sprayed over it to form a sugar syrup. The syrup is heated in the dilution tank. From the dilution tank the syrup flows to the preparation tank where more heating and mixing are accomplished. From the preparation tank the syrup flows to the blending tank. Phosphoric acid is added to the syrup as it flows to the blending tank. In the blending tank lime is added. This treatment with acid, lime, and heat serves two purposes. The first is that of clarification, that is, the treatment causes the coagulation and precipitation of the no-sugar organics. The second purpose is to eliminate the coloration of the raw sugar. From the blending tank the syrup continues to the process.

The following variables are thought to be important to control.

1. Temperature in the dilution tank.

2. Temperature in the preparation tank.

3. Density of the syrup leaving the preparation tank.

4. Level in preparation tank.

5. Level in 50% acid tank. The level in the 75% acid tank can be assumed constant.

6. The strength of the 50% acid. The strength of the 75% acid can be assumed constant.

7. The flow of syrup and 50% acid to the blending tank.

8. The pH of the solution in the blending tank.

9. Temperature in the blending tank.

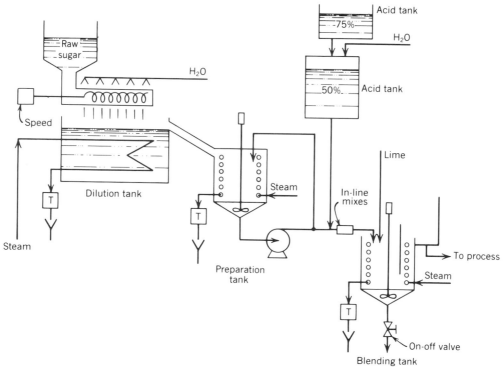

Figure B-4. Sugar refining process.

10. The blending tank requires only a high-level alarm.

The flowmeters used in this process are magnetic flowmeters. The density unit used in the sugar industry is °Brix, which is roughly equivalent to the percentage of sugar solids in the solution by weight.

Design the control systems necessary to control all of the above variables. How would you control the production rate? Briefly discuss your design. Show the action of control valves and controllers and also show the necessary alarms.

Case V. CO₂ Removal from Synthesis Gas

Consider the process shown in Fig. B-5 for removing CO_2 from synthesis gas. The plant treats 1646.12 MSCFH of feed gas. The feed gas will be supplied at 1526°F and 223 psig with the following composition:

	Volume (%)
Hydrogen	50.29
Nitrogen	0.16
Carbon dioxide	5.60
Carbon monoxide	9.94
Methane	2.62
Water vapors	31.39

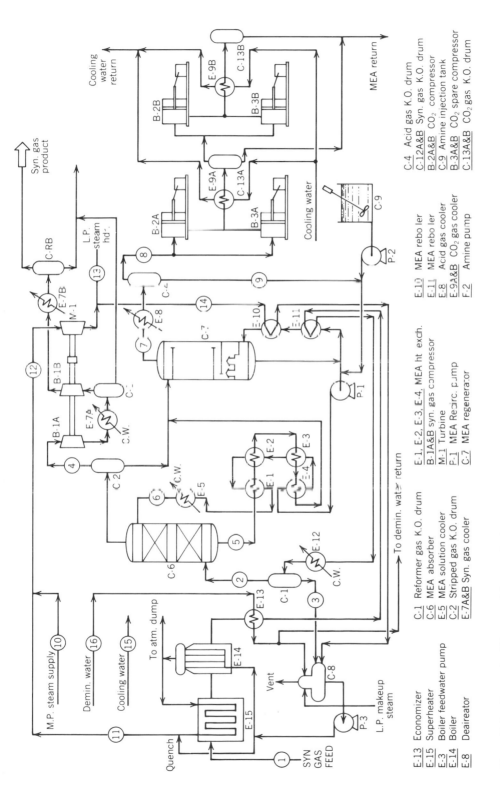

Figure B-5. Process for CO₂ removal from synthesis gas.

The products from this plant will be:

1. Synthesis gas at 115°F and 600 psig with a maximum of 50 volumetric ppm CO_2.

2. CO_2 at 115°F and 325 psig.

The process is as follows. The feed gas enters the plant at 1526°F and 223 psig. The gas must be cooled to 105°F prior to entering the absorber for CO_2 removal. This cooling is done in four stages. First, the feed gas passes through a superheater (E-15) and a boiler (E-14). The heat removed produces 27,320 lb/hr medium-pressure steam. Second, the feed gas passes through an economizer (E-13) heating the demineralized water prior to deaeration. Third, the feed gas passes through a reboiler (E-11) where the feed gas provides 84% of the reboiler duty under full operation. Finally, the feed gas is cooled in the feed gas heat exchanger (E-12) by plant cooling water. By means of these four stages, 75% of the Btu's removed from the feed gas is recovered for process heating requirements.

The cooled gases enter the absorber (C-6) in countercurrent fashion where monoethanolamine (MEA) strips the CO_2 from the gases. The remaining gases are compressed in a two-stage centrifugal compression (B-1A and B). The compressor is driven by a steam turbine (M-1). Interstage and exit cooling of the gases is provided (E-7A and B) with "knockout" drums (C-12A and B) to separate out any condensation.

The compressor operates on medium-pressure steam of which 68% is provided by the feed gas steam boiler. Ninety-three percent of the outlet low-pressure steam is available for plant use. The remaining 7% provides heat to one of the reboilers (E-10) of the regenerator column (C-7).

The CO_2 is carried with the MEA to the regenerator (C-7), where it is separated from the MEA. The regenerator is operated at low pressure and high temperature. This causes the CO_2 to be released with water vapor out the top of the tower while the lean MEA is recirculated (P-1) back to the absorber. Before this MEA enters the top of the absorber, it passes through four heat exchangers (E-1, E-2, E-3, and E-4) where it is cross-exchanged with the bottoms of the absorber. These four exchangers recover 8.8 MM Btu/hr. From these exchangers the MEA passes through another cooler (E-5) and finally enters the absorber.

The CO_2 gases from the regenerator are compressed in a two-stage compressor (B-2 and 3) powered by electric motors. Interstage and exit coolers (E-9A and B) are provided along with knock out drums (C-13A and B).

Table B-1 gives the conditions of the streams numbered in the flow diagram.

The process engineer believes that the following variables must be controlled:

1. Temperature of superheated steam leaving E-15.

2. Pressure of superheated steam produced in E-14/E-15.

3. Level in boiler drum.

4. Pressure in deaerator.

5. Level in deaerator.

6. Flow of low-pressure makeup steam to deaerator.

7. Temperature of feed gas leaving E-13 going to reboiler E-11.

8. Temperature of feed gas into absorber C-6.

Table B-1

Stream / Component	1 Feed gas	2 Absorber feed	3 Gas knock-out drum	4 Syn gas to compressor	5 Absorber Btums	6 MEA return	7 Stripper ovhd	8 CO$_2$ to Compressor
H$_2$, MSCFH	676.91	676.91		676.34	0.57		0.57	0.57
N$_2$, MSCFH	2.22	2.22		2.22				
CO$_2$, MSCFH	75.36	75.36		0.04	118.36	43.02	75.32	75.32
CO, MSCFH	133.80	133.80		133.74	0.06		0.06	0.06
CH$_4$, MSCFH	35.33	35.33		35.30	0.03		0.03	0.03
H$_2$O, gpm	422.50							
H$_2$O Vap/stm, MSCFH		4.61	51.19	3.68				
MEA solution, MSCFH					274.50	274.50	129.93	4.64
Total mass flow, lb/hr	43889.05	24041.00	15553.00	15261.00	147034.00	138254.00	13594.00	8919.00
Temperature, °F	1526	105	105	100	160	100	217	106
Pressure, psig	223	207	207	205	207	206	11	5

Table B-1 (Continued)

Component	Stream 9 Stripper Ko drum Btm	10 M. press. steam	11 M.P. steam from boiler	12 M.P. steam to turbine	13 Low-press. steam	14 Low-press. steam	15 Cooling water	16 Demin. water
H_2, MACFH								
N_2, MSCFH								
CO_2, MSCFH								
CO, MSCFH								
CH_4, MSCFH								
H_2O, gpm	9.35						4513.00	55.74
H_2O vap/stm, MSCFH								
MEA solution, MSCFH								
Total mass flow, lbs/hr	4675.00	12719.00	27320.00	40039.00	32747.00	7292.00	22565.00	27868.00
Temperature, °F	106	600	600	600	3780	370	86	60
Pressure, psig	5	400	400	400	75	75	28	100

9. Flow of MEA to absorber C-6.

10. Temperature in the bottom third of C-6.

11. Level in regenerator C-7.

12. Temperature in bottom of regenerator C-7.

13. Temperature of CO_2 gases leaving E-8.

14. Pressure in the regenerator area.

15. Temperature of synthesis gas between the two stages of compressor and exit temperature.

16. Pressure of the synthesis gas leaving the compressor B-1.

17. Interstage and exit temperature of CO_2 gases through compressor B-2.

18. Level in *all* knockout drums.

These may not be all the necessary control loops for smooth operation; however, they are the first ones proposed by the process engineer. You may propose your own.

(a) Design all of the above control loops and any other you may feel necessary. Specify the valves' action and controllers' action.

(b) Specify the necessary check valves, block valves, and alarms and process indicators as you deem necessary.

Case VI. Sulfuric Acid Process

Figure B-6 shows a simplified flow diagram for the manufacture of sulfuric acid (H_2SO_4).

Sulfur is loaded into a melting tank where it is kept in the liquid state. From this tank it goes to a burner where it is reacted with the oxygen in the air to produce SO_2 by the following reaction:

$$S_{(l)} + O_{2(g)} \rightarrow SO_{2(g)}$$

From the burner the gases are then passed through a waste heat boiler where the heat of reaction of the above reaction is recovered by producing steam. From the boiler the gases are then passed through a four-stage catalytic converter (reactor). In this converter the following reaction takes place:

$$SO_{2(g)} + 1/2\, O_{2(g)} \rightleftarrows SO_{3(g)}$$

From the converter, the gases are then sent to an absorber column where the SO_3 gases are absorbed by dilute H_2SO_4 (93%). The water in the dilute H_2SO_4 reacts with the SO_3 gas producing H_2SO_4:

$$H_2O_{(l)} + SO_{3(g)} \rightarrow H_2SO_{4(l)}$$

The liquid leaving the absorber, concentrated H_2SO_4 (98%), goes to a circulation tank where it is diluted back to 93% using H_2O. Part of the liquid from this tank is then used as the absorbing medium in the absorber.

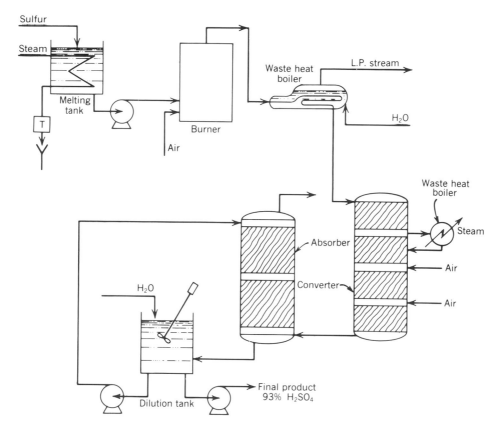

Figure B-6. Sulfuric acid process.

A. The following variables are thought to be important to control.
  1. Level in the melting tank.
  2. Temperature of sulfur in the melting tank.
  3. Air to the burner.
  4. Level of water in the waste-heat boiler.
  5. The concentration of SO_3 in the gas leaving the absorber.
  6. Concentration of H_2SO_4 in the dilution tank.
  7. Level in the dilution tank.
  8. Temperature of the gases entering the first stage of the converter.
  Design the necessary control systems to accomplish the above. Be sure to show all of the instrumentation specifying the action of valves and controllers. Briefly discuss your design.

B. How would you set the production rate for this plant?

APPENDIX
C
Sensors, Transmitters, and Control Valves

This appendix presents some of the hardware necessary to build control systems and is closely related to Chapter 5. Some of the most common sensors—pressure, flow, level, and temperature—are examined as well as two different types of transmitters, one pneumatic and the other electronic. The appendix ends with a presentation of the different types of control valves and a presentation of additional considerations when sizing control valves.

PRESSURE SENSORS[1,2,3,4]

The most common pressure sensor is the *Bourdon tube*, developed by the French engineer Eugene Bourdon. The Bourdon tube, shown in Fig. C-1, is basically a piece of tubing in the form of a horseshoe with one end sealed and the other connected to the pressure source. Since the cross section of the tube is elliptical or flat, as pressure is applied the tubing tends to straighten, and when the pressure is released the tubing returns to its original form so long as the elastic limit of the material of the tubing was not exceeded. The amount of straightening the tubing undergoes is proportional to the applied pressure. So if the open end of the tubing is fixed, then the closed end can be connected to a pointer to indicate pressure or to a transmitter to generate a pneumatic or electrical signal.

The pressure range that can be measured by the Bourdon tube depends on the wall thickness and the material of the tubing. An extended Bourdon tube in the form of a helical spiral was developed to permit additional motion of the sealed end. This element, called the *helix*, is shown in Fig. C-2. The helix can handle pressure ranges of about 10:1 with an accuracy of ± 1% of the calibrated span.[1] Another common type of Bourdon tube is the *spiral* element, shown in Figs. C-2d and C-3.

Another type of pressure sensor is the *bellows*, shown in Figs. C-2c and C-4, which looks like a corrugated capsule made up of a somewhat elastic material such as stainless steel, brass, etc. On increasing pressure the bellows expands (or contracts), and on decreasing pressure they contract (or expand). The amount of expansion or contraction is proportional to the applied pressure. Similar to the bellows is the *diaphragm* sensor, shown in Figs. C-2b and C-5. As the process pressure increases, the center of the

545

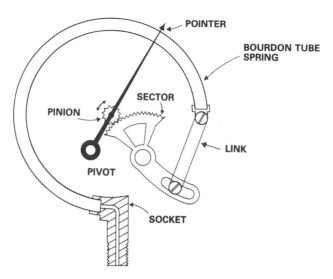

Figure C-1.   Simple Bourdon tube. (Courtesy of the Instrument Society of America.)

diaphragm moves away from the pressure. The amount of motion is proportional to the applied pressure.

FLOW SENSORS[2,4,5,6,7]

Flow is one of the two most commonly sensed process variables, the other being temperature; consequently, many different types of flow sensors have been developed. This section stresses the most used and mentions some others. Table C-1[6] shows several characteristics of some common sensors.

(a)

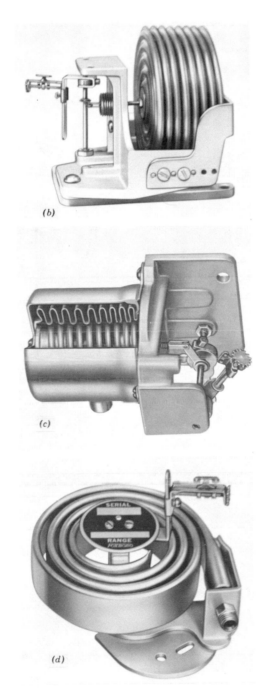

Figure C-2. Types of Bourdon tubes. (Courtesy of the Foxboro Co.) (*a*) Helical. (*b*) and (*c*) Bellows. (*d*) Spiral.

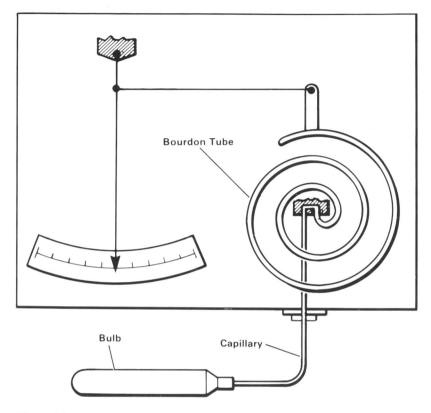

Figure C-3.   Spiral Bourdon tube. (Courtesy of Taylor Instrument Co.)

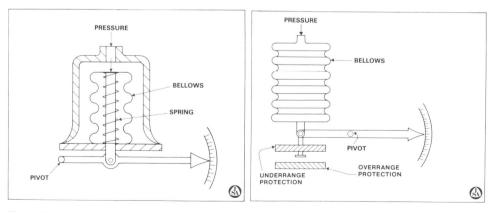

Figure C-4.   Bellows pressure sensor. (Courtesy of the Instrument Society of America.)

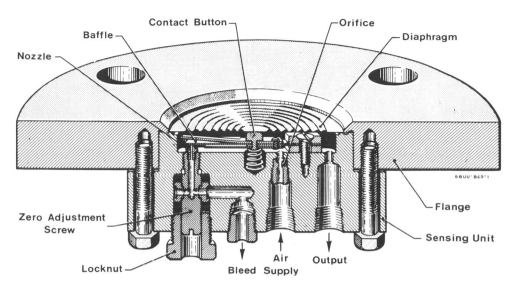

Figure C-5. Diaphragm pressure sensor. (Courtesy of Taylor Instrument Co.)

Probably the most popular flow sensor is the *orifice meter*, which is a flat disk with a machined hole as shown in Fig. C-6. The disk is inserted in the process line perpendicular to the fluid motion with the intention of producing a pressure drop, ΔP. This pressure drop across the orifice is proportional to the volumetric flow rate through the orifice. Accurate orifice meter flow equations are complex and are presented in many fine references;[8,9,10] however, most installations probably use the following simple equation:

$$q = C \sqrt{\frac{\Delta P}{\rho}}$$ (C-1)

where

q = flow rate
ΔP = pressure drop across orifice
C = orifice coefficient
ρ = fluid density

The previously given references also show how to size the required orifice diameter. Most orifice diameters vary between 10% and 75% of the pipe diameter.

The pressure drop across the orifice is usually measured with:

1. Flange taps, shown in Fig. C-7, are the most common. Their technique consists of measuring the pressure drop across the flanges holding the orifice in the process line.

2. Vena contracta taps, shown in Fig. C-8, indicate the largest pressure drop.

3. Other types include radius taps, corner taps, and line taps. These are not as popular as the two previous ones.

Table C-1. Characteristics of Typical Flow Sensors

Primary element	Type of fluid	Pressure loss[a]	Flow rangeability	Error	Upstream piping[b]	Viscosity Effect	Cost	Type of readout
Concentric orifice	Liquid, gas, and steam	50–90%	3:1	3/4%	10–30D	High	Low	Square root
Segmental orifice	Liquid slurries	60–100%	3:1	2½%	10–30D	High	Low	Square root
Eccentric orifice	Liquid–gas comb.	60–100%	3:1	2%	10–30D	High	Low	Square root
Quadrant-edged orifice	Viscous liquids	45–85%	3:1	1%	20–500	Low	Medium	Square root
Segmental wedge	Slurries and viscous liquids	30–80%	3:1	1%	10–30D	Low	High	Square root
Venturi tube	Liquid and gas	10–20%	3:1	1%	5–10D	Very high	Very high	Square root
Dall tube	Liquids	5–10%	3:1	1%	5–10D	High	High	Square root
Flow nozzle	Liquid, gas, and steam	30–70%	3:1	1½%	10–30D	High	Medium	Square root
Elbow meter	Liquid	None	3:1	1%	30D	Negligible	Medium	Square root

Meter	Fluid	Pressure / Range	Rangeability	Accuracy	Upstream piping[b]	Pressure loss[a]	Cost	Output
Rotameter	All fluids	1–200"WG	10:1	2%	None	Medium	Medium	Linear
V-notch weir	Liquids	None	30:1	4%	None	Negligible	Medium	5/2
Trapezoidal weir	Liquids	None	10:1	4%	None	Negligible	Medium	3/2
Parshall flume	Liquid slurries	None	10:1	3%	None	Negligible	High	3/2
Magnetic flow meter	Liquid slurries	None	30:1	1%	None	None	High	Linear
Turbine meter	Clean liquids	0–7 psi	14:1	1/2%	5–10D	High	High	Linear
Pitot tube	Liquids	None	3:1	1%	20–30D	Low	Low	Square root
Pitot venturi	Liquids and gases	None	3:1	1%	20–30D	High	Low	Square root
Positive displacement	Liquids	0–15 psi	10:1	1/2–2%	None	None	High	Linear totalization
Swirlmeter	Gases	0–2 psi	10:1 to 100:1	1%	10D	None	High	Linear
Vortex shedding	Liquids and gases	0–6 psi 0–5"WF	30:1 to 100:1	1/4%	15–30D	Minimum Reynolds No. 10,000	High	Linear
Ultrasonic	Liquids	None			None	None	High	Linear

[a] Pressure loss percentages are stated as a percent of differential pressure produced.

[b] Upstream piping is stated in the number of straight pipe diameters required preceding the primary element.

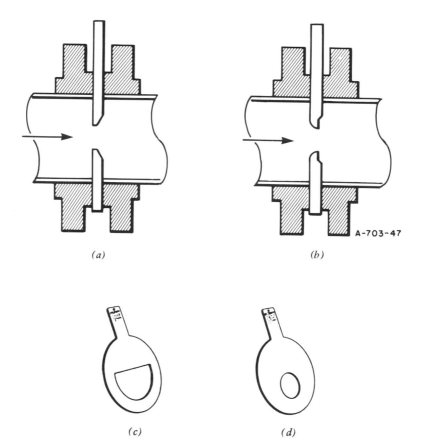

Figure C-6. Schematic of orifice meter. (*a*) Sharp edge. (*b*) Quadrant edge. (Courtesy of Taylor Instrument Co.) (*c*) Segmental edge. (*d*) Eccentric edge. (Courtesy of Foxboro Co.)

The tap upstream from the orifice is called the high-pressure tap and the one downstream from the orifice is called the low-pressure tap. Most tap diameters vary between ¼ and ¾ in. The pressure drop sensed will be a function of tap location as well as flow rate.

Several things must be stressed about the use of orifice meters to measure flows. The first is that the output signal from the orifice/transmitter combination is the pressure drop across the orifice, *not* the flow. A differential pressure sensor, Fig. C-9, is used to measure the pressure drop across the orifice. Equation C-1 shows that this pressure drop is related to the square of the flow, or

$$\Delta P = \left(\frac{\rho}{C^2}\right) q^2 \tag{C-2}$$

Consequently, if the flow is desired, then the square root of the pressure drop must be obtained. Chapter 8 presents the instrumentation required to do so. However, several manufacturers offer the option of installing a square root extraction unit within the

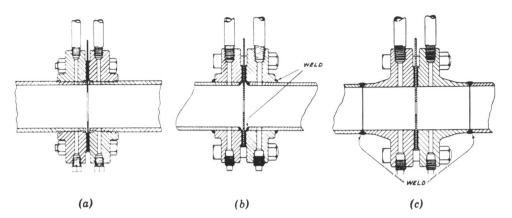

Figure C-7. Flange taps. (*a*) Threaded union. (*b*) Slip-on union. (*c*) Welding-neck union. (Courtesy of the Foxboro Co.)

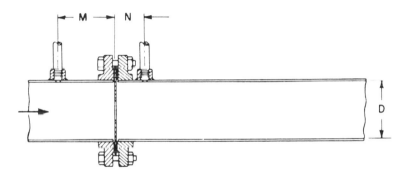

Dimension *M*: anywhere between ½ and 2 pipe diameters above the upstream face of the orifice plate

Orifice to Pipe Diameter Ratio	Location of Downstream Pressure Tap (*N*) (Pipe-Diameters)		
d/D	*Minimum*	*Mean*	*Maximum*
0.2	0.37	0.85	1.30
0.3	0.44	0.80	1.15
0.4	0.47	0.73	1.00
0.5	0.47	0.66	0.84
0.6	0.42	0.57	0.70
0.7	0.35	0.45	0.55
0.8	0.25	0.33	0.41

Figure C-8. Vena contracta taps. (Courtesy of Foxboro Co.)

transmitter. In this case, the output signal from the transmitter is linearly related to the volumetric flow. The second thing that must be stressed is that not all of the pressure drop measured by the taps is lost by the process fluid. A certain amount is recovered by the fluid, in the next few pipe diameters, as it reestablishes its flow regime. Finally, the rangeability of the orifice meter, the ratio of the maximum measurable flow to the

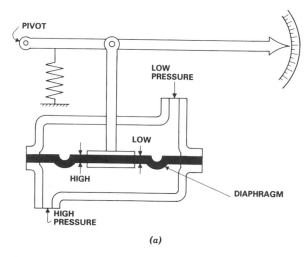

(a)

Figure C-9. Differential pressure sensor and transmitter. *(a)* (Courtesy of the Instrument Society of America.)

Figure C-9. *(Continued)* *(b)* (Courtesy of Fischer & Porter.)

minimum measurable flow, is about 3:1, as indicated in Table C-1. This is important to know as it would indicate the expected accuracy when running the process at low or high loads.

    There are several possible causes that prevent the use of orifice sensors. Among such causes are not enough available pressure to provide pressure drop, as in the case of gravity flow, and the flow of corrosive fluids, fluids with suspended solids that may plug the orifice, or fluids close to their saturated vapor pressure that may flash when subjected to a drop in pressure. These cases require the use of other sensors to measure flow.

Figure C-9. (*Continued*) (*c*) (Courtesy of Rosemount, Inc.)

Another common type of sensor is the *magnetic flowmeter* shown in Fig. C-10. The operating principle of this element is Faraday's Law; that is, as a conductive material (a fluid) moves at right angles through a magnetic field, it induces a voltage. The voltage created is proportional to the intensity of the magnetic field and to the velocity of the fluid. If the intensity of the magnetic field is constant, the voltage is then only proportional to the velocity of the fluid. Furthermore, the velocity measured is the average velocity and thus this sensor can be used for both regimes, laminar and turbulent. During the calibration of this flowmeter, the cross-sectional area of the pipe is taken into consideration so that the electronics associated with the meter can calculate the volumetric flow. Thus, the output is linearly related to the volumetric flow rate.

Since the magnetic flowmeter does not restrict flow, it is a zero pressure drop device suitable for measuring gravity flow, slurry flows, and flow of fluids close to their vapor pressure. However, the fluid must have a minimum required conductivity of about 10 μohm/cm^2, making the meter unsuitable for the measurement of both gases and hydrocarbon liquids.

Table C-1 shows that the rangeability of the magnetic flowmeter is 30:1, which is significantly greater than that of orifice meters; however, their cost is also greater. The cost differential increases as the size of the process pipe increases.

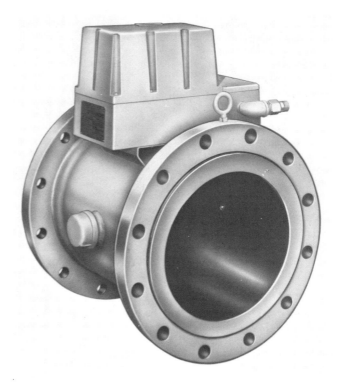

Figure C-10.   Magnetic flowmeter. (Courtesy of Fischer & Porter.)

Finally, an important consideration in the application and maintenance of magnetic flowmeters is coating of the electrodes. This coating represents another electrical resistance resulting in erroneous readings. Manufacturers offer techniques such as ultrasonic cleaners to maintain clean electrodes.

Another important flowmeter is the *turbine meter* shown in Fig. C-11. This meter is one of the most accurate of the commercially available flowmeters. Its working principle consists of a rotor that is caused to spin by the fluid velocity. The rotation of the blades is detected by a magnetic pickup coil that emits pulses the frequency of which is proportional to the volumetric flow rate. This pulse is equally converted to a 4–20 mA signal so that it can be used with standard electronic instrumentation. The converter, or transducer, is usually an integral part of the meter. The most common problem associated with turbine meters is the bearings, which require clean fluids with some lubricating properties.

We have briefly discussed three of the most common flowmeters in use in the process industries. There are many other types ranging from rotameters, flow nozzles, venturi tubes, pitot tubes, and Annubars, which have been used for many years, to more recent developments such as vortex-shedding meters, ultrasonic meters, and swirlmeters. Lack of space does not permit a discussion of these meters. The reader is directed to the many fine references given at the beginning of this section for discussion of these meters.

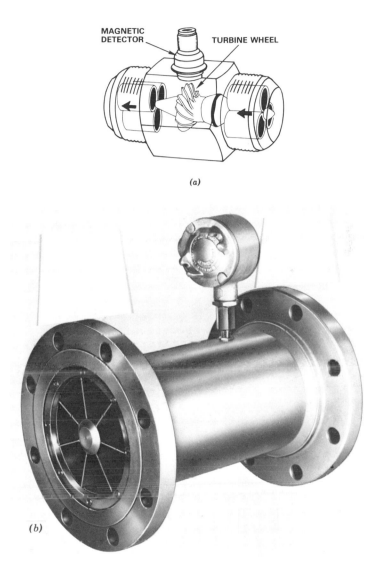

(a)

(b)

Figure C-11. Turbine flowmeter. (*a*) (Courtesy of the Instrument Society of America.) (*b*) (Courtesy of Fischer & Porter.)

LEVEL SENSORS[2,5,6,11]

The three most important level sensors are the differential pressure, float, and air bubbler. The *differential pressure* method consists of sensing the difference in pressure between the pressure at the bottom of a liquid and that above the liquid level. This differential pressure is caused by the hydrostatic head developed by the liquid level. This sensor is shown in Fig. C-12. The side that senses the pressure at the bottom of the liquid is referred to as the high-pressure side and the one that senses the pressure above the liquid level is referred to as the low-pressure side. With the knowledge of the differential pressure and of the density of the liquid, the level can be obtained. Fig. C-12 shows the installation of the differential pressure sensor in open and closed vessels. If the vapors above the liquid level are noncondensable, then the low-pressure piping, also known as the wet leg, can be empty. However, if the vapors are likely to condense, then the wet leg must be filled with a suitable seal liquid. If the density of the liquid varies, then some compensation technique must be employed.

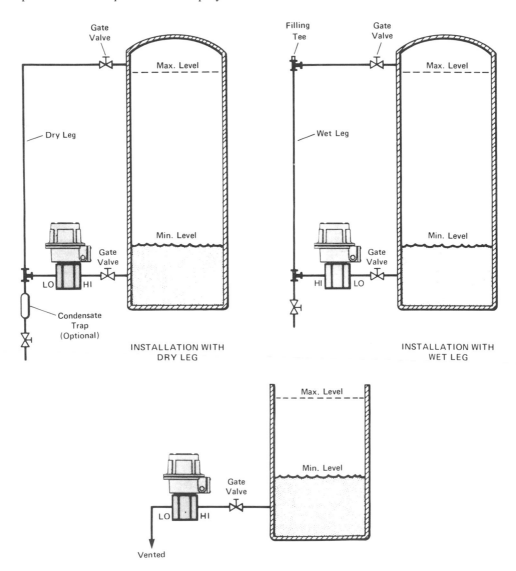

Figure C-12. Differential pressure transmitters installed in closed and in open vessels. (Courtesy of Taylor Instrument Co.)

The *float sensor* detects the change in buoyant force on a body immersed in the liquid. This sensor is generally installed in an assembly mounted externally to the vessel as shown in Fig. C-13. The force required to keep the float in place, which is proportional to the liquid level, is then converted to a signal by the transmitter. This type of sensor is less expensive than most other level sensors; however, their major disadvantage resides in their inability to change their zero and span. To change the zero requires relocation of the complete housing.

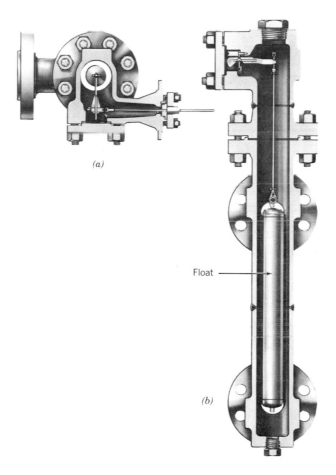

Figure C-13.   Level float sensor. (Courtesy of Taylor Instrument Co.) (*a*) Top view. (*b*) Front view.

The *bubbler sensor* is another type of hydrostatic pressure sensor. It consists, as shown in Fig. C-14, of an air or inert gas pipe immersed in the liquid. The air or inert gas flow through the pipe is regulated to produce a continuous stream of bubbles. The pressure required to produce this continuous stream is a measure of the hydrostatic head or liquid level.

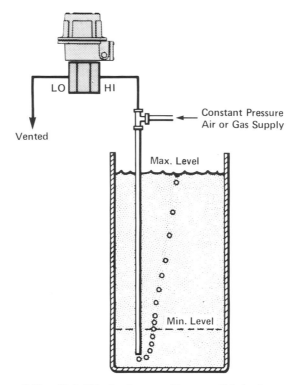

Figure C-14.   Air bubbler level sensor. (Courtesy of Taylor Instrument Co.)

There are some newer methods to measure level in tanks. Some of these are capacitance gauges, ultrasonic systems, and nuclear radiation systems. The last two sensors are also used to measure level of solid material. The references cited at the beginning of this section are recommended for further reading.

TEMPERATURE SENSORS[*,2,3,6]

Temperature, along with flow, is the most frequently measured variable in the process industries today. A simple reason is that very few physical phenomena are not affected by it. Temperature is often used to infer other process variables. Two of the most common examples are in distillation columns and in chemical reactors. In distillation columns temperature is commonly used to infer the purity of one of the exit streams. In chemical reactors temperature is used as an indication of the extent of reaction or conversion.

Because of the many effects produced by temperature, numerous devices have been developed to measure it. With a few exceptions, the devices fall into four general clas-

---

*This section is partially written from notes developed by Dr. L. A. Scott for his undergraduate instrumentation courses at the University of South Florida.

sifications as shown in Table C-2. Quartz thermometers, pyrometric cones, and specialized paints are some of the sensors that do not fit into the classifications shown in Table C-2. Table C-3[6] shows some characteristics of typical sensors.

Liquid-in-glass thermometers indicate temperature change caused by the difference between the temperature coefficient of expansion for glass and the liquid employed. Mercury and alcohol are the most widely used liquids. Mercury-in-glass thermometers made from ordinary glass are useful between $-35°F$ and $600°F$. The lower limit is due to the freezing point of mercury and the upper limit to its boiling point. By filling the space above the mercury with an inert gas, usually nitrogen, to prevent boiling, the useful range may be extended to $950°F$. Such thermometers usually bear the inscription "nitrogen filled." For temperatures below the freezing point of mercury ($-38°F$) another liquid must be employed. Alcohol is the most widely used fluid for temperatures down to $-80°F$, pentane for temperatures to $-200°F$, and toluene for temperatures below $230°F$.

The *bimetallic strip thermometer* works on the principle that metals expand with temperature and that the expansion coefficients are not the same for all metals. Figure C-15 shows a typical bimetallic strip thermometer. The temperature-sensitive element is a composite of two different metals fastened together into a strip. One metal has a high thermal expansion coefficient and the other metal has a low thermal expansion coefficient. A common combination is invar (64% Fe, 36% Ni), which has a low coefficient and another nickel-iron alloy that has a high coefficient. Usually the expansion with temperature is low and this is the reason for having the bimetal strip wound in the form of a spiral. As the temperature increases, the spiral will tend to bend toward the side of the metal with the low thermal coefficient.

Fig. C-16 shows the elements of a typical *filled system thermometer*. Temperature variations cause the expansion or contraction of the fluid in the system, which is sensed by the Bourdon spring and transmitted to an indicator or transmitter. Because of the design simplicity, reliability, relatively low cost, and inherent safety, these elements are

Table C-2. Popular Sensors for Temperature Measurement

I. Expansion thermometer
   A. Liquid-in-glass thermometer
   B. Solid-expansion thermometers (bimetallic strip)
   C. Filled-system thermometers (pressure thermometers)
      1. Gas-filled
      2. Liquid-filled
      3. Vapor-filled

II. Resistance-sensitive devices
   A. Resistance thermometers
   B. Thermistors

III. Thermocouples

IV. Noncontact methods
   A. Optical pyrometers
   B. Radiation pyrometers
   C. Infrared techniques

Table C-3. Characteristics of Typical Temperature Sensors

	Range, °F	Accuracy, °F	Advantages	Disadvantages
Glass-stem thermometers	Practical: −200 to 600 Extreme: −321 to 1100	0.1–2.0	Low cost Simplicity Long life	Difficult to read Only local measurement, no automatic control or recording capability
Bimetallic thermometers	Practical: −80 to 800 Extreme: −100 to 1000	1.0–20	Less subject to breakage Dial reading Less costly than thermal or electrical	Less accurate than glass stem thermometer Changes calibration with rough handling
Filled thermal elements	Practical: −300 to 1000 Extreme: −450 to 1400	±0.5–2.0% of full scale	Simplicity No auxiliary power needed Sufficient response times	Larger bulb size than electrical systems, and greater minimum spans Bulb to readout distance is maximum of 50–200 ft Factory repair only
Resistance thermometer	−430 to 1800	0.1 (best)	System accuracy Low spans (10°F) available Fast response, small size	Self-heating may be a problem Long-term drift exceeds that of thermocouple Some forms expensive, difficult to mount
Thermocouples	−440 to 5000	0.2 (best)	Small size, low cost Convenient mount Wide range	Not as simple as direct-reading thermometers Cold working of wires can affect calibration; 70°F nominal minimum span
Radiation pyrometer	0 to 7000	±0.5–1.0% of full scale	No physical contact Wide range, fast response Measure small target or average over large area	More fragile than other electrical devices Nonlinear scale, relatively wide span required
Thermistors	−150 to 600	0.1 (best)	Small size, fast response Good for narrow spans Low cost, stable No cold junction	Very nonlinear response Stability above 600°F is a problem; not suitable for wide spans High resistance makes system prone to pick up noise from power lines

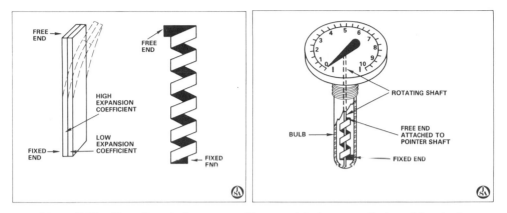

Figure C-15. Bimetallic strip thermometer. (Courtesy of the Instrument Society of America.)

popular in the process industries. The Scientific Apparatus Manufacturers' Association (SAMA) has established four major classes, with subclassifications, of filled systems. Table C-4 lists these classifications. The most significant differences between the classifications are the fluid used and the compensation for temperature difference between the bulb, the capillary, and the Bourdon spring. Fig. C-17 shows several of these filled systems. For a more extensive description of these systems the reader is referred to References 2 and 3.

Resistance thermometer devices (RTDs) are elements based upon the principle that the electrical resistance of pure metals increases with an increase in temperature. Since measurements of electrical resistance can be made with high precision, this also provides a very accurate way to make temperature measurements. The most commonly used metals are platinum, nickel, tungsten, and copper. Figure C-18 shows a schematic of a typical RTD.

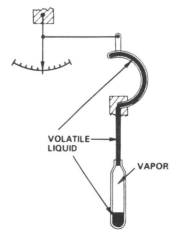

Figure C-16. Typical filled system thermometer. (Courtesy of the Instrument Society of America.)

Table C-4. Scientific Apparatus Manufacturers' Association Classification of Filled System Thermometers

Class	Filling	Characteristics
I	Liquid other than mercury	Uncompensated
IA	Liquid other than mercury	Case and capillary compensated
IB	Liquid other than mercury	Case compensated
IIA	Vapor	For bulb above ambient applications
IIB	Vapor	For bulb below ambient applications
IIC	Vapor	For bulb either above or below ambient Large bulb used
IID	Vapor	For bulb either above or below ambient Nonvolatile liquid used for transmission
IIIA	Gas	Case and capillary compensated
IIIB	Gas	Case compensated
VA	Mercury	Case and capillary compensated
VB	Mercury	Case compensated

Note: There is no SAMA classification IV.

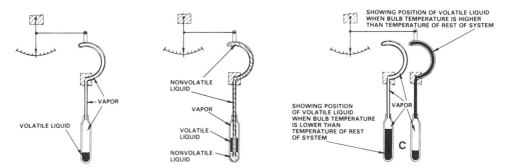

Figure C-17. Other types of filled system thermometers. (Courtesy of the Instrument Society of America.)

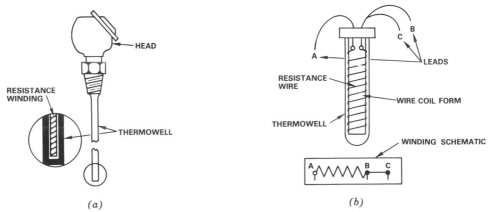

Figure C-18. Schematic of a resistance thermometer device. (*a*) Assembly. (*b*) Components. (Courtesy of the Instrument Society of America.)

A Wheatstone bridge is usually used for the resistance, and consequently also the temperature, reading. Figure C-19 shows the schematics of the two- and three-wire bridges used.

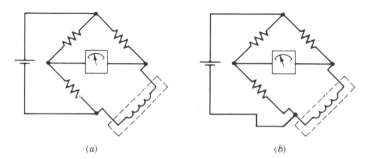

(a)   (b)

Figure C-19.   (a) Two-wire Wheatstone bridge. (b) Three-wire Wheatstone bridge.

Thermistor elements detect very sensitive temperature changes. Thermistors are made of a sintered combination of ceramic material and some kind of semiconducting metallic oxide such as nickel, manganese, copper, titanium, or iron. Thermistors have a very high negative, or sometimes positive, temperature coefficient of resistivity. Figure C-20 shows some typical thermistors. The Wheatstone bridges shown in Fig. C-19 are usually used to measure the resistance and, therefore, also temperature. Some of the advantages are their small size and low cost. Their main disadvantages lie in their nonlinear temperature versus resistance relationship and in the fact that they usually require shielded power lines.

LEADS   LEADS

SEMICONDUCTING MATERIAL   SEMICONDUCTING MATERIAL

Figure C-20.   Typical thermistor construction. (Courtesy of the Instrument Society of America.)

The last temperature element that we will discuss is the *thermocouple*, probably the best known industrial temperature sensor. The thermocouple works on a principle discovered by T. J. Seebeck in 1821. The Seebeck effect or Seebeck principle states that an electric current flows in a circuit of two dissimilar metals if the two junctions are at different temperatures. Figure C-21 shows a schematic of a simple circuit. M_1 and M_2 are the two metals, T_H is the temperature being measured and T_C is the temperature of what is usually known as the cold, or reference, junction. The voltage produced by this thermoelectric effect depends on the temperature difference between the two junctions and the metals used. Figure C-22 shows some voltages generated by typical metals. A more realistic schematic of a measuring circuit is shown in Fig. C-23. The most common type of thermocouples are platinum-platinum/rhodium alloy, copper-constantan, iron-constantan, chromel-alumel and chromel-constantan. Figure C-24 (b) shows an assembly of an industrial thermocouple setup. The protecting tube, also called a thermowell, is not necessary in all installations. This thermowell tends to slow down the response of the sensor system. For a more detailed discussion of thermocouples References 2 and 3 are strongly recommended.

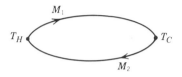

Figure C-21.   Single thermocouple circuit.

DEG F	0	2	4	6	8	10
−100	−2.699	−2.736	−2.773	−2.810	−2.867	−2.883
10	−0.478	−0.435	−0.392	−0.349	−0.305	−0.262
100	1.520	1.566	1.611	1.657	1.703	1.748

(a)

DEG F	0	2	4	6	8	10
−100	−2.581	−2.616	−2.650	−2.665	−2.719	−2.753
10	−0.467	−0.425	−0.383	−0.341	−0.299	−0.256
100	1.518	1.565	1.611	1.450	1.705	1.752

(b)

DEG F	0	2	4	6	8	10
−100	−3.492	−3.541	−3.590	−3.639	−3.688	−3.737
10	−0.611	−0.556	−0.501	−0.445	−0.390	−0.334
100	1.942	2.000	2.058	2.117	2.175	2.233

(c)

Figure C-22.   Voltages generated (millivolts) for different types of thermocouples. (Reference junction at 32°F.) (a) Type K: Nickel-chromium vs. nickel-aluminum (chromel-alumel). (b) Type T: Copper vs. copper-nickel (copper-constantan). (c) Type J: Iron vs. copper-nickel (iron-constantan).

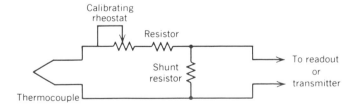

Figure C-23.   Thermocouple measuring circuit.

COMPOSITION SENSORS[2,25,26,27,28]

Another important class of sensors are composition sensors. These sensors are used in environmental and product quality measurement and control. There are many different types of measurement sensors such as density, viscosity, chromatography, pH, and ORP. Because of lack of space we cannot present these sensors; however, we want to make the reader aware of their importance. References are given for interesting reading and learning.

TRANSMITTERS

This section presents an example of a pneumatic transmitter and one of an electrical transmitter. The objective is to present the reader with the working principles of these typical transmitters. As mentioned earlier in this appendix, the purpose of a transmitter is to convert the output from the sensor to a signal strong enough so that it can be transmitted to a controller or any other receiving device. In general, most transmitters can be divided into force-balance types and motion-balance types. These are the most common types of transmitters and are used extensively in industry.

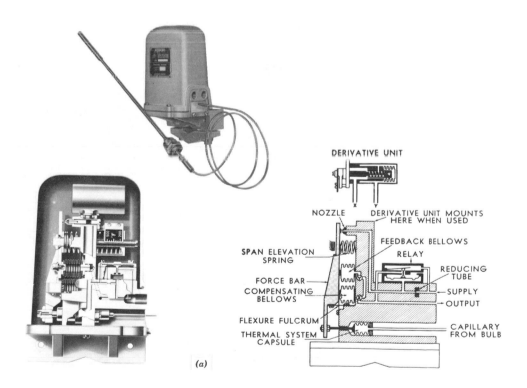

(a)

Figure C-24. *(a)* Assembly of gas-filled bulb. (Courtesy of the Foxboro Co.)

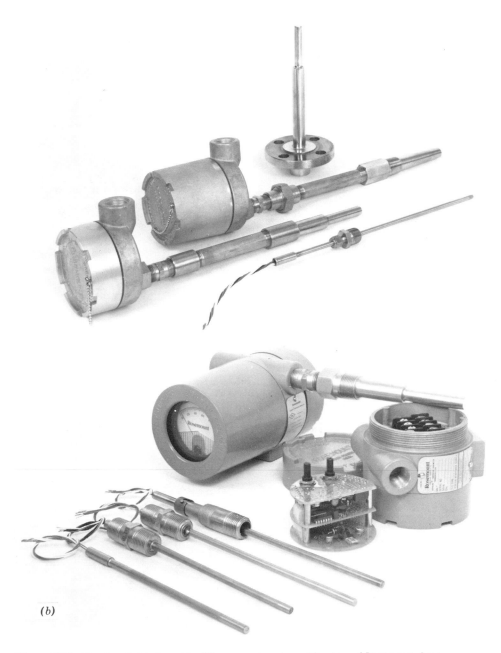

(b)

Figure C-24. (*Continued*) (b) Assembly of thermocouple systems. (Courtesy of Rosemount, Inc.)

Pneumatic Transmitter

All pneumatic transmitters use a flapper-nozzle arrangement to produce an output signal proportional to the output from the sensor. A pneumatic differential pressure transmitter,[12] which is a force-balance type transmitter, will be used to show the working principles. This transmitter is shown in Fig. C-25.

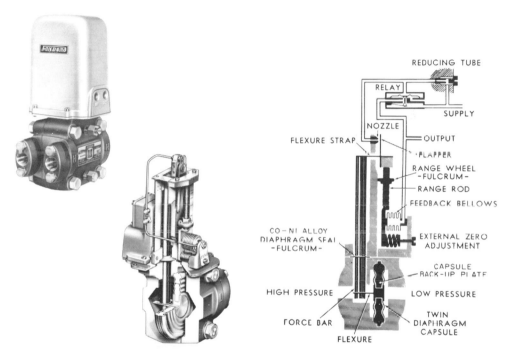

Figure C-25. Pneumatic differential pressure transmitter. (Courtesy of the Foxboro Co).

The twin diaphragm capsule is the sensor. It senses the difference in pressure between the high- and low-pressure sides. Previously, we learned that this type of sensor is used to measure liquid level and flow. The diaphragm is connected to a force bar by a flexure. The force bar is connected to the body of the transmitter by a stainless steel diaphragm. This diaphragm serves as a seal to the measuring cavity and also as a positive fulcrum for the force bar. The top of the force bar is connected by a flexure strap to a range rod. This range rod has a range wheel that also serves as a fulcrum. A feedback bellows and a zero adjustment are located in the bottom part of the range rod. Above the range rod a flapper-nozzle arrangement and a pneumatic relay are located. As seen, the flapper is connected to the force bar–range rod combination.

As the diaphragm capsule senses a difference in pressure, this creates a tension or force on the lower end of the force bar. To be more specific, you may assume that the pressure on the high side increases, creating a pulling force on the force bar. This force results in a motion at the outer end of the bar, causing the flapper to move closer to the nozzle. In this case the output of the relay increases and this increases the force that the feedback bellows exerts on the range rod. This force balances the force of the differential pressure across the diaphragm capsule. These balanced forces result in an output signal from the transmitter that is proportional to the difference in pressure.

The recommended supply pressure to most pneumatic instruments is between 20 and 25 psig. This will ensure proper performance at the 15 psig output level. The calibration of these instruments requires the adjustment of the zero and span (or range). In the instrument shown in Fig. C-25 this is done with the external zero adjustment screw and with the range wheel.

The preceding paragraphs have described the working principle of a typical pneumatic instrument. As mentioned at the beginning, all pneumatic instruments use some kind of flapper-nozzle arrangement to produce an output signal. This is a reliable and simple technique that has proven very successful for many years.

Electronic Transmitter

Figure C-26 shows a simplified diagram of an electronic differential pressure transmitter.[13] This transmitter is a motion-balance type transmitter and will be used to illustrate the working principles of typical electronic instrumentation.

An increase in differential pressure, acting on the measuring element diaphragms, develops a force that moves the lower end of the force beam to the left. This motion of the force beam is transferred to the strain gage force unit through the connecting wire. The strain gage force unit contains four strain gages connected in a bridge configuration. Movement of the force beam causes the strain gages to change resistance. This change in resistance produces a differential signal that is proportional to the input differential pressure. This differential signal is applied to the inputs of the input amplifier. One side of the signal is applied directly to the inverting input of the amplifier, while the other side is applied to the noninverting input through the zero network. This zero network provides the zero adjustment for the transmitter.

The signal from the input amplifier drives the output current regulator. The current regulator controls the transmitter output current through the span network and the output current sense circuit. The span network provides the span adjustment for the transmitter. The signal from the span network is fed back to the input circuit through a buffer amplifier and is used to control the gain of the input circuit. If the transmitter output current increases above 20 mA D.C., the voltage across the output current sense resistor will turn on the output current limiter, which will limit the output.

TYPES OF CONTROL VALVES[14,15,16,17,18,19,20]

There are many different types of control valves on the market. Almost every other month a "new improved" control valve appears. Consequently, it is difficult to classify them; however, we will classify them into two broad categories: reciprocating stem and rotating stem.

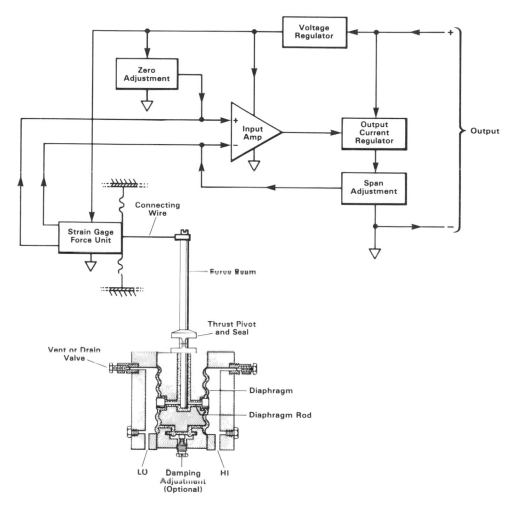

Figure C-26.   Electronic differential pressure transmitter. (Courtesy of Taylor Instrument Co.)

Reciprocating Stem

Figure C-27 shows a typical reciprocating stem control valve. This particular valve is called a *single-seated sliding stem globe valve*. *Globe valves* are a family of valves characterized by a closure member that travels in a line perpendicular to the valve seat. They are used primarily for throttling purposes and general flow control. Figure C-27 also shows in detail the different components of the valve. The valve is shown to be divided into two general areas: the *actuator* and the *body*. The actuator is the part of the valve that converts the energy input to the valve into mechanical motion to increase or decrease the flow restriction. Figure C-28 shows a *double-seated sliding stem globe valve*. Double-seated valves can handle high process pressure with a standard actuator. However,

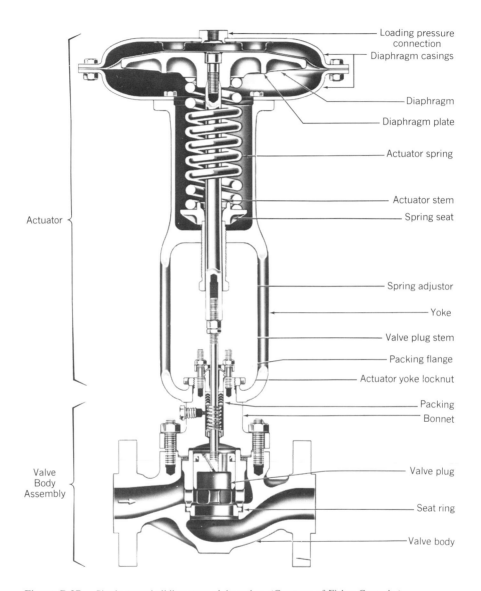

Loading pressure
connection

Diaphragm casings

Diaphragm

Diaphragm plate

Actuator spring

Actuator stem

Spring seat

Spring adjustor

Yoke

Valve plug stem

Packing flange

Actuator yoke locknut

Packing

Bonnet

Valve plug

Seat ring

Valve body

Actuator

Valve
Body
Assembly

Figure C-27. Single-seated sliding stem globe valve. (Courtesy of Fisher Controls.)

if tight shut-off is required, single-seated valves are usually used. Double-seated valves tend to have greater leakage when closed than single-seated valves.

Another type of body in common use is the *split-body valve* shown in Fig. C-29. This type of body is frequently used in process lines where frequent changes of plug and seat are required to prevent corrosion.

Cage valves, shown in Fig. C-30, have hollow plugs with internal passages.

Three-way valves, shown in Fig. C-31, are also a reciprocating type. Three-way valves can be either diverging or converging and, consequently, they can either split one

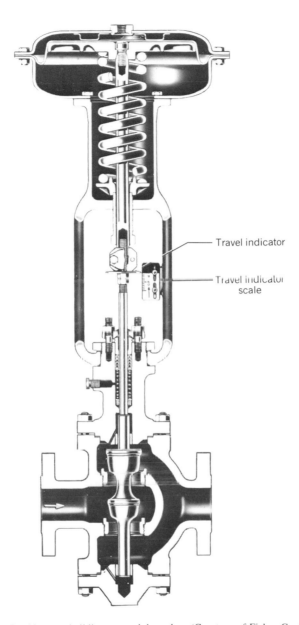

Travel indicator

Travel indicator
scale

Figure C-28. Double-seated sliding stem globe valve. (Courtesy of Fisher Controls.)

stream into two other streams or blend two streams into only one. They are commonly used for control purposes.

　There are some other types of reciprocating stem control valves. Most of them are used in specialized services. Some of these are the *Y-style valve*, which is commonly used in molten metal or cryogenic service. The *pinch or diaphragm valve* consists of some kind of flexure, such as a diaphragm, that can be moved together or open or close

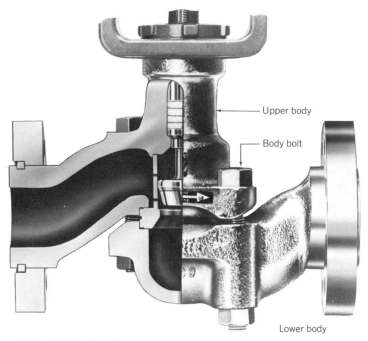

Figure C-29.   Split-body valve. (Courtesy of Masoneilan Division, McGraw-Edison Co.)

the area of flow. They are commonly used for highly corrosive fluids, slurries, and high-viscosity liquids, as well as in some food processing operations, such as the making of beer and wine. The *gate valve* is another type of reciprocating stem valve. It is mainly used for fully open or fully closed services, however, and is not commonly used in throttling services.

Rotating Stem

There are several very popular types of rotating stem valves. One of the most common is the *butterfly valve*, shown in Fig. C-32. These valves consist of a disk rotating about a shaft. They require minimum space for installation and provide high-capacity flow with minimum pressure drop. They are used in low-pressure services. Conventional disks provide throttling control for up to 60 degree rotation. New patented disks allow throttling control for a full 90 degrees rotation.

Another common rotating stem valve is the *ball valve* (Fig. C-33). These valves also provide high-capacity flow with minimum pressure drop. They are commonly used to handle slurries or fibrous materials. They have low leakage tendency and are small in size.

A very brief introduction to several types of control valves has been presented. However, these are by no means the only control valves, nor are they the only types of

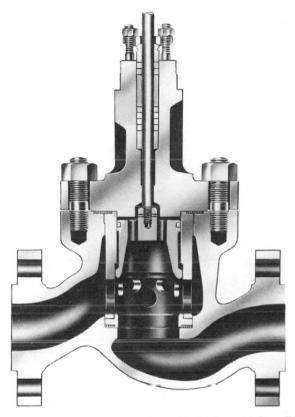

Figure C-30.   Cage valve. (Courtesy of Masoneilan Division, McGraw-Edison Co.)

valves. There are a great number of valves available to meet the requirements for specialized services as well as safety and other types of regulation.

CONTROL VALVE ACTUATOR

As previously defined, the actuator is the part of the valve that converts the energy input, either pneumatic or electrical, into mechanical motion to open or close the valve.

Pneumatically Operated Diaphragm Actuator

These are the most common actuators in the process industries. Figure C-34 shows a typical diaphragm actuator. These actuators consist of a flexible diaphragm placed between two casings. One of the chambers resulting from this arrangement must be made pressure tight. The force generated within the actuator is opposed by a "range" spring. The controller air signal goes into the pressure-tight chamber, and an increase or decrease in air pressure produces a force that is used to overcome the force of the actuator's range spring and the forces within the valve body.

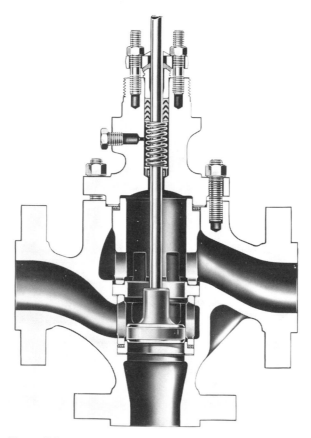

Figure C-31.   Three-way valve. (Courtesy of Fisher Controls.)

The action of the valve, FC or FO, is determined by the actuator. Figure C-34*a* shows a fail closed or air-to-open valve. Figure C-34*b* shows a fail open or air-to-close valve.

The size of the actuator depends on the process pressure against which it must move the stem and on the air pressure available. The most common air pressure range is 3 to 15 psig but ranges of 6 to 30 psig and 3 to 27 psig are also used. These diaphragm actuators are simple in construction and also dependable and economical. Equations are provided by manufacturers to size actuators.

Piston Actuator

Piston actuators are normally used when maximum thrust output is required along with fast response. This usually occurs when working against high process pressure. These actuators operate using a high air pressure supply, up to 150 psig. The best designs are double acting to give maximum thrust in both directions.

Figure C-32. Butterfly valve. (Courtesy of Fisher Controls.)

Electrohydraulic and Electromechanical Actuators

Not as commonly used as the two previous types, electrohydraulic and electromechanical actuators will become more common with the increasing use of electrical control signals. Shown in Figs. C-35 and C-36, they require electric power to the motor and an electric signal from the controller.

In this family of actuators, probably the most common one is the solenoid actuator. A solenoid valve can be used to actuate a double-acting piston actuator. By making or breaking an electric current signal, the solenoid switches the output of a connected hydraulic pump to either above or below the actuator piston. Accurate control of valve position can be obtained with this unit.

Manual-Handwheel Actuator

These actuators are used where automatic control is not required. They are available for reciprocating stem and rotary stem. Figure C-37 shows a typical handwheel actuator.

Figure C-33.   Ball valve with positioner. (Courtesy of Masoneilan Division, McGraw-Edison Co.)

CONTROL VALVE ACCESSORIES

There are a number of devices, called accessories, that usually go along with control valves. This section presents a brief introduction to some of the most common of these accessories.

Positioners

A positioner is a device that acts very much like a controller. Its job is to compare the signal from the controller with the valve stem position. If the stem is not where the controller wants it to be positioned, the positioner adds or exhausts air from the valve until the correct valve position is obtained. That is, when it is important to position the valve's stem accurately, a positioner is normally used. Figure C-38 shows a valve with a positioner. The figure shows the bar-linkage arrangement by which the positioner senses the stem position. Another positioner is shown in fig. C-33.

The use of positioners tends to minimize the effects of:

1.   Lag in large-capacity actuators.

2.   Stem friction due to tight stuffing boxes.

3.   Friction due to viscous or gummy fluids.

4.   Process line pressure changes.

A positioner is recommended when the response of the positioner-valve combination is much faster than the process itself. Some control loops for which this is the case are temperature, liquid level, concentration, and gas flow loops. Some fast loops with which the use of positioners may be discouraged are liquid flow and liquid pressure loops.

The positioners are usually pneumatic or electropneumatic. The pneumatic positioners accept a pneumatic signal as input and output a pneumatic signal to the valve. The electropneumatic positioners accept an electrical signal (1–5 V, 4–20 mA, or 10–50 mA) as input and output a pneumatic signal to the valve. Positioners can also be used to choose the action of control valves (FC or FO).

Boosters

Boosters, also called air relays, are used on valve actuators to speed up the response of the valve to a changing signal from a low-output-capacity pneumatic controller or transducer. It may also be noticed that for fast responding control loops, such as liquid flow

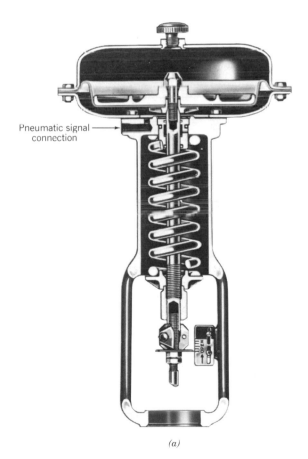

Pneumatic signal connection

(a)

Figure C-34a. Diaphragm actuator. fail closed, air-to-open. (Courtesy of Fisher Controls.)

(b)

Figure C-34*b.*   Diaphragm actuator: fail open, air-to-close. (Courtesy of Fisher Controls.)

or liquid pressure, with which the use of positioners is discouraged, the use of boosters may be the proper choice.[14]

Boosters also have several other possible uses:

1. Amplify a pneumatic signal. Some typical amplification ratios are 1:2 and 1:3

2. Reduce a pneumatic signal. Typical ratios are 5:1, 2:1, and 3:1

3. Reversing a pneumatic signal. This means that as the input signal increases, the output signal decreases. When the input signal is 3 psig, the output is 15 psig. When the input signal is 15 psig, the output signal is 3 psig.

Limit Switches

These limit switches are mounted on the side of the valves and are triggered by the position of the stem. These switches are usually used to drive alarms, solenoid valves, lights, or any other such device.

Figure C-35. Electrohydraulic actuator mounted on a rotary valve. (Courtesy of Fisher Controls.)

CONTROL VALVES—ADDITIONAL CONSIDERATIONS

This section presents a number of additional considerations when sizing and choosing a control valve. Thus, this section complements Section 5-2.

Figures C-39a through C-39d show examples of manufacturer catalogs (Masoneilan and Fisher Controls). Once the C_V coefficient has been calculated using the equations presented in Chapter 5, these figures are used to determine valve size.

Viscosity Corrections

Equation (5-1) does not take into consideration the effect of liquid viscosity in calculating the C_V coefficient. For most liquid service the viscosity correction can be ignored; however, in some other instances this can lead to sizing errors.

Masoneilan[15] proposes to calculate a turbulent C_V and a laminar C_V and then to use the larger value as the required C_V.

Figure C-36. Electromechanical actuator mounted on a butterfly valve. (Courtesy of Fisher Controls.)

Figure C-37. Manual actuator. (Courtesy of Fisher Controls.)

Figure C-38. Positioner installed on a valve. (Courtesy of Taylor Co.)

Turbulent Flow

$$C_V = q\sqrt{\frac{G_f}{\Delta P}} \tag{C-3a}$$

Laminar Flow

$$C_V = 0.072\left(\frac{\mu q}{\Delta P}\right)^{2/3} \tag{C-3b}$$

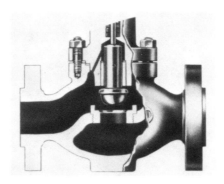

20000 Series

ANSI Class 150-600 (Sch. 40)

Nominal Trim Size	¼	⅜	½	¾	1	1½	2	3	4	6	8	10
Orifice Dia. (in.)	.250	.375	.500	.750	.812	1.250	1.625	2.625	3.500	5.000	6.250	8.000

Valve Size (in.)	Reduced Trim					Full Capacity Trim						
¾	1.7	3.7	6.4	11								
1	1.7	3.7	6.4	11	12							
1½	1.7	3.8	6.6	12	13	25						
2	1.7	3.8	6.7	13	19*	26	46					
3				14		31	47	110				
4						32	49	113	195			
6							53	126	208	400		
8								133	224	415	640	
10									233	442	648	1000

Orifice Diameter .994　　　　　　　　　　　　　　　　　　　　　　　　　**Refer to Bulletin 334E**

(a)

Figure C-39a.　Example of Masoneilan's valve catalog. (Courtesy of Masoneilan Division, McGraw-Edison Co.)

where

μ = viscosity, centipoises

Fisher Controls[16] has developed a nomograph and procedure that provide a correction factor, F_V, that can be applied to the standard C_V coefficient to determine a corrected coefficient, C_{Vr}. Figure C-40 shows the nomograph with the instructions on how to use it. Once the correction factor is obtained then, the corrected coefficient is calculated as follows:

$$C_{Vr} = F_V C_V \qquad (C-4)$$

37000 Series
MiniTork

Critical Flow Factor – C_f
Flow in either direction — .65

ANSI Class 150-300 (2″ to 12″)
ANSI Class 150 (14″ to 24″)

Valve Size (in.)	$\dfrac{\text{Line Size}}{\text{Valve Size}} = 1.0$
2	54
3	180
4	390
6	1150
8	2050
10	3200
12	4600
14	5600
16	7400
18	9500
20	11800
24	17200

Refer to Bulletin 352E

38000 Series

Critical Flow Factor – C_f — .60
ANSI Class 150

Valve Size (in.)	C_v
2	130
3	265
4	460
6	1025
8	1775
10	2500
12	4400

Refer to Bulletin 371E

Figure C-39b.   Example of Masoneilan's valve catalog. (Courtesy of Masoneilan Division, McGraw-Edison Co.)

FLOW COEFFICIENTS   For additional body information see *Bulletin 51.1:EB(10)*

Equal Percentage

Equal Percentage Characteristic

Coeffi-cients	Body Size, In.	Port Diameter, In.	Total Travel, In.	Valve Opening—Percent of Total Travel										K_m and C_f
				10	20	30	40	50	60	70	80	90	100	
C_v (Liquid)	1	1	3/4	.834	1.22	1.79	2.54	3.60	5.46	8.04	10.4	11.3	12.0	.92
	1-1/2	1-1/2	3/4	1.18	1.76	2.63	3.58	5.17	7.63	10.5	17.0	22.8	28.9	.85
	2	2	1-1/8	2.18	3.35	4.55	6.40	9.10	13.5	22.0	34.6	44.5	50.0	.94
	3	3	1-1/2	4.13	7.07	10.6	15.8	23.6	39.0	62.4	92.4	112	120	.83
	4	4	2	5.54	9.39	14.4	20.5	29.1	41.0	61.9	108	155	183	.86
	1*	3/4	3/4	.361	.597	.878	1.38	1.98	3.02	4.51	6.30	7.37	7.84	.97
	1-1/2*	3/4	3/4	.354	.559	.874	1.34	2.02	3.01	4.61	6.78	8.54	9.71	.96
		1	3/4	.803	1.13	1.61	2.50	3.58	4.95	8.20	12.3	15.7	16.9	.93
	2*	1	3/4	.772	1.17	1.65	2.47	3.53	5.38	8.20	11.2	13.7	14.9	.93
		1-1/2	3/4	1.11	1.64	2.42	3.46	4.86	7.22	12.0	17.4	24.0	29.1	.83
	3*	1-1/2	3/4	1.17	1.75	2.46	3.62	4.66	7.06	11.0	17.9	25.4	31.5	.79
		2	1-1/8	2.01	3.14	4.57	6.39	9.01	12.8	22.0	37.1	51.8	60.9	.84
	4*	2	1-1/8	2.18	3.27	4.61	6.28	8.78	12.8	23.1	40.2	53.8	64.1	.83
		3	1-1/2	3.47	6.12	9.66	14.8	22.6	37.1	62.2	92.5	117	136	.84
C_g (Gas)	1	1	3/4	27.0	37.5	53.0	77.6	112	162	242	344	434	482	40.1
	1-1/2	1-1/2	3/4	40.7	56.5	81.3	112	162	221	312	514	784	992	34.3
	2	2	1-1/8	72.9	111	149	207	293	392	639	1100	1580	1800	36.0
	3	3	1-1/2	127	219	317	471	706	1160	1970	3000	3870	4280	35.7
	4	4	2	176	293	445	619	838	1170	1810	3300	5960	6650	36.3
	1*	3/4	3/4	12.9	19.9	27.0	42.4	59.9	91.0	136	201	265	331	42.2
	1-1/2*	3/4	3/4	12.3	18.2	27.9	42.2	62.8	94.3	143	210	290	367	37.8
		1	3/4	20.4	32.0	48.4	73.4	108	162	244	358	463	617	36.5
	2*	1	3/4	27.1	38.5	53.0	77.4	111	162	236	348	466	582	39.1
		1-1/2	3/4	37.0	54.5	80.8	110	152	202	316	495	778	973	33.4
	3*	1-1/2	3/4	38.8	56.4	80.1	113	146	227	328	527	780	1040	33.0
		2	1-1/8	66.3	102	142	196	271	388	622	1060	1590	2040	33.5
	4*	2	1-1/8	65.1	107	147	193	281	386	617	1050	1530	2100	32.8
		3	1-1/2	114	200	305	450	704	1100	1850	2950	3960	4920	36.2
C_s (Steam)	1	1	3/4	1.35	1.88	2.65	3.88	5.60	8.10	12.1	17.2	21.7	24.1	40.1
	1-1/2	1-1/2	3/4	2.04	2.83	4.07	5.60	8.10	11.1	15.6	25.7	39.2	49.6	34.3
	2	2	1-1/8	3.65	5.55	7.45	10.4	14.7	19.6	32.0	55.0	79.0	90.0	36.0
	3	3	1-1/2	6.35	10.7	15.9	23.6	35.3	58.0	98.5	150	194	214	35.7
	4	4	2	8.80	14.7	22.3	31.0	41.9	58.5	90.5	165	298	333	36.3
	1*	3/4	3/4	.645	.995	1.35	2.12	3.00	4.55	6.80	10.1	13.3	16.6	42.2
	1-1/2*	3/4	3/4	.615	.910	1.40	2.11	3.14	4.72	7.15	10.5	14.5	18.4	37.8
		1	3/4	1.02	1.60	2.42	3.67	5.40	8.10	12.2	17.9	23.2	30.9	36.5
	2*	1	3/4	1.36	1.93	2.65	3.87	5.55	8.10	11.8	17.4	23.3	29.1	39.1
		1-1/2	3/4	1.85	2.73	4.04	5.50	7.60	10.1	15.8	24.8	38.9	48.7	33.4
	3*	1-1/2	3/4	1.94	2.82	4.01	5.65	7.30	11.4	16.4	26.4	39.0	52.0	33.0
		2	1-1/8	3.32	5.10	7.10	9.80	13.6	19.4	31.1	53.0	79.5	102	33.5
	4*	2	1-1/8	3.26	5.35	7.35	9.65	14.1	19.3	30.9	52.5	76.5	105	32.8
		3	1-1/2	5.70	10.0	15.3	22.5	35.2	55.0	92.5	148	198	246	36.2

*This column lists the K_m values for the C_v coefficients and the C_f values for the C_g and C_s coefficients at 100% travel.
• Restricted Trim

Figure C-39c. Example of Fisher's valve catalog: design EAB, specs C and D, equal percentage valve plug. (Courtesy of Fisher Controls.)

Flashing and Cavitation

The presence of either flashing or cavitation in a control valve can have significant effects on the operation of the valve and in its sizing procedure. It is important to understand the meaning and significance of these two phenomena. Figure C-41 shows the pressure profile of a liquid flowing through a restriction (possibly a control valve).

To maintain steady-state flow, the velocity of the liquid must increase as the cross-sectional area for flow decreases. The liquid velocity reaches its maximum at a point just past the minimum cross-sectional area (the port area for a control valve). The point of maximum velocity is called the "vena contracta." At this point the liquid also experiences the lowest pressure. What happens is that the increase in velocity (kinetic energy) is accompanied by a decrease in "pressure energy." Energy is transformed from one form to another.

FLOW COEFFICIENTS

For additional body information see *Bulletin 51.1:EB(10)*

Linear   Linear Characteristic

Coefficients	Body Size In.	Port Diameter In.	Total Travel In.	Valve Opening—Percent of Total Travel										Km* and Cf
				10	20	30	40	50	60	70	80	90	100	
Cv (Liquid)	1	1	3/4	2.01	3.52	4.81	5.96	7.46	8.93	10.1	11.0	11.8	12.4	.90
	1-1/2	1-1/2	3/4	4.20	7.93	11.7	15.6	19.7	23.7	27.1	30.7	32.8	33.6	.87
	2	2	1-1/8	5.96	11.7	17.6	23.6	29.5	35.7	42.0	48.6	51.0	51.4	.96
	3	3	1-1/2	16.7	32.2	47.2	63.4	78.0	91.2	102	112	117	120	.85
	4	4	2	20.5	38.2	55.3	60.3	96.5	124	151	173	191	201	.84
	1-1/2*	1	3/4	1.90	3.32	4.80	5.93	7.58	9.00	10.6	12.6	14.7	16.2	.90
	2*	1	3/4	1.81	3.28	4.76	6.24	7.75	9.30	10.8	12.3	13.8	15.1	.90
		1-1/2	3/4	4.39	7.93	11.7	15.5	19.3	22.9	27.0	30.2	33.0	35.1	.81
	3*	1-1/2	3/4	4.24	7.97	11.7	15.4	19.3	23.2	27.2	31.3	35.1	38.5	.80
		2	1-1/8	5.63	11.4	17.7	24.0	30.8	37.6	47.5	57.7	65.1	68.7	.81
	4*	2	1-1/8	5.45	11.3	17.7	24.6	31.9	39.9	52.0	59.7	69.7	76.8	.81
		3	1-1/2	14.8	30.1	44.9	59.8	75.3	91.9	109	125	133	136	.87
Cg (Gas)	1	1	3/4	66.6	113	159	197	247	294	334	376	430	469	37.8
	1-1/2	1-1/2	3/4	131	251	371	507	633	762	1010	1120	1190	1190	35.4
	2	2	1-1/8	173	365	553	753	960	1180	1440	1660	1840	1920	37.4
	3	3	1-1/2	512	999	1520	2040	2510	3000	3440	3850	4150	4380	36.4
	4	4	2	610	1210	1830	2490	3150	3870	4970	6290	7090	7480	37.2
	1-1/2*	1	3/4	61.6	00.7	147	195	242	289	335	395	492	573	35.4
	2*	1	3/4	55.9	99.8	146	195	242	291	343	388	473	570	37.7
		1-1/2	3/4	129	249	367	497	614	739	864	980	1110	1240	35.3
	3*	1-1/2	3/4	122	247	362	489	614	742	864	986	1130	1280	33.2
		2	1-1/8	172	327	520	763	951	1180	1450	1790	2180	2420	35.2
	4*	2	1-1/8	184	357	557	761	974	1220	1520	1840	2200	2590	33.7
		3	1-1/2	482	963	1470	1960	2440	2920	3460	4060	4710	5150	37.9
Cs (Steam)	1	1	3/4	3.33	5.65	7.95	9.85	12.4	14.7	18.8	21.5	23.5	23.7	35.4
	1-1/2	1-1/2	3/4	6.55	12.6	18.6	25.4	31.7	38.1	44.4	50.5	56.0	59.5	35.4
	2	2	1-1/8	8.65	18.3	27.7	37.7	48.0	59.0	72.0	83.0	92.0	96.0	37.4
	3	3	1-1/2	25.6	50.0	76.0	102	126	150	172	193	208	219	36.4
	4	4	2	30.5	60.5	91.5	125	158	194	249	315	355	374	37.2
	1-1/2*	1	3/4	2.58	4.99	7.35	9.75	12.1	14.5	16.8	19.8	24.6	28.7	35.4
	2*	1	3/4	2.80	4.99	7.30	9.75	12.1	14.6	17.2	19.4	22.7	39.5	37.7
		1-1/2	3/4	6.48	12.5	18.4	24.9	30.7	37.0	43.2	49.0	55.5	62.0	35.3
	3*	1-1/2	3/4	6.10	12.4	18.1	24.6	30.7	37.1	43.2	49.3	56.5	64.0	33.2
		2	1-1/8	8.60	16.4	26.0	38.2	47.6	59.0	72.5	89.5	109	121	35.2
	4*	2	1-1/8	9.20	17.9	27.9	38.1	48.7	61.0	76.0	92.0	110	130	33.7
		3	1-1/2	24.1	48.2	73.5	98.0	122	146	173	203	236	258	37.9

*This column lists the Km values for the Cv coefficients and the Cf values for the Cg and Cs coefficients at 100% travel. • Restricted Trim

Figure C-39d. Example of Fisher's valve catalog: design EB, specs C and D, linear valve plug. (Courtesy of Fisher Controls.)

As the liquid passes the vena contracta, its velocity decreases and in so doing it recovers part of its pressure. Valves such as butterfly valves, ball valves, or most rotary valves have a high-pressure recovery characteristic. Most reciprocating stem valves show a low-pressure recovery characteristic. The flow path through these reciprocating stem valves is more tortuous than through rotary valves. Therefore, the liquid experiences more pressure drop through these valves than through rotary type valves.

Looking again at Fig. C-41, let us suppose that the vapor pressure of the liquid at the flowing temperature is P_V. When the pressure of the liquid falls below this pressure, P_V, some of the liquid starts changing phase from the liquid phase to the vapor phase, that is, the liquid flashes. *Flashing* can produce serious erosion damage to the valve plug and seat, as shown in Fig. C-42.

Aside from the physical damage to the valve, flashing tends to lower the flow capacity of the valve. As bubbles start forming, this tends to cause a "crowding condition" at the valve, which limits the flow. Furthermore, this crowding condition may get bad enough to "choke" the flow through the valve. That is, beyond this choked condition, increases in pressure drop across the valve will not result in an increased flow. It is important to recognize that the valve equation, Eq. (5-3), does not describe this condition.

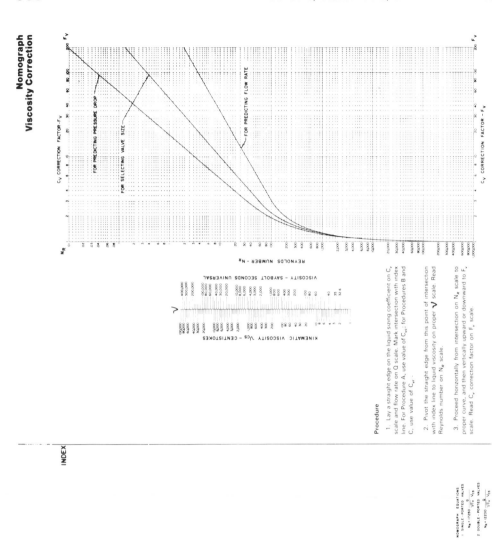

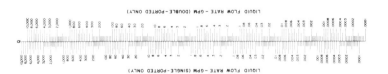

Figure C-40. Nomograph for viscosity correction. (Courtesy of Fisher Controls.)

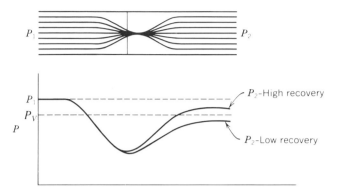

Figure C-41.   Pressure profile of a liquid across a restriction.

Figure C-42.   Damage to valve plug due to flashing. (Courtesy of Masoneilan Division, McGraw-Edison Co.)

As the pressure drop increases, the equation will predict higher flow rates. This relationship is shown graphically in Fig. C-43 along with the choked flow condition.

Notice from this figure that it is important for the engineer to know what maximum pressure drop, $\Delta P_{\max}$, is effective in producing flow. The manufacturers have chosen to indicate this maximum pressure drop by giving an equation for ΔP_{allow}. At higher pressure drops than this ΔP_{allow}, choked flow results. ΔP_{allow} is very much a function of not only the fluid but also of the type of valve. Several manufacturers have conducted research to obtain data and develop an equation for prediction of ΔP_{allow}. Masoneilan[18] proposes the following equation:

$$\Delta P_{\text{allow}} = C_f^2 \Delta P_S \tag{C-5}$$

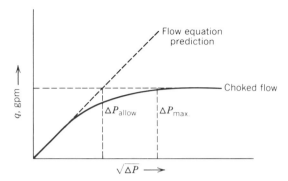

Figure C-43.   Choked flow condition.

And

$$\Delta P_S = P_1 - \left(0.96 - 0.28 \sqrt{\frac{P_V}{P_C}}\right) P_V \tag{C-6}$$

or, if $P_V < 0.5 P_1$,

$$\Delta P_S = P_1 - P_V \tag{C-7}$$

where

P_V = vapor pressure of liquid in psia
C_f = critical flow factor[15]
P_C = critical pressure of liquid in psia

The critical flow factor, C_f, is shown in Fig. C-44 for different types of valves. These values are the result of flow tests performed on the valves.

Fisher Controls[16] proposes the following equation:

$$\Delta P_{\text{allow}} = K_m (P_1 - r_C P_V) \tag{C-8}$$

where

K_m = valve recovery coefficient[16]
r_C = critical pressure ratio[16]

The K_m coefficient depends on the type of valve and is also a result of flow tests. Figures C-39c and C-39d show in the last column values of K_m for the particular type of valve. The r_C term is determined from Fig. C-45.

If the pressure recovery experienced by the liquid is enough to raise the pressure above the vapor pressure of the liquid, then the vapor bubbles will start collapsing, or imploding. This implosion is called "cavitation." The energy released during cavitation will produce noise, as if gravel were flowing through the valve,[14] and will tear away the material of the valve. Figure C-46 shows typical damage produced by cavitation. Certainly, high-pressure recovery, rotary stem valves tend to experience cavitation more often than low-pressure recovery, reciprocating stem valves.

Tests have shown that for low-pressure recovery valves, choked flow and cavitation occur nearly at the same ΔP and, consequently, Eqs. (C-5) and (C-8) can also be used to calculate the pressure drop at which cavitation starts. For high-pressure recovery valves, cavitation can occur at pressure drops below ΔP_{allow}. For these types of valves Masoneilan[15] proposes the following equation:

$$\Delta P_{cavitation} = K_C(P_1 - P_V) \tag{C-9}$$

Valve Type	Trim Size	Flow To	C_f (F_L)	K_c*	C_{fr} (F_{LP}) D/d = 1.5 or greater	X_T
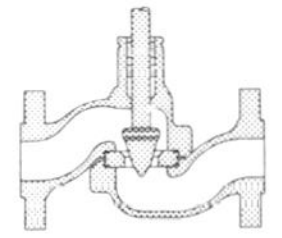 *Split Body Globe Valves*	A	Close → Open ←	.80 .75	.51 .46	.77 .72	.54 .47
	B	Close → Open ←	.80 .90	.52 .65	.80 .89	.54 .68
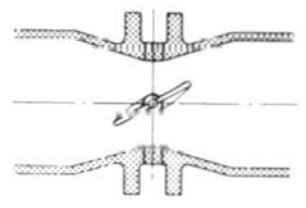 *37000 Series*	A	Flow in Either Direction	.05	.32	.00	.35
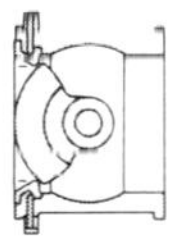 *Control Ball Valve*	A	Open →	.60	.24	.55	.30
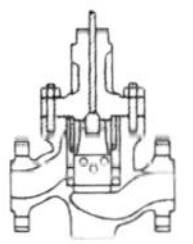 *40000 Series Balanced*	1½"- 4"	Close →	.94	.71	.87	.74
	6"-16"	Close →	.92	.68	.89	.71
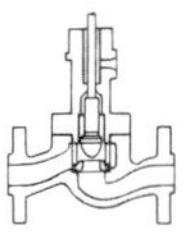 *40000 Series Unbalanced*	A	Open ←	.90	.65	.79	.68
	B	Open ←	.90	.65	.86	.68
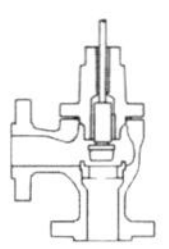 *70000 Series*	A	Close → Open ←	.81 .89	.53 .64	.78 .85	.55 .67
	B	Close → Open ←	.80† .90	.52† .65	.80 .90	.54 .68

(A) Full capacity trim, orifice dia. ≈ .8 valve size.
(B) Reduced capacity trim 50% of (A) and below.
†With Venturi Liner $C_f = 0.50$, $K_c = 0.19$

Note $X_t = 0.84C_f^2$

Figure C-44. Critical-flow factor, C_f, at full opening. (Courtesy of Masoneilan Division, McGraw-Edison Co.)

Valve Type	Trim Size	Flow To	C_f (F_L)	K_c*	C_{fr} (F_{LP}) $D/d = 1.5$ or greater	X_T
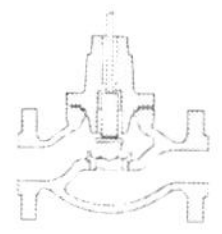 20000 Series	A	Close → / Open ←	.85 .90	.58 .65	.81 .86	.61 .68
	B	Close → / Open ←	.80 .90	.52 .65	.80 .90	.54 .68
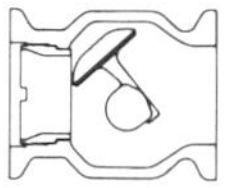 Camflex Valve	A	Close ← / Open →	.68 .85	.35 .60	.65 .80	.39 .61
	B	Close ← / Open →	.70 .88	.39 .62	.70 .87	.41 .65
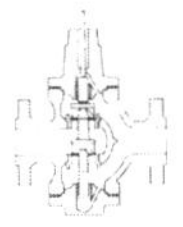 10000 Series	A	Contoured V-Port	.90 .98	.70 .80	.86 .94	.68 .81
	B	Contoured V-Port	.80 .95	.31 .73	.80 .94	.54 .76

Note $X_T = 0.84C_f^2$

Figure C-44. (*Continued*)

where

K_C = coefficient of incipient cavitation, shown in Fig. C-44

Fisher Controls[19] also proposes the same equation and their K_C term is shown in Fig. C-47.

Special anticavitation trims are produced by valve manufacturers that tend to increase the K_C term of the valve and therefore the pressure drop at which cavitation will occur.

SUMMARY

The purpose of this brief appendix was to introduce the reader to some of the most common instrumentation used for process control loops. The instrumentation shown included some of the hardware necessary for the measurement (M) of process variables (primary elements) such as flow, pressure, temperature, and pressure. Two types of transmitters were also presented and their working principles discussed. Finally, some very common types of valves (final control elements) used to take action (A) were also presented along with their flow characteristics.

As mentioned in the first sentence of this summary, this has been a brief appendix. It is impossible to present and discuss in this book all of the details related to the different types of instruments; however, there are entire handbooks and a very exhaustive collection of articles available for this purpose. The reader is referred to the fine references given at the end of this appendix. In addition to the many different types of instruments available today, every month new, ''improved'' types of primary elements, transmitters, and final control elements are introduced on the market. In the primary elements area new sensors

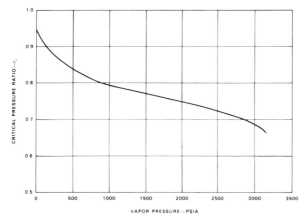

Critical Pressure of Various
Fluids, Psia*

Ammonia	1636
Argon	705.6
Butane	550.4
Carbon Dioxide	1071.6
Carbon Monoxide	507.5
Chlorine	1118.7
Dowtherm A	465
Ethane	708
Ethylene	735
Fluorine	808.5
Helium	33.2
Hydrogen	188.2
Hydrogen Chloride	1198
Isobutane	529.2
Isobutylene	580
Methane	673.3
Nitrogen	492.4
Nitrous Oxide	1047.6
Oxygen	736.5
Phosgene	823.2
Propane	617.4
Propylene	670.3
Refrigerant 11	635
Refrigerant 12	596.9
Refrigerant 22	716
Water	3206.2

Use this curve for water. Enter on the abscissa at the water vapor pressure at the valve inlet. Proceed vertically to intersect the curve. Move horizontally to the left to read the critical pressure ratio, r_c, on the ordinate.

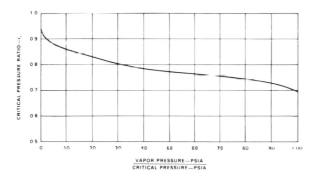

Use this curve for liquids other than water. Determine the vapor pressure/critical pressure ratio by dividing the liquid vapor pressure at the valve inlet by the critical pressure of the liquid. Enter on the abscissa at the ratio just calculated and proceed vertically to intersect the curve. Move horizontally to the left and read the critical pressure ratio, r_c, on the ordinate.

Figure C-45. Fisher's critical pressure ratio. (Courtesy of Fisher Controls.)

that can measure difficult variables, such as concentration, more exactly, more repeatably, and faster are developed constantly. In the transmitter area the latest word is "smart transmitters."

These are today's transmitters that with the aid of microprocessors will be able to present the information to the controllers in a more understandable manner. The final control elements present another very active area of research. Not only are pneumatic control valves continually being upgraded but now electric actuators are being developed and improved to allow interfacing with other electronic components such as controllers and computers. In addition, other final control elements, such as drivers for variable-

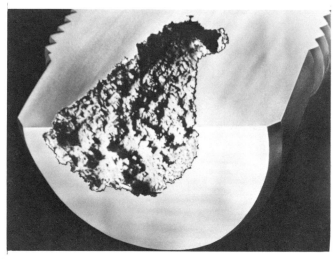

Figure C-46. Damage due to cavitation. (Courtesy of Masoneilan Division, McGraw-Edison Co.)

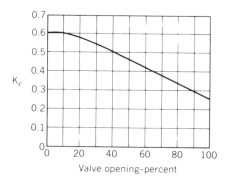

Figure C-47. Fisher's coefficient of incipient cavitation. (Courtesy of Fisher Controls.)

speed pumps and fans, are being developed. The impetus behind this development is energy conservation. The lack of space in this book prevents a presentation on the feasibility and justification of the use of these variable-speed pumps and fans for flow throttling; however, the reader is referred to References 21, 22, 23, and 24 for a presentation on this subject.

Certainly, the last paragraph should have indicated to the reader that there is a lot of research being conducted, principally by manufacturers in the instrumentation area, that should result in better measurement and control. This is one reason why process control is such a dynamic field.

REFERENCES

1. Ryan, J. B., "Pressure Control," *Chemical Engineering*, February 3, 1975.

2. Liptak, Bela G., ed., *Instrument Engineers' Handbook, Vol. I, Process Measurement*, Chilton Book Co., New York.

3. Considine, D. M., Ed., *Process Instruments and Controls Handbook*, McGraw-Hill Book Co., New York, 1957.

4. Considine, D. M., Ed., *Handbook of Instrumentation and Controls*, McGraw-Hill Book Co., New York, 1961.

5. Smith, C. L., "Liquid Measurement Technology," *Chemical Engineering*, April 3, 1978.

6. Zientara, Dennis E., "Measuring Process Variables," *Chemical Engineering*, September 11, 1972.

7. Kern, R., "How to Size Flowmeters," *Chemical Engineering*, March 3, 1975.

8. "Handbook Flowmeter Orifice Sizing," Handbook No. 10B9000, Fischer Porter Co., Warminster, Pa.

9. Spink, L. K., "Principles and Practice of Flow Meter Engineering," The Foxboro Co., Foxboro, Mass.

10. Kern, R., "Measuring Flow in Pipes with Orifices and Nozzles," *Chemical Engineering*, February 3, 1975.

11. Wallace, L. M., "Sighting in on Level Instruments," *Chemical Engineering*, Feb. 16, 1976.

12. "Process Control Instrumentation," The Foxboro Co. Publication 105A-15M-4/71, Foxboro, Mass.

13. Taylor Instrument Co., "Differential Pressure Transmitter Manual, IB-12B215."

14. "Control Valve Handbook," Fisher Controls Co., Marshalltown, Iowa.

15. "Masoneilan Handbook for Control Valve Sizing," Masoneilan International, Inc., Norwood, Mass.

16. "Fisher Catalog 10," Fisher Controls Co., Marshalltown, Iowa.

17. Casey, J. A., and D. Hammitt, "How to Select Liquid Flow Control Valves," *Chemical Engineering*, April 3, 1978.

18. Chalfin, S., "Specifying Control Valves," *Chemical Engineering*, Oct. 14, 1974.

19. Kern, R., "Control Valves in Process Plants," *Chemical Engineering*, April 14, 1975.

20. Hutchison, J. W., Ed., "ISA Handbook of Control Valves," Instrument Society of America."

21. Fischer, K. A., and D. J. Leigh, "Using Pumps for Flow Control" Instruments and Control Systems, March 1983.

22. Jarc, D. A., and J. D. Robechek, "Control Pumps with Adjustable Speed Motors Save Energy," Instruments and Control Systems, May 1981.

23. Pritchett, D. H., "Energy-Efficient Pump Drive," *Chemical Engineering Progress*, Oct. 1981.

24. Baumann, H. D., "Control Valve vs. Variable-Speed Pump," *Chemical Engineering*, June 24, 1981.

25. Utterback, V. C., "Online Process Analyzers," *Chemical Engineering*, June 21, 1976.

26. Foster, R. "Guidelines for Selecting Outline Process Analyzers," *Chemical Engineering*, March 17, 1975.

27. Ottmers, D. M., et al., "Instruments for Environmental Monitoring," *Chemical Engineering*, October 15, 1979.

28. Creason, S. C., "Selection and Care of pH Electrodes," *Chemical Engineering*, October 23, 1978.

APPENDIX
D
Computer Program to Solve for Roots of Polynomials

The program listed in Fig. D-1 can be used directly to find all of the roots, real and complex, of a polynomial. It uses subroutine MULLER, which calculates the roots by Müller's method of Lagrangian quadratic interpolation. For a complete description of Müller's method the reader should consult the original reference by Müller[1] or the text by Conte and de Boor.[2]

While the main program and subroutine POLY are specific for polynomial root solving, subroutine MULLER is general and can solve for real and complex roots of any nonlinear function, not necessarily a polynomial. It uses subroutine DEFLAT to numerically reduce the polynomial or function, as in Eq. (2-51), an operation known as *deflation*. This ensures that the same root will not be found more than once except for the case of repeated roots.

The main program accepts the degree of the polynomial and the polynomial coefficients, in ascending powers of x, and passes them to subroutine POLY through a COMMON statement. It also accepts the maximum number of iterations allowed per root, the relative error tolerance, and the initial approximations to the roots, and passes them to subroutine MULLER through its arguments. Müller[1] recommends that the initial approximations to the roots be set to zero so that the roots are found in order of increasing magnitude. Although this results in the most accurate approximations to the roots, they are not always found in increasing order.

The relative error tolerance is defined in this program as the maximum difference allowed between two successive approximations of a root divided by the magnitude of the root. For example, a value of 10^{-6} ensures that the two successive approximations agree to 6 *significant digits*, independent of the magnitude of the root.

Input data.

First Line

NDEG,   MAXIT,   RTOL

```
C
C             MAIN PROGRAM TO FIND THE ROOTS OF A POLYNOMIAL BY MULLER'S
C             METHOD
C
C          VARIABLES IN INPUT DATA
C
C          VARIABLE   TYPE   DIMENSION   DESCRIPTION
C
C          NDEG        I        -        DEGREE OF THE POLYNOMIAL
C          MAXIT       I        -        MAXIMUM NUMBER OF ITERATIONS
C          RTOL        R        -        RELATIVE ERROR TOLERANCE ON ROOTS
C          A           R      NDEG+1     POLYNOMIAL COEFFICIENTS ORDERED
C                                        IN ASCENDING POWERS OF X
C          XS          C       NDEG      INITIAL APPROXIMATIONS TO THE ROOTS
C
C          SUBPROGRAMS CALLED
C
C             MULLER  - TO COMPUTE ROOTS OF POLYNOMIAL BY MULLER'S METHOD
C
          COMPLEX XS(20)
          COMMON NDEG, A(21)
          EXTERNAL POLY
C
C          READ AND PRINT INPUT DATA
C
   10     WRITE ( 6, 100 )
  100     FORMAT(' ENTER NDEG, MAXIT, RTOL' )
          READ( 5, * )   NDEG, MAXIT, RTOL
          IF( NDEG .EQ. 0 ) STOP
          NC = NDEG + 1
C
          WRITE( 6, 110 )
  110     FORMAT(' ENTER POLYNOMIAL COEFFS., ASCENDING POWERS OF X' )
          READ ( 5, * ) (A(I),I=1,NC)
          WRITE( 6, 120 )
  120     FORMAT(' ENTER INITIAL APPROXIMATIONS TO THE ROOTS' )
          READ ( 5, * ) (XS(I),I=1,NDEG)
C
          WRITE ( 6, 130 ) NDEG, RTOL, (A(I),I=1,NC)
          WRITE ( 4, 130 ) NDEG, RTOL, (A(I),I=1,NC)
          WRITE ( 6, 140 ) (XS(I),I=1,NDEG)
          WRITE ( 4, 140 ) (XS(I),I=1,NDEG)
C
  130     FORMAT('1',10X,'MULLERS  METHOD TO FIND THE ROOTS OF A POLY',
         1          'NOMIAL OF DEGREE',I3,
         2        //11X,'RELATIVE ERROR TOLERANCE',1PG10.2,5X
         3        //11X,'THE COEFFICIENTS OF THE POLYNOMIAL ARE (CONSTANT ',
         4          'TERM FIRST)'//(13X,5G13.6))
  140     FORMAT(//11X,'THE INITIAL APPROXIMATIONS TO THE ROOTS ARE'
         1        //18X,'REAL',8X,'IMAGINARY'/(/12X,1P2G15.6))
C
C          EVALUATE ROOTS OF POLYNOMIAL BY MULLER'S METHOD
C
          CALL MULLER( NDEG, XS, MAXIT, RTOL, POLY )
C
C          PRINT RESULTS
C
          WRITE ( 6, 150 ) (XS(I),I=1,NDEG)
          WRITE ( 4, 150 ) (XS(I),I=1,NDEG)
  150     FORMAT(//11X,'THE FINAL APPROXIMATIONS TO THE ROOTS ARE'
         1        //18X,'REAL',8X,'IMAGINARY'/(/12X,1P2G15.6))
          GOTO 10
          END
C*
C*------------------------------------------------------------------------
```

Figure D-1.   Program to calculate all of the roots, real and complex, of a polynomial.

```
C*
      SUBROUTINE POLY( X, P )
C
C         PURPOSE - TO EVALUATE A POLYNOMIAL OF NTH DEGREE
C                         FOR SUBROUTINE MULLER
C
C         METHOD  - NESTED MULTIPLICATION (SYNTHETIC DIVISION)
C
C         VARIABLES IN COMMON
C
C         NDEG          DEGREE OF THE POLYNOMIAL
C         A(NDEG+1)     COEFFICIENTS OF THE POLYNOMIAL ORDERED IN
C                         ASCENDING POWERS OF X
C
C         VARIABLES IN ARGUMENT LIST
C
C         VARIABLE  TYPE  I/O  DESCRIPTION
C
C            X        C    I   COMPLEX VALUE AT WHICH THE
C                                 POLYNOMIAL IS TO BE EVALUATED
C            P        C    O   COMPLEX VALUE OF THE POLYNOMIAL
C
      COMPLEX X, P, B
      COMMON  NDEG, A(21)
C
      NC = NDEG + 1
      B  = A(NC)
      DO 10 I=1,NDEG
         K = NC - I
  10     B = B * X + A(K)
      P = B
C
      RETURN
      END
C*
C*----------------------------------------------------------------------
C*
      SUBROUTINE MULLER( N, XS, MAXIT, RTOL, FUNCTN )
C
C         PURPOSE - TO COMPUTE APPROXIMATIONS OF THE ROOTS OF
C                         NONLINEAR EQUATIONS
C
C         REF:  CONTE AND DE BOOR, "ELEMENTARY NUMERICAL ANALYSIS",
C               3RD ED., MCGRAW-HILL, NEW YORK, 1980, PP. 120-127.
C
C         METHOD:  MULLER'S METHOD OF QUADRATIC INTERPOLATION
C
C         VARIABLES IN ARGUMENT LIST
C
C         VARIABLE  TYPE  I/O  DIMENSION  DESCRIPTION
C
C            N       I    I       -       NUMBER OF ROOTS TO BE COMPUTED
C            XS      C    I/O      N       ARRAY CONTG. THE INITL. APPRS.
C                                           AND RETG. THE FINAL VALUES
C            MAXIT   I    I       -       MAXIMUM NUMBER OF ITERATIONS
C            RTOL    R    I       -       RELATIVE ERROR TOLERANCE ON
C                                           THE ROOTS
C            FUNCTN  E    I       -       NAME OF SUBROUTINE USED TO
C                                           EVALUATE THE FUNCTION
C
C         SUBPROGRAMS CALLED
C
C            FUNCTN - TO EVALUATE THE NONLINEAR FUNCTION
C            DEFLAT - TO DEFLATE THE FUNCTION
```

Figure D-1. (*Continued*)

```
C
      COMPLEX XS(N),    H,   X,   FX,     F, DFP, FP, LAMBDA, DF, DFPL,
     C        DELTA, TOP,  G, RAD,  BOT
C
C         GUARD AGAINST ZERO OR NEGATIVE ERROR TOLERANCE
C
      ETOL   = AMAX1( RTOL, 1.E-6 )
C
C         START OF LOOP TO EVALUATE N ROOTS
C
      DO 100 I=1,N
         NEVAL = 0
C
C            EVALUATE AND DEFLATE FUNCTION AT FIRST APPROXIMATION
C
  10     H       = 0.5
         X       = XS(I) + H
         NEVAL   = NEVAL + 1
         CALL FUNCTN( X, FX )
         CALL DEFLAT( I, X, FX, F, XS, IFLG )
         IF( IFLG .NE. 0 ) GOTO 10
C
C            EVALUATE AND DEFLATE FUNCTION AT SECOND APPROXIMATION
C
         DFP     = F
         X       = XS(I) - H
         NEVAL = NEVAL + 1
         CALL FUNCTN( X, FX )
         CALL DEFLAT( I, X, FX, F, XS, IFLG )
         IF( IFLG .NE. 0 ) GOTO 10
C
C            INITIALIZATION OF ITERATIVE CALCULATION
C
         FP      = F
         DFP     = FP - DFP
         X       = XS(I)
         LAMBDA = - 0.5
         NEV     = NEVAL + 1
C
C            ITERATIVE CALCULATION OF THE ROOT
C
         DO 30 NEVAL=NEV,MAXIT
C
            CALL FUNCTN( X, FX )
            CALL DEFLAT( I, X, FX, F, XS, IFLG )
            IF( IFLG .NE. 0 ) GOTO 10
C
C               COMPUTE NEXT ESTIMATE OF ROOT
C
            DF     = F - FP
            DFPL   = DFP * LAMBDA
            DELTA  = 1. + LAMBDA
            TOP    = - 2. * F * DELTA
            G      = ( DELTA + LAMBDA ) * DF - LAMBDA * DFPL
            RAD    = CSQRT( G**2 + 2. * TOP * LAMBDA * ( DF - DFPL) )
            BOT    = G + RAD
            IF( REAL(G) * REAL(RAD) + AIMAG(G) * AIMAG(RAD) .LT. 0.)
     *         BOT = G - RAD
            LAMBDA = TOP
            IF( CABS(BOT) .NE. 0. ) LAMBDA = TOP / BOT
C
            FP = F
            DFP = DF
```

Figure D-1. (*Continued*)

```
              H   = H * LAMBDA
              X   = X + H
C
C             CHECK FOR CONVERGENCE
C
              IF( CABS(H)  .LT. ( ETOL * CABS(X) ) ) GOTO 100
C
 30           CONTINUE
C
C             REACHED MAXIMUM NUMBER OF ITERATIONS
C
         WRITE( 6, 110 ) NEVAL, I, X, F
         WRITE( 4, 110 ) NEVAL, I, X, F
 110 FORMAT(//11X,'FAILED TO CONVERGE AFTER',I3,1X,'EVALUATIONS',
    *' FOR ROOT',I3//11X,'ROOT',1P2G15.6,5X,'F(ROOT)',2G15.6)
C
 100     XS(I) = X
C
     RETURN
C
     END
C*
C*-----------------------------------------------------------------------
C*
     SUBROUTINE DEFLAT( I, X, FX, FDEF, XS, IFLG )
C
C        PURPOSE - TO DEFLATE THE FUNCTION F SO THAT THE SAME
C                  ROOT IS NOT ENCOUNTERED MORE THAN ONCE
C
C        VARIABLES IN ARGUMENT LIST
C
C        VARIABLE  TYPE  I/O  DIMENSION  DESCRIPTION
C
C           I       I    I       -       NUMBER OF ROOT SOUGHT
C           X       C    I       -       CURRENT APPROXIMATION
C           FX      C    I       -       FUNCTION VALUE AT X
C           FDEF    C    O       -       DEFLATED VALUE AT X
C           XS      C    I       I       PREVIOUS ROOTS FOUND
C           IFLG    I    O       -       FLAG SET TO ONE IF CURRENT
C                                        ROOT MATCHES PREVIOUS ROOT
C
     COMPLEX X, FX, FDEF, XS(1), BOT
C
     IFLG = 0
     FDEF = FX
     IF( I .LT. 2 ) RETURN
C
C        DEFLATION OF F(X)
C
     DO 10 K=2,I
        BOT    = X - XS(K-1)
        IF( CABS(BOT) .LT. 1E-20 ) GOTO 20
 10     FDEF = FDEF / BOT
     RETURN
C
C        CURRENT APPROXIMATION MATCHES A ROOT.  MODIFY AND SET IFLAG.
C
 20  XS(I) = X + 0.001
     IFLG  = 1
     RETURN
C
     END
```

Figure D-1. (*Continued*)

where

NDEG is the degree of the polynomial
MAXIT is the maximum number of iterations allowed per root
RTOL is the relative error tolerance

Second Line

$$A(I), \qquad I = 1, \text{NDEG} + 1$$

where

$A(I)$ are the coefficients of the polynomial, arranged in ascending powers of x, that is,

$$p(x) = A(1) + A(2)x + A(3)x^2 + \ldots + A(\text{NDEG} + 1)x^{\text{NDEG}}$$

Third Line

$$XS(I), \qquad I = 1, \text{NDEG}$$

where

$XS(I)$ are the initial approximations to the roots

After each problem is solved, the program returns to the start to read data for another case. *To stop execution* enter a value of *zero* for NDEG.

Format. The format of the input data is free, that is, the numbers for each line are entered and separated by blanks or commas. The program is written with prompts so that it can be used interactively from a time-sharing terminal.

Results. The results consist of an echo of the input data and the final approximations to the roots, given in two columns for the real and imaginary parts, respectively.

Caution: A small imaginary part is usually printed for roots that are real. This is due to numerical inaccuracies, and the imaginary part should be discarded if it is several orders of magnitude smaller than the real part. Similarly, the real part should be discarded if it is several orders of magnitude smaller than the imaginary part.

Output Unit Assignments. Because of the time-sharing feature, the program prints each line of output twice, once on unit 6, presumed to be the terminal, and once on unit 4, for printing on paper. For batch use, one of the two WRITE statements should be removed from each pair.

Example The input data for finding the roots of the following polynomial equation:

$$s^5 + 4s^4 + 16s^3 + 25s^2 + 48s + 36 = 0$$

with 50 iterations allowed per root, a relative error tolerance of 10^{-5}, and zero initial approximations, are

First line: 5, 50, 1E-4
Second line: 36, 48, 25, 16, 4, 1
Third line: (0,0), (0,0), (0,0), (0,0), (0,0)

The roots are known to be

$$-1, i2, -i2, -\frac{3}{2} + i\frac{3\sqrt{3}}{2}, -\frac{3}{2} - i\frac{3\sqrt{3}}{2}.$$

The output of the program for this example is given in Fig. D-2.

```
MULLERS  METHOD TO FIND THE ROOTS OF A POLYNOMIAL OF DEGREE  5

RELATIVE ERROR TOLERANCE    1.0E-05

THE COEFFICIENTS OF THE POLYNOMIAL ARE (CONSTANT TERM FIRST)
     36.0000      48.0000      25.0000      16.0000      4.0000
     1.00000

THE INITIAL APPROXIMATIONS TO THE ROOTS ARE
        REAL          IMAGINARY
       .0             .0
       .0             .0
       .0             .0
       .0             .0
       .0             .0

THE FINAL APPROXIMATIONS TO THE ROOTS ARE
        REAL          IMAGINARY
     -1.00000        7.99139E-13
      8.22254E-07   -2.00000
      3.95433E-07    2.00000
     -1.50000       -2.59808
     -1.50000        2.59808
```

Figure D-2. Sample results of Müller's program.

REFERENCES

1. Müller, D. E., "A Method for Solving Algebraic Equations Using an Automatic Computer," *Mathematical Tables and Other Aids to Computation*, Vol. 10, 1956, pp. 208–215.

2. Conte, S. D. and C. de Boor, *Elementary Numerical Analysis*, 3rd ed., McGraw-Hill, New York, 1980, pp. 120–127.

INDEX

Accumulation, 449
Action, 3
Actuators, 575
Antoine equation, 46
Argument, of complex number, 53
Arrhenius equation, 46
Automatic controller position, 179
Automatic process control, 2
  objective, 3
  reasons, 8
Auxiliary variable, 471

Balance equations, 449
Balances, independent, 450
Batch switch, 253
Block diagrams, 76
  rules, 77
Bode plot, 300
  from pulse test, 335
Boiler control, 358, 384
Boosters, 579
Boundary conditions, 468
Bourdon, Eugene, 545
Bourdon tube, 545
Bubbler sensor, 559

Canned subroutines, 492
Cascade control, 362
Characteristic equation, 37, 181
Closed-loop control, 4
Closed-loop response specification,
  238
Closed-loop transfer function, 179
Column pressure, 459–460
Complex conjugate roots, 28
Complex differentiation theorem, 16
Complex number, 52
  argument, 53
  conjugate, 53
  magnitude, 53
  operations, 54
  polar notation, 53
Complex plane, 53
Complex translation theorem, 17

Computer control, tuning, 234–236
Computer program, *see* Program
Computer simulation, 471–491
Computer time, 478
Computing blocks, 344
Computing relays, 344
Condenser accumulator model, 460–
  463
Condenser model, 458–460
Conformal mapping, 322
Conjugate, of complex number, 54
Conservation equations, 449
Control, 4
  boiler, 358, 384
  cascade, 80, 362
  cross-limiting, 360
  feedback, 4
  feedforward, 6, 369
  multivariable, 396
  override, 392
  ratio, 354
  regulatory, 4
  selective, 392
  servo, 4
Controlled variable, 3
Controller, 2, 3, 154
  action, 157
  auto/manual, 154
  derivative time, 165
  digital, 168
  feedback, 154
  gain, 159
  local/remote, 155
  offset, 159
  proportional, 159
  proportional band, 161
  proportional-derivative (PD), 168
  proportional-integral (PI), 163
  proportional-integral-derivative
    (PID), 165
  rate time, 165
  reset rate, 164
  reset time, 163
  reset wind-up, 169, 249–253

607

Controller (*Continued*)
   synthesis, 237
   tuning, 4, 210–237, 239–249
Controller modes, selection, 239–245
Control loop program, 502
Control loop simulation, 500
Control system, 2
Control volume, 449

Dahlin response, 238
Dahlin tuning, 245
Dahlin tuning parameter, 239
Damping ratio, 123
Dead time, 16, 33, 83
   estimation, 217–220
   effect on stability, 208
   simulation, 496
Dead time program, 502
Decay ratio, 125
Decision, 3
Decoupler, 420
Deflation, 597
Density, of ideal gas, 50
Dependent variable, 22
Derivative mode, 165
   program, 502
   selection, 243–245
   simulation, 500
Deviation variable, 43
Differential equation, general form,
   471
Differential equations, solution, 22
Differential pressure meter, 557
Differential pressure transmitter, 569
   electronic, 570
   pneumatic, 569
Dirac delta function, 12
Direct substitution, 205
Discretization, 469
Distillation model, 450–464
Distributed control, 234
Distributed parameter model, 449, 464
Disturbance, 3

input, 228
response, 228
tuning, 230
Domain, 23
Dominant eigenvalue, 475, 506
Dominant time constant, 204
Duration of simulation runs, 475

Eigenvalue, 37, 277
   dominant, 475, 506
   estimation, 507
Entry points, 484
Equilibrium relationships, 454
Error integral criteria, 226
Error integral tuning, 230–231
Error tolerance, 597
Euler integration, 475
   stability, 514
Evaporator control, 518
Explicit integration, 512

Factoring polynomials, 25
Feedback control, 5, 177
   program, 502
   simulation, 500
   synthesis, 237
Feedforward control, 369
Final control element, 2, 3
Final time, 475
Final value theorem, 18
Finite differences, 469
First-order lag, 67
   process, 67
   program, 502
   simulation, 493
First-order lead, 127–128
First-order-plus-dead-time, *see*
   FOPDT
Fit 1 of step response, 217
Fit 2 of step response, 219
Fit 3 of step response, 219
Float sensor, 559
FOPDT, 97

fits, of step response, 217–220
model, 215
process, 241
response, 218
Forcing function, 22, 65
FORTRAN program, *see* Program
Fourier transform, 331
numerical evaluation, 333–336
of rectangular pulse, 332–333
Francis's weir formula, 454
Frequency, 126
cyclical, 126
natural, 126
radian, 126
Frequency response, 292
amplitude ratio, 295
Bode plot, 300
Bode stability criterion, 310
gain margin, 318
magnitude ratio, 295
Nichols plots, 328
Nyquist stability criterion, 324
phase angle, 295
phase margin, 318
polar plots, 320
Function deflation, 597
Furnace model, 464–468

Gain, 70
estimation, 217
Gain margin, 318
Gear, C. W., 515

Hougen, Joel O., 329, 336

IAE, 227
tuning, 230–231, 244
Ideal gas density, 50
Imaginary number, 52
Implicit integration, 515
Implicit modified Euler, 515
Impulse, 12
Independent variable, 22

Initial conditions:
for furnace, 468
for startup, 463
at steady state, 463, 474
Initial time, 475
Initial value theorem, 18
Input variable, 22
to distillation column, 464
to furnace, 468
Instrumentation labels, 527
Instrumentation symbols, 527
Integration, numerical, 475–491
implicit, 515
main program, 486
Integration interval, 475
selection of, 478
Interacting systems, 92, 111
Inverse response, 390
Inversion of Laplace transform, 24
ISE, 227
tuning, 230–231
ITAE, 228
tuning, 230–231
Iteration, 38
ITSE, 228

Lag:
first-order, 67
second-order, 107
third-order, 109
Laplace domain, 23
Laplace transform, 10
of derivatives, 14
of integrals, 15
inversion, 24–37
linearity, 14
properties, 13–18
solution procedure, 22–37, 42–43
table, 14
usefulness, 23
Lead, first-order, 127, 128
Lead/lag unit, 128, 374, 378–379
program, 502

Lead (*Continued*)
  simulation, 497
Left-hand plane, 199
Level control model, 457, 461
Limit switches, 580
Linear equation, 43
Linearization, 43, 86
  base value, 44
  of multivariable functions, 48
  of single variable functions, 45
  by Taylor series expansion, 86
Loop gain, 203
Lopez, A. M., 229
Lumped parameter model, 449
Luyben, W. L., 336

Magnetic flowmeter, 555
Magnitude, of complex number, 53
Manipulated variable, 3
Manual controller position, 179
Marginal stability, 205
Martin, Jacob, 224, 248
Mathematical modeling, 450
Measurement, 3
Microprocessor control, 245
Minimal phase systems, 310
Model:
  of accumulator drum, 460–463
  of column pressure, 459–460
  of condenser, 458–460
  development, 450
  of distillation column, 450–464
  of distillation tray, 452–455
  distributed, 449, 464
  first-order, 215
  of furnace, 464–468
  of heat radiation, 467
  of level controller, 457, 461
  lumped, 449
  of reactor, 471–475
  second-order, 215
  of steam chest, 457
Modeling, 8, 450
Modified Euler integration, 482

implicit, 515
subroutine, 485
Modular programming, 484
Moore, C. F., 236
Muller's method, 38, 597
  subroutine, 599
Multiple-effect evaporator, 520
Multivariable control, 396
  decoupling, 420
  Interaction Index, 416
  interaction and stability, 419
  Niederlinski theorem, 410
  pairing, 407
  relative gain matrix, 409
Murphree efficiency, 453
Murrill, P. W., 226

Nested multiplications, 39
Newton-Bairstow method, 38
Newton's method, 39
Nichols plots, 328
Nisenfeld, E., 416
Noninteracting systems, 104
Nonlinearity, 88
Nonminimal phase systems, 310
Numerical integration subroutines,
    491–492
Nyquist stability criterion, 324

Offset, 159
  calculation of, 187–191
On-line tuning, 211–214
Open-loop control, 3
Orifice meter, 549
Oscillation period, 125
Output function, 22
Override control, 392
Overshoot, 125

Pade approximation, 208–243
Partial differential equations, 467
  solution, 469–470
Partial fractions expansion, 24–37
Period of oscillation, 125

Perturbation variable, 43
Phase margin, 318
pH control, 517
Physical properties, 454
PID controller, 165
  program, 502
  simulation, 498
  tuning, 212, 225, 230–231, 245
Polar notation, 53
Polar plots, 320
Pole, 276
Polynomial:
  factoring, 25
  roots, 25
Polynomial roots, 25, 597
  program, 598
Positioners, 578
Pressure, in distillation column, 459–460
Principle of superposition, 76
Process:
  characterization, 211, 214
  coupled, 92
  first-order, 67
  gain, 70, 217
  higher-order, 104
  interacting, 92, 111
  noninteracting, 104
  nonlinearity, 88
  personality, 6
  reaction curve, 216
  stable, 75
Process control, 1
  objective, 3
  reasons, 8
Process testing:
  pulse, 329–336
  step, 216
Program:
  for control loop, 502
  for general integration, 486
  for modified Euler, 485
  for Muller's method, 599
  for polynomial roots, 598

for reactor, 480, 487
for Runge-Kutta-Simpson, 490
for tank drainage, 511
Proportional mode, function, 246
Pulse testing, 329–336
  amplitude, 330
  duration, 330
  of integrating process, 335–336

QDR *see* Quarter-decay-ratio tuning
Quadratic factor, 30
Quarter-decay-ratio tuning, 212, 225
Quasi-steady-state assumption, 506

Radians, 30
Radiation model, 467
Range, 136
Ratio control, 354–362
Reactor model, 471–475
Reactor program, 480, 487
Real differentiation theorem, 14
Real integration theorem, 15
Real translation theorem, 16
Reboiler model, 455–458
Rectangular pulse, 11, 330
Reduced polynomial, 39
Regulator, 228
Regulatory control, 4
Relative error tolerance, 597
Relative volatility, 46
Repeated roots, 31
Reset feedback, 253, 473
  program, 502
  simulation, 499
Reset windup, 169
  prevention of, 249
Response, 93, 122
  critically damped, 124
  inverse, 390
  overdamped, 124
  underdamped, 123
Rise time, 125
Root locus, 277–292
  angle criterion, 283

Root locus (*Continued*)
  magnitude criterion, 283
  rules for plotting, 282
Roots of polynomials, 25
  complex conjugate, 28
  computation of, 597
  determination, 38
  repeated, 31
  unrepeated real, 26
Round-off error, 478
Routh's test, 200
Rovira, A. A., 231, 234
Runge-Kutta-Simpson integration, 489
  subroutine, 490

Sampled-data controller, 234
Sample time, 235
Saturation, 249
Schultz, G., 416
S domain, 23
Second-order lag, 107
  simulation, 494
Second-order-plus-dead-time, *see*
  SOPDT model
Second-order system response, 123
  critically damped, 124
  overdamped, 124
  underdamped, 123
Selective control, 392–396
Self-regulating process, 198
Sensor, 2, 3, 545
  composition, 567
  flow, 546
  gain, 137
  level, 557
  pressure, 545
  range, 136
  span, 136
  temperature, 560
  zero, 136
Servo control, 4
Servo-regulator, 228
Set point, 3
  input, 228
  tuning, 231

Settling time, 125
Shinskey, F. G., 418
Signal flow graphs, 399
Signals, 4
  digital, 4
  electrical, 4
  pneumatic, 4
Simulation, 9, 471–491
  of control loop, 500
  of dead time, 496
  of derivative unit, 500
  dynamic, 448
  of first-order lag, 493
  of lead-lag unit, 497
  of PID controller, 498
  of reset feedback, 499
  of second-order lag, 494
  of tank drainage, 508–512
Simulation languages, 492–493
Simulation runs:
  duration, 475
  types of, 463
Sine function, 13
Sinusoidal testing, 293
Slave controller, 228
Smith, Cecil L., 219, 226
SOPDT model, 215
Span, 136
Stability, 37
  criterion, 199
  of feedback loop, 198
  of numerical integration, 514
State variables, 463
Steady-state gain, 70
Steam chest model, 457
Step function, 11
Step testing, 216–217
  FOPDT fits, 217–220
  rationale, 228
Stiffness, 505
  reduction, 506
  source, 505
Subroutine:
  for control loop, 502
  for Muller's method, 599

for reactor, 487
for Runge-Kutta-Simpson, 490
for tank drainage, 511
Subroutines, for numerical integration, 492
Superposition, principle, 76
Symbols, for instrumentation, 527
Synthesis of controllers, 237
Synthesis tuning, 245
Synthetic division, 39
Systems:
   first-order, 67
   higher-order, 104
   interacting, 92, 111
   minimal phase, 310
   noninteracting, 104
   nonminimal phase, 310

Tank drainage simulation, 508–512
Taylor series expansion, 45, 48, 86
Thermistor, 565
Thermocouple, 565
Thermometers:
   bimetallic strip, 561
   filled systems, 561
   liquid-in-glass, 561
   resistance devices (RTD), 563
Time constant, 68
   characteristic, 123
   effective, 114
   estimation, 217–220
Time delay, see Dead time
Time domain, 23
Time response, display of, 479
Time units, 475
Tolerance, 597
Transducer, 4
Transfer function, 24, 66, 74
   closed-loop, 79, 179
   first-order, 67
   open-loop, 278
   pole, 276
   properties, 75
   second-order, 107
   third-order, 109

   zero, 276
Transmitter, 2, 3, 567
   electronic, 570
   gain, 137
   pneumatic, 569
   range, 136
   span, 136
   zero, 136
Transportation lag, see Dead time
Trapezoidal rule, 482
Triple-effect evaporator, 520
Truncation error, 478
Tuning:
   computer controllers, 234–236
   for disturbances, 230
   of feedback controllers, 210
   for 5% overshoot, 245
   for minimum error integral, 230–231
   for quarter-decay ratio, 211
   sampled-data controllers, 234
   for set-point change, 231
   by synthesis, 245
   Ziegler-Nichols, 211, 225
Turbine meter, 556
Two-point fit of step response, 219

Ultimate frequency, 205
Ultimate gain, 198, 211
Ultimate period, 205, 211
Unit impulse, 12
Unit step, 11
Unity feedback loop, 187
Unrealizable controller, 241
Upset, 3

Valves, 138–154, 570–594
   action, 138
   actuators, 575
   boosters, 579
   cavitation, 590
   critical flow, 141
   flashing, 586
   flow characteristics, 147
   gain, 152
   limit switches, 580

Valves (*Continued*)
  positioner, 578
  pressure drop, 144
  rangeability, 143
  sizing, 139
  types, 570
  viscosity correlation, 581
Variable:
  controlled, 3
  dependent, 22
  deviation, 43–45, 66
  independent, 22

input, 65
manipulated, 3
output, 65
perturbation, 43
responding, 65
state, 463

Zero:
  of transfer function, 276
  of transmitter, 136
Ziegler-Nichols tuning, 211, 225